Amitsur Centennial Symposium

Shimshon Avraham Amitsur (1921–1994)

CONTEMPORARY MATHEMATICS

800

Israel Mathematical Conference Proceedings

Amitsur Centennial Symposium

Amitsur Centennial Symposium
November 1–4, 2021
Virtual and the Israel Institute for Advanced Studies (IIAS)
The Hebrew University of Jerusalem
Jerusalem, Israel

Avinoam Mann
Louis H. Rowen
David J. Saltman
Aner Shalev
Lance W. Small
Uzi Vishne
Editors

AMS American Mathematical Society

אוניברסיטת בר־אילן
Bar-Ilan University

2020 *Mathematics Subject Classification.* Primary 11Exx, 14Exx, 15Axx, 16Kxx, 16Nxx, 16Rxx, 17Dxx, 19Dxx, 20Cxx, 20Gxx.

Library of Congress Cataloging-in-Publication Data

Names: Mann, Avinoam, 1937–2023, editor. | Amitsur, S. A. (Shimshon A.), 1921–1994, honoree.
Title: Amitsur Centennial Symposium, November 1–4, 2021, virtual and the Israel Institute for Advanced Studies (IIAS), The Hebrew University of Jerusalem, Jerusalem, Israel / Avinoam Mann, Louis H. Rowen, David J. Saltman, Aner Shalev, Lance W. Small, Uzi Vishne, editors.
Description: Providence, Rhode Island : American Mathematical Society ; Ramat Gan, Israel : Bar-Ilan University, [2024] | Series: Contemporary mathematics, 0271-4132 ; volume 800 | Israel Mathematical Conference proceedings | Includes bibliographical references.
Identifiers: LCCN 2023044737 | ISBN 9781470475550 (paperback) | 9781470476137 (ebook)
Subjects: LCSH: Algebra–Congresses. | Rings (Algebra)–Congresses. | Group theory–Congresses. | AMS: Algebraic geometry – Birational geometry. | Linear and multilinear algebra; matrix theory – Basic linear algebra. | Associative rings and algebras – Division rings and semisimple Artin rings. | Associative rings and algebras – Radicals and radical properties of rings. | Associative rings and algebras – Rings with polynomial identity. | Nonassociative rings and algebras – Other nonassociative rings and algebras. | K-theory – Higher algebraic K-theory. | Group theory and generalizations – Representation theory of groups. | Group theory and generalizations – Linear algebraic groups and related topics.
Classification: LCC QA150 .A55 2024 | DDC 512–dc23/eng/20240201
LC record available at https://lccn.loc.gov/2023044737

Contemporary Mathematics ISSN: 0271-4132 (print); ISSN: 1098-3627 (online)

DOI: https://doi.org/10.1090/conm/800

∞ The paper used in this book is acid-free and falls within the guidelines established to ensure permanence and durability.
Visit the AMS home page at https://www.ams.org/

10 9 8 7 6 5 4 3 2 1 29 28 27 26 25 24

Contents

Preface

Born in Jerusalem in 1921, Shimshon Avraham Amitsur was a pioneer in several branches of algebra, and the leading algebraist in Israel for several decades. The **Amitsur Centennial Symposium** took place on November 1–4, 2021, partly physically at the Hebrew University of Jerusalem amidst Covid restrictions, and partly online. This event was the largest, so far, in the Amitsur Memorial Symposium series, hosted annually by different institutions in Israel since Amitsur's passing away in 1994.

This volume presents 15 contributions by invited speakers of the Centennial Symposium, coming from areas of algebra on which Professor Amitsur had a fundamental impact: group theory, structure theory of rings, central simple algebras, and nonassociative algebra.

The volume also includes a biography of Amitsur, written by Avinoam Mann for Amitsur's Selected Papers (AMS, 2001). Mann, who was one of Amitsur's first PhD students, and a world figure in group theory, passed away in March 2023, after the work on this volume was largely completed. His paper titled "Challenges for 21st century group theory" opens the volume.

We acknowledge with thanks the support provided for the Conference by the Hebrew University of Jerusalem, the Israel Institute of Advanced Studies, and Bar-Ilan University. We note that the production of the volume is partially supported by the Israeli Science Foundation grant #1994/20." Finally, we thank Miriam Beller, who served as scientific editor.

The Editors

Conference Program

Monday, 1 November

12:00-12:50 *Registration and Lunch at the Israel Institute of Advanced Study (IIAS)*

13:00-13:50 *Visit to the Newton Manuscripts (National Library of Israel)*

13:55-14:00 *Greetings by Yitzhak Hen (IIAS Director)*

14:00-14:40 **Yoav Segev** (Ben-Gurion University of the Negev): A characterization of the quaternions using commutators

14:40-15:20 **Eliyahu Matzri** (Bar-Ilan University): On the symbol length of symbols in Galois cohomology

15:20-16:00 **Adam Chapman** (Academic College, Tel-Aviv Yafo): The essential dimension of central simple algebras in the bad characteristic case

16:00-16:20 *Coffee Break*

16:20-17:00 **Uriya First** (University of Haifa): Azumaya algebras admitting just one type of involution

17:00-17:40 **Amiram Braun** (University of Haifa): p-groups and the polynomial ring of invariants question

17:40-18:20 **Michael Larsen** (Indiana University, Bloomington): Almost polynomial identities

18:30-19:30 *Reception with the Amitsur family*

Tuesday, 2 November

15:00-15:40 **Sean Eberhard** (University of Cambridge): Probabilistically nilpotent groups of class two

15:50-16:30 **Chloe Perin** (The Hebrew University of Jerusalem): Homogeneity in torsion-free hyperbolic groups

16:40-17:20 **Pham Huu Tiep** (Rutgers University): Word maps on simple groups: width and distribution

17:30-18:10 **Anna Erschler** (École Normale Supérieure): Poisson-Furstenberg boundary and Liouville property of linear finitely generated groups

18:20-19:00 **Raman Parimala** (Emory University): Pencils of quadric period index questions for hyperelliptic curves

19:10-19:50 **Robert Guralnick** (University of Southern California): Generation theorems for finite groups

Wednesday, 3 November

15:00-15:40 **Agata Smoktunowicz** (University of Edinburgh): From Amitsur's conjectures on nil rings to braces, pre-Lie algebras and Hopf-Galois extensions

15:50-16:30 **Alex Lubotzky** (The Hebrew University of Jerusalem): Locally testable good codes

16:40-17:20 **Sue Sierra** (University of Edinburgh): Affine Lie algebras and just-infinite growth

17:30-18:10 **Igor Rapinchuk** (Michigan State University): Groups with good reduction and the genus problem

18:20-19:00 **Jason Bell** (University of Waterloo): Specht's theorem, commutativity theorems, and decision procedures

19:10-19:50 **Danny Krashen** (Rutgers University): Splitting fields of central simple algebras

Thursday, 4 November

15:00-15:30 *Reminiscences and personal memories of Shimshon Amitsur*

15:40-16:00 **Alexei Belov** (Bar-Ilan University): Gelfand conjecture and the method of proof of Specht's problem

16:00-16:20 **Eugene Plotkin** (Bar-Ilan University): Geometric equivalence of algebras

16:20-16:40 **Leonid Bokut** (Sobolev Institute of Mathematics): Gröbner-Shirshov bases theory and word problems for metatrivial Poisson algebras

16:40-17:20 **Miriam Cohen** (Ben-Gurion University of the Negev): Representations of the quantum double of quasi-triangular Hopf algebras

17:30-18:10 **Claudio Procesi** (Universitá degli Studi di Roma): Tensor fundamental theorems of Invariant Theory

18:20-19:00 **Be'eri Greenfeld** (UC San Diego): How do algebras grow?

19:10-19:50 **Efim Zelmanov** (UC San Diego): Brackets and algebras

Contemporary Mathematics
Volume **800**, 2024
https://doi.org/10.1090/conm/800/16048

Shimshon Avraham Amitsur (1921–1994)

Avinoam Mann

Shimshon Avraham Amitsur was born, under the name Shimshon Kaplan on 26 August 1921, in Jerusalem, and passed away on 5 September 1994, in the same city. He was survived by his wife Sarah, his children Hanna, Michal, and Eli, and several grandchildren.

Several years after his birth his family moved to Tel Aviv, where he attended a commercial school. His talents were noticed by his teachers, and his principal, S. Maharshak, a well known writer of high school textbooks in mathematics, arranged a fund for his studies at the Hebrew University. These studies were interrupted by his army service during World War II and during Israel's War of Independence. His letters to Levitzki show how he kept up his interest in mathematics even during this service. He received his M.Sc. degree in 1946, and his Ph.D. in 1950, both under the supervision of Professor Jacob Levitzki. Around that time he changed his name to the Hebrew one Amitsur. Levitzki's report on the M.Sc. thesis is titled "A Report on the Thesis of Shimshon Amitsur", but in the report itself Levitzki forgot the change of name and refers to Mr. Kaplan. Amitsur's appreciation of his great teacher was often witnessed by all his friends, and he was a very worthy successor to Levitzki in their jointly loved area, ring theory. Teacher and student received in 1953, for their joint work, the first Israel Prize in exact sciences, the most distinguished prize in Israel. Later Amitsur won several other prizes, and in 1990 was awarded the degree of Doctor of Philosophy (Honoris Causa) from Ben-Gurion University of the Negev.

The early death of Levitzki in 1956, at age 52, left Amitsur as the leading algebraist in Israel, a position he held indisputably until his death, even after his retirement in 1989. This retirement, marked by a two week ring theory conference organised by the Hebrew University and Bar-Ilan University, did not stop his mathematical productivity in the least. His influence on Israeli mathematics was enormous, by no means restricted only to algebra, let alone ring theory. Witness his students: the first one worked in non-associative algebra, the second in group theory, another one in category theory, and one of the last ones in Lie superalgebras. This diversity resulted from Amitsur's willingness to give his students free rein to choose their area. As one of these students, I can testify that, while he let me work in my preferred area, he was still very much interested in all that I did, monitoring

This biography appeared in the *Selected Papers of S. A. Amitsur*, American Mathematical Society, 2001. Originally published by the Israel Journal of Mathematics (Mann, Avinoam, "Amitsur (1921–1994)," vol. 96 (1996), ix–xxvii). Reprinted here with permission.

it closely and giving valuable advice. He was always inquiring about my progress; it was very difficult to be slack while working under Shimshon. On the other hand, he would also tell me about his own progress, seeming, to my surprise, to regard me as an equal, and was always ready to share his knowledge. In all of this he set a remarkable personal example, no less instructive than direct mathematical advice.

Of course he also had many students in ring theory. Moreover, many ring theorists throughout the world, while not his students nominally, feel themselves so morally. For nearly fifty years he influenced the ring theory world greatly, through example, suggestion, and personal contact.

Amitsur was one of the leaders of the Einstein Institute of Mathematics at the Hebrew University, contributing much to its development and seeing to it that it kept its high level. He was also very active in the life of the university. Though he avoided offices such as dean etc., he was a member of many committees, and his voice was often heard in faculty meetings. He was a member of the Israel Academy of Science, in which he fulfilled many important duties, e.g. Head of the Section for Experimental Sciences. He was one of the founders and editors of the Israel Journal of Mathematics. He was also an editor of many other mathematical journals, and the mathematical editor of the Hebrew Encyclopedia (in that office he replaced the late A. A. Fraenkel).

He was a member of the Department of Science Teaching at the university, and had several students in that area. For many years he was the motivating force in developing new programs for the teaching of mathematics in Israeli high schools. He will be missed in this as in all his other activities.

Contemporary Mathematics
Volume **800**, 2024
https://doi.org/10.1090/conm/800/16049

Challenges for 21st century group theory

Avinoam Mann

ABSTRACT. We suggest some directions for future research in group theory.

I list here some of the directions in which I would like group theory to develop. I challenge other mathematicians to produce their lists. Despite the title, this paper is meant to be also read by people who are not specialists in group theory; hence I expand on some details which group theorists may consider superfluous.

1. The Inverse Galois Problem.
2. CFSG
3. Grand Unified Theory
4. Analytic Group Theory
5. Classical Problems

1. The Inverse Galois Problem

Is every finite group the Galois group of some extension of the rationals? In other words: is every finite group the Galois group of some equation $a_nx^n + a_{n-1}x^{n-1} + \ldots + a_0 = 0$, with integer coefficients?

It may be claimed that this is not a group theoretical problem, but rather a field theoretical or a number theoretical problem. That may be so. The fact is that progress in finite group theory, such as CFSG (to be discussed in the next section) brought about also progress in that problem, with some of that progress made by group theorists. Thus I consider it a proper problem for workers in group theory, not to mention its great interest. The study of equations starts as early as in early Mesopotomian mathematics, nearly four thousands years ago. Thus it may be thought of as the oldest problem in mathematics. But as long as the inverse problem is still open, we cannot claim to fully understand these, the simplest of equations.

2. The Classification of the Finite Simple Groups (CFSG)

That classification is one of the great achievements of 20th century mathematics (actually, it continued to the early years of the 21st century). The term 'simple

---

2020 *Mathematics Subject Classification.* 20-02.

Key words and phrases. Groups.

Professor Avinoam Mann passed away in March 2023. He was a Professor Emeritus in the Einstein Institute of Mathematics at The Hebrew University of Jerusalem.

group' was coined by Galois, at the same time as the term 'group'. Galois also gave the first (infinitely many) non-trivial examples. More examples were constructed throughout the 19th century; they consist of several infinite families, and were summarized by L.E. Dickson in his book *Linear Groups* [**5**]. Dickson himself went on to produce two further families. That was in the early years of the 20th century, and then the matter laid dormant till the mid century, when C. Chevalley and others discovered more families. At the same time began the effort of the classification, aimed at producing a list of all finite simple groups. As mentioned above, that effort succeeded, after about half a century. This is the result of joint work by many mathematicians, who produced dozens of papers, containing thousands of pages. The most prominent name in that effort is John G. Thompson, and, in the later stages, Michael Aschbacher. We should mention also Richard Brauer, as one of the initiators and encouragers of that undertaking, and Michio Suzuki, who both characterized many of the known families, as well as discovering new ones. The final theorem states that finite simple groups consist of the following four types:

1. Groups of prime order.
2. The alternating groups A_n, for $n \geq 5$.
3. Groups of Lie type.
4. Twenty six further groups, known as 'sporadic'.

We are going to elaborate on the latter two types. The name 'of Lie type' will be explained in the next section. For now, let F be any field, and $n \geq 2$ an integer. We consider the *general linear group* $G := GL(n, F)$ of all invertible $n \times n$ matrices over F. Equivalently, we may consider the isomorphic group $GL(V)$ of all invertible linear transformation of an n-dimensional vector space V over F. That group has two natural normal subgroups. The first is the *special linear group* $SL(n, F)$ of all matrices of determinant 1. Since this is the kernel of the determinant homomorphism, the factor group over it is isomorphic to the multiplicative group of F. The second normal subgroup is Z, the centre of G, which consists of all non-zero scalar matrices. The factor group G/Z is isomorphic to a group of collineations of the $(n-1)$-dimensional projective space over F; hence it is termed the *general projective group* $PGL(n, F)$, and its subgroup, the image of the special linear group, is the *special projective group* $PSL(n, F)$. This latter group, the special projective one, is simple, with two exceptions when $n = 2$ and F has only 2 or 3 elements. In particular, when the field F is finite, hence its order is p^k, for some prime p, we obtain a finite simple group, often denoted by $PSL(n, p^k)$. This is a typical simple group of Lie type. Then there are subgroups of that group which preserve some non-degenerate bilinear form (u, v) on V. Three types of bilinear forms are relevant here: Symmetric forms, in which $(u, v) = (v, u)$, give rise to the *orthogonal groups*, skew-symmetric forms, $(u, v) = -(v, u)$, to the *symplectic groups*, and Hermitian forms, $(u, v) = (v, u)^\sigma$, to the *unitary groups*. Here σ is an automorphism of order 2 of F, e.g., complex conjugation if $F = \mathbb{C}$, the complex numbers. To obtain simple groups, we usually have to take in these groups the elements of determinant 1, and divide by the centre, which consists of scalar matrices. There are complications in the case of the orthogonal groups, in which the determinant has to be replaced by the so-called *spinorial norm*, and in characteristic 2, when symmetric and skew-symmetric are the same, so the first one is replaced by non-zero quadratic forms, which are the function $Q(u) = (u, u)$, and the skew-symmetric forms are replaced by alternate ones, i.e., satisfying $(u, u) = 0$. Note that non-degenerate skew-symmetric (or alternate) forms exist only when n is even, and, in the finite field case, F has an automorphism of order 2 iff $|F| = p^k$ with k even. Also, in the finite field case, there are two types of symmetric forms, giving rise to two families of simple groups, while there is only one type of symplectic or unitary groups (for the symplectic group, this is true for any field, but the situation for the orthogonal and unitary group is much more complicated for infinite fields).

These four types of groups are known as *the classical groups*. Next, there are five families of *exceptional groups*. These are denoted by E_6, E_7, E_8, F_4, and G_2. They can also be considered as groups of matrices. They are defined for any field, but only in one dimension, peculiar for each family. Finally, there are the *twisted groups*, which are essentially the sets of fixed points of some of the above groups under some special automorphism. For example, the unitary groups are a twisted form of the general linear ones. These automorphisms may exist only for some special groups, e.g., a family known as the *Suzuki groups* is defined only for fields of order 2^{2k+1}, for some k (and also for some infinite fields of characteristic 2). This ends the list of groups of Lie type, which comprise most of the finite simple groups.

There remain the sporadic groups. These are the twenty six groups that do not fit in the previous families. The first five, known as the *Mathieu groups* $M_{11}, M_{12}, M_{22}, M_{23}$, and M_{24},, were already discovered in the late 19th century, as interesting permutation groups. Recall that a permutation group G on a set S is *k-transitive* if, given any two ordered k-tuples of distinct elements from S, say $s_1, \ldots, s_k$ and $t_1, \ldots, t_k$, there is an element $x \in G$ such that $s_i^x = t_i$. It is clear that if $|S| = n$, then the symmetric group S_n is the only n-transitive group on S, and also the only $(n-1)$-transitive group. And S_n and A_n are the only $(n-2)$-transitive groups. Apart from these two, it is not easy to find multiply transitive groups, i.e., with $k \geq 2$. As one example, for any field F, consider the group of all functions $ax+b$, with $a, b \in F$ and $a \neq 0$. This is a 2-transitive group on F. There are several more families of 2-transitive groups, e.g., the aforementioned Suzuki groups were discovered first as certain 2-transitive groups. Then it was noticed that they are also a twisted form of some symplectic groups. There are also infinite families of triply transitive groups ($k = 3$). The Mathieu groups are permutation groups on the indicated number of letters, with M_{12} and M_{24} 5-transitive. Then M_{11} and M_{23}, which are the stabilizers of a single letter in the previous groups, are 4-transitive, and M_{22}, the stabilizer of a letter in M_{23}, is 3-transitive. As a corollary of CFSG, all 2-transitive finite groups were determined, and it turns out that, apart from the symmetric and alternating groups, the two groups M_{12} and M_{24} are the only 5-transitive groups, and with the addition of M_{11} and M_{23}, we have the only 4-transitive groups. There are no 6-transitive groups (the situation for infinite groups is very different). The other sporadic groups were discovered since the mid 20th century, during the effort of the classification. They arise sometimes as automorphism groups of strange combinatorial structures, and sometimes as exotic extensions of other groups. For a more detailed description of the simple groups, see [**15**].

As we have said, the proof of CFSG occupies hundreds of papers spanning thousands of pages. Given that, two things are clear. First, no single person has read all these papers, let alone checked them for correctness. Second, it is impossible to write that amount of mathematics without some errors creeping in. Thus, the belief in the classification theorem is to some extent an act of faith. On the other hand, CFSG is such a useful tool, that we would not like to give it up. To remedy that situation, the late Daniel Gorenstein initiated writing a series of books, which will contain the proof of the classification, omitting steps that were found to be unnecessary, and carefully checking everything. This is being carried out now by Ron Solomon, Richard Lyons, and Inna Capdeboscq. It is expected that the series will consist of 12 or 13 books, of which 9 have already been published. In parallel, there is a programme called Revisionism, which aims at simplifying some of the proof, by using alternative methods. This still keeps the overall strategy of the proof, and, even if successful, it is still going to be very long and demanding.

Thus the challenge is simple: find an approach to CFSG, which will enable mathematicians to go through it, without devoting about half of their lifetime for that. The problem is that we have no idea where to look for such an approach. Since the proof of the CFSG, a lot of research has been done on the finite simple groups, investigating their representations, subgroups, etc. Hopefully, this will lead to some insight about the correct way to go about the classification.

3. Grand Unified Theory

CFSG is not the first classification of simple groups. It was preceded by the determination, by W. Killing and Elie Cartan, of the simple Lie groups, around the turn of the 20th century, and, by C. Chevalley, of the simple algebraic groups, in the middle of the same century. Let us review these.

Recall that a *topological group* is a group, which is also a topological space, in which the group operations, multiplication and inversion, are continuous. An *n-manifold* is a topological space in which, first, each point has a neighbourhood homeomorphic to Euclidean space E^n. Moreover, given two points x and y, and a point z lying in the intersection of the two Euclidean neighbourhoods of x and y, the point z can be described by two systems of coordinates, one in each neighbourhood. It is required that the passage between these two systems is given by continuous functions. A *Lie group* is a topological group which, topologically, is an n-manifold, and in which the above passage functions, as well as the group operations, are not only continuous, but analytic. Among the subgroups of Lie groups we consider only the closed ones. These groups were introduced by Sophus Lie, in an attempt to develop for differential equations a theory similar to the Galois theory for polynomial equations, and they occur everywhere, both in mathematics and in its applications to physics, chemistry, etc. If we replace in the definition the Euclidean space E^n by complex space $\mathbb{C}^n$, we obtain *analytic groups.*

The simple analytic groups consist of four families and five more *exceptional groups.* The four families, denoted by A_n, B_n, C_n, and D_n, include linear, orthogonal, and symplectic groups, defined over the complex field. The simple Lie groups include some more, e.g., the unitary groups, which are not analytic, because conjugation is not an analytic function. See also the previous section for some remarks regarding the exceptional groups.

An *algebraic group* is a group which is also a finite dimensional algebraic variety, in which the group operations are rational. We consider only subgroups which are themselves algebraic varieties, i.e., they are closed in the Zarisky topology. We are looking for simple algebraic groups over algebraically closed fields. In characteristic 0, the list is identical to that of the simple Lie groups. In positive characteristic, there was a problem: it was not known if groups analogous to the exceptional Lie groups exist. Over finite fields, two of the exceptional groups were already constructed by L.E. Dickson, who did not explain how he discovered them. C. Chevalley managed to construct all of them, over any field, starting from the Lie algebras associated to the relevant groups. This constructed some of the groups of Lie type discussed in the previous section; then the twisted ones were added. Chevalley was now able to show that also in positive characteristic the list of the simple algebraic groups (over algebraically closed fields) is identical to the list of simple Lie groups.

We thus have three classification theorems. In each of them we start from a similar premise, we are interested in simple groups with some extra structure. And we obtain a similar answer, the list consists essentially of the groups of Lie type. Naturally there are some variations in each case. In the finite group case, these are the occurrence of the alternating and sporadic groups. When we have such a situation of similarity, we look for a unified frame. This is what I mean by 'a grand unified theory'. A point of view that will explain the similarity of the three theorems, and, hopefully, would point the way to a unified proof. As in the previous section, we have no idea of where to look for such a theory.

It is possible that there will be a fourth theorem to add to the above three (that was pointed out to me by G. Cherlin), the determination of the simple groups of finite Morley rank. I am not going to give the exact definition of finite Morley rank, just remark that it is a model theoretic analogue of finite dimensionality. It is conjectured that the simple groups of finite Morley rank are the simple algebraic groups over algebraically closed fields. That conjecture is still an active area of research, and, if proved, it will be the fourth classification theorem in our list.

4. Analytic Group Theory

By the title of this section, I mean the application of the powerful methods of mathematical analysis to group theory. I will mention some instances, but, as I will explain, I would like to see many more.

An area where it seems natural to apply such methods is *asymptotic group theory*, an area which was developed in recent decades. The most known branch of this is centered on the notion of *growth*. Given a finitely generated group G, a system $\{s_1, \ldots, s_d\}$ of generators for it, and an element $x \in G$, we write x as a product of the generators and their inverses, and among all such expressions, we choose the ones with a minimal number of factors. This number is termed the *length* $l(x)$ of x. We write $a_n(G)$ and $s_n(G)$ for the number of elements of length n, and of length at most n, respectively. These numbers measure *the growth* of G. If G is finite, then eventually we have $a_n = 0$ and $s_n = |G|$. Our main interest is in infinite groups. It is easy to see that the growth is at most exponential, indeed $a_n \leq 2d(2d-1)^{n-1}$, with equality iff G is either a free group of rank d, or the free product of $2d$ groups of order 2. On the other hand, since G is infinite, there must be an element of each length, therefore $a_n \geq 1$. Moreover, there exists an element of size $2n$, and writing it as xy, with $l(x) = l(y) = n$, we see that $x \neq y^{-1}$; therefore, $a_n \geq 2$ and $s_n \geq 2n - 1$, with equality iff G is infinite cyclic ($d = 1$) or infinite dihedral ($d = 2$). It is not difficult to see that if G is abelian, then the growth functions are polynomial in n. It is more difficult to establish that if G is nilpotent, then the growth is *polynomial*, meaning that there exist numbers A and s, such that $s_n(G) \leq An^s$. A seminal result of M. Gromov is that the converse holds: G has polynomial growth iff it is virtually nilpotent, i.e., contains a nilpotent subgroup of finite index. Gromov's proof applies the solution of Hilbert's 5th problem, so it is an instance of the use of analysis in group theory, and the same holds for alternative proofs that were found, which employ the concept of harmonic functions on groups, which is modeled on the analytic notion of harmonic functions. But it seems that more applications are called for. Since the growth is always at most exponential, it is natural to consider the generating functions $A(x) = \sum a_n x^n$ and $S(x) = \sum s_n x^n$ (here $A(x) = (1-x)S(x)$). What can we tell about these functions, and what do they tell us about G? In some cases these functions satisfy interesting functional equations. On the other hand, it was first observed by J.P. Serre that for some groups the values of these functions are related to the so called *Euler characteristic* of the group. These type of results are among the 'many more' that I have mentioned above.

Gromov's theorem shows that most groups have super-polynomial growth. R.I. Grigorchuk was the first to construct groups of *intermediate growth*, meaning that the growth is super-polynomial but sub-exponential (no inequality $s_n(G) \geq Ac^n$, with $c > 1$, holds). By now, many such groups are known, with various types of growth. The groups of sub-exponential growth (including polynomial growth) can be characterized as the groups for which the power series $A(x)$ and $S(x)$ above have radius of convergence 1, while for exponential growth that radius is smaller (for more details on growth, see [7] and [**12**], and Ch. 15 of the latter for the generating functions).

Another branch of asymptotic group theory is *subgroup growth*. This starts from the fact, that if G is finitely generated, and n is some natural number, then G contains only finitely many subgroups of index n. Thus it is natural to consider the number $a_n(G)$ of such subgroups (not to be confused with the notation $a_n(G)$ in the previous paragraph). The maximal value of these numbers is obtained when $G = F^d$ is a free group, and if it has rank d, then this number is asymptotically $n \cdot n^{d-1}$, which is super-exponential. This time we can say that the analytic applications came first. Indeed, while these numbers were considered already in the late 19th century, and the asymptotic formula above is deduced from a recursion formula for $a_n(F^d)$ due to M. Hall, the present interest arose when D. Segal had the idea to define, analogously to the Dedekind ζ-function of an algebraic number fields, the Dirichlet series $\zeta_G(s) = \sum \frac{a_n(G)}{n^s}$ (in the Dedekind function one counts only ideals). For this series to converge, $a_n(G)$ has to grow polynomially. This happens, e.g., when G is nilpotent, and for that case many interesting results were derived, including functional equations and more. M.P.F. du Sautoy and F. Gruenwald have produced an analytic continuation for the ζ-function of finitely generated nilpotent groups, and applied this to obtain results about the numbers $a_n(G)$. This is exactly the kind of analytic methods to which I referred above.

The groups for which the subgroup growth is polynomial were determined by A. Lubotzky, A. Mann and D. Segal. For all that, see [**11**]. We should remark that, since we are considering subgroups of finite index, the discussion is usually limited to groups in which the intersection of all these subgroups is the identity. Such groups are called *residually finite*, and they include free groups, as well as finitely generated nilpotent groups. In a residually finite group, the intersection of all normal subgroups of finite index is also trivial; therefore, the groups embeds in the Cartesian product of all its finite homomorphic images.

Several variations on that ζ-function were suggested, e.g., counting only normal subgroups. Also, one can count other objects, e.g., conjugacy classes of a given size, or irreducible representations of a given degree, where we have to assume that these numbers are finite. We will mention one more suggestion for a ζ-function, one which does not involve counting.

As noted above, a residually finite group G is a subgroup of a Cartesian product of finite groups, say C. We give each of these finite factors the discrete topology, and give C the product topology. If G is infinite, then C is not discrete, but it is compact, by Tychonoff's theorem. If G happens to coincide with its closure in C, it is termed a *profinite group*. Equivalent definitions of a profinite group are as inverse limit of finite groups, and as a topological group which is compact, Hausdorff, and totally disconnected (the only connected subsets are single points). In any case, the closure of G in C is termed the *profinite closure* of G.

It is natural, when considering topological groups, and in particular profinite ones, to consider only closed subgroups. However, a deep theorem of N. Nikolov–D.Segal states that in a finitely generated profinite group, all subgroups of finite index are closed (the proof employs CFSG, among other tools). Since below we will be concerned with finite index subgroups of finitely generated groups, we do not have to worry about the closure of these subgroups.

A compact topological group G carries a finite *Haar measure*, i.e., a measure which is invariant under translations, a subset A and its translates xA and Ax have the same measure, and G has finite measure. We can then normalize that measure so that G has measure 1, and is thus a probability space. This can be done in a unique way. If G can be generated, as a topological group, by k elements, then the set of all ordered k-tuples which generate it is closed in G^k, therefore measurable (in the product measure, which is the normalized Haar measure on G^k), say it has measure $P(G,k)$. The author suggested calling a profinite group for which $P(G,k) > 0$, for some k, a *positively finitely generated* group **(PFG)**. This terminology extends to residually finite groups, by considering their profinite closures. Soluble groups, and arithmetic groups with the congruence subgroups property, are **PFG**, while free groups of rank at least 2 are not. There are many more examples.

In particular, consider $\mathbf{Z}$, the infinite cyclic group. For it, we have $P(\mathbf{Z},1) = 0$, but $P(\mathbf{Z},k) = \frac{1}{\zeta(k)}$ for $k \geq 2$, where $\zeta(k)$ is the usual number theoretical ζ-function. This and other examples led the author to make the following

CONJECTURE. *If G is a* **PFG** *group, then the positive values of $P(G,k)$ can be interpolated to an analytic function $f(z)$.*

The conjecture holds, e.g., for soluble groups, for arithmetic groups with the congruence subgroup property, and many more. The reciprocal of $f(z)$, when it exists, is then also a candidate for a ζ-function. It has come to be known as the *probabilistic ζ-function.* The author suggested two ways to define it, one as a Dirichlet series, and another as an infinite product of Dirichlet polynomials. The latter way shows that this ζ-function has a factorization similar to the Euler factorization of the ordinary ζ-function, with the factors indexed by the finite simple groups. To prove the conjecture, we have to show that the series, or the product, converges.

There are several characterizations of **PFG** groups. The first relates them to subgroup growth. Let us write $m_n(G)$ for the number of maximal subgroups of G of index n. We term a group a *polynomial maximal subgroup growth* **(PMSG)** if these numbers grow polynomially. A. Mann–A. Shalev proved, that a finitely generated profinite groups is **PFG** iff it is **PMSG**. While the direction from **PMSG** to **PFG** of this equivalence is easy, the other direction requires CFSG and various other tools, both from group theory and from probability.

Another characterization is by the multiplicities of chief factors. This is due to A. Jaikin-Zapirain and L. Pyber, who established other interesting properties of these groups. The proof that soluble groups are **PFG** is by showing that they are **PMSG**, and a similar argument shows that in a finitely generated soluble groups the number of all subgroups grows at most exponentially, $a_n(G) \leq An^s$ (recall that in non-abelian free groups that growth is super-exponential). We term such groups as **ESP** (*exponential subgroup property*). Jaikin-Zapirain and Pyber characterized also **ESP** groups, and that implies that **PFG** are **ESP**, but the two properties are not equivalent.

It is then natural to consider for **ESP** groups the generating function $\sum a_n(G)x^n$, and for **PFG** groups the function $\sum \frac{m_n(G)}{n^s}$. The ground is rich for analytic considerations.

5. Classical Problems

Let us start with the Burnside problem. This is the question: do there exist finitely generated infinite torsion groups? The positive answer was given by Golod-Shafarevich. It is of interest to note that their proof employs analytic tools, though by now purely algebraic proofs are available. There remained the question: can such groups have the orders of its elements bounded? equivalently, can they have a finite exponent, i.e., there exists a number n such that all elements x of the group satisfy $x^n = 1$? If $n = 2$, such groups are abelian and finite. Burnside proved that they are also finite if $n = 3$. Much later, Sanov showed the same for $n = 4$, and M. Hall for $n = 6$. These are the only exponents for which finitely generated groups of that exponent are known to be finite. In particular, the case of exponent 5 is still wide open. And if any one wants to find more such exponents, my feeling is that it is better to try exponents such as 8, 9, or 12, rather than 5.

The existence of finitely generated infinite groups of finite exponent was announced by P.S. Novikov in 1959, who claimed it for exponents at least 72. The proof was published much later, in a joint work by Novikov and S.I. Adian (1968), and only for odd exponents at least 4381. Later S. Ivanov and I. Lysyonok proved it also for even exponents n, provided that $n \geq 8000$ and is divisible by 16. By now the size of odd exponents has been reduced to 101, so there is still a big gap between the finite and infinite cases. The question then is, for which exponents finitely generated groups of that exponents are finite? This is probably very hard, and we would like at least to see a narrowing of the gap.

To quote one of the reviews on papers dealing with the Burnside problem: the constructions of Golod and Shafarevich, or even of Novikov and Adian, do not end the story, they are just one more contribution to the question: what makes a group finite? Here is a still unanswered question from the first edition (1965) of the Kourovka notebook of unsolved problems in group theory: is a residually finite group with the maximum condition virtually polycyclic? Here we recall that a group is *polycyclic*, if it has a normal series with cyclic factors. Equivalently, it is soluble and satisfies the maximum condition. Virtually polycyclic groups are residually finite.

These questions regard finitely generated groups. What if we narrow the field, and look only at finitely presented groups, which number only countably many, against continuously many finitely generated ones? These are wide open: do there exist finitely presented infinite torsion groups? Do there exist infinite finitely presented groups satisfying the minimum condition? or even both chain conditions? Note that the minimum condition implies torsion.

Another question regarding finitely presented groups is: do there exist such groups of intermediate growth? The association with previous questions is natural, if we recall that the Grigorchuk group of intermediate growth was first presented as another, simply constructed, finitely generated infinite torsion group (of infinite exponent).

The factor group $B(r,n) = F^r/(F^r)^n$ of the free group of rank r by its subgroup generated by nth powers is called the *Burnside Group* of rank r and exponent n. It is known that if $B(2,n)$ is finite, then so are all Burnside groups of the same exponent. In that case, $B(r,n)$ is the biggest finite group of exponent n which can be generated by r elements. But we know already that for most exponents, these groups are infinite. The *Restricted Burnside Problem* asks: among the *finite* groups of exponent n and r generators, does there exist a biggest one (and then all others are homomorphic images of it). A seminal paper of P. Hall and G. Higman shows, among other things, that CFSG reduces that question to the case when n is a prime power. Higman and A.I. Kostrikin gave a positive answer for $n = 5$, with the latter also dealing with $n = 7$. Then Kostrikin gave a positive answer for all prime exponents, and finally E.I. Zelmanov settled the general case (see [**14**]). The question there arises: how big is that biggest finite group?

Upper bounds for its order were first given by S.I. Adian and A.A. Razborov. The following upper bound was given by D. Groves and M.R. Vaughan-Lee, improving an earlier one by Zelmanov and Vaughan-Lee:

$$|B(d,n)| \leq d^{d^{\cdot^{\cdot^{\cdot^{d}}}}}$$

where d occurs in the above tower n^{4^n} times (the Vaughan-Lee–Zelmanov bound was similar, but the height of the tower was n^{n^n}). If these bounds look terrifying, remember that they are very significant improvements on the previous ones. When $n = p$ is prime, the height of the tower can be reduced to 2^p, (3^p if $p \leq 23$), but this does not really console us. However, the following much better bounds holding for small values of n promise some hope (see [**14**] and the papers quoted there). In these bounds, n is either a prime power, so $B(d,n)$ is a p-group; or, $n = 6$.

$$|B(d,2)| = 2^d;$$
$$|B(d,3)| = 3^{d+\binom{d}{2}+\binom{d}{3}};$$
$$|B(2,4)| = 2^{12}, \quad |B(3,4)| = 2^{69}, \quad |B(4,4)| = 2^{422}, \quad |B(5,4)| = 2^{2728},$$
$$|B(d,4)| = 2^k \qquad \text{where } 2^{2d-1} \leq k < \tfrac{1}{2}(4+2^{3/2})^d;$$
$$|B(2,5)| = 5^{34}, \quad |B(3,5)| = 5^{2282};$$
$$|B(d,6)| = 2^a 3^b \qquad \text{for } a = 1+(d-1)3^{d+\binom{d}{2}+\binom{d}{3}} \text{ and } b = 1+(d-1)2^d;$$
$$|B(2,7)| = 7^{20416}.$$

So it seems possible that for a fixed prime power n, a bound for the class which is polynomial in d holds. However, it was pointed out by M.F. Newman that a bound in terms of n must be of a much bigger order of magnitude. Indeed, if $n = p^e$, we construct a group of exponent n as follows. Let F be a free group on d generators. Start with the largest factor group F/N_1 of exponent p. This has order at least p^d (the order of an abelian group of exponent p with d generators). The kernel N_1 is also a free group, with $1 + (d-1)|F : N_1|$ generators. Now take the largest factor group N_1/N_2 of exponent p, etc. After e steps we get a factor group F/N_e of exponent n with d generators and order at least $p^{p^{\cdot^{\cdot^{\cdot^{d}}}}}$, where the exponential tower has height e. Thus "toweric" functions do give the right order of magnitude, but there is still a very significant difference in the tower height between the upper and lower bounds. This is especially clear for $n = p$. For that case an improved lower bound was given by S.I. Adyan and N.N. Repin, who have shown that (for $p \geq 5$) $cl(B(2,p)) \geq 2^{2^{cp}}$ for some constant c (e.g., $c = 2/31$ is adequate.). A bound c for the class implies a bound n^{d^c} for the order.

The gap between the upper and lower bounds is still enormous, so we ask for reasonable estimates for the order of $B(d,n)$, and if n is a prime power, also for the nilpotency class and derived length.

In the investigation of the restricted Burnside problem, Engel identities play a significant role. We say that a group G, or a Lie ring L, are *n-Engel*, if they satisfy the identity $[x, y, \ldots, y] = 1$, where y occurs n times. If G satisfies such an identity for each pair x, y, but without a restriction on the number of occurrences of y, we say that G is *an Engel group*. Burnside first noted that groups of exponent 3 are 2-Engel, and applied this in his proof that groups of exponent three are locally finite. Generally, if G has exponent p, its associated Lie ring satisfies the $(p-1)$-st Engel identity, a fact that was instrumental in Kostrikin's solution of the restricted Burnside problem for exponent p. There is no similar result for non-prime exponent, but Zelmanov's results do imply that n-Engel Lie algebras are locally nilpotent (for characteristic 0, even nilpotent). Clearly, nilpotent groups of class c are c-Engel. The question corresponding to Burnside's problem here is: *are n-Engel groups locally nilpotent?* The answer is yes for $n \leq 4$ [**8**], and A. Juhasz and E. Rips have constructed n-Engel groups that are not locally nilpotent for $n \geq 20$. Their proof is as yet available only in recorded video talks, not in a written form.

As for other problems, it is still open if there exist finitely presented Engel groups that are not locally nilpotent. All these questions regarding finitely presented groups are believed to be very hard.

To be more specific about small n, 2-Engel groups are nilpotent of class at most three [**9**], and satisfy $exp(\gamma_3(G)) = 3$; so if they contain no elements of order 3, they are of class two [**3**].[1] This applies in particular for p-groups, if $p \neq 3$. This phenomenon of "exceptional primes" is typical.

Thus, 3-Engel groups satisfy $\gamma_5(G)^{20} = 1$, so if G contains no elements of order 2 or 5, it is nilpotent of class at most four. But in general a 3-Engel group need not be nilpotent, unless it is finitely generated. If it has d generators, it is nilpotent of class at most $2d-1$; this bound is best possible, but if G contains no elements of order 5, the class is at most $d+2$. Finally, G^5 is soluble, of length at most four (for these and further results, see [**6**] and papers quoted there).

[1] These results were reproved by F.W. Levi [**10**], and are often attributed to him. Indeed, [**9**] has some minor inaccuracies.

As for 4-Engel groups, they are nilpotent of class seven at most, provided that they do not have elements of order 2, 3, or 5. If they do not have elements of order 2 or 5, they need not be nilpotent, but they are soluble, of derived length at most six (see [**1**] and papers quoted there). Two-generated 4-Engel groups have nilpotence class at most four, and a d-generated 4-Engel group has class at most $4d$ [**13**].

For $n \geq 5$, we have to add to the Engel condition the assumption that G is locally nilpotent. Zelmanov proved that a torsion free locally nilpotent n-Engel group is nilpotent, with a bound $c(n)$, depending on n, for the class. It can be deduced from this, using general model theoretic principles, that there exists a set of *exceptional primes*, i.e., a finite set Π of primes, such that a locally nilpotent p-group is nilpotent, with the same bound on the class, provided that $p \notin \Pi$. This is equivalent to $\gamma_{c(n)+1}(G)$ being a Π-group. In [**2**], a stronger statement is proved: locally nilpotent n-Engel groups are actually (of finite exponent, say e)-by-(nilpotent of class c); and they are also (nilpotent of class c)-by-(of exponent e). Here both c and e depend on n. In particular, if G is a p-group and p does not divide e, then G is nilpotent. In [**4**], it is shown that c can be chosen so that even $G/Z_c(G)$ has bounded exponent. The obvious question is to find good estimates for c and e as functions of n.

References

[1] A. Abdollahi and G. Traustason, *On locally finite p-groups satisfying an Engel condition*, Proc. Amer. Math. Soc. **130** (2002), no. 10, 2827–2836, DOI 10.1090/S0002-9939-02-06421-3. MR1908264

[2] R. G. Burns and Y. Medvedev, *A note on Engel groups and local nilpotence*, J. Austral. Math. Soc. Ser. A **64** (1998), no. 1, 92–100. MR1490149

[3] W. Burnside, *On groups in which every two conjugate operators are permutable*, Proc. London Math. Soc. **35** (1902), 28-37.

[4] P. G. Crosby and G. Traustason, *A remark on the structure of n-Engel groups*, Comm. Algebra **39** (2011), no. 11, 3998–4001, DOI 10.1080/00927872.2010.514874. MR2855107

[5] L. E. Dickson, *Linear groups: With an exposition of the Galois field theory*, Dover Publications, Inc., New York, 1958. MR104735

[6] N. D. Gupta and M. F. Newman, *Third Engel groups*, Bull. Austral. Math. Soc. **40** (1989), no. 2, 215–230, DOI 10.1017/S0004972700004329. MR1012830

[7] P. de la Harpe, *Topics in geometric group theory*, Chicago Lectures in Mathematics, University of Chicago Press, Chicago, IL, 2000. MR1786869

[8] G. Havas and M. R. Vaughan-Lee, *4-Engel groups are locally nilpotent*, Internat. J. Algebra Comput. **15** (2005), no. 4, 649–682, DOI 10.1142/S0218196705002475. MR2160572

[9] C. Hopkins, *Finite Groups in which Conjugate Operations are Commutative*, Amer. J. Math. **51** (1929), no. 1, 35–41, DOI 10.2307/2370559. MR1506699

[10] F. W. Levi, *Groups in which the commutator operation satisfies certain algebraic conditions*, J. Indian Math. Soc. (N.S.) **6** (1942), 87–97. MR7417

[11] A. Lubotzky and D. Segal, *Subgroup growth*, Progress in Mathematics, vol. 212, Birkhäuser Verlag, Basel, 2003, DOI 10.1007/978-3-0348-8965-0. MR1978431

[12] A. Mann, *How groups grow*, London Mathematical Society Lecture Note Series, vol. 395, Cambridge University Press, Cambridge, 2012. MR2894945

[13] G. Traustason, *Locally nilpotent 4-Engel groups are Fitting groups*, J. Algebra **270** (2003), no. 1, 7–27, DOI 10.1016/j.jalgebra.2003.07.012. MR2015928

[14] M. Vaughan-Lee, *The restricted Burnside problem*, 2nd ed., London Mathematical Society Monographs. New Series, vol. 8, The Clarendon Press, Oxford University Press, New York, 1993. MR1364414

[15] R. A. Wilson, *The finite simple groups*, Graduate Texts in Mathematics, vol. 251, Springer-Verlag London, Ltd., London, 2009, DOI 10.1007/978-1-84800-988-2. MR2562037

Contemporary Mathematics
Volume **800**, 2024
https://doi.org/10.1090/conm/800/16050

Local systems and Suzuki groups

Levent Alpöge, Nicholas M. Katz, Gabriel Navarro, E. A. O'Brien, and Pham Huu Tiep

ABSTRACT. We study geometric monodromy groups $G_{\mathrm{geom},\mathcal{F}_q}$ of the local systems $\mathcal{F}_q$ on the affine line over $\mathbb{F}_2$ of rank $D = \sqrt{q/2}(q-1)$, $q = 2^{2n+1}$, constructed in N. Katz [*Exponential sums, Ree groups and Suzuki groups: conjectures*, Exp. Math. **28** (2019), 49-56.]. The main result of the paper shows that $G_{\mathrm{geom},\mathcal{F}_q}$ is either the Suzuki simple group ${}^2B_2(q)$, or the special linear group SL_D. We also show that $\mathcal{F}_8$ has geometric monodromy group ${}^2B_2(8)$, and arithmetic monodromy group $\mathrm{Aut}({}^2B_2(8))$ over $\mathbb{F}_2$, thus establishing Katz's Conjecture 2.2 in the above cited paper in the case $q = 8$.

CONTENTS

2020 *Mathematics Subject Classification.* Primary 11T23; Secondary 20C15, 20C33, 20D06, 20G05.

Key words and phrases. Local systems, airy sheaves, monodromy groups, Suzuki simple groups.

The first author gratefully acknowledges the support of the NSF (grant DMS-2002109) and the Society of Fellows.

The third author gratefully acknowledges the support of the Grant PID2019-103854GB-I00 funded by MCIN/AEI/10.13039/501100011033.

The fourth author gratefully acknowledges the support of the Marsden Fund of New Zealand via grant UOA 107.

The fifth author gratefully acknowledges the support of the NSF (grants DMS-1840702 and DMS-2200850), the Simons Foundation, and the Joshua Barlaz Chair in Mathematics.

The authors are grateful to the referee for careful reading and helpful comments on the paper.

Introduction

In an earlier paper [**26**], one of us, inspired by a paper [**12**] of Gross, defined, for each $n \geq 1$, a local system on $\mathbb{A}^1/\mathbb{F}_2$ of rank $2^n(2^{2n+1}-1)$, whose geometric monodromy group was conjectured to be the Suzuki group $\mathrm{Sz}(q) := {}^2B_2(q)$, $q = 2^{2n+1}$, in one of its two lowest dimensional nontrivial irreducible representations. These representations, complex conjugates of each other, are of dimension $2^n(2^{2n+1}-1)$ and have traces in $\mathbb{Z}[i]$, the Gaussian integers.

The definition involved p-Witt vectors of length 2 for $p = 2$, with values in $\mathbb{F}_2$-algebras. We identify $W_2(\mathbb{F}_2)$ with $\mathbb{Z}/4\mathbb{Z}$, by the map $[a, b] \mapsto a^2 + 2b$, with the usual convention that we first lift a, b to $\mathbb{Z}$ and then reduce $a^2 + 2b$ modulo 4. We take the additive character of $\mathbb{Z}/4\mathbb{Z}$ given by $n \mapsto i^n$, and view it as the additive character $\psi_2 : W_2(\mathbb{F}_2) \to \mu_4(\mathbb{Z}[i])$ given by

$$\psi_2([a, b]) := i^{a^2+2b}.$$

Attached to a Witt vector of length 2 with coefficients in $\mathbb{F}_2[x]$, say $[a(x), b(x)]$, is the Artin-Schreier-Witt sheaf $\mathcal{L}_{\psi_2([a(x),b(x)])}$ on $\mathbb{A}^1/\mathbb{F}_2$. It is lisse of rank one, and its trace function is given as follows. For $k/\mathbb{F}_2$ a finite extension, and $x \in k$,

$$\mathrm{Trace}(\mathrm{Frob}_{x,k}|\mathcal{L}_{\psi_2([a(x),b(x)])}) = \psi_2(\mathrm{Trace}_{W_2(k)/W_2(\mathbb{F}_2)}([a(x), b(x)])).$$

If instead we take $a(x), b(x) \in k_0[x]$ for some finite extension $k_0/\mathbb{F}_2$, then the Artin-Schreier-Witt sheaf $\mathcal{L}_{\psi_2([a(x),b(x)])}$ is lisse of rank one on $\mathbb{A}^1/k_0$, and for k/k_0 a finite extension and $x \in k$, the trace of $\mathrm{Frob}_{x,k}$ is given by the same formula (which only makes sense when k is an extension of k_0).

The unique nontrivial additive character ψ of $\mathbb{F}_2$ is related to ψ_2 by the formula

$$\psi(b) = \psi_2([0, b]);$$

this is simply the identity $(-1)^b = i^{2b}$.

Under Witt vector addition, $[a, b] = [a, 0]+[0, b]$. Thus we have the factorization

$$\mathcal{L}_{\psi_2([a(x),b(x)])} = \mathcal{L}_{\psi_2([a(x),0])} \otimes \mathcal{L}_{\psi(b(x))}.$$

When both $a(x)$ and $b(x)$ are polynomials of degree prime to $p = 2$, $\mathcal{L}_{\psi_2([a(x),0])}$ has Swan conductor $\mathsf{Swan}_\infty = p\deg(a(x))$, while $\mathcal{L}_{\psi(b(x))}$ has $\mathsf{Swan}_\infty = \deg(b(x))$. So in this case, $\mathcal{L}_{\psi_2([a(x),b(x)])}$ has $\mathsf{Swan}_\infty = \max(p\deg(a(x)), \deg(b(x)))$.

Quite generally, for an Artin-Schreier-Witt sheaf $\mathcal{L} := \mathcal{L}_{\psi_2([a(x),b(x)])}$ on $\mathbb{A}^1/k_0$ with Swan conductor $\mathsf{Swan}_\infty = n \geq 2$, its Fourier transform $\mathrm{FT}_\psi(\mathcal{L})$ on $\mathbb{A}^1/k_0$ is an *Airy sheaf* in the sense of Šuch [**44**]. It is lisse of rank $n-1$, and all its ∞-slopes are $\frac{n}{n-1}$. It is pure of weight one. Its trace function is given as follows: for k/k_0 a finite extension, and $t \in k$,

$$\mathrm{Trace}(\mathrm{Frob}_{t,k}|\mathrm{FT}_\psi(\mathcal{L})) = -\sum_{x\in k} \psi_2\big(\mathrm{Trace}_{W_2(k)/W_2(\mathbb{F}_2)}([a(x), b(x) + tx])\big).$$

Some key facts about Airy sheaves and their monodromy groups are due to Šuch [**44**], and are fundamental for the investigations reported on here.

We now turn to the Suzuki "candidates" of [**26**]. Here

$$n \geq 1\ , q_0 := 2^n,\ q := 2q_0^2,\ t(q) := q + 1 - 2q_0.$$

We take the Witt vector

$$\left[x^{t(q)}, \sum_{i=1}^{n} x^{(1+2^i)t(q)}\right],$$

form the Artin-Schreier-Witt sheaf

$$\mathcal{L} := \mathcal{L}_{\psi_2([x^{t(q)}, \sum_{i=1}^{n} x^{(1+2^i)t(q)}])},$$

form its $\mathrm{FT}_\psi(\mathcal{L})$, and twist by the constant field twist $(\frac{1}{1-(-1)^n i})^{\deg}$, to arrive at the local system $\mathcal{F}_q$ on $\mathbb{A}^1/\mathbb{F}_2$, whose trace function is given as follows. For $k/\mathbb{F}_2$ a finite extension, and $t \in k$,

$$\begin{aligned}&\mathrm{Trace}(\mathrm{Frob}_{t,k}|\mathcal{F}_q)\\ &= \frac{-1}{\big(1-(-1)^n i\big)^{\deg(k/\mathbb{F}_2)}} \sum_{x \in k} \psi_2\Big(\mathrm{Trace}_{W_2(k)/W_2(\mathbb{F}_2)}\Big(\Big[x^{t(q)}, \sum_{i=1}^{n} x^{(1+2^i)t(q)} + tx\Big]\Big)\Big).\end{aligned}$$

A key fact about $\mathcal{F}_q$ is that the input Witt vector $[x^{t(q)}, \sum_{i=1}^{n} x^{(1+2^i)t(q)}]$ is a function of $x^{t(q)}$. So in the trace formula for $\mathcal{F}_q$, if we restrict to $t \neq 0$ and make the substitution $x \mapsto x/t$, the formula becomes

$$\begin{aligned}&\mathrm{Trace}(\mathrm{Frob}_{t,k}|\mathcal{F}_q)\\ &= \frac{-1}{\big(1-(-1)^n i\big)^{\deg(k/\mathbb{F}_2)}} \sum_{x \in k} \psi_2\Big(\mathrm{Trace}_{W_2(k)/W_2(\mathbb{F}_2)}\Big(\Big[\frac{x^{t(q)}}{t^{t(q)}}, \sum_{i=1}^{n} \frac{x^{(1+2^i)t(q)}}{t^{(1+2^i)t(q)}} + x\Big]\Big)\Big),\end{aligned}$$

which is a function of $t^{t(q)}$. Thus $\mathcal{F}_q|\mathbb{G}_m$ has a descent to a lisse sheaf $\mathcal{G}_q$ on $\mathbb{G}_m/\mathbb{F}_2$ whose trace function is given by

$$\begin{aligned}&\mathrm{Trace}(\mathrm{Frob}_{t,k}|\mathcal{G}_q)\\ &= \frac{-1}{\big(1-(-1)^n i\big)^{\deg(k/\mathbb{F}_2)}} \sum_{x \in k} \psi_2\Big(\mathrm{Trace}_{W_2(k)/W_2(\mathbb{F}_2)}([\frac{x^{t(q)}}{t}, \sum_{i=1}^{n} \frac{x^{(1+2^i)t(q)}}{t^{(1+2^i)}} + x])\Big),\end{aligned}$$

and such that under the $t(q)$ Kummer pullback,

$$[t(q)]^\star \mathcal{G}_q = \mathcal{F}_q|\mathbb{G}_m.$$

For the case $n = 1$, i.e., for ${}^2B_2(8)$, we prove in Theorem 6.1 that $\mathcal{F}_8$ has the predicted geometric monodromy group $G_{\mathrm{geom},\mathcal{F}_8} = {}^2B_2(8)$ and arithmetic monodromy group $G_{\mathrm{arith},\mathcal{F}_8,\mathbb{F}_2} = \mathrm{Aut}({}^2B_2(8))$ over $\mathbb{F}_2$, and thus establish [**26**, Conjecture 2.2] in full in this case. For each $n \geq 1$, we show that $G_{\mathrm{geom},\mathcal{F}_q}$ is either ${}^2B_2(q)$ or SL_D for $D = \mathrm{rank}(\mathcal{F}_q) = q_0(q-1)$. A (huge) calculation of fourth moment for $\mathcal{F}_8$ shows that $G_{\mathrm{geom},\mathcal{F}_8}$ cannot be SL_{14}. It remains an open problem to prove (or disprove) that each $\mathcal{F}_q$ has $G_{\mathrm{geom},\mathcal{F}_q} = {}^2B_2(q)$ when $q \geq 32$.

In fact, we show in Theorem 8.4 that we have such a dichotomy of possible geometric monodromy groups G_{geom}, either ${}^2B_2(q)$ or SL_D, $D = q_0(q-1)$, for a general class of local systems of "the same shape" as $\mathcal{F}_q$ (see Remark 8.6 and Theorem 9.18 for examples of such local systems with $G_{\mathrm{geom}} = \mathrm{SL}_D$). Key to our investigation is the fact that these sheaves all satisfy the condition (**S**+) of [**30**, Definition 1.2]. Somewhat to our surprise, the most difficult part of establishing condition (**S**+) was to show the sheaves in question are geometrically primitive, i.e., that the representation of their G_{geom} is not induced. Our proof utilizes the existence of primitive prime divisors of the integer $t(q)$, cf. Theorem 3.4. We give a second proof which will be useful in future studies of geometric monodromy groups of quite general Airy sheaves. Condition (**S**+) implies that for each of these sheaves, either G_{geom} is a finite, almost quasisimple group, or has G°_{geom} acting irreducibly. This initial dichotomy, plus a substantial group-theoretic analysis, in which Theorem 4.5 plays a key role, is what leads to the ${}^2B_2(q)/\mathrm{SL}_D$ dichotomy. Pursuing the study of primitive prime divisors, we prove in Theorems 3.4, 3.7, and Corollary 3.5 the existence of primitive prime divisors in the orders of maximal tori of the Suzuki-Ree groups ${}^2B_2(q)$, ${}^2G_2(q)$, and ${}^2F_4(q)$. We also extend in §7 the classification of low-dimensional representations of classical groups in characteristic $p \geq 0$, beyond the bounds in [**32**] and [**33**]. These results will be useful in other situations as well. Finally, the structure of arithmetic monodromy groups, assuming finiteness, is determined in Theorem 9.14.

1. Descents of Suzuki candidates and moment calculations

For each odd power $q = 2^{2n+1}$ of 2, starting with $q = 8$, the finite simple group ${}^2B_2(q)$ has two (complex conjugate) lowest dimensional nontrivial irreducible representations, of dimension $d(q) := q_0(q-1)$, with $q_0 := 2^n$. In each case, we have a factorization

$$1 + d(q) = (q_0 + 1)t(q) \quad \text{with } t(q) := q + 1 - 2q_0.$$

In [**26**], for each such q there is proposed an Airy sheaf in the sense of Šuch [**44**], call it $\mathcal{F}_q$ on $\mathbb{A}^1/\mathbb{F}_2$ which is lisse of rank $d(q)$ and with $\mathsf{Swan}_\infty(\mathcal{F}_q) = 1 + d(q)$. Its $I(\infty)$-representation is irreducible of dimension $d(q)$, with all ∞-slopes $\frac{1+d(q)}{d(q)}$. Moreover, $\mathcal{F}_q$ is given [**26**, Section 4] with an explicit descent to a lisse sheaf $\mathcal{G}_q$ on $\mathbb{G}_m/\mathbb{F}_2$ whose Kummer pullback by $t(q)^{\text{th}}$ power is (the restriction to $\mathbb{G}_m/\mathbb{F}_2$ of) $\mathcal{F}_q$:

$$[t(q)]^\star \mathcal{G}_q \cong \mathcal{F}_q|\mathbb{G}_m.$$

Thus $\mathcal{G}_q$ is lisse on $\mathbb{G}_m/\mathbb{F}_2$, tame at 0 with $I(0)$-representation a direct sum of Kummer characters of order dividing $t(q)$, and whose $I(\infty)$-representation is irreducible of dimension $d(q)$, with all ∞-slopes $\frac{q_0+1}{d(q)} = \frac{q_0+1}{q_0(q-1)}$.

Next we give a slight improvement of [**28**, Theorem 6.5].

THEOREM 1.1. *Let $\mathcal{H}_0$ be a lisse sheaf on $\mathbb{G}_m/\mathbb{F}_q$ which is tame at 0 and pure of weight zero. Let a, b be nonnegative integers, and consider the moment $M_{a,b}$. Denote by*

$$\mathcal{H}_0^{a,b} := \mathcal{H}_0^{\otimes a} \otimes (\mathcal{H}_0^\vee)^{\otimes b}.$$

Denote by A, B, C, D the following constants:

$$\begin{aligned}
C &:= \textit{dimension of the space of } I(0)\textit{-invariants in } \mathcal{H}_0^{a,b},\\
D &:= \textit{dimension of the space of } I(\infty)\textit{-invariants in } \mathcal{H}_0^{a,b},\\
B &:= \mathsf{Swan}_\infty(\mathcal{H}_0^{a,b}) + M_{a,b},\\
A &:= B + M_{a,b} - C - D.
\end{aligned}$$

Then we have the following estimate:

$$\left|\frac{1}{q-1}\sum_{u\in\mathbb{F}_q^\times} \mathrm{Trace}(\mathrm{Frob}_{u,\mathbb{F}_q}|\mathcal{H}_0^{a,b})\right| \leq \frac{q}{q-1}M_{a,b} + \frac{A\sqrt{q}}{q-1} + \frac{B-A}{q-1}.$$

PROOF. For any lisse sheaf $\mathcal{F}$ on $\mathbb{G}_m$, the Lefschetz trace formula gives

$$\sum_{u\in\mathbb{F}_q^\times} \mathrm{Trace}(\mathrm{Frob}_{u,\mathbb{F}_q}|\mathcal{F})$$
$$= \mathrm{Trace}\big(\mathrm{Frob}_{\mathbb{F}_q}|H^2_c(\mathbb{G}_m/\overline{\mathbb{F}_q},\mathcal{F})\big) - \mathrm{Trace}\big(\mathrm{Frob}_{\mathbb{F}_q}|H^1_c(\mathbb{G}_m/\overline{\mathbb{F}_q},\mathcal{F})\big).$$

If $\mathcal{F}$ is pure of weight zero, then H^2_c is pure of weight 2, and H^1_c is mixed of weight ≤ 1; indeed,

$$H^1_c = H^1_c(\mathrm{wt}=1)\oplus H^1_c(\mathrm{wt}\le 0).$$

Thus, for $\mathcal{F}$ pure of weight zero,

$$\left|\sum_{u\in\mathbb{F}_q^\times} \mathrm{Trace}(\mathrm{Frob}_{u,\mathbb{F}_q}|\mathcal{F})\right| \le qh^2_c + \sqrt{q}h^1_c(\mathrm{wt}=1) + h^1_c(\mathrm{wt}\le 0).$$

When $\mathcal{F}$ is tame at 0, the Euler-Poincaré formula gives

$$\mathsf{Swan}_\infty(\mathcal{F}) = h^1_c - h^2_c.$$

To compute the dimension of $H^1_c(\mathrm{wt}=1)$, we use the fact that for the inclusion $j:\mathbb{G}_m\subset\mathbb{P}^1$, the group $H^1(\mathbb{P}^1/\overline{\mathbb{F}_q},j_\star\mathcal{F})$ is pure of weight one. We exploit this by looking at the short exact sequence of sheaves on $\mathbb{P}^1/\overline{\mathbb{F}_q}$ given by

$$0\to j_!\mathcal{F}\to j_\star\mathcal{F}\to(\mathcal{F}^{I(0)})_0\oplus(\mathcal{F}^{I(\infty)})_\infty\to 0,$$

where the last two summands are skyscraper sheaves at 0 and ∞. The group

$$H^2(\mathbb{P}^1/\overline{\mathbb{F}_q},j_\star\mathcal{F}) = H^2_c(\mathbb{G}_m/\overline{\mathbb{F}_q},\mathcal{F})$$

is the Tate-twisted group of π_1^{geom} co-invariants in $\mathcal{F}$, but as $\mathcal{F}$ is pure, the action of π_1^{geom} is semisimple, so this is also the (Tate-twisted) group of π_1^{geom} invariants. The group

$$H^0(\mathbb{P}^1/\overline{\mathbb{F}_q},j_\star\mathcal{F}) = H^0(\mathbb{G}_m/\overline{\mathbb{F}_q},\mathcal{F})$$

is the space of π_1^{geom} invariants in $\mathcal{F}$, so has dimension $h^0=h^2_c$. Consider the long exact sequence

$$0\to H^0(\mathbb{P}^1/\overline{\mathbb{F}_q},j_\star\mathcal{F})\to\mathcal{F}^{I(0)}\oplus\mathcal{F}^{I(\infty)}\to H^1_c(\mathbb{G}_m/\overline{\mathbb{F}_q},\mathcal{F})\to H^1(\mathbb{P}^1/\overline{\mathbb{F}_q},j_\star\mathcal{F}_0)\to 0.$$

Apply it with $\mathcal{F}$ taken to be $\mathcal{H}_0^{a,b}$. Then $h^0=h^2_c$ is $M_{a,b}$, and the Euler-Poincaré formula gives

$$h^1_c = \mathsf{Swan}_\infty(\mathcal{H}_0^{a,b}) + M_{a,b}.$$

Thus we have the equalities

$$h^1_c(\mathrm{wt}=1) = h^1_c + h^0 - \dim\mathcal{F}^{I(0)} - \dim\mathcal{F}^{I(\infty)} = A,$$

and

$$h^1_c(\mathrm{wt}\le 0) = B - A.$$

Then the estimate

$$\left|\sum_{u\in\mathbb{F}_q^\times} \mathrm{Trace}(\mathrm{Frob}_{u,\mathbb{F}_q}|\mathcal{F})\right| \le qh^2_c + \sqrt{q}h^1_c(\mathrm{wt}=1) + h^1_c(\mathrm{wt}\le 0)$$

becomes

$$\left|\sum_{u\in\mathbb{F}_q^\times} \mathrm{Trace}(\mathrm{Frob}_{u,\mathbb{F}_q}|\mathcal{H}_0^{a,b})\right| \le qM_{a,b} + A\sqrt{q} + (B-A).$$ □

We will now apply this estimate to the descent $\mathcal{G}_8$ to $\mathbb{G}_m/\mathbb{F}_2$ of the lisse sheaf $\mathcal{F}_8$ on $\mathbb{A}^1/\mathbb{F}_2$. Recall from [**26**, §2] that $\mathcal{F}_8$ is the Fourier transform of the Artin-Schreier-Witt sheaf $\mathcal{L}_{\psi_2([x^5,x^{15}])}$, with a constant field twist by $\frac{1}{1+i}$. Thus for $k/\mathbb{F}_2$ a finite extension, and $t \in k$,

$$\mathrm{Trace}(\mathrm{Frob}_{t,k}|\mathcal{F}_8) = -\left(\frac{1}{1+i}\right)^{\deg(k/\mathbb{F}_2)} \sum_{x\in k} \psi_2(\mathrm{Trace}_{W_2(k)/W_2(\mathbb{F}_2)}([x^5, x^{15} + tx])).$$

The descent $\mathcal{G}_8$ is the lisse sheaf on $\mathbb{G}_m/\mathbb{F}_2$ whose trace function is given as follows. For $k/\mathbb{F}_2$ a finite extension, and $t \in k^\times$,

$$\mathrm{Trace}(\mathrm{Frob}_{t,k}|\mathcal{G}_8) = -\left(\frac{1}{1+i}\right)^{\deg(k/\mathbb{F}_2)} \sum_{x\in k} \psi_2(\mathrm{Trace}_{W_2(k)/W_2(\mathbb{F}_2)}([x^5/t, x^{15}/t^3+x])).$$

Thus one visibly has, for $t \in k^\times$, the identity

$$\mathrm{Trace}(\mathrm{Frob}_{t,k}|\mathcal{F}_8) = \mathrm{Trace}(\mathrm{Frob}_{t^5,k}|\mathcal{G}_8),$$

simply by the substitution $x \mapsto x/t$ in the formula for $\mathcal{F}_8$.

In any extension field $k/\mathbb{F}_2$ such that $\gcd(5, \#k^\times) = 1$, the map $t \mapsto t^5$ is bijective on $k^\times$. Such $k/\mathbb{F}_2$ are precisely those whose degree over $\mathbb{F}_2$ is not divisible by 4. For such a $k/\mathbb{F}_2$, the traces $\mathrm{Trace}(\mathrm{Frob}_{t,k}|\mathcal{G}_8)$ as t runs over $k^\times$ are precisely the traces $\mathrm{Trace}(\mathrm{Frob}_{t,k}|\mathcal{F}_8)$ as t runs over $k^\times$. An extensive calculation shows that over $\mathbb{F}_{2^{18}}^\times$, the seven traces which occur, with their multiplicities, are

- $-2i$, multiplicity 16256,
- -2, multiplicity 4095,
- -1, multiplicity 52429,
- 0, multiplicity 112347,
- 1, multiplicity 60495,
- $2i$, multiplicity 16512,
- 14, multiplicity 9.

[We have not been able to find any conceptual explanation for the multiplicities of these traces, let alone for the fact that all traces are algebraic integers. Had we known that the traces of all Frobenii over all finite extensions of $\mathbb{F}_2$ are algebraic integers, we would have been able to conclude that $\mathcal{G}_8$ has finite geometric monodromy group G_{geom}, and the proof of the main result Theorem 6.1 would have been much simpler.]

This computation was carried out using MAGMA 2.26-6 [**3**] at the University of Auckland. It exploited a new feature of MAGMA which supports running tasks in parallel on multiple processors: each task performed the necessary computation for a given $t \in k^\times$. Using 55 3.0GHz processors, the computation completed in 5.75 days, taking about 7500 hours of CPU time.

The empirical $M_{2,2}$ for $\mathcal{G}_8$ over $\mathbb{F}_{2^{18}}^\times$ is thus approximately

$$3.99963378766551080898593515753.$$

Applying Theorem 1.1, we find

COROLLARY 1.2. *For the lisse sheaf $\mathcal{G}_8$ on $\mathbb{G}_m/\mathbb{F}_2$, $M_{2,2} > 2$.*

PROOF. The sheaf $\mathcal{F}_8$ is a geometrically irreducible Airy sheaf, lisse on $\mathbb{A}^1/\mathbb{F}_2$ of rank 14, whose $I(\infty)$ representation is irreducible, with all slopes 15/14. Its descent $\mathcal{G}_8$ is thus lisse and geometrically irreducible on $\mathbb{G}_m$, its $I(0)$ representation is the direct sum with multiplicities of the characters of order dividing 5, and its $I(\infty)$ representation is irreducible, with all slopes $3/14$.

Let us denote by

$$\mathcal{K} := (\mathcal{G}_8)^{\otimes 2} \otimes (\mathcal{G}_8^\vee)^{\otimes 2}.$$

In the proof of Theorem 1.1, the estimate

$$\left|\sum_{t\in\mathbb{F}_q^\times} \mathrm{Trace}(\mathrm{Frob}_{t,\mathbb{F}_q}|\mathcal{K})\right| \leq qh_c^2 + \sqrt{q}h_c^1(\mathrm{wt}=1) + h_c^1(\mathrm{wt}\leq 0)$$

can first be weakened to

$$\left|\sum_{x\in\mathbb{F}_q^\times} \mathrm{Trace}(\mathrm{Frob}_{u,\mathbb{F}_q}|\mathcal{K})\right| \leq qh_c^2 + \sqrt{q}h_c^1(\mathrm{wt}=1) + h_c^1.$$

The equality

$$h_c^1(\mathrm{wt}=1) = \mathsf{Swan}_\infty(\mathcal{K}) + 2M_{2,2} - \dim \mathcal{K}^{I(0)} - \dim \mathcal{K}^{I(\infty)}$$

can be weakened to

$$h_c^1(\mathrm{wt}=1) \leq \mathsf{Swan}_\infty(\mathcal{K}) + M_{2,2} - \dim \mathcal{K}^{I(0)},$$

simply because in the exact sequence, the space H^0 of global invariants of $\mathcal{K}$ injects into the space $\mathcal{K}^{I(\infty)}$ of $I(\infty)$-invariants.

Thus $\mathcal{K}$ is lisse on $\mathbb{G}_m$, its $I(0)$ representation is the direct sum with multiplicities of the characters of order dividing 5, and all its $I(\infty)$ slopes are $\leq 3/14$. So we have the crude estimate

$$\mathsf{Swan}_\infty(\mathcal{K}) \leq \mathrm{rank}(\mathcal{K})(\text{biggest slope}) \leq 14^4(3/14) = 8232.$$

The $I(0)$ representation of $\mathcal{K}$ is End(End(the $I(0)$-representation of $\mathcal{G}_8$)).

We now turn to the $I(0)$ representation of $\mathcal{G}_8$. The action of $I(0)$ is through μ_5. So in terms of a fixed character χ of order 5, it is a direct sum

$$a\mathbb{1} + b\chi + c\chi^2 + d\chi^3 + e\chi^4,$$

with non-negative integers a, b, c, d, e which sum to 14. We also know from Deligne's "independence of ℓ" result [**11**, Theorem 9.8] or from Serre-Tate [**43**, Theorem 2(ii)], that the character of this representation of $I(0)$ has values in the field $\mathbb{Q}(i)$ (because $\mathcal{G}_8$ is part of a compatible system over $\mathbb{Q}(i)$). On the other hand, this character has values in $\mathbb{Q}(\zeta_5)$. But the intersection $\mathbb{Q}(i) \cap \mathbb{Q}(\zeta_5)$ is just $\mathbb{Q}$, so the trace has values in $\mathbb{Q}$. In other words, the quantity

$$a + b\zeta_5 + c\zeta_5^2 + d\zeta_5^3 + e\zeta_5^4$$

lies in $\mathbb{Q}$. This in turn forces $b = c = d = e$, and so $a + b\zeta_5 + b\zeta_5^2 + b\zeta_5^3 + b\zeta_5^4 = a - b$. But $a + 4b = 14$, so there are only four possibilities for the character, namely,

$$2\mathbb{1}+3\chi+3\chi^2+3\chi^3+3\chi^4,\ ,6\mathbb{1}+2\chi+2\chi^2+2\chi^3+2\chi^4, 10\mathbb{1}+1\chi+1\chi^2+1\chi^3+1\chi^4,\ 14\mathbb{1}.$$

More intrinsically, let us denote by Reg the regular representation of μ_5:

$$\mathsf{Reg} = \mathbb{1} + \chi + \chi^2 + \chi^3 + \chi^4.$$

Then the $I(0)$ representation of $\mathcal{G}_8$ is one of

$$-\mathbb{1} + 3\mathsf{Reg},\ 4\mathbb{1} + 2\mathsf{Reg},\ 9\mathbb{1} + \mathsf{Reg},\ 14\mathbb{1}.$$

Because the character takes real (in fact integer) values, $\dim \mathcal{K}^{I(0)}$ is the coefficient of $\mathbb{1}$ in the fourth power of the character. For each of our four candidates, this is easily computed by hand, because $\mathsf{Reg}^2 = 5\mathsf{Reg}$, $\mathsf{Reg}^3 = 5^2\mathsf{Reg}$, $\mathsf{Reg}^4 = 5^3\mathsf{Reg}$. The least of the possible values of $\dim \mathcal{K}^{I(0)}$ is 7684, the value attained by the candidate $-\mathbb{1} + 3\mathsf{Reg}$. So the weakened estimate becomes

$$\left| \sum_{t \in \mathbb{F}_{2^{18}}^\times} \mathrm{Trace}(\mathrm{Frob}_{t,\mathbb{F}_{2^{18}}}|\mathcal{K}) \right| \le 2^{18} M_{2,2} + (8232 + M_{2,2} - 7684)2^9 + (8232 + M_{2,2}).$$

Dividing by $2^{18} - 1$, we get

$$3.999 \le \frac{2^{18}}{2^{18}-1} M_{2,2} + \frac{(548 + M_{2,2})2^9}{2^{18}-1} + \frac{8232 + M_{2,2}}{2^{18}-1}.$$

Here $2^{18} - 1 = 262143$, and $2^9/(2^{18}-1) \le (2^9+1)/(2^{18}-1) = 1/(2^9-1) = 1/511$.

Thus if $M_{2,2}$ were 2, we would get

$$3.999 \le (1 + (1/262143))2 + 550/511 + 8332/262143,$$

which is nonsense. □

2. Background results on determinants, rationality, and slopes

THEOREM 2.1. *Let p be a prime, $k/\mathbb{F}_p$ a finite extension, $\nu \ge 1$ an integer, and $\mathcal{F}$ a lisse $\overline{\mathbb{Q}_\ell}$ sheaf on a smooth, geometrically connected scheme U/k which is pure of weight zero and part of a compatible system of lisse sheaves on U whose trace functions take values in $\mathbb{Q}(\zeta_{p^\nu})$. Denote by $G_{\mathrm{geom}} \le G_{\mathrm{arith}}$ the geometric and arithmetic monodromy groups of $\mathcal{F}$. Then we have the following results:*

(i) $\det(G_{\mathrm{arith}})$ *has finite order dividing $2p^\nu$ if p is odd, dividing 2^ν if $p = 2$.*
(ii) $\det(G_{\mathrm{geom}})$ *has finite order dividing $2p^\nu$ if p is odd, dividing 2^ν if $p = 2$.*
(iii) *Suppose that U is a dense open set of $\mathbb{P}^1$, and that at each point $x \in \mathbb{P}^1(\overline{k}) \setminus U(\overline{k})$, all $I(x)$-slopes of $\mathcal{F}$ are < 1. Then $\det(G_{\mathrm{geom}})$ has order dividing 2, and is trivial if $p = 2$.*

PROOF. (i) For any finite extension L/k and any point $t \in U(L)$, $\det(\mathrm{Frob}_{t,L}|\mathcal{F})$ lies in $\mathbb{Q}(\zeta_{p^\nu})$. It has absolute value 1 at every archimedean place of $\mathbb{Q}(\zeta_{p^\nu})$ (purity of weight zero), and is a λ-adic unit at every finite place λ of residue characteristic $\neq p$ (being part of a compatible system). Because $\mathbb{Q}(\zeta_{p^\nu})$ has a unique place above p, the product formula tells us that this determinant is a unit at all finite places. Thus it is a root of unity in $\mathbb{Q}(\zeta_{p^\nu})$, so of order dividing $2p^\nu$ for p odd, and of order dividing 2^ν if $p = 2$.

As G_{geom} is a subgroup of G_{arith}, we trivially obtain (ii).

To prove (iii), the slope hypotheses imply that det as a character of G_{geom} has slope < 1, hence 0, at each missing point, and thus is everywhere tame, and hence (being on a dense open set of $\mathbb{P}^1$) has order prime to p. □

THEOREM 2.2. *Let p be a prime, $k/\mathbb{F}_p$ a finite extension, $\nu \ge 1$ an integer, and $\mathcal{F}$ a lisse $\overline{\mathbb{Q}_\ell}$ sheaf on a smooth, geometrically connected scheme U/k which is pure of weight zero and part of a compatible system of lisse sheaves on U whose trace functions take values in a number field E. Suppose further that there exists a proper smooth curve with geometrically connected fibres*

$$\pi : \mathcal{C} \to U,$$

a finite group Γ of automorphisms of $\mathcal{C}/U$, and a linear character

$$\chi : \Gamma \to E^\times$$

such $\mathcal{F}$ is isomorphic to the χ-component of $R^1\pi_\star(\overline{\mathbb{Q}_\ell})$. Then we have the following results:

(i) *For any finite extension L/k and any point $t \in U(L)$, the action of $\mathrm{Frob}_{t,L}|\mathcal{F}$ is semisimple.*

(ii) *If U is a curve, with complete nonsingular model X, then for each point $x \in X(\overline{k}) \setminus U(\overline{k})$, the character of the action of the inertia group $I(x)$ acting on $\mathcal{F}$ has values in K.*

PROOF. The semisimplicity of Frobenii on H^1 of curves goes back to Weil. By Serre-Tate [**43**, Theorem 2(ii)], the character of $I(x)$ on $R^1\pi_\star(\overline{\mathbb{Q}_\ell})$ has values in $\mathbb{Z}$. When we project the $I(x)$ action onto the χ component, the character of the resulting $I(x)$ action has values in $\mathbb{Q}(\chi)$, a subfield of K. □

REMARK 2.3. Theorem 2.2 applies to the Airy sheaves of Šuch [**44**] and any of their descents, where the family of curves in question is a family of Artin-Schreier-Witt coverings of $\mathbb{A}^1$ (compactified by adding its one point at ∞).

Next we record a general result on G_{geom} of lisse sheaves on open sets of $\mathbb{A}^1$:

THEOREM 2.4. *Let U be a dense open set of $\mathbb{A}^1/\overline{\mathbb{F}_p}$, and $\mathcal{F}$ a lisse $\overline{\mathbb{Q}_\ell}$-sheaf on U, with $\ell \neq p$. Suppose all the ∞-slopes of $\mathcal{F}$ are at most σ, for some $0 < \sigma < 1$. Suppose the geometric monodromy group $G = G_{\mathrm{geom}}$ of $\mathcal{F}$ admits a representation $\Phi : G \to \mathrm{GL}_d(\mathbb{F})$ over some algebraically closed field $\mathbb{F}$ of characteristic $\neq p$, of dimension $d < 1/\sigma$. Then Φ is tame at ∞.*

PROOF. The hypothesis that all ∞-slopes of $\mathcal{F}$ are $\leq \sigma$ is that for all $y > \sigma$, the upper numbering subgroup $I(\infty)^y$ acts trivially on $\mathcal{F}$, i.e., dies in G_{geom}, and hence dies under Φ. Thus all ∞-slopes of $\mathcal{F}$ are $\leq \sigma$, and hence $\mathsf{Swan}_\infty(\Phi) \leq d\sigma < 1$. As Swan conductors are non-negative integers, $\mathsf{Swan}_\infty(\Phi) = 0$. This means $P(\infty)$ acts trivially in Φ, i.e., Φ is tame at ∞. □

Now we prove a generalization of [**30**, Theorem 4.16].

THEOREM 2.5. *Let U be a dense open set of $\mathbb{A}^1/\overline{\mathbb{F}_p}$, $\mathcal{F}$ a lisse $\overline{\mathbb{Q}_\ell}$-sheaf on U, with $\ell \neq p$, and let G be the geometric monodromy group of $\mathcal{H}$. Suppose that the following hold:*

(a) *All ∞-slopes of $\mathcal{H}$ are at most σ for some $0 < \sigma < 1$, and $\mathcal{H}$ is not tame at ∞;*

(b) *G is a finite almost quasisimple group: $S \lhd G/\mathbf{Z}(G) \leq \mathrm{Aut}(S)$ for some finite non-abelian simple group S;*

(c) *For some normal subgroup R of $G/\mathbf{Z}(G)$ containing S, R admits either a faithful d-dimensional linear representation*

$$\Phi : R \to \mathrm{GL}_d(\mathbb{F}),$$

or an e-dimensional projective representation

$$\Psi : R \to \mathrm{PGL}_e(\mathbb{F})$$

which is nontrivial over S, over some algebraically closed field $\mathbb{F}$ of characteristic $\neq p$.

Then

$$1/\sigma \leq d \cdot [G/\mathbf{Z}(G) : R] \leq d \cdot |\mathrm{Out}(S)|,$$

respectively,

$$1/\sigma \le (e^2-1)\cdot[G/\mathbf{Z}(G):R] \le (e^2-1)\cdot|\mathrm{Out}(S)|.$$

PROOF. Note that the given Ψ is faithful. Indeed, $\mathrm{Ker}(\Psi) \lhd R$ does not contain S, so it intersects S trivially by simplicity of S. Because both S and $\mathrm{Ker}(\Psi)$ are normal in R, the commutator $[S, \mathrm{Ker}(\Psi)] \subset S \cap \mathrm{Ker}(\Psi) = 1$. Thus $\mathrm{Ker}(\Psi) \le \mathbf{C}_R(S) \le \mathbf{C}_{\mathrm{Aut}(S)}(S) = 1$. Hence R is embedded in $\mathrm{PGL}(U)$, where $U = \mathbb{F}^e$. Composing this embedding with the faithful action of $\mathrm{PGL}(U)$ on $\mathrm{End}^0(U) = \mathrm{End}(U)/\text{scalars}$, we obtain a faithful action of R on a module of dimension $\le e^2-1$. Thus it suffices to prove the bound $1/\sigma \le d\cdot[G/\mathbf{Z}(G):R]$ when $\Phi : R \to \mathrm{GL}(V)$ is given.

So assume the contrary, i.e., $\Phi : R \to \mathrm{GL}(V)$ is faithful for some V with $\dim(V) = d$, but

$$1/\sigma > d\cdot[G/\mathbf{Z}(G):R]. \tag{2.5.1}$$

Let $\tilde{V}$ denote the $\bar{G}$-module $\mathrm{Ind}_R^{\bar{G}}(V)$ for $\bar{G} := G/\mathbf{Z}(G)$. Note that $\bar{G}$ acts faithfully on $\tilde{V}$. Indeed, let $K \lhd \bar{G}$ denote the kernel of the action of $\bar{G}$ on $\tilde{V}$. By the construction of V as the induced representation, the R-module $\tilde{V}$ contains V as a submodule. But S acts faithfully on V, hence $S \cap K = 1$. As $S \lhd \bar{G}$, it follows that $[S, K] = 1$, and so

$$K \le \mathbf{C}_{\bar{G}}(S) \le \mathbf{C}_{\mathrm{Aut}(S)}(S) = 1.$$

We also note that

$$\dim(\tilde{V}) = [\bar{G}:R]\cdot\dim(V) = d\cdot[\bar{G}:R] < 1/\sigma$$

by (2.5.1).

Now view $\tilde{V}$ as a representation of G, of dimension $< 1/\sigma$. By Theorem 2.4, this representation is tame at ∞. Thus the image Q in G of $P(\infty)$ acts trivially on $\tilde{V}$. But $G/\mathbf{Z}(G)$ acts faithfully on $\tilde{V}$. Therefore Q lands in $\mathbf{Z}(G)$. Recall that $I(\infty)$ has finite image J in G, and J/Q is cyclic. As $Q \le \mathbf{Z}(J)$, it follows that J is abelian. Thus all simple J-summands in $\mathcal{H}$ are one-dimensional, and at least one of them is wild, as $\mathcal{H}$ is not tame at ∞. Each one-dimensional wild component has Swan a strictly positive integer, which is also its slope, contradicting the hypothesis that all ∞-slopes of $\mathcal{H}$ are < 1. □

3. Primitive prime divisors for Suzuki-Ree groups

The order of a finite group of Lie type $G(\mathbb{F}_q)$ over a field $\mathbb{F}_q$ is usually a product of a power of $q = p^f$ (p the defining characteristic) and the values at q of cyclotomic polynomials $\Phi_m(q)$ for various m. In a number of problems on $G(\mathbb{F}_q)$, the existence of primitive prime divisors $\mathrm{ppd}(q, m)$ or $\mathrm{ppd}(p, mf)$ for certain m was helpful. Recall [**50**] that for $a, m \in \mathbb{Z}_{\ge 2}$, a *primitive prime divisor* $\ell = \mathrm{ppd}(a, m)$ is a prime divisor of $a^m - 1$ that does not divide $\prod_{i=1}^{m-1}(a^i - 1)$; such a prime divisor always exists unless $(a, m) = (2, 6)$ or $m = 2$ and $a + 1$ is a 2-power. For the Suzuki-Ree groups ${}^2B_2(q)$ with $q = 2^n$, $2 \nmid n \ge 3$, ${}^2G_2(q)$ with $q = 3^n$, $2 \nmid n \ge 3$, and ${}^2F_4(q)$ with $q = 2^n$, $2 \nmid n \ge 3$, some factor $\Phi_m(q)$ of $|G(\mathbb{F}_q)|$ decomposes further into values at $\sqrt{q}$ of polynomials over $\mathbb{Z}[\sqrt{2}]$ or $\mathbb{Z}[\sqrt{3}]$. More precisely,

$$\Phi_4(q) = q^2 + 1 = (q - \sqrt{2q} + 1)(q + \sqrt{2q} + 1)$$

for ${}^2B_2(q)$,

$$\Phi_6(q) = q^2 - q + 1 = (q - \sqrt{3q} + 1)(q + \sqrt{3q} + 1)$$

for ${}^2G_2(q)$, and

$$\Phi_{12}(q) = q^4 - q^2 + 1 = (q^2 - q\sqrt{2q} + q - \sqrt{2q} + 1)(q^2 + q\sqrt{2q} + q + \sqrt{2q} + 1)$$

for ${}^2F_4(q)$. In applications, it is desirable to prove that these factors also possess primitive prime divisors $\mathrm{ppd}(p, 4n)$, respectively $\mathrm{ppd}(p, 6n)$, $\mathrm{ppd}(p, 12n)$, whose existence does not follow from [**50**]. The main results of this section establish the existence of such prime divisors for Suzuki-Ree groups.

3A. Almost equidistribution of coprime integers in congruence classes. For $n \in \mathbb{Z}_{\geq 1}$, let $\phi(n)$ denote the Euler function of n, let $\mu(n)$ denote the Möbius function of n, and let $\omega(n)$ denote the number of distinct prime divisors of n (not counting multiplicities). First we prove the following.

PROPOSITION 3.1. *Let $m, n \in \mathbb{Z}_{\geq 1}$ be coprime integers. For any integer $0 \leq a \leq m-1$, the number N_a of integers $1 \leq k \leq n$ such that $\gcd(k, n) = 1$ and $k \equiv a \pmod{m}$ satisfies*

$$\left| N_a - \frac{\phi(n)}{m} \right| < 2^{\omega(n)}.$$

PROOF. For $k \in \mathbb{Z}_{\geq 1}$, define $F(k) := 0$ if $\gcd(k, n) > 1$ and $F(k) := 1$ if $\gcd(k, n) = 1$. By [**37**, Theorem 4.7], $F(k) = \sum_{d \mid \gcd(k,n)} \mu(d)$. Now

$$\begin{aligned} N_a &= \sum_{1 \leq k \leq n,\ k \equiv a \pmod{m}} F(k) \\ &= \sum_{1 \leq k \leq n,\ k \equiv a \pmod{m}} \ \sum_{d \mid \gcd(k,n)} \mu(d) = \sum_{d \mid n} \mu(d) N(a, d), \end{aligned} \tag{3.1.1}$$

with

$$N(a, d) := \sum_{1 \leq k \leq n,\ k \equiv a \pmod{m},\ d \mid k} 1.$$

If $d|n$, then $\gcd(d, m) = 1$, so we can find $1 \leq e \leq m-1$ such that $de \equiv 1 \pmod{m}$. Now write every $1 \leq k \leq n$ with $d|k$ as $k = dl$ with $1 \leq l \leq n/d$. Then the condition that $k \equiv a \pmod{m}$ is equivalent to $l \equiv ea \pmod{m}$, in which case we can write $l = s + mi$ with $0 \leq r \leq m-1$, $ea \equiv r \pmod{m}$, and $i \in \mathbb{Z}$. To count the number $N(a, d)$ of i occurring, write $n/d = qm + r$ with $0 \leq r \leq m-1$ and $q \in \mathbb{Z}_{\geq 0}$. Certainly, every $0 \leq i \leq q-1$ works, but neither $i = -1$ nor $i = q+1$ can occur. It follows that

$$n/md - 1 < q \leq N(a, d) \leq q + 1 < n/md + 1. \tag{3.1.2}$$

We also note by [**37**, (4.1)] that $\phi(n) = \sum_{d|n} \mu(d) n/d$ and that $\sum_{d|n} |\mu(d)|$ is the number of square-free divisors of n and hence equals to $2^{\omega(n)}$. Combining with (3.1.1) and (3.1.2), this yields

$$|N_a - \frac{\phi(n)}{m}| = \left| \sum_{d|n} \mu(d)(N(a, d) - n/md) \right| < \sum_{d|n} |\mu(d)| = 2^{\omega(n)}. \qquad \square$$

We will also need the following analogue of Proposition 3.1:

PROPOSITION 3.2. *Let $n \in \mathbb{Z}_{\geq 1}$ be an odd integer divisible by 3. For any integer $0 \leq a \leq 11$ coprime to 3, the number N_a of integers $1 \leq k \leq n$ such that* $\gcd(k,n) = 1$ *and* $k \equiv a \pmod{12}$ *satisfies*

$$\left| N_a - \frac{\phi(n)}{8} \right| < 2^{\omega(n)-1}.$$

PROOF. As in the proof of Proposition 3.2, we have

$$\begin{aligned}(3.2.1) \qquad N_a &= \sum_{1\leq k\leq n,\ k\equiv a \pmod{12}} F(k) = \sum_{1\leq k\leq n,\ k\equiv a \pmod{12}} \ \sum_{d\mid \gcd(k,n)} \mu(d) \\ &= \sum_{d\mid n} \mu(d) N(a,d),\end{aligned}$$

with

$$N(a,d) := \sum_{1\leq k\leq n,\ k\equiv a \pmod{12},\ d\mid k} 1.$$

Now, if $d|n$ but $3|d$, then $N(a,d) = 0$ since $3 \nmid a$. Hence,

$$(3.2.2) \qquad N_a = \sum_{d\mid n,\ 3\nmid d} \mu(d) N(a,d).$$

Next, $\mu(d) = 0$ if $9|d$, and $\mu(d) = -\mu(d/3)$ if $3|d$. It follows from [**37**, (4.1)] that

$$(3.2.3) \qquad \frac{\phi(n)}{n} = \sum_{d\mid n} \frac{\mu(d)}{d} = \sum_{d\mid n,\ 3\nmid d} \frac{\mu(d)}{d} - \sum_{3d'=d\mid n, 3\nmid d'} \frac{\mu(d')}{3d'} = \frac{2}{3} \sum_{d\mid n,\ 3\nmid d} \frac{\mu(d)}{d}.$$

If $d|n$ and $3 \nmid d$, then $\gcd(d, 12) = 1$, and so we can find $1 \leq e \leq 11$ such that $de \equiv 1 \pmod{12}$. The proof of (3.1.2) repeated verbatim shows that

$$n/12d - 1 < q \leq N(a,d) \leq q + 1 < n/12d + 1.$$

Combining with (3.2.2) and (3.2.3), this yields

$$\left| N_a - \frac{\phi(n)}{8} \right| = \left| \sum_{d\mid n,\ 3\nmid d} \mu(d)(N(a,d) - n/12d) \right| < \sum_{d\mid n,\ 3\nmid d} |\mu(d)| = 2^{\omega(n)-1}. \qquad \square$$

3B. Primitive prime divisor for Suzuki groups. We make the choice $\sqrt{2} > 0$. For odd $n \in \mathbb{Z}_{\geq 1}$ and $a = 1, 3$, set

$$P_{2,a}(n) := \prod_{1\leq k<8n,\ \gcd(k,n)=1,\ k\equiv a \pmod 8} \left(3 - 2\sqrt{2}\cos\frac{k\pi}{4n} \right).$$

PROPOSITION 3.3. *If $2 \nmid n \geq 2603$ and $a = 1, 3$, then $P_{2,a}(n) > 2n$.*

PROOF. (i) First we note that

$$(3.3.1) \qquad \phi(n) \geq \max\left(2^{2.2\omega(n)}, n^{6/7}\right)$$

when $2 \nmid n$ and $n \neq 1, 3, 9, 15, 21, 33, 45, 75, 105, 165, 195$. Indeed, suppose $s := \omega(n) \geq 1$ and write $n = \prod_{i=1}^{s} p_i^{a_i}$ for some prime divisors $2 < p_1 < p_2 < \ldots < p_s$ of n. If $p_1 \geq 5$, then

$$\frac{n}{\phi(n)} = \prod_{i=1}^{s} \frac{p_i}{p_i - 1} \leq (5/4)^s < 5^{s/7} \leq n^{1/7},$$

and so $\phi(n) > n^{6/7}$. If $p_1 \geq 7$, then

$$\phi(n) = \prod_{i=1}^{s} p_i^{a_i}(1 - 1/p_i) \geq 6^s > 2^{2.5s}.$$

If $p_1 = 5$ and $s \geq 2$, then

$$\phi(n) = \prod_{i=1}^{s} p_i^{a_i}(1 - 1/p_i) \geq 4 \cdot 6^{s-1} > 2^{2.25s}.$$

If $p_1 = 5$ and $s = 1$, but $n \neq 5$, then $\phi(n) \geq 20 > 2^{4s}$.

In the rest of the proof of (3.3.1), we may assume that $p_1 = 3$. First suppose that $s \geq 4$. As $(11^6/10^7)^{s-3} < (2 \cdot 4 \cdot 6)^7/(3 \cdot 5 \cdot 7)^6$,

$$\frac{n}{\phi(n)} = \prod_{i=1}^{s} \frac{p_i}{p_i - 1} \leq \frac{3}{2} \cdot \frac{5}{4} \cdot \frac{7}{6} \cdot \left(\frac{11}{10}\right)^{s-3} < \left(3 \cdot 5 \cdot 7 \cdot 11^{s-3}\right)^{1/7} < n^{1/7}.$$

Also, $\phi(n) \geq 2 \cdot 4 \cdot 6 \cdot 10^{s-3} > 2^{2.23s}$.

Next suppose that $s = 3$. If $p_3 \geq 11$, then

$$\frac{n}{\phi(n)} = \prod_{i=1}^{s} \frac{p_i}{p_i - 1} \leq \frac{3}{2} \cdot \frac{5}{4} \cdot \frac{11}{10} < (3 \cdot 5 \cdot 11)^{1/7} < n^{1/7}.$$

If $p_3 < 11$ then $(p_1, p_2, p_3) = (3, 5, 7)$; thus if $n \neq 105$ then $n > 3 \cdot 105$ and so

$$n/\phi(n) = (3 \cdot 5 \cdot 7)/(2 \cdot 4 \cdot 6) < n^{1/7}.$$

Also, if $n \neq 105, 165, 195$, then $\phi(n) \geq 2 \cdot 6 \cdot 10 > 2^{2.2s}$.

Suppose now that $s = 2$. If $p_2 \geq 13$, then $n/\phi(n) \leq (3/2) \cdot (13/12) < (3 \cdot 13)^{1/7} \leq n^{1/7}$. If $p_3 = 11$ and $n \neq 33$, then $n/\phi(n) = (3/2) \cdot (11/10) < (3 \cdot 3 \cdot 11)^{1/7} \leq n^{1/7}$. If $p_3 = 7$ and $n \neq 21$, then $n/\phi(n) = (3/2) \cdot (7/6) < (3 \cdot 3 \cdot 7)^{1/7} \leq n^{1/7}$. If $p_3 = 5$ and $n \neq 15, 45, 75$, then $n/\phi(n) = (3/2) \cdot (5/4) < (6 \cdot 3 \cdot 5)^{1/7} < n^{1/7}$. Also, if $n \neq 15, 21, 33, 45$, then $\phi(n) \geq 24 > 2^{2.2s}$.

Finally, assume that $s = 1$, and $n = 3^a$ with $a \geq 3$. Then $n/\phi(n) = 3/2 < 3^{a/7} = n^{1/7}$, and $\phi(n) \geq 18 > 2^{4s}$, completing the proof of (3.3.1).

(ii) Now assume that $n \geq 2602$. By (3.3.1) $m := \phi(n) \geq n^{6/7} > 846$, so $m \geq 847$, and $2^{\omega(n)} \leq m^{5/11}$.

Fix $a \in \{1, 3\}$ and let $S_j := \{(j-1)n \leq k < jn \mid \gcd(k, 8n) = 1, k \equiv a \pmod 8\}$ for $0 \leq j \leq 7$. For each j, observe that $k \in S_j$ if and only if $0 \leq k' := k - jn < n$ is coprime to n and $k' \equiv a - jn \pmod 8$. By Proposition 3.1,

$$|S_j| = |\{0 \leq k' < n \mid \gcd(k, n) = 1, k' \equiv a - jn \pmod 8\}|$$

satisfies $\phi(n)/8 - 2^{\omega(n)} < |S_j| < \phi(n)/8 + 2^{\omega(n)}$. By the above,

$$m/8 - m^{5/11} < |S_j| < m/8 + m^{5/11}. \tag{3.3.2}$$

Now, if $k \in S_0 \cup S_7$, then $3 - 2\sqrt{2}\cos(k\pi/4n) \geq 3 - 2\sqrt{2}$. If $k \in S_1 \cup S_6$, then $3 - 2\sqrt{2}\cos(k\pi/4n) \geq 1$. If $k \in S_2 \cup S_5$, then $3 - 2\sqrt{2}\cos(k\pi/4n) \geq 3$. Finally, if $k \in S_3 \cup S_4$, then $3 - 2\sqrt{2}\cos(k\pi/4n) \geq 5$. It follows from (3.3.2) that

$$P_{2,a}(n) \geq (3 - 2\sqrt{2})^{m/4 + 2m^{5/11}} \cdot 15^{m/4 - 2m^{5/11}} = A^{m/4} B^{-2m^{5/11}}$$

with $A := 15(3 - 2\sqrt{2})$ and $B := 15(3 + 2\sqrt{2})$.

Setting $f(t) := A^{t/4}B^{-2t^{5/11}}t^{-7/6}$,

$$g(t) := \log f(t) = (t/4)\log(A) - 2t^{5/11}\log(B) - (7/6)\log(t).$$

Now $g'(t) = \log(A)/4 - \log(B)/(1.1t^{6/11}) - (7/6t)$ is increasing, so $g'(t) \geq g'(847) > 0.13$ when $t \geq 847$. It follows that $g(m) \geq g(847) > 0.75$, and so $f(m) = \exp(g(m)) > 2.11$. Thus, for $m = \phi(n) \geq 847$

$$P_{2,a}(n) \geq f(m)m^{7/6} > 2m^{7/6} > 2n,$$

as desired. □

As we will see in part (iii) of the proof of the following theorem, Proposition 3.3 actually holds for all odd $n \geq 7$.

THEOREM 3.4. *Let $q = 2^n$ with $2 \nmid n$. If $n \geq 7$ then $t(q) = t^-(q) := q - \sqrt{2q} + 1$ is divisible by a primitive prime divisor* ppd$(2, 4n)$ *of $q^2 + 1$. In all cases, $t^+(q) := q + \sqrt{2q} + 1$ is divisible by a primitive prime divisor* ppd$(2, 4n)$ *of $q^2 + 1$.*

PROOF. (i) The second statement is obvious for $n \leq 5$; also note that $t(2) = 1$, $t(8) = 5$. Henceforth we may assume $n \geq 7$. Consider the sets

$$\begin{aligned} A_j &:= \{1 \leq k \leq 4n \mid \gcd(k, 2n) = 1,\ k \equiv j \pmod 8\}, \\ B_j &:= \{4n + 1 \leq k \leq 8n \mid \gcd(k, 2n) = 1, k \equiv j \pmod 8\} \end{aligned}$$

for $j = 1, 3, 5, 7$. Then the map $k \mapsto 8n - k$ yields bijections

$$A_1 \longleftrightarrow B_7,\ A_7 \longleftrightarrow B_1,\ A_3 \longleftrightarrow B_5,\ A_5 \longleftrightarrow B_3,$$

and the map $k \mapsto k + 4n$ yields bijections

$$A_1 \longleftrightarrow B_5,\ A_5 \longleftrightarrow B_1,\ A_3 \longleftrightarrow B_7,\ A_7 \longleftrightarrow B_3.$$

It follows that

$$|A_1| = |A_3| = |B_5| = |B_7|,\ |A_5| = |A_7| = |B_1| = |B_3|. \tag{3.4.1}$$

Since $\sqcup_{j=1,3,5,7}(A_j \sqcup B_j) = \{1 \leq k < 8n \mid \gcd(k, 8n) = 1\}$, we now see that

$$|A_j| + |B_j| = \phi(8n)/4 = \phi(n) \tag{3.4.2}$$

for each $j = 1, 3, 5, 7$.

(ii) Make the choice of $\zeta := \zeta_{8n} = \exp(\pi i/4n)$ and consider the cyclotomic polynomial

$$\Phi_{8n}(X) = \prod_{1 \leq k < 8n,\ \gcd(k,2n)=1} (X - \zeta^k).$$

If $1 \leq k < 4n$ then $(X - \zeta^k)(X - \zeta^{k+4n}) = (X - \zeta^k)(X + \zeta^k) = X^2 - \zeta_{4n}^k$. It follows that

$$\Phi_{8n}(X) = \prod_{1 \leq k < 4n,\ \gcd(k,2n)=1} (X^2 - \zeta_{4n}^k) = \Phi_{4n}(X^2).$$

In particular,

$$\Phi_{4n}(2) = \Phi_{8n}(\sqrt{2}). \tag{3.4.3}$$

Setting

$$\Phi_{8n,a}(X) := \prod_{1 \leq k < 8n,\ \gcd(k,2n)=1,\ k \equiv \pm a \pmod 8} (X - \zeta^k), \tag{3.4.4}$$

for $a = 1, 3$, we have $\Phi_{8n}(X) = \Phi_{8n,1}(X)\Phi_{8n,3}(X)$. Next, since

$$(\sqrt{2} - \zeta^k)(\sqrt{2} - \zeta^{8n-k}) = 3 - 2\sqrt{2}\cos\frac{k\pi}{4n}$$

for each $1 \leq k < 8n$, using the bijection $k \mapsto 8n - k$ in (3.4.1)

$$P_{2,a}(n) = \Phi_{8n,a}(\sqrt{2}), \tag{3.4.5}$$

for $a = 1, 3$. Using (3.4.3) also

$$P_{2,1}(n)P_{2,3}(n) = \Phi_{4n}(2). \tag{3.4.6}$$

(iii) To show that $P_{2,1}(n)$ and $P_{2,3}(n)$ are integers, we use $\gcd(8, n) = 1$ to write $1 = ns + 8t$ for some $s, t \in \mathbb{Z}$ with $\gcd(s, 8) = \gcd(t, n) = 1$, and set $\zeta = \zeta_{8n} = \alpha\beta$ with $\alpha := \zeta^{ns}$, a 8th root of unity, and $\beta := \zeta^{8t}$, an nth root of unity. When k runs over $A_1 \sqcup B_1$, $\zeta^k = \alpha^k\beta^k = \alpha\beta^k$, and $k \pmod n$ runs over units in $\mathbb{Z}/n\mathbb{Z}$, each at most once. Using (3.4.2), we see that each unit is met exactly once. Repeating the same argument for $k \in A_7 \sqcup B_7$, we get

$$\Phi_{8n,1}(X) := \prod_{l \text{ unit mod } n} (X - \alpha\beta^l)(X - \alpha^{-1}\beta^l) = \prod_{l \text{ unit mod } n} (X^2 - (\alpha + \alpha^{-1})\zeta_n^l X + \zeta_n^{2l}).$$

Note that $\alpha + \alpha^{-1} = \epsilon\sqrt{2}$ for some $\epsilon = \pm 1$. It follows from (3.4.5) that

$$P_{2,1}(n) = \Phi_{8n,1}(\sqrt{2}) = \prod_{l \text{ unit mod } n} (2 - 2\epsilon\zeta_n^l + \zeta_n^{2l})$$

is the norm $\mathrm{Norm}_{\mathbb{Q}(\zeta_n)/\mathbb{Q}}$ of the algebraic integer $2 - 2\epsilon\zeta_n + \zeta_n^2$, hence it is an integer. The same arguments show that $P_{2,3}(n)$ is the norm $\mathrm{Norm}_{\mathbb{Q}(\zeta_n)/\mathbb{Q}}$ of the algebraic integer $2 + 2\epsilon\zeta_n + \zeta_n^2$, hence it is an integer.

Using this norm interpretation for $P_{2,1}(n)$ and $P_{2,3}(n)$, a calculation with MAGMA shows that $P_{2,a}(n) > 2n$ for odd integers $7 \leq n \leq 2601$. Together with Proposition 3.3, this shows that

$$P_{2,a}(n) > 2n \tag{3.4.7}$$

for $a = 1, 3$ and odd $n \geq 7$.

(iv) We also set

$$f^-(X) := X^{2n} - \sqrt{2}X^n + 1,\ f^+(X) := X^{2n} + \sqrt{2}X^n + 1,$$

so that

$$t^-(q) = f^-(\sqrt(2)),\ t^+(q) = f^+(\sqrt(2)),\ f^-(X)f^+(X) = X^{4n} + 1.$$

Certainly, any root of $X^{4n} + 1$ is ζ^k for some odd integer $1 \leq k < 8n$. If $k \equiv \pm 1 \pmod 8$, then

$$f^-(\zeta^k) = \exp(k\pi i/2) + 1 - \sqrt{2}\exp(k\pi i/4) = 0.$$

Similarly, if $k \equiv \pm 3 \pmod 8$, then

$$f^+(\zeta^k) = \exp(k\pi i/2) + 1 + \sqrt{2}\exp(k\pi i/4) = 0.$$

It follows that

$$f^-(X) = \prod_{1 \leq k < 8n,\ k \equiv \pm 1 \pmod 8} (X - \zeta^k),\ f^+(X) = \prod_{1 \leq k < 8n,\ k \equiv \pm 3 \pmod 8} (X - \zeta^k).$$

Comparing to (3.4.4) and (3.4.5), we see that

$$f^-(\sqrt{2})/P_{2,1}(n) = \prod_{1\le k<8n,\ \gcd(k,2n)>1,\ k\equiv\pm1 \pmod 8} (\sqrt{2}-\zeta^k)$$

is an algebraic integer. But $f^-(\sqrt{2}) = t^-(q)$ is an integer, and $P_{2,1}(n)$ is an integer by (iii). Hence $P_{2,1}(n)$ divides $t^-(q)$. Similarly, $P_{2,3}(n)$ divides $t^+(q)$.

(v) By (3.4.7), $P_{2,1}(n) > n$. Consider any prime divisor ℓ of $P_{2,1}(n)$, which then divides $t^-(q)$ by the result of (iv), and divides $\Phi_{4n}(2)$ by (3.4.6). Suppose that ℓ is not a primitive prime divisor of $2^{4n}-1$. By [**34**, Satz 1] (cf. [**42**, Proposition 2]) $\ell|n$, and moreover $\ell^2 \nmid \Phi_{4n}(2)$. It follows that the ℓ-part of $P_{2,1}(n)$ is ℓ. Hence, if $t^-(q)$ is not divisible by any primitive prime divisor of $2^{4n}-1$, then $P_{2,1}(n)$ divides n, a contradiction.

The proof for $t^+(q)$ is entirely similar. □

COROLLARY 3.5. *Let $q = 2^n$ with $2\nmid n$. If $n \ge 3$ then $\Phi''_{24} := q^2 - q\sqrt{2q} + q - \sqrt{2q} + 1$ is divisible by a primitive prime divisor* ppd$(2, 12n)$ *of $q^4 - q^2 + 1$. In all cases, $\Phi'_{24} := q^2 + q\sqrt{2q} + q + \sqrt{2q} + 1$ is divisible by a primitive prime divisor* ppd$(2, 12n)$ *of $q^4 - q^2 + 1$.*

PROOF. Note that if $n = 1$ then $\Phi''_{24} = 1$ and $\Phi'_{24} = 13$. Assume that $n \ge 3$. By Theorem 3.4,

$$t^-(q^3) = q^3 - q\sqrt{2q} + 1 = (q + \sqrt{2q} + 1)(q^2 - q\sqrt{2q} + q - \sqrt{2q} + 1) = t^+(q)\Phi''_{24}$$

is divisible by a primitive prime divisor $\ell_1 = \mathrm{ppd}(2, 12n)$. Since $t^+(q)|(q^2+1)$ and $\ell_1|(q^4 - q^2 + 1)$, we see that $\ell_1 \nmid t^+(q)$, and so $\ell_1|\Phi''_{24}$. The argument for Φ'_{24} is similar, using

$$t^+(q^3) = q^3 + q\sqrt{2q} + 1 = (q - \sqrt{2q} + 1)(q^2 + q\sqrt{2q} + q + \sqrt{2q} + 1) = t^-(q)\Phi'_{24}.$$

□

3C. Primitive prime divisor for Ree groups. The results in this subsection are not needed for the rest of the paper, however they will be used elsewhere [**31**]. We make the choice $\sqrt{3} > 0$. For odd $n \in \mathbb{Z}_{\ge 1}$ and $a = 1, 5$, set

$$P_{3,a}(n) := \prod_{1\le k<12n,\ \gcd(k,n)=1,\ k\equiv a \pmod{12}} \left(4 - 2\sqrt{3}\cos\frac{k\pi}{6n}\right).$$

PROPOSITION 3.6. *If $2\nmid n \ge 3$ and $a = 1, 5$, then $P_{3,a}(n) > 2n$.*

PROOF. A computation with Mathematica shows that $P_{3,a}(n) > 2n$ when $3 \le n \le 353$. Now assume that $n \ge 354$. By (3.3.1) $m := \phi(n) \ge n^{6/7} > 153$, so $m \ge 154$, and $2^{\omega(n)} \le m^{5/11}$.

Fix $a \in \{1, 5\}$ and let $R_j := \{(j-1)n \le k < jn \mid \gcd(k, 12n) = 1, k \equiv a \pmod{12}\}$ for $0 \le j \le 11$. For each j, observe that $k \in R_j$ if and only if $0 \le k' := k - jn < n$ is coprime to n and $k' \equiv a - jn \pmod{12}$. If $3 \nmid n$, then according to Proposition 3.1,

$$|R_j| = |\{0 \le k' < n \mid \gcd(k, n) = 1, k' \equiv a - jn \pmod{12}\}|$$

satisfies $\phi(n)/12 - 2^{\omega(n)} < |R_j| < \phi(n)/12 + 2^{\omega(n)}$. On the other hand, if $3|n$ then $3 \nmid (a - jn)$, so by Proposition 3.2,

$$|R_j| = |\{0 \le k' < n \mid \gcd(k, n) = 1, k' \equiv a - jn \pmod{12}\}|$$

satisfies $\phi(n)/8 - 2^{\omega(n)-1} < |R_j| < \phi(n)/8 + 2^{\omega(n)-1}$.

Setting $b := 12$ when $3 \nmid n$ and $b := 8$ when $3|n$, by the above consideration, now

(3.6.1) $$m/b - m^{5/11} < |R_j| < m/b + m^{5/11}.$$

Now, if $k \in S_0 \cup S_{11}$, then $4 - 2\sqrt{3}\cos(k\pi/6n) \geq 4 - 2\sqrt{3}$. If $k \in S_1 \cup S_{10}$, then $4 - 2\sqrt{3}\cos(k\pi/6n) \geq 1$. If $k \in S_2 \cup S_9$, then $4 - 2\sqrt{3}\cos(k\pi/6n) \geq 4 - \sqrt{3}$. If $k \in S_3 \cup S_8$, then $4-2\sqrt{3}\cos(k\pi/6n) \geq 4$. If $k \in S_4 \cup S_7$, then $4-2\sqrt{3}\cos(k\pi/6n) \geq 4+\sqrt{3}$. Finally, if $k \in S_5 \cup S_6$, then $4-2\sqrt{3}\cos(k\pi/6n) \geq 7$. It follows from (3.6.1) that

$$\begin{aligned} P_{3,a}(n) &\geq \left(4 - 2\sqrt{3}\right)^{2m/b+2m^{5/11}} \cdot \left((4-\sqrt{3}) \cdot 4 \cdot (4+\sqrt{3}) \cdot 7\right)^{2m/b-2m^{5/11}} \\ &= A^{2m/b}B^{-2m^{5/11}} \geq A^{m/6}B^{-2m^{5/11}} \end{aligned}$$

with $A := 364(4-2\sqrt{3})$ and $B := 364/(4-2\sqrt{3})$.

Setting $f(t) := A^{t/6}B^{-2t^{5/11}}t^{-7/6}$,

$$g(t) := \log f(t) = (t/6)\log(A) - 2t^{5/11}\log(B) - (7/6)\log(t).$$

Now $g'(t) = \log(A)/6 - \log(B)/(1.1t^{6/11}) - 7/6t$ is increasing, so $g'(t) \geq g'(154) > 0.48$ when $t \geq 154$. It follows that $g(m) \geq g(154) > 0.74$, and so $f(m) = \exp(g(m)) > 2.09$. Thus, for $m = \phi(n) \geq 154$

$$P_{3,a}(n) \geq f(m)m^{7/6} > 2m^{7/6} > 2n,$$

as desired. □

THEOREM 3.7. *Let $q = 3^n$ with $2 \nmid n$. If $n \geq 3$, then $t(q) = t^-(q) := q - \sqrt{3q} + 1$ is divisible by a primitive prime divisor* ppd$(3, 6n)$ *of $q^2 - q + 1$. In all cases, $t^+(q) := q + \sqrt{3q} + 1$ is divisible by a primitive prime divisor* ppd$(3, 6n)$ *of $q^2 - q + 1$.*

PROOF. (i) Note that $t(3) = 1$. Henceforth we will assume $n \geq 3$. Consider the sets

$$\begin{aligned} A_j &:= \{1 \leq k \leq 6n \mid \gcd(k, 12n) = 1, \ k \equiv j \ (\text{mod } 12)\}, \\ B_j &:= \{6n + 1 \leq k \leq 12n \mid \gcd(k, 12n) = 1, k \equiv j \ (\text{mod } 12)\} \end{aligned}$$

for $j = 1, 5, 7, 11$. Then the map $k \mapsto 12n - k$ yields bijections

$$A_1 \longleftrightarrow B_{11}, \ A_{11} \longleftrightarrow B_1, \ A_5 \longleftrightarrow B_7, \ A_7 \longleftrightarrow B_5,$$

and the map $k \mapsto k + 6n$ yields bijections

$$A_1 \longleftrightarrow B_7, \ A_7 \longleftrightarrow B_1, \ A_5 \longleftrightarrow B_{11}, \ A_{11} \longleftrightarrow B_5.$$

It follows that

(3.7.1) $$|A_1| = |A_5| = |B_7| = |B_{11}|, \ |A_7| = |A_{11}| = |B_1| = |B_5|.$$

Since $\sqcup_{j=1,5,7,11}(A_j \sqcup B_j) = \{1 \leq k < 12n \mid \gcd(k, 12n) = 1\}$, we now see that

(3.7.2) $$|A_j| + |B_j| = \phi(12n)/4 = \phi(3n)/2$$

for each $j = 1, 5, 7, 11$.

(ii) Make the choice of $\zeta := \zeta_{12n} = \exp(\pi i/6n)$ and consider the cyclotomic polynomial

$$\Phi_{12n}(X) = \prod_{1 \leq k < 12n, \ \gcd(k,12n)=1} (X - \zeta^k).$$

If $1 \le k < 6n$ then $(X-\zeta^k)(X-\zeta^{k+6n}) = (X-\zeta^k)(X+\zeta^k) = X^2 - \zeta_{6n}^k$. It follows that

$$\Phi_{12n}(X) = \prod_{1\le k<6n,\ \gcd(k,6n)=1} (X^2 - \zeta_{6n}^k) = \Phi_{6n}(X^2).$$

In particular,

$$\Phi_{6n}(3) = \Phi_{12n}(\sqrt{3}). \tag{3.7.3}$$

Setting

$$\Phi_{12n,a}(X) := \prod_{1\le k<12n,\ \gcd(k,12n)=1,\ k\equiv\pm a \pmod{12}} (X-\zeta^k), \tag{3.7.4}$$

for $a = 1, 5$, we have $\Phi_{12n}(X) = \Phi_{12n,1}(X)\Phi_{12n,5}(X)$. Next, since

$$(\sqrt{3}-\zeta^k)(\sqrt{3}-\zeta^{12n-k}) = 4 - 2\sqrt{3}\cos\frac{k\pi}{6n}$$

for each $1 \le k < 12n$, using the bijection $k \mapsto 12n - k$ in (3.7.1)

$$P_{3,a}(n) = \Phi_{12n,a}(\sqrt{3}), \tag{3.7.5}$$

for $a = 1, 5$. Using (3.7.3) also

$$P_{3,1}(n)P_{3,5}(n) = \Phi_{6n}(3). \tag{3.7.6}$$

(iii) Here we show that $P_{3,1}(n)$ and $P_{3,5}(n)$ are integers. Clearly, they are algebraic integers in $\mathbb{Q}(\zeta)$. Hence it suffices to show that each of them is fixed by any Galois automorphism $\sigma : \zeta \mapsto \zeta^l$, $\gcd(l, 12n) = 1$. First consider the case $j \equiv \pm 1 \pmod{12}$. Then σ fixes $\sqrt{3} = \zeta_{12} + \zeta_{12}^{-1}$, and fixes each of the sets $C := \sqcup_{j=1,11}(A_j \cup B_j)$ and $D := \sqcup_{j=1,11}(A_j \cup B_j)$ modulo $12n$. Since

$$P_{3,1}(n) = \prod_{k\in C}(\sqrt{3}-\zeta^k),\ P_{3,5}(n) = \prod_{k\in D}(\sqrt{3}-\zeta^k),$$

it follows that σ fixes each of $P_{3,1}(n)$ and $P_{3,5}(n)$. Now assume that $j \equiv \pm 5 \pmod{12}$. Then σ sends $\sqrt{3}$ to $-\sqrt{3}$ and ζ^k to ζ^{kl}, and thus

$$\sigma(\sqrt{3}-\zeta^k) = -\sqrt{3}-\zeta^{kl} = -(\sqrt{3}-\zeta^{6n+kl}).$$

Note that modulo $12n$, when k runs over C, $6n+kl$ runs over C, covering each element of C exactly once. Also, $|C| = \phi(3n)$ by (3.7.2), and so $|C|$ is even. It follows that σ sends $P_{3,1}(n)$ to $(-1)^{|C|}P_{3,1}(n) = P_{3,1}(n)$, and similarly σ fixes $P_{3,5}(n)$.

(iv) We also set

$$f^-(X) := X^{2n} - \sqrt{3}X^n + 1,\ f^+(X) := X^{2n} + \sqrt{3}X^n + 1,$$

so that

$$t^-(q) = f^-(\sqrt{(3)}),\ t^+(q) = f^+(\sqrt{(3)}),\ f^-(X)f^+(X) = X^{4n} - X^{2n} + 1.$$

Certainly, any root of $X^{4n} - X^{2n} + 1$ is ζ^k for some integer $1 \le k < 12n$ coprime to 6. If $k \equiv \pm 1 \pmod{12}$, then

$$f^-(\zeta^k) = \exp(k\pi i/3) + 1 - \sqrt{3}\exp(k\pi i/6) = 0.$$

Similarly, if $k \equiv \pm 5 \pmod{12}$, then

$$f^+(\zeta^k) = \exp(k\pi i/3) + 1 + \sqrt{3}\exp(k\pi i/6) = 0.$$

It follows that

$$f^-(X) = \prod_{1\le k<12n,\ k\equiv\pm1\ (\mathrm{mod}\ 12)} (X-\zeta^k),\ f^+(X)$$
$$= \prod_{1\le k<12n,\ k\equiv\pm5\ (\mathrm{mod}\ 12)} (X-\zeta^k).$$

Comparing to (3.7.4) and (3.7.5), we see that

$$f^-(\sqrt{3})/P_{3,1}(n) = \prod_{1\le k<8n,\ \gcd(k,12n)>1,\ k\equiv\pm1\ (\mathrm{mod}\ 12)} (\sqrt{3}-\zeta^k)$$

is an algebraic integer. But $f^-(\sqrt{3}) = t^-(q)$ is an integer, and $P_{3,1}(n)$ is an integer by (iii). Hence $P_{3,1}(n)$ divides $t^-(q)$. Similarly, $P_{3,5}(n)$ divides $t^+(q)$.

(v) By Proposition 3.6, $P_{3,1}(n) > n$. Consider any prime divisor ℓ of $P_{3,1}(n)$, which then divides $t^-(q)$ by the result of (iv), and divides $\Phi_{6n}(3)$ by (3.7.6). Since $\ell|(q^2-q+1)$, $\ell \neq 2,3$. Suppose that ℓ is not a primitive prime divisor of $3^{6n}-1$. Again by [**34**, Satz 1] $\ell|3n$, whence $\ell|n$ as $\ell\ge 5$, and moreover $\ell^2 \nmid \Phi_{6n}(3)$. It follows that the ℓ-part of $P_{3,1}(n)$ is ℓ. Hence, if $t^-(q)$ is not divisible by any primitive prime divisor of $3^{6n}-1$, then $P_{3,1}(n)$ divides n, a contradiction.

The proof for $t^+(q)$ is entirely similar. □

3D. Primitive prime divisors for Suzuki-Ree groups: another approach. For $\delta\in\mathbb{Z}^+$, $\alpha\in(\mathbb{Z}/\delta)^\times$, and $x\in\mathbb{R}^+$ with $x\ge 1$, let

$$f_n^{(\alpha \bmod \delta)}(x) := \prod_{a\in(\mathbb{Z}/\delta n)^\times : a\equiv\pm\alpha\ (\mathrm{mod}\ \delta)} (x-\zeta_{\delta n}^a).$$

Note that for such δ,α, and x, $f_n^{(\alpha \bmod \delta)}(x)\in\mathbb{R}^+$ by pairing a with $-a$. Note also that

$$f_n^{(1 \bmod 8)}(\sqrt{2}) = P_{2,1}(n) = \Phi_{8n,1}(\sqrt{2}),$$
$$f_n^{(3 \bmod 8)}(\sqrt{2}) = P_{2,3}(n) = \Phi_{8n,3}(\sqrt{2}),$$
$$f_n^{(1 \bmod 12)}(\sqrt{3}) = P_{3,1}(n) = \Phi_{12n,1}(\sqrt{3}),$$
$$f_n^{(5 \bmod 12)}(\sqrt{3}) = P_{3,5}(n) = \Phi_{12n,5}(\sqrt{3})$$

in the notation above.

LEMMA 3.8. *Let $n\in\mathbb{Z}^+$. Let $x\ge 1$. Then*

$$f_n^{(\alpha \bmod \delta)}(x) \ge f_n^{(\alpha \bmod \delta)}(1)\cdot\left(\frac{x+1}{2}\right)^{\frac{2\cdot\phi(\delta n)}{\phi(\delta)}}.$$

PROOF. The claim is evident for $x=1$, and we will show that

$$\frac{f_n^{(\alpha \bmod \delta)}(x)}{(x+1)^{\frac{2\cdot\phi(\delta n)}{\phi(\delta)}}}$$

is increasing in x. Indeed

$$\frac{f_n^{(\alpha \bmod \delta)}(x)}{(x+1)^{\frac{2\cdot\phi(\delta n)}{\phi(\delta)}}} = \prod_{a\in(\mathbb{Z}/\delta n)^\times :\, a\equiv\pm\alpha\ (\mathrm{mod}\ \delta)} \frac{|x-\zeta_n^a|}{x+1},$$

and it suffices to show each factor is increasing in x. But for all $z \in S^1$, denoting by $\Re(z)$ the real part of z,

$$\frac{|x-z|^2}{(x+1)^2} = \frac{x^2 - 2x \cdot \Re z + 1}{x^2+2x+1} = 1 - \frac{2x}{(x+1)^2} \cdot (1+\Re z),$$

which is increasing in x because $\frac{x}{(x+1)^2} = \frac{1}{x+1} - \frac{1}{(x+1)^2}$ decreases in x when $x \geq 1$. □

For $a, b \in \mathbb{Z}^+$, write $\gcd(a, b^\infty) := \prod_{p|b} p^{v_p(a)}$, and $\operatorname{rad}(a) := \prod_{p|a} p$.

LEMMA 3.9. *Let $n \in \mathbb{Z}^+$. Let $m := \frac{n}{\gcd(n,\delta^\infty)}$. Then*

$$f_n^{(\alpha \bmod \delta)}(1) = \prod_{S \subseteq \{p|m\}} \left|1 - \zeta_\delta^{\alpha \cdot \prod_{p\in S} p^{-1} \pmod{\delta}}\right|^{2\cdot(-1)^{\#|S|}}.$$

PROOF. We first claim that if $\operatorname{rad}\left(\frac{n}{\gcd(n,\delta^\infty)}\right) = \operatorname{rad}\left(\frac{n'}{\gcd(n',\delta^\infty)}\right)$, then $f_n^{(\alpha \bmod \delta)}(1) = f_{n'}^{(\alpha \bmod \delta)}(1)$. This follows by repeatedly applying the following. If $p|n$ is such that $k := v_p(\delta n) \geq 2$, then, writing $\delta n =: p^k \cdot s$, because $X^p - Y^p = \prod_{b\in\mathbb{F}_p}(X - \zeta_p^b \cdot Y)$ as elements of $\mathbb{Z}[\zeta_p][X,Y]$,

$$\begin{aligned} f_n^{(\alpha \bmod \delta)}(1) &= \prod_{a\in(\mathbb{Z}/p^{k-1}s)^\times : a\equiv\pm\alpha \pmod{\delta}} \prod_{b\in\mathbb{F}_p} (1 - \zeta_{p^k\cdot s}^{a+p^{k-1}\cdot s\cdot b}) \\ &= \prod_{a\in(\mathbb{Z}/p^{k-1}s)^\times : a\equiv\pm\alpha \pmod{\delta}} (1 - \zeta_{p^{k-1}\cdot s}^{a}). \end{aligned}$$

Therefore without changing $f_n^{(\alpha \bmod \delta)}(1)$, we may assume without loss of generality that n is squarefree and such that $\gcd(n,\delta) = 1$.

Now if $p|n$, writing $\delta n =: p\cdot n_0$ and letting $p', n_0' \in \mathbb{Z}$ be such that $pp' + n_0 n_0' = 1$,

$$\begin{aligned} f_n^{(\alpha \bmod \delta)}(1) &= \prod_{a\in(\mathbb{Z}/n_0)^\times : a\equiv\pm\alpha \pmod{\delta}} \prod_{b\in\mathbb{F}_p^\times} (1 - \zeta_{pn_0}^{app' + bn_0n_0'}) \\ &= \prod_{a\in(\mathbb{Z}/n_0)^\times : a\equiv\pm\alpha \pmod{\delta}} \prod_{b\in\mathbb{F}_p^\times} (1 - \zeta_{n_0}^{ap'} \cdot \zeta_p^{bn_0'}) \\ &= \prod_{a\in(\mathbb{Z}/n_0)^\times : a\equiv\pm\alpha \pmod{\delta}} \prod_{b\in\mathbb{F}_p^\times} (1 - \zeta_{n_0}^{ap'} \cdot \zeta_p^{b}) \end{aligned}$$

via $b \mapsto p \cdot b$. Now we apply the identity $\frac{X^p - Y^p}{X-Y} = \prod_{b\in\mathbb{F}_p^\times}(X - \zeta_p^b \cdot Y)$ to find:

$$\begin{aligned} f_n^{(\alpha \bmod \delta)}(1) &= \prod_{a\in(\mathbb{Z}/n_0)^\times : a\equiv\pm\alpha \pmod{\delta}} \frac{(1-\zeta_{n_0}^a)}{(1-\zeta_{n_0}^{ap'})} \\ &= \frac{f_{\frac{n}{p}}^{(\alpha \bmod \delta)}(1)}{f_{\frac{n}{p}}^{(\alpha p' \bmod \delta)}(1)}. \end{aligned}$$

Since the lemma is evident for $n = 1$, by induction on the number of prime factors of n, we find that

$$f_n^{(\alpha \bmod \delta)}(1) = \prod_{S \subseteq \{\ell | \frac{n}{p}\}} \left(\frac{\left|1 - \zeta_\delta^{\alpha \cdot \prod_{\ell \in S} \ell^{-1} \pmod{\delta}}\right|}{\left|1 - \zeta_\delta^{\alpha \cdot p^{-1} \cdot \prod_{\ell \in S} \ell^{-1} \pmod{\delta}}\right|} \right)^{2 \cdot (-1)^{\#|S|}}$$
$$= \prod_{S \subseteq \{\ell | n\}} \left|1 - \zeta_\delta^{\alpha \cdot \prod_{\ell \in S} \ell^{-1} \pmod{\delta}}\right|^{2 \cdot (-1)^{\#|S|}},$$

and we are done. □

COROLLARY 3.10. *Let $n \in \mathbb{Z}^+$ with $n \geq 3$ be odd. Then*

$$f_n^{(1 \bmod 8)}(1) = \begin{cases} 1 & \exists p | n : p \equiv \pm 1 \pmod{8}, \\ (1 + \sqrt{2})^{-2^{\omega(n)}} & \textit{else} \end{cases}$$

and

$$f_n^{(3 \bmod 8)}(1) = \begin{cases} 1 & \exists p | n : p \equiv \pm 1 \pmod{8}, \\ (1 + \sqrt{2})^{2^{\omega(n)}} & \textit{else}, \end{cases}$$

whence

$$f_n^{(1 \bmod 8)}(\sqrt{2}) \geq \left(\frac{\sqrt{2} + 1}{2} \right)^{2 \cdot \phi(n)} \cdot \begin{cases} 1 & \exists p | n : p \equiv \pm 1 \pmod{8}, \\ (1 + \sqrt{2})^{-2^{\omega(n)}} & \textit{else} \end{cases}$$

and

$$f_n^{(3 \bmod 8)}(\sqrt{2}) \geq \left(\frac{\sqrt{2} + 1}{2} \right)^{2 \cdot \phi(n)} \cdot \begin{cases} 1 & \exists p | n : p \equiv \pm 1 \pmod{8}, \\ (1 + \sqrt{2})^{2^{\omega(n)}} & \textit{else}. \end{cases}$$

PROOF. If there is a $p | n$ with $p \equiv \pm 1 \pmod{8}$, then pair $S - \{p\}$ with $S \cup \{p\}$ in the conclusion of Lemma 3.9 to see that $f_n^{(1 \bmod 8)}(1) = f_n^{(3 \bmod 8)}(1) = 1$. Otherwise every $p | n$ is such that $p \equiv \pm 3 \pmod{8}$, so that by Lemma 3.9

$$f_n^{(1 \bmod 8)}(1) = \left(\frac{|1 - \zeta_8|}{|1 - \zeta_8^3|} \right)^{2^{\omega(n)}} = (1 + \sqrt{2})^{-2^{\omega(n)}}.$$

By the same reasoning

$$f_n^{(3 \bmod 8)}(1) = \left(\frac{|1 - \zeta_8^3|}{|1 - \zeta_8|} \right)^{2^{\omega(n)}} = (1 + \sqrt{2})^{2^{\omega(n)}}.$$

We are done by Lemma 3.8. □

COROLLARY 3.11. *Let $n \in \mathbb{Z}^+$ with $n \geq 3$ be odd. Let $m := \frac{n}{\gcd(n, 6^\infty)}$. Then*

$$f_n^{(1 \bmod 12)}(1) = \begin{cases} 1 & \exists p | m : p \equiv \pm 1 \pmod{12}, \\ (2 + \sqrt{3})^{-2^{\omega(m)}} & \textit{else} \end{cases}$$

and

$$f_n^{(5 \bmod 12)}(1) = \begin{cases} 1 & \exists p | m : p \equiv \pm 1 \pmod{12}, \\ (2 + \sqrt{3})^{2^{\omega(m)}} & \textit{else}, \end{cases}$$

whence

$$f_n^{(1 \bmod 12)}(\sqrt{3}) \geq \left(\frac{\sqrt{3}+1}{2}\right)^{\frac{\phi(12n)}{2}} \cdot \begin{cases} 1 & \exists p|m : p \equiv \pm 1 \pmod{12}, \\ (2+\sqrt{3})^{-2^{\omega(m)}} & \textit{else} \end{cases}$$

and

$$f_n^{(5 \bmod 12)}(\sqrt{3}) \geq \left(\frac{\sqrt{3}+1}{2}\right)^{\frac{\phi(12n)}{2}} \cdot \begin{cases} 1 & \exists p|m : p \equiv \pm 1 \pmod{12}, \\ (2+\sqrt{3})^{2^{\omega(m)}} & \textit{else.} \end{cases}$$

PROOF. If there is a $p|m := \frac{n}{\gcd(n,6^\infty)}$ with $p \equiv \pm 1 \pmod{12}$, then pair $S - \{p\}$ with $S \cup \{p\}$ in the conclusion of Lemma 3.9 to see that $f_n^{(1 \bmod 12)}(1) = f_n^{(5 \bmod 12)}(1) = 1$. Otherwise every $p|m$ is such that $p \equiv \pm 5 \pmod{12}$, so that by Lemma 3.9

$$f_n^{(1 \bmod 12)}(1) = \left(\frac{|1-\zeta_{12}|}{|1-\zeta_{12}^5|}\right)^{2^{\omega(m)}} = (2+\sqrt{3})^{-2^{\omega(m)}}.$$

By the same reasoning

$$f_n^{(5 \bmod 12)}(1) = \left(\frac{|1-\zeta_{12}^5|}{|1-\zeta_{12}|}\right)^{2^{\omega(m)}} = (2+\sqrt{3})^{2^{\omega(m)}}.$$

We are done by Lemma 3.8. □

4. Action on 2-groups and primitivity of local systems

We begin this section with a group theoretic lemma which will be used in the proof of primitivity given in Theorem 4.5.

LEMMA 4.1. *Suppose A is a cyclic group of prime order p that acts faithfully on a finite q-group G, where $p \neq q$ are primes. Let n be the order of q modulo p. Let $\chi \in \mathrm{Irr}(G)$ be A-invariant and faithful, and write $\chi(1) = q^a$. Then $n \leq 2a$.*

PROOF. We argue by induction on $|G|$. Let $C = \mathbf{C}_G(A) < G$. Let $C \leq N < G$ be maximal A-invariant in G. Since G is nilpotent, $N \lhd G$. Also, G/N is an irreducible A-module. Notice that A cannot act trivially on G/N, because $G = [G,A]C$, by coprime action. Hence, G/N is a faithful irreducible $\mathbb{F}_q[A]$-module. By [**35**, Example 2.7], say, $|G/N| = q^n$. Let $\theta \in \mathrm{Irr}(N)$ be A-invariant under χ; such exists by [**20**, Theorem 13.27]. Now, since G/N is an abelian chief factor of the semidirect product GA and χ is GA-invariant, by the Isaacs "going down" theorem [**20**, Theorem 6.18], either $\chi_N = \theta$, or $\chi_N = e\theta$ with $e^2 = |G/N|$, or $\theta^G = \chi$.

In the third case, $\chi(1) = q^n\theta(1) = q^{n+b} = q^a$, for some $b \geq 0$. Thus $n \leq n+b = a \leq 2a$. In the second case, q^n is a square, and that $\chi(1)/\theta(1) = q^{n/2}$. Then $\chi(1) = q^{\frac{n}{2}+c} = q^a$, for some $c \geq 0$. Then $a = \frac{n}{2}+c$, and again $n \leq 2a$.

In the first case, $\chi_N = \theta \in \mathrm{Irr}(N)$. If A does not act trivially on N, then A acts faithfully on N. Since θ is A-invariant and faithful, then we are done by the inductive hypothesis. Otherwise, A acts trivially on N, and therefore $N = \mathbf{C}_G(A)$ (because N is maximal A-invariant). By [**39**, Lemma 2.1], $[G,A] \subseteq \mathrm{Ker}(\chi)$. Since χ is faithful, A acts trivially on G. But this cannot happen. □

We will now introduce a general class of Airy sheaves $\mathcal{F}(q,f)$ that includes the sheaves $\mathcal{F}_q$. Recall that $q = 2^{2n+1} = 2q_0^2$ and $t(q) := q - 2q_0 + 1$. Let $k_0/\mathbb{F}_2$ be a finite extension, and $f(x) \in k_0[x]$ a polynomial of degree $(1+q_0)t(q)$. Form the Artin-Schreier-Witt lisse sheaf

$$\mathcal{L}(q,f) := \mathcal{L}_{\psi_2([x^{t(q)},f(x)]}$$

on $\mathbb{A}^1/k_0$. Then $\mathcal{F}(q,f)$ is the constant field twisted Fourier transform

$$\mathcal{F}(q,f) := \mathrm{FT}_\psi(\mathcal{L}(q,f)) \otimes (1-(-1)^n i)^{-\deg}. \tag{4.1.1}$$

The trace function of $\mathcal{F}(q,f)$ at $t \in k$, k/k_0 a finite extension, is

$$t \mapsto \frac{-1}{\left((1-(-1)^n i\right)^{\deg(k/\mathbb{F}_2)}} \sum_{x\in k} \psi_2(\mathrm{Trace}_{W_2(k)/W_2(\mathbb{F}_2)}([x^{t(q)}, f(x)+tx])). \tag{4.1.2}$$

Assume in addition that

$$f(x) = f_1(x^{t(q)}) \tag{4.1.3}$$

for some polynomial $f_1(x) \in k_0[x]$ of degree $1+q_0$. If we make the substitution $x \mapsto x/t$, then (4.1.2) for $t \in k^\times$ becomes

$$t \mapsto \frac{-1}{\left(1-(-1)^n i\right)^{\deg(k/\mathbb{F}_2)}} \sum_{x\in k} \psi_2\big(\mathrm{Trace}_{W_2(k)/W_2(\mathbb{F}_2)}([x^{t(q)}/t^{t(q)}, f_1(x^{t(q)}/t^{t(q)})+x])\big),$$

For

$$t(q) = rs,$$

we get a descent $\mathcal{G}(q,f,r)$ of $\mathcal{F}(q,f)$ to $\mathbb{G}_m/k_0$ whose trace function is now

$$t \mapsto \frac{-1}{\left(1-(-1)^n i\right)^{\deg(k/\mathbb{F}_2)}} \sum_{x\in k} \psi_2\big(\mathrm{Trace}_{W_2(k)/W_2(\mathbb{F}_2)}([x^{t(q)}/t^s, f_1(x^{t(q)}/t^s)+x])\big),$$

and whose Kummer pullback by r^{th} power is (the restriction to $\mathbb{G}_m/k_0$ of) $\mathcal{F}(q,f)$:

$$[r]^\star\mathcal{G}(q,f,r) \cong \mathcal{F}(q,f)|\mathbb{G}_m.$$

We next give a key lemma of Šuch, which is proved in [**44**, Proposition 11.1] but not stated there explicitly.

LEMMA 4.2 (Šuch). *Let $\mathcal{F}$ be an Airy sheaf of rank $n \geq 2$ on $\mathbb{A}^1/\overline{\mathbb{F}_p}$, i.e., $\mathcal{F}$ is the Fourier transform $\mathrm{FT}(\mathcal{L})$ of a lisse, rank one sheaf $\mathcal{L}$ on $\mathbb{A}^1/\overline{\mathbb{F}_p}$ with* $\mathsf{Swan}_\infty(\mathcal{L}) = n+1$. *If $\mathcal{F}$ is induced, then it is induced from an Artin-Schreier covering of $\mathbb{A}^1/\overline{\mathbb{F}_p}$. In particular, if $\mathcal{F}$ is induced, then it is induced from a normal subgroup of index p of its G_{geom}.*

PROOF. Let us recall the argument, which is contained in the proof of [**44**, Proposition 11.1]. If $\mathcal{F}$ is induced, then it is $g_\star\mathcal{H}$, in which case

$$\mathrm{End}(\mathcal{F}) = (g_\star\mathcal{H}) \otimes (g_\star\mathcal{H}^\vee) \supseteq g_\star(\mathcal{H}\otimes\mathcal{H}^\vee) \supseteq g_\star\mathbb{1},$$

and hence $\mathrm{End}^0(\mathcal{F}) \supseteq g_\star\mathbb{1}/\mathbb{1}$. But $g_\star\mathbb{1}/\mathbb{1}$ has rank $< n$, and all its slopes are $\leq (n+1)/n = 1+1/n$. But each slope of $g_\star\mathbb{1}/\mathbb{1}$ has denominator in lowest terms at most the rank of $g_\star\mathbb{1}/\mathbb{1}$, which is $< n$. Therefore each slope, being at most $1+1/n$, is in fact ≤ 1. Then by [**44**, Corollary 3.3], $g_\star\mathbb{1}/\mathbb{1}$ is the direct sum of various $\mathcal{L}_{\psi(ax)}$.

The rest of the argument is given in the first eight lines of the second paragraph of the proof of [**44**, Proposition 11.1]. □

We will need the following general result, a slight generalization of [**30**, Theorem 4.6].

THEOREM 4.3. *Let $\mathcal{F}$ be a semisimple lisse sheaf on $\mathbb{G}_m/\overline{\mathbb{F}_p}$ which is tame at 0. Denote by J the image of $I(\infty)$ in $G := G_{\text{geom},\mathcal{F}}$. Then G is the Zariski closure of the normal subgroup of G generated by all G-conjugates of J.*

PROOF. Denote by G_∞ this Zariski closure. Then G is reductive, and hence its quotient G/G_∞ is reductive. It suffices to show that every irreducible representation of G/G_∞ is trivial. But a d-dimensional irreducible representation is a lisse sheaf of rank d on $\mathbb{G}_m/\overline{\mathbb{F}_p}$ which is tame at 0 (because $P(0)$ dies in G) and lisse at ∞. By multiplicative inversion, this is an irreducible local system on $\mathbb{A}^1/\overline{\mathbb{F}_p}$ which is tame at ∞, so a representation of $\pi_1^{\text{tame}}(\mathbb{A}^1/\overline{\mathbb{F}_p})$, which is the trivial group. □

We will also need a very special case (about subgroups of index 2) of (the second part of) the following proposition. Since we cannot find a reference for it, we give a proof.

PROPOSITION 4.4.

(i) *Let G be a Lie group and H an abstract subgroup of finite index in G. Then H is closed.*
(ii) *Let G be a reductive linear algebraic group over an algebraically closed field k, and let H be an abstract subgroup of finite index in G. Then H is Zariski closed.*

PROOF. (a) We give a proof of statement (i), which is due to Jason DeVito, posted on `MathStackExchange`.

(a1) First we show that if G is connected, then G is generated by its divisible subgroups. Indeed, there is an open set $U \subseteq G$ containing the identity such that $U \subseteq \exp(\mathcal{L})$ for the Lie algebra $\mathcal{L}$ of G. For any u in U, if $u = \exp(X)$, then u lies in the divisible subgroup $\{\exp(tX) \mid t \in \mathbb{R}\}$ of G. Since G is generated by U, the claim follows.

(a2) Next we show that if G is connected and $H \leq G$ is of finite index, then $H = G$. Indeed, by considering the action via left translation of G on the finite set G/H of left H-cosets, we see that H contains a normal subgroup $K \lhd G$ of finite index. By (i), G, and so G/K, is generated by its divisible subgroups. But G/K is finite, so $K = G$.

(a3) In the general case, consider the map $\iota : G^\circ/(H \cap G^\circ) \to G/H$ defined by $x(H \cap G^\circ) \mapsto xH$. Then ι is injective, and so $H \cap G^\circ$ has finite index in G°. By (a2), $H \geq G^\circ$, and so H is a union of a finite number of G°, each of which is closed. Hence H is closed.

(b) For statement (ii), the argument in (a3) shows that it suffices to prove that if G is connected reductive and H has finite index in G, then $H = G$. The argument in (a2) allows us to further assume that $H \lhd G$. Recall [**5**, p. 16] that $G = T[G,G]$, where $T = \mathbf{Z}(G)^\circ$ is a central torus of G, and the derived subgroup $[G,G]$ is a central product $G_1 \circ \ldots \circ G_n$ of simple groups. In particular, for each i, $H \cap G_i$ is a normal (abstract) subgroup of finite index of G_i. As G_i is generated by (unipotent) root subgroups, [**48**, Main Theorem] implies that $H \cap G_i$ is either equal to G_i or contained in $\mathbf{Z}(G_i)$. The finite index assumption now ensures that $H \geq G_i$ for all i, and thus $H \geq [G,G]$. Next, $H \cap T$ is a normal (abstract) subgroup of finite index say m in T. In particular, H contains t^m for all $t \in T$. But the torus T is $(k^\times)^r$ for some r, so $k = \overline{k}$ implies that any element in T is an mth power, and hence $H \geq T$, completing the proof. [The referee kindly pointed out a much simpler argument as follows. If $H \lhd G$ has index N in G then G contains all powers g^N, $g \in G$. But G, being connected reductive, is generated by its maximal tori, and the map $g \mapsto g^N$ is surjective on each maximal torus. Hence $H = G$.] $\square$

As pointed out by the referee, in both of the cases of Proposition 4.4, once H is closed, it is also open (indeed, $G \smallsetminus H$ is a disjoint union of a finite number of cosets gH, each being closed, and so it is closed).

THEOREM 4.5. *Let $q = 2^{2n+1}$ with $n \in \mathbb{Z}_{\geq 1}$ and $n \neq 2$. Under the assumption (4.1.3), the group $G_{\text{geom},\mathcal{F}(q,f)}$ of the Airy sheaf $\mathcal{F}(q,f)$ has no subgroups of index 2. As a consequence, the Airy sheaf $\mathcal{F}(q,f)$ is not geometrically induced, and hence none of the sheaves $\mathcal{G}(q,f,r)$ is geometrically induced.*

PROOF. Assume to the contrary that the Airy sheaf $\mathcal{F}(q,f)$ is induced. Then, by Lemma 4.2, the underlying representation V of the geometric monodromy group $G := G_{\text{geom}}$ of $\mathcal{F}(q,f)$ is induced from a subgroup G_1 of G of index 2.

(i) For each divisor $r > 1$ of $t(q)$, consider the descent $\mathcal{G}(q,f,r)$. Because $\mathcal{F}(q,f)$ is lisse at 0, the image of $I(0)$ in the geometric monodromy group H of $\mathcal{G}(q,f,r)$ is the cyclic group $\mu_r(\overline{\mathbb{F}_2})$ of order r. Because the ∞-slopes of $\mathcal{F}(q,f)$ are $\dfrac{(2^n+1)t(q)}{(2^n+1)t(q)-1}$, if $r > 1$, then the ∞-slopes of $\mathcal{G}(q,f,r)$ are $(1/r)\dfrac{(2^n+1)t(q)}{(2^n+1)t(q)-1}$ < 1. It then follows from [**30**, Proposition 4.2] that H is equal to the Zariski closure H_0^{Zar} in H of the normal closure H_0 in H of the image of $I(0)$.

We next show that for $r > 1$, H has no subgroup of index 2. We argue by contradiction. Suppose that H has a subgroup H_1 of index 2. Since every H-conjugate of the image of $I(0)$ has odd order r, all such conjugates are contained in H_1, and thus $H_0 \leq H_1$. But H_1 is closed in H by Proposition 4.4(ii), so $H = H_0^{\text{Zar}} \leq H_1$, and hence $H = H_1$, a contradiction. We have shown that H has no subgroup of index 2.

(ii) We also know that G is a normal subgroup of H of index dividing r, simply because by the rth power map, $\mathbb{G}_m/\overline{\mathbb{F}_2}$ becomes a finite étale Galois covering of itself with cyclic group of order r, and G and H are respectively the Zariski closure of the images of $\pi_1(\mathbb{G}_m/\overline{\mathbb{F}_2})$ and of a normal subgroup of cyclic index r of that group. Now if $H = G$, then the existence of G_1 leads to a contradiction by (i). If we now take r to be a prime dividing $t(q)$, then $H/G \cong C_r$.

Let J and Q denote the images of $I(\infty)$ and of $P(\infty)$ in H. By Theorem 4.3, H is equal to the Zariski closure H_∞^{Zar} in H of the normal closure H_∞ in H of J. Suppose for the moment that $r \nmid |J|$. Then every H-conjugate of J has order coprime to r, and so they are all contained in G, and thus $H_\infty \leq G$. But G is closed in H, so $H_\infty^{\text{Zar}} \leq G$, and hence $H = G$, a contradiction.

We have shown that r divides $|J|$. Recall that $J = Q \rtimes C$, with C a cyclic $2'$-group that permutes cyclically and transitively the $q-1$ simple Q-submodules V_i in V, each of dimension $q_0 = 2^n$. As Q is a 2-group, C is of order divisible by r.

(iii) Since $n \neq 2$, by Theorem 3.4 the integer $t(q) = q - \sqrt{2q} + 1$ admits a prime divisor

$$r = \text{ppd}(2, 4(2n+1)). \tag{4.5.1}$$

At this point, we take for r a ppd$(2, 4(2n+1))$ which divides $t(q)$. Fix $c \in C$ of order r. Since $r|t(q)$, $r \nmid (q-1)$, and so c stabilizes each of the subspace V_i, and certainly normalizes Q. For each i, let Φ_i denote the representation of $\langle Q, c\rangle$ on V_i. We claim that $\Phi_i(c)$ centralizes $\Phi_i(Q)$. Otherwise, $\Phi_i(c)$ acts faithfully on $\Phi_i(Q)$, of prime order r, and $\Phi_i(Q)$ is faithful and irreducible of degree 2^n. Hence by Lemma 4.1, the order of 2 modulo r is at most $2n$, which contradicts (4.5.1).

Therefore, for each i, $\Phi_i(c) = \alpha_i \cdot \mathrm{Id}_{V_i}$ for some $\alpha_i \in \mathbb{C}^\times$. But the cyclic group C permutes the V_i's transitively and $c \in C$, so $\alpha_1 = \ldots = \alpha_{q-1} =: \alpha$, i.e., c acts on V as $\alpha \cdot \mathrm{Id}_V$. As $|c| = r$, $\alpha \neq 1$ is a primitive r^{th} root of unity. Thus we may assume that $c \in J$ has trace $\dim(V) \cdot \zeta_r$. On the other hand, by Theorem 2.2 and (4.1.2), the trace of every element in J belongs to $\mathbb{Q}(i)$. Thus $\zeta_r \in \mathbb{Q}(i)$, a contradiction since r is an odd prime.

(iii) Since the irreducible representation V of $G = G_{\mathrm{geom}}$ of $\mathcal{F}(q,f)$ is not induced, for any $r|t(q)$, the representation of G_{geom} of $\mathcal{G}(q,f,r)$ on V, which contains G, is not induced. □

5. Condition (S+) and autoduality for Airy sheaves

In this section, we continue to consider Airy sheaves $\mathcal{F}(q,f)$ of the same general shape (4.1.1), but we consider them only geometrically, i.e., as lisse sheaves on $\mathbb{A}^1/\overline{\mathbb{F}_2}$. The key insight on which the results of this section are based is due to Šuch, cf. [**44**, Proposition 11.1].

LEMMA 5.1 (Šuch). *Let $\mathcal{F}$ be an Airy sheaf of rank $D \geq 2$ on $\mathbb{A}^1/\overline{\mathbb{F}_p}$. Let $\mathcal{H}$ be a direct factor of* $\mathrm{End}(\mathcal{F})$ *of rank $r < D$. Then $\mathcal{H}$ is a direct sum of $\mathcal{L}_{\psi(ax)}$ for various $a \in \overline{\mathbb{F}_p}$.*

PROOF. The ∞-slopes of $\mathcal{F}$ are all $1 + 1/D$. All slopes of $\mathrm{End}(\mathcal{F})$ are therefore $\leq 1 + 1/D$. Thus all slopes of $\mathcal{H}$ are $\leq 1 + 1/D$. But each slope of $\mathcal{H}$, written in lowest terms, has denominator $\leq r$. Therefore $\mathcal{H}$ has all slopes ≤ 1. Now take an irreducible constituent $\mathcal{K}$ of $\mathcal{H}$. Its Fourier transform, i.e., $t \mapsto H_c(A^1/\overline{\mathbb{F}_p}, \mathcal{K} \otimes \mathcal{L}_{\psi(tx)})$, is perverse irreducible on $A^1/\overline{\mathbb{F}_p}$, so is either a single delta function δ_a or is the extension by direct image of a lisse sheaf on a dense open set. But on a dense open set of the t-line, $\mathcal{K} \otimes \mathcal{L}_{\psi(tx)}$ has all ∞-slopes 1, so $H^2_c(A^1/\overline{\mathbb{F}_p}, \mathcal{K} \otimes \mathcal{L}_{\psi(tx)}) = 0$ and by the Euler-Poincaré formula $\chi_c(A^1/\overline{\mathbb{F}_p}, \mathcal{K} \otimes \mathcal{L}_{\psi(tx)}) = 0$, so $\mathrm{FT}(\mathcal{K})$ is punctual, hence a single δ_a. This means in turn that each irreducible constituent of $\mathcal{H}$ is an $\mathcal{L}_{\psi(ax)}$. Because $\mathcal{F}$ is irreducible, $\mathrm{End}(\mathcal{F})$ is completely reducible, hence $\mathcal{H}$ is completely reducible, and so it is the sum of its irreducible constituents. □

THEOREM 5.2. *Let $\mathcal{F}$ be an Airy sheaf of rank at least* 2 *on $\mathbb{A}^1/\overline{\mathbb{F}_p}$. Then the following conditions are equivalent:*

(i) *$\mathcal{F}$ is geometrically induced.*
(ii) $\mathrm{End}(\mathcal{F})$ *contains a summand $\mathcal{L}_{\psi(ax)}$ for some $a \neq 0$ in $\overline{\mathbb{F}_p}$.*
(iii) *There exists a geometric isomorphism $\mathcal{F} \cong \mathcal{F} \otimes \mathcal{L}_{\psi(-ax)}$ for some $a \neq 0$ in $\overline{\mathbb{F}_p}$.*

PROOF. Let us denote by n the rank of $\mathcal{F}$. If $\mathcal{F}$ is induced, then it is $g_\star\mathcal{H}$, in which case

$$\mathrm{End}(\mathcal{F}) = (g_\star\mathcal{H}) \otimes (g_\star\mathcal{H}^\vee) \supseteq g_\star(\mathcal{H} \otimes \mathcal{H}^\vee) \supseteq g_\star\mathbb{1},$$

and hence $\mathrm{End}^0(\mathcal{F}) \supseteq g_\star \mathbb{1}/\mathbb{1}$. But $g_\star \mathbb{1}/\mathbb{1}$ has rank less than n. So by Lemma 5.1, $g_\star \mathbb{1}/\mathbb{1}$ is the direct sum of various $\mathcal{L}_{\psi(ax)}$. As $\mathrm{End}^0(\mathcal{F})$ does not contain the trivial sheaf (by irreducibility of $\mathcal{F}$), we find that $\mathrm{End}^0(\mathcal{F})$, and hence $\mathrm{End}(\mathcal{F})$, contains some $\mathcal{L}_{\psi(ax)}$ for some nonzero $a \in \overline{\mathbb{F}_2}$.

Conversely, suppose $\mathrm{End}(\mathcal{F})$ contains a summand $\mathcal{L}_{\psi(ax)}$ for some $a \neq 0$ in $\overline{\mathbb{F}_p}$. Thus there is a nonzero geometric homomorphism from $\mathcal{F}$ to $\mathcal{F} \otimes \mathcal{L}_{\psi(-ax)}$. But source and target are geometrically isomorphic, so every nonzero homomorphism is an isomorphism. Thus we have a geometric isomorphism

$$\mathcal{F} \cong \mathcal{F} \otimes \mathcal{L}_{\psi(-ax)}.$$

That $\mathcal{F}$ is geometrically induced is the special case, where G is G_{geom} for $\mathcal{F} \oplus \mathcal{L}_{\psi(-ax)}$, V is $\mathcal{F}$, L is $\mathcal{L}_{\psi(-ax)}$, N is p, and $\mathbb{E}$ is $\overline{\mathbb{Q}_\ell}$, of the following general statement in characteristic zero representation theory. □

THEOREM 5.3. *Over an algebraically closed field $\mathbb{E}$ of characteristic zero, let V be a finite dimensional irreducible representation of dimension $d \geq 2$ of a group G, and L a one-dimensional representation of G which, viewed as a linear character χ of G, has finite order $N > 1$. Suppose that $V \cong V \otimes L$ as representations of G. Denote by $G_0 := \mathrm{Ker}(\chi)$ the kernel of χ, so that $G_0 \lhd G$ with G/G_0 cyclic of order N. Then V is induced from a representation of a subgroup H with $G_0 \leq H < G$.*

PROOF. We first reduce to the case when N is prime. If $V \cong V \otimes L$, then by induction $V \cong V \otimes L^{\otimes n}$ for every integer n. Let r be a prime dividing N. Replacing L by $L^{\otimes(N/r)}$ and $\mathrm{Ker}(\chi)$ by the overgroup $K := \mathrm{Ker}(\chi^{(N/r)})$, we are reduced to the case when $N = r$ is prime.

Let U be a simple summand of $V|_K$, which is semisimple since $K \lhd G$. By Clifford theory, cf. [**7**, (50.5)], V is induced so long as $V|_K$ is not isotypic. Hence, if V is not induced, then $V|_K \cong eU := \underbrace{U \oplus U \oplus \ldots \oplus U}_{e \text{ times}}$ for some $e \in \mathbb{Z}_{\geq 1}$. Note that $\mathrm{Ind}_K^G(\mathbb{E}) \cong \oplus_{i=1}^r L^{\otimes i}$, so by Frobenius reciprocity, cf. [**8**, (10.20)],

$$\mathrm{Ind}_K^G(V|_K) \cong \mathrm{Ind}_K^G\big((V|_K) \otimes \mathbb{E}\big) \cong V \otimes \mathrm{Ind}_K^G(\mathbb{E}) \cong \oplus_{i=1}^r (V \otimes L^{\otimes i}) \cong rV.$$

Again by Frobenius reciprocity [**8**, (10.8)],

$$e^2 = \dim \mathrm{Hom}_K(V|_K, V|_K) = \dim \mathrm{Hom}_G\big(V, \mathrm{Ind}_K^G(V|_K)\big) = \dim \mathrm{Hom}_G(V, rV) = r,$$

a contradiction as r is a prime. □

The following general result is stated for ease of later reference.

THEOREM 5.4. *Let $\mathcal{F}$ be a lisse irreducible $\overline{\mathbb{Q}_\ell}$-sheaf of rank $d \geq 2$ on a smooth, geometrically connected $X/\overline{\mathbb{F}_p}$. Suppose that $\mathrm{End}^0(\mathcal{F})$ contains a rank one summand $\mathcal{L}$. Then $\mathcal{F}$ is induced.*

PROOF. Because $\mathcal{L}$ occurs in $\mathrm{End}^0(\mathcal{F})$, it cannot be trivial (as by irreducibility of $\mathcal{F}$, $\mathrm{End}(\mathcal{F})$ contains $\mathbb{1}$ exactly once). Exactly as in the proof of the implication (ii) $\Longrightarrow$ (iii) in Theorem 5.2, we infer that $\mathcal{F} \cong \mathcal{F} \otimes \mathcal{L}^{-1}$. Taking determinants of this isomorphism, we find that $\mathcal{L}^{\otimes d}$ is trivial. Thus $\mathcal{L}$ has finite order $N > 1$ as a character of $\pi_1(X)$. Now apply Theorem 5.3. □

THEOREM 5.5. *Let $k_0/\mathbb{F}_2$ be a finite extension, and $f(x) \in k_0[x]$ a polynomial of degree $(1+q_0)t(q)$. Then the Airy sheaf $\mathcal{F}(q, f)$ introduced in (4.1.1) is geometrically primitive, i.e., is not geometrically induced.*

PROOF. (i) In view of Theorem 5.2, it suffices to show that there is no geometric isomorphism from $\mathcal{F}(q,f)$ to $\mathcal{F}(q,f)\otimes\mathcal{L}_{\psi(ax)}$ for any $a\neq 0$ in $\overline{\mathbb{F}_p}$. We argue by contradiction. Because $\mathcal{F}(q,f)$ was geometrically a Fourier transform, so is $\mathcal{F}(q,f)\otimes\mathcal{L}_{\psi(ax)}$; the effect of tensoring with $\mathcal{L}_{\psi(ax)}$ after FT is the same as translating additively by $-a$ before FT:

$$\mathcal{F}(q,f)\otimes\mathcal{L}_{\psi(ax)}=\mathrm{FT}([x\mapsto x-a]^{\star}\mathcal{L}(q,f)).$$

Since FT is invertible, an isomorphism of $\mathcal{F}(q,f)$ with $\mathcal{F}(q,f)\otimes\mathcal{L}_{\psi(ax)}$ gives an isomorphism of $\mathcal{L}(q,f)$ with its additive translate by $-a$:

$$\mathcal{L}_{\psi_2([x^{t(q)},f(x)])}\cong\mathcal{L}_{\psi_2([(x-a)^{t(q)},f(x-a)])}.$$

We will show no such isomorphism exists.

(ii) Suppose that $n\geq 2$. In this case, we argue as follows. If two lisse rank one sheaves are isomorphic, then so are their tensor squares.

Quite generally, addition of Witt vector of length 2 over an $\mathbb{F}_2$-algebra is given by

$$[a,b]+[A,B]=[a+A,B+b+aA],\ [a,b]+[a,b]=[0,a^2].$$

Thus

$$\mathcal{L}^{\otimes 2}_{\psi_2([g(x),f(x)])}=\mathcal{L}_{\psi_2([g(x),f(x)]+[g(x),f(x)])}=\mathcal{L}_{\psi_2([0,g(x)^2])}=\mathcal{L}_{\psi(g(x)^2)}\cong\mathcal{L}_{\psi(g(x))}.$$

So we would have a geometric isomorphism

$$\mathcal{L}_{\psi(x^{t(q)}))}\cong\mathcal{L}_{\psi((x-a)^{t(q)}))},$$

or equivalently a geometric isomorphism

$$\mathcal{L}_{\psi((x-a)^{t(q)}-x^{t(q)})}\cong\overline{\mathbb{Q}_\ell}.$$

Now we use the explicit shape of $t(q)=q+1-2q_0=2q_0^2-2q_0+1=(q_0-1)(2q_0)+1$.

The key point is that $t(q)=1+dQ$ with $d\geq 2$ prime to $p=2$ (here $d=q_0-1$) and Q a strictly positive power of $p=2$ (here $Q=2q_0$). Then

$$\begin{aligned}(x-a)^{1+qD}&=(x-a)(x^Q-a^Q)^d\\&=(x-a)\big(x^{dQ}-da^Qx^{(d-1)Q}+(\text{terms of degree }\leq(d-2)Q)\big)\\&=x^{1+dQ}-da^Qx^{1+(d-1)Q}+(\text{terms of degree }\leq 1+(d-2)Q)\\&\quad+(\text{polynomial in }x^Q\text{ of degree }d).\end{aligned}$$

Thus, up to Artin-Schreier equivalence,

$$\begin{aligned}&(x-a)^{1+dQ}-x^{1+dQ}\\&=-da^Qx^{1+(d-1)Q}+(\text{terms of degree }\leq 1+(d-2)Q)+(\text{a term of degree }d).\end{aligned}$$

Thus $\mathcal{L}_{\psi((x-a)^{1+dQ}-x^{1+dQ})}$ has $\mathsf{Swan}=1+(d-1)Q>0$, so is not geometrically trivial.

(iii) We now turn to the case $n=1$, which requires a more delicate analysis. What makes the $n\geq 2$ case argument work is that $q_0-1=2^n-1$ has $n\geq 2$ binary digits. In the $n=1$ case, $t(8)=5$, so the above argument would involve examining

$$(x-a)^5-x^5=(x-a)(x^4-a^4)-x^5=ax^4-a^4x+a^5,$$

but this is Artin-Schreier equivalent to $(a^{1/4}-a^4)x$, which for $a\in\mu_{15}$ is Artin-Schreier trivial.

Thus we must look instead at

$$\mathcal{L}_{\psi_2([(x-a)^5, f(x-a)])} \otimes \mathcal{L}^{-1}_{\psi_2([x^5, f(x)])} = \mathcal{L}_{\psi_2([(x-a)^5, f(x-a)]-[x^5, f(x)])}.$$

Here $-[a, b] = [a, b - a^2]$, implying $-[x^5, f(x)] = [x^5, f(x) - x^{10}]$, so this is

$$\mathcal{L}_{\psi_2([(x-a)^5+x^5, f(x-a)+f(x)-x^{10}+x^5(x-a)^5])}.$$

But $f(x)$ has degree 15, say $f(x) = bx^{15} + cx^{14} + dx^{13} +$ lower terms with $b \neq 0$. So under Artin-Schreier equivalence

$$f(x) \equiv bx^{15} + dx^{13} + \text{lower terms}.$$

Similarly, under Artin-Schreier equivalence

$$f(x - a) \equiv b(x - a)x^{15} + d(x - a)^{13} + \text{lower terms}.$$

Thus under Artin-Schreier equivalence

$$f(x - a) + f(x) - x^{10} + x^5(x - a)^5 \equiv b(x - a)^{15} - bx^{15} + \text{lower terms}.$$

The difference

$$b(x - a)^{15} - bx^{15} = -abx^{14} + ba^2x^{13} + \text{lower terms}$$

is thus Artin-Schreier equivalent to

$$ba^2x^{13} + \text{lower terms}.$$

Now view $\mathcal{L}_{\psi_2([(x-a)^5+x^5, f(x-a)+f(x)-x^{10}+x^5(x-a)^5])}$ as the tensor product (using $[a, b] = [a, 0] + [0, b]$)

$$\mathcal{L}_{\psi_2([(x-a)^5+x^5, 0]} \otimes \mathcal{L}_{\psi(f(x-a)+f(x)-x^{10}+x^5(x-a)^5)}.$$

The first factor has $\mathsf{Swan}_\infty \leq 2 \times 4$, while the second factor, which is $\mathcal{L}_{\psi(\text{a polynomial of degree 13})}$, has $\mathsf{Swan}_\infty = 13$, and hence their tensor product has $\mathsf{Swan}_\infty = 13$, so is not geometrically trivial. □

Recall that a lisse sheaf $\mathcal{H}$ is *Lie-irreducible* if, in the underlying representation of its G_{geom}, the identity component G°_{geom} acts irreducibly. It is *Lie-self-dual* if the given representation of G_{geom}, when restricted to G°_{geom}, is self-dual.

THEOREM 5.6. *Suppose that a lisse sheaf $\mathcal{F}(q, f)$ as in* (4.1.1) *is Lie-irreducible and Lie-self-dual. Then $\mathcal{F}(q, f)$ is self-dual.*

PROOF. The two sheaves $\mathcal{F}(q, f)$ and its dual $\mathcal{F}(q, f)^\vee$ are irreducible representations of G_{geom} whose restrictions to the identity component G°_{geom} are isomorphic. So the two, as representations of G_{geom}, differ by a linear character of $G_{\text{geom}}/G^\circ_{\text{geom}}$. Thus

$$\mathcal{F}(q, f)^\vee \cong \mathcal{F}(q, f) \otimes \mathcal{L}$$

for some lisse, rank one $\mathcal{L}$ on $\mathbb{A}^1$. Both $\mathcal{F}(q, f)^\vee$ and $\mathcal{F}(q, f)$ have all ∞-slopes $1 + 1/\operatorname{rank}(\mathcal{F}(q, f)) < 2$. Therefore, $\mathsf{Swan}_\infty(\mathcal{L}) \leq 1$ (otherwise $\mathcal{F}(q, f)^\vee$ would have all ∞-slopes ≥ 2).

If $\mathsf{Swan}_\infty(\mathcal{L}) = 0$, then $\mathcal{L}$ is lisse on $\mathbb{A}^1$ and tame at ∞, so geometrically trivial, and $\mathcal{F}(q, f)$ is geometrically self-dual. If $\mathsf{Swan}_\infty(\mathcal{L}) = 1$, then $\mathcal{L}$ is $\mathcal{L}_{\psi(ax)}$ for some nonzero $a \in \overline{\mathbb{F}_p}$. We will show that this case cannot arise.

Recall that $\mathcal{F}(q, f) := \mathrm{FT}_\psi(\mathcal{L}_{\psi_2([x^{t(q)}, f(x)])})$. In general, the interaction of FT with duality is a geometric isomorphism

$$D(\mathrm{FT}_\psi(\mathcal{H})) \cong \mathrm{FT}_{\overline{\psi}}(D\mathcal{H})).$$

In characteristic 2, where ψ takes values ± 1, we have $\overline{\psi} = \psi$. Thus

$$\mathcal{F}(q,f)^{\vee} \cong \mathrm{FT}_{\psi}(\mathcal{L}_{\psi_2(-[x^{t(q)},f(x)])}) = \mathrm{FT}_{\psi}(\mathcal{L}_{\psi_2([x^{t(q)},f(x)+(x^{t(q)})^2])})$$

while

$$\mathcal{F}(q,f) \otimes \mathcal{L}_{\psi(ax)} \cong \mathrm{FT}_{\psi}([x \mapsto x-a]^{\star}\mathcal{L}_{\psi_2([x^{t(q)},f(x)])} = \mathrm{FT}_{\psi}(\mathcal{L}_{\psi_2([(x-a)^{t(q)},f(x-a)])}).$$

By Fourier inversion, this is equivalent to a geometric isomorphism

$$\mathcal{L}_{\psi_2([x^{t(q)},f(x)+(x^{t(q)})^2])} \cong \mathcal{L}_{\psi_2([(x-a)^{t(q)},f(x-a)])}.$$

We first treat the case $n \geq 2$. Already the tensor squares of these two lisse rank one sheaves are not geometrically isomorphic, by the identical argument used to treat the case $n \geq 2$ in the proof of Theorem 5.5.

In the case $n = 1$, we must show that the lisse rank one sheaf

$$\mathcal{L}_{\psi_2([(x-a)^s,f(x-a)])} \otimes (\mathcal{L}_{\psi_2([x^5,f(x)+(x^5)^2])})^{-1}$$

is not geometrically trivial. The second tensor factor is

$$(\mathcal{L}_{\psi_2([x^5),f(x)+(x^5)^2])})^{-1} = \mathcal{L}_{\psi_2(-[x^5,f(x)+(x^5)^2])} = \mathcal{L}_{\psi_2([x^5,f(x)])}.$$

So we must show that

$$\begin{aligned}\mathcal{L}_{\psi_2([(x-a)^5,f(x-a)])} \otimes \mathcal{L}_{\psi_2([x^5),f(x)])} &= \mathcal{L}_{\psi_2([(x-a)^5,f(x-a)]+[x^5,f(x)])}\\ &= \mathcal{L}_{\psi_2([(x-a)^5+x^5,f(x-a)+f(x)+x^5(x-a)^5])}\\ &= \mathcal{L}_{\psi_2([(x-a)^5+x^5,0])} \otimes \mathcal{L}_{\psi(f(x-a)+f(x)+x^5(x-a)^5)}\end{aligned}$$

is not geometrically trivial. Exactly as in the proof of the $n = 1$ case of Theorem 5.5, the first factor has $\mathsf{Swan}_{\infty} \leq 2\times 4 = 8$, while the second factor has $\mathsf{Swan}_{\infty} = 13$. □

THEOREM 5.7. *No sheaf $\mathcal{F}(q,f)$ introduced in* (4.1.1) *is geometrically self-dual.*

PROOF. Recall that $\mathcal{F}(q,f) := \mathrm{FT}_{\psi}(\mathcal{L}_{\psi_2([x^{t(q)},f(x)])})$. In general, the interaction of the Fourier transform FT with the duality functor $D(\cdot)$ is a geometric isomorphism

$$D(\mathrm{FT}_{\psi}(\mathcal{H})) \cong \mathrm{FT}_{\overline{\psi}}(D\mathcal{H}).$$

In characteristic 2, where ψ takes values ± 1, we have $\overline{\psi} = \psi$. Therefore, $\mathcal{F}(q,f)$ is self-dual if and only if $\mathcal{L}_{\psi_2([x^{t(q)},f(x)])}$ is self-dual. Its dual is $\mathcal{L}_{\psi_2(-[x^{t(q)},f(x)])}$. We have the Witt vector addition law $[x,y] = [x,0]+[0,y]$, and, for Witt vectors over $\mathbb{F}_2$-algebras, $-[x,y] = [x,y+x^2]$. So the dual of $\mathcal{L}_{\psi_2([x^{t(q)},f(x)])}$ is $\mathcal{L}_{\psi_2([x^{t(q)},f(x)+(x^{t(q)})^2])}$, and the asserted isomorphism is

$$\mathcal{L}_{\psi_2([x^{t(q)},0)])} \otimes \mathcal{L}_{\psi(f(x))} \cong \mathcal{L}_{\psi_2([x^{t(q)},0)])} \otimes \mathcal{L}_{\psi(f(x)+(x^{t(q)})^2)}.$$

This holds if and only if there is an isomorphism

$$\mathcal{L}_{\psi(f(x))} \cong \mathcal{L}_{\psi(f(x)+(x^{t(q)})^2)},$$

or equivalently if $\mathcal{L}_{\psi((x^{t(q)})^2)}$ is geometrically trivial. By the Artin-Schreier reduction, we have $\mathcal{L}_{\psi((x^{t(q)})^2)} \cong \mathcal{L}_{\psi(x^{t(q)})}$. The latter sheaf is not geometrically trivial, because $x^{t(q)}$ is a polynomial of odd degree $t(q)$ and so the sheaf has $\mathsf{Swan}_{\infty}(\mathcal{L}_{\psi(x^{t(q)})}) = t(q) \neq 0$. □

COROLLARY 5.8. *Suppose that the sheaf $\mathcal{F}(q,f)$ introduced in* (4.1.1) *is Lie-irreducible. Then it is not Lie-self-dual.*

PROOF. Combine Theorem 5.6 and Theorem 5.7. □

THEOREM 5.9. *The sheaf $\mathcal{F}(q,f)$ introduced in* (4.1.1) *is tensor indecomposable.*

PROOF. We will show that $\mathcal{F}(q,f)$ is tensor indecomposable as a representation of $\pi_1 := \pi_1(\mathbb{A}^1/\overline{\mathbb{F}_p})$. Because π_1 has cohomological dimension ≤ 1 (this is true for the π_1 of any smooth, connected affine curve over an algebraically closed field), the argument of [**27**, Corollary 10.4] shows that if $\mathcal{F}(q,f)$ is tensor decomposable, then it is linearly tensor decomposable, i.e., we have an isomorphism of local systems on $\mathbb{A}^1/\overline{\mathbb{F}_p}$,

$$\mathcal{F}(q,f) \cong \mathcal{A} \otimes \mathcal{B},$$

with both $\mathrm{rank}(\mathcal{A}), \mathrm{rank}(\mathcal{B}) \geq 2$. To fix ideas, suppose $\mathrm{rank}(\mathcal{A}) \leq \mathrm{rank}(\mathcal{B})$. Then

$$\begin{aligned}\mathrm{End}(\mathcal{F}(q,f)) &= \mathrm{End}(\mathcal{A}) \otimes \mathrm{End}(\mathcal{B}) \\ &= (\mathbb{1} + \mathrm{End}^0(\mathcal{A})) \otimes (\mathbb{1} + \mathrm{End}^0(\mathcal{B})) \text{ contains } \mathbb{1} + \mathrm{End}^0(\mathcal{A}).\end{aligned}$$

Thus $\mathrm{End}^0(\mathcal{F}(q,f))$ contains $\mathrm{End}^0(\mathcal{A})$ as a direct factor. Now $\mathrm{End}^0(\mathcal{A})$ has rank less than

$$(\mathrm{rank}(\mathcal{A}))^2 \leq \mathrm{rank}(\mathcal{A})\,\mathrm{rank}(\mathcal{B}) = \mathrm{rank}(\mathcal{F}(q,f)).$$

So by Lemma 5.1, $\mathrm{End}^0(\mathcal{F}(q,f))$ contains some $\mathcal{L}_{\psi(ax)}$ with $a \neq 0$ ($a \neq 0$ because $\mathrm{End}^0(\mathcal{F}(q,f))$ only contains $\mathbb{1}$ once, by irreducibility of $\mathcal{F}$). The proof of Theorem 5.5 shows this is impossible. □

LEMMA 5.10. *For every integer $n \geq 2$, the integer $2^n - 1$ is never a perfect power x^m with $x \in \mathbb{Z}$ and $m \geq 2$.*

PROOF. We argue by contradiction. If $2^n - 1 = x^m$, then x is odd and $x^m = 2^n - 1 \equiv 3 \pmod 4$. Thus m is odd, and hence $2^n = x^m + 1$ is divisible by $x+1$. The quotient $\frac{x^m+1}{x+1} > 1$ is the alternating sum of m powers of the odd integer x, so is itself odd. Thus $\frac{x^m+1}{x+1}$ is an odd divisor of 2^n, the desired contradiction. □

COROLLARY 5.11. *For $n \geq 1$, no lisse sheaf of rank $2^n(2^{2n+1}-1)$ can be tensor induced. In particular, $\mathcal{F}(q,f)$ is not tensor induced.*

THEOREM 5.12. *The local systems $\mathcal{F}(q,f)$ on $\mathbb{A}^1/\overline{\mathbb{F}_2}$ introduced in* (4.1.1) *all satisfy the condition* (**S**+) *of* [**30**, Definition 1.2].

PROOF. First, by [**44**, Proposition 7.4] the underlying representation V of G_{geom} of $\mathcal{F}(q,f)$ is irreducible. That $\mathcal{F}(q,f)$ is primitive, tensor indecomposable, and not tensor induced is the content of Theorem 5.5, Theorem 5.9 and Corollary 5.11. That $\det(\mathcal{F}(q,f))$ has finite order results from the fact that $\mathcal{F}(q,f)$ began life over a finite subfield of $\overline{\mathbb{F}_p}$ (in fact any subfield containing the coefficients of f). □

THEOREM 5.13. *For the local system $\mathcal{F}(q,f)$ on $\mathbb{A}^1/\overline{\mathbb{F}_2}$ introduced in* (4.1.1), *every irreducible constituent of $\mathrm{End}^0(\mathcal{F}(q,f))$ has dimension $\geq \mathrm{rank}(\mathcal{F}(q,f))$. In particular, if $G_{\mathrm{geom},\mathcal{F}(q,f)}$ is not finite, then $G^\circ_{\mathrm{geom},\mathcal{F}(q,f)}$ is a simple algebraic group of dimension $\geq \mathrm{rank}(\mathcal{F}(q,f))$.*

PROOF. By Lemma 5.1, any irreducible constituent of dimension $< D := \mathrm{rank}(\mathcal{F}(q,f))$ is a single $\mathcal{L}_{\psi(ax)}$, while Theorem 5.5 shows that $\mathrm{End}^0(\mathcal{F}(q,f))$ contains no $\mathcal{L}_{\psi(ax)}$. Because $\mathcal{F}(q,f)$ satisfies condition (**S**+) by Theorem 5.12, if $G_{\mathrm{geom},\mathcal{F}(q,f))}$ is not finite, then its identity component $G^\circ_{\mathrm{geom},\mathcal{F}(q,f))}$ is a simple algebraic group. In that case, $\mathrm{Lie}(G^{\circ}_{\mathrm{geom},\mathcal{F}(q,f))})$ is an irreducible constituent of $\mathrm{End}^0(\mathcal{F}(q,f))$, so has dimension $\geq D$. □

THEOREM 5.14. *Consider the sheaf $\mathcal{F}(q,f)$ in* (4.1.1) *subject to the condition* (4.1.3). *Then, for each $r|t(q)$, the descent $\mathcal{G}(q,f,r)$ satisfies the condition* (**S**+) *of* [**30**, Definition 1.2].

PROOF. Because $G_{\mathrm{geom},\mathcal{F}(q,f)}$ is a subgroup of $G_{\mathrm{geom},\mathcal{G}(q,f,r)}$ of finite index, the fact that $\mathcal{F}(q,f)$ satisfies condition (**S**+), see Theorem 5.12, implies that $\mathcal{G}(q,f,r)$ does as well. [Condition (**S**+) holds for (G,V) if it holds for (H,V) with H a subgroup of G of finite index.] □

We also record the following:

THEOREM 5.15. *Under the condition* (4.1.3), *if the sheaf $\mathcal{F}(q,f)$ in* (4.1.1) *is Lie-irreducible, then it is not Lie-self-dual.*

PROOF. This is just a restatement of Corollary 5.8, under the more restrictive hypotheses of (4.1.3). Given Theorem 5.7, it suffices to show that if $\mathcal{F}(q,f)$ is Lie-irreducible and Lie-self-dual, then it is self-dual.

However, there is a simpler proof of this last fact, using the descent $\mathcal{G}(q,f) := \mathcal{G}(q,f,t(q))$ in Theorem 5.14. Exactly as in the proof of Theorem 5.6, we find an isomorphism

$$\mathcal{G}(q,f)^{\vee} \cong \mathcal{G}(q,f) \otimes \mathcal{L}$$

for some lisse $\mathcal{L}$ on $\mathbb{G}_m$ of rank one. Because $\mathcal{G}(q,f)^{\vee}$and $\mathcal{G}(q,f)$ are both tame at 0 and with all ∞-slopes < 1, $\mathcal{L}$ is tame on $\mathbb{G}_m$, hence a Kummer sheaf $\mathcal{L}_{\chi}$. Because $\mathcal{G}(q,f)^{\vee}$and $\mathcal{G}(q,f)$ both have $I(0)$ representations which are sums of characters of order dividing $t(q)$, χ is a ratio of characters of order dividing $t(q)$, so χ has order dividing $t(q)$. Pulling back this isomorphism by $t(q)^{\mathrm{th}}$ power, we get an isomorphism $\mathcal{F}(q,f)^{\vee} \cong \mathcal{F}(q,f)$. □

6. A local system for the Suzuki group $^2B_2(8)$

Now we can prove the first main result of the paper, which establishes [**26**, Conjecture 2.2] in the case $q = 8$:

THEOREM 6.1. *Let $q = 8$. Both the local systems $\mathcal{F}_q$ and $\mathcal{G}_q$ have geometric monodromy group $G_{\mathrm{geom}} \cong {}^2B_2(8)$ in one of its irreducible representations of degree* 14. *Over $\mathbb{F}_2$, the local systems $\mathcal{F}_q$ and $\mathcal{G}_q$ have arithmetic monodromy group $G_{\mathrm{arith}} \cong \mathrm{Aut}({}^2B_2(8))$.*

PROOF. (a) Let G, respectively H, denote the geometric monodromy group of $\mathcal{F}_q$, respectively of $\mathcal{G}_q$. Similarly, let G_{arith}, respectively H_{arith}, denote the arithmetic monodromy group of $\mathcal{F}_q$, respectively of $\mathcal{G}_q$, over $\mathbb{F}_2$. We will use the fact that $f := \mathrm{Frob}_{1,\mathbb{F}_2}$ has order 15 and trivial determinant. Indeed, a MAGMA calculation shows that

$$\mathrm{Trace}(\mathrm{Frob}_{1,\mathbb{F}_{2^{15}}}|\mathcal{G}_q) = \mathrm{Trace}(\mathrm{Frob}_{1,\mathbb{F}_{2^{15}}}|\mathcal{F}_q) = 14.$$

By Theorem 2.2, $\mathrm{Frob}_{1,\mathbb{F}_{2^{15}}}|\mathcal{G}_q$ is semisimple, hence (being of weight zero in a 14-dimensional representation) is the identity. Therefore, $\det(\mathrm{Frob}_{1,\mathbb{F}_{2^{15}}}|\mathcal{G}_q))$ is a root of unity of order dividing both 4 (by Theorem 2.1) and 15, so this determinant is trivial. But $H_{\mathrm{arith}} = \langle f, H\rangle$, and H has trivial determinant (by Theorem 2.1), and hence H_{arith} also has trivial determinant. As $G \lhd H$ and $G_{\text{arith}} \leq H_{\text{arith}}$ are subgroups, both G and G_{arith} have trivial determinants.

Recall that G is normal in H of index dividing $t(q) = 5$. By Theorem 5.12, H satisfies (**S**+). Since the rank of the sheaves is 14, not a prime power, by [**30**, Lemma 1.1] this implies that either the identity component H° of H is a simple algebraic group that acts irreducibly on the underlying representation V of H, or H is a finite almost quasisimple group with $L := H^{(\infty)}$ also acting irreducibly on V.

(b) Consider the former case. Note that if H° is classical of rank r, then either it is of type A and $r \leq 13$, or $r \leq 7$ by [**32**, Proposition 5.4.11]. Using the tables of [**33**], we see that H° is of type SL_2, SL_{14}, Sp_4, Sp_6, Sp_{14}, SO_{14}, or G_2. Since $[H : G] < \infty$, $G^\circ = H^\circ$.

Suppose first that $H^\circ = \mathrm{SL}_{14}(\mathbb{C})$. In this case, V is just the natural module for H°; hence $M_{2,2}(V)$ takes the smallest possible value 2 for all H°, H, G°, and G. But this contradicts Corollary 1.2.

Next suppose that $G^\circ = H^\circ$ is of type SL_2, Sp_4, Sp_6, Sp_{14}, SO_{14}, or G_2. In all these cases, the G°-module V is self-dual, and this contradicts Corollary 5.8.

(c) We have shown that H is finite, and we are in the almost quasisimple case. In this case, H_{arith} is also finite, and $H_{\text{arith}}^{(\infty)} = H^{(\infty)} = G_{\text{arith}}^{(\infty)} = G^{(\infty)} = L$. Let φ denote the character of H_{arith} in the underlying representation V. Then the formula for the trace function of $\mathcal{G}_q$ and the existence of an element with trace $2\zeta_4$ show that

$$\mathbb{Q}(\varphi) = \mathbb{Q}(\zeta_4). \tag{6.1.1}$$

Since any element z of $\mathbf{C}_{H_{\text{arith}}}(L) = \mathbf{Z}(H_{\text{arith}})$ acts as a root of unity γ on V, this implies that $\gamma^4 = 1$. Since H_{arith} has trivial determinant, $\gamma^{14} = 1$ and thus $\gamma = \pm 1$. It follows that

$$|\mathbf{C}_{H_{\text{arith}}}(L)| = |\mathbf{Z}(H_{\text{arith}})| \leq 2. \tag{6.1.2}$$

Now, using (6.1.1) and the fact that L acts irreducibly on $V = \mathbb{C}^{14}$, the classification results of [**16**] show that $L = \mathrm{PSL}_2(13)$, $\mathrm{SL}_2(13)$, A_7, A_8, $\mathrm{SU}_3(3)$, $G_2(3)$, $2 \cdot J_2$, A_{15}, or ${}^2B_2(8)$.

In all but the last case, $\varphi|_L$ is real-valued; furthermore, $|\mathrm{Out}(L)| \leq 2$. As shown in the proof of Theorem 4.5, G has no subgroups of index 2, so G can induce only inner automorphisms of L. But $G_{\text{arith}} = \langle G, f \rangle$, implying $G_{\text{arith}}/G \hookrightarrow C_{15}$; so the same conclusion holds for G_{arith}, and hence $G_{\text{arith}} = \mathbf{C}_{G_{\text{arith}}}(L)L = \mathbf{Z}(G_{\text{arith}})L$, with $|\mathbf{Z}(G_{\text{arith}}| \leq 2$ by (6.1.2). On the other hand, as we saw in the computation for the proof of Corollary 1.2, some $g \in G_{\text{arith}}$ has trace $2\zeta_4$ on V. Write $g = zh$ with $z \in \mathbf{Z}(G_{\text{arith}})$ and $h \in L$. It follows that $2\zeta_4 = \varphi(g) = \pm\varphi(h)$, a contradiction since $\varphi(h) \in \mathbb{R}$.

We have therefore shown that $L = {}^2B_2(8)$. As $|\mathrm{Out}(L)| = 3$, (6.1.2) implies that $|H/L|$ divides 6. On the other hand, H is the normal closure of the image of order 5 of $I(0)$, so we conclude that $H = L$. Since $L \lhd G \lhd H$, we also get $G = L$.

Recalling again that $G_{\text{arith}}/G \hookrightarrow C_{15}$, we now see from (6.1.2) that $\mathbf{Z}(G_{\text{arith}}) = 1$ and thus $G_{\text{arith}} \hookrightarrow \mathrm{Aut}(L) \cong L \rtimes C_3$. A MAGMA computation shows the existence of a Frobenius element with trace ζ_4, and this implies that $G_{\text{arith}} > L$, whence $G_{\text{arith}} = \mathrm{Aut}(L)$. Since H_{arith} is generated over H by $f = \mathrm{Frob}_{1,\mathbb{F}_2}$, an element of order 15, $\mathbf{Z}(H_{\text{arith}}) = 1$ by (6.1.2). Thus $H_{\text{arith}} \hookrightarrow \mathrm{Aut}(L)$, and so $H_{\text{arith}} = \mathrm{Aut}(L)$ as $H_{\text{arith}} \geq G_{\text{arith}}$. $\square$

7. Low-dimensional representations of classical groups

In this section, we will extend the classification results obtained in [**32**, Proposition 5.4.11] and [**33**, Theorem 5.1]. Even though the intended applications in the paper only need the complex case of these results, we establish them in the modular case, which is interesting in its own right.

Let $\mathbb{F}$ be an algebraically closed field of characteristic $p \geq 0$ and let G be a simple, simply connected, classical algebraic group of rank r over $\mathbb{F}$. Fixing a maximal torus in G, we consider the set of simple roots $\{\alpha_1, \dots, \alpha_r\}$ and the corresponding set of fundamental weights $\{\varpi_1, \dots, \varpi_r\}$ (in the ordering of [**40**]). Then the set

$$\Lambda^+ = \left\{ \sum_{i=1}^{r} a_i \varpi_i \mid a_i \in \mathbb{Z}, a_i \geq 0 \right\}$$

of dominant weights admits the partial ordering $\succ$ where $\lambda \succ \mu$ precisely when $\lambda - \mu = \sum_{i=1}^{r} k_i \alpha_i$ for some non-negative integers k_i. As usual, W denotes the Weyl group. If $\lambda \in \Lambda^+$, let $L(\lambda)$ denote the irreducible $\mathbb{F}G$-module with highest weight λ.

We will rely on the following two results.

THEOREM 7.1 ([**41**]). *Let G be a simple, simply connected algebraic group in characteristic $p > 0$. If the root system of G has different root lengths, then we assume that $p \neq 2$, and if G is of type G_2, then we also assume that $p \neq 3$. Let λ be a restricted dominant weight. Then the set of weights $\Pi(\lambda)$ of the irreducible G-module $L(\lambda)$ is the union of the W-orbits of dominant weights μ with $\lambda \succ \mu$.* □

LEMMA 7.2 ([**17**, Lemma 10.3B]). *Let $\lambda = \sum_{i=1}^{r} a_i \varpi_i$ be a dominant weight. Then the stabilizer of λ in the Weyl group is the Young subgroup generated by the reflections ρ_i along the simple roots α_i for which $a_i = 0$.* □

Our first result treats groups of type A and includes a strengthening for SL_{22}:

THEOREM 7.3. *Let $G = \mathrm{SL}_n(\mathbb{F})$ with $n = r + 1 \geq 8$, and let $L(\lambda)$ be an irreducible $\mathbb{F}G$-representation, which is restricted if $p = \mathrm{Char}(\mathbb{F}) > 0$. Suppose that $\dim L(\lambda) \leq M$, where $M := \binom{n}{4}$ if $n \neq 22$ and $M := 8176$ if $n = 22$. Then $\lambda = 0$, $a\varpi_1$ or $a\varpi_r$ with $1 \leq a \leq 3$, $\varpi_1 + \varpi_r$, ϖ_2 or ϖ_{r-1}, ϖ_3 or ϖ_{r-2}, ϖ_4 or ϖ_{r-3}, $\varpi_1 + \varpi_2$ or $\varpi_{r-1} + \varpi_r$, $2\varpi_1 + \varpi_r$ or $\varpi_1 + 2\varpi_r$, and $\varpi_2 + \varpi_r$ or $\varpi_1 + \varpi_{r-1}$.*

PROOF. Write the highest weight λ as $\sum_{i=1}^{r} a_i \varpi_i$ with $a_i \in \mathbb{Z}_{\geq 0}$.

(a) First suppose that there is some weight $\mu = \sum_{i=1}^{r} b_i \varpi_i \in \Lambda^+$ with $\lambda \succ \mu$ and $b_j \neq 0$ for some $s + 1 \leq j \leq r - s$, where

$$s := 3 \text{ if } n \neq 22 \text{ and } s := 4 \text{ if } n = 22.$$

By Theorem 7.1, $\Pi(\lambda)$ contains the W-orbit $\mathcal{O}(\mu)$ of μ. By Lemma 7.2,

$$\mathrm{Stab}_W(\mu) \leq \langle \rho_1, \dots, \rho_{j-1}, \rho_{j+1}, \dots, \rho_r \rangle = \mathsf{S}_j \times \mathsf{S}_{n-j}$$

(where $\rho_i = (i, i+1) \in W = \mathsf{S}_n$). Hence

$$\dim L(\lambda) \geq |\mathcal{O}(\mu)| \geq [\mathsf{S}_n : (\mathsf{S}_j \times \mathsf{S}_{n-j})] = \binom{n}{j} \geq \binom{n}{s+1} \geq M,$$

a contradiction if $n = 22$, or if $5 \leq j \leq r - 4$. Suppose $j \in \{4, r - 3\}$ and $n \neq 22$. Then $\dim L(\lambda) = |\mathcal{O}(\mu)| = \binom{n}{4}$, showing μ is the unique dominant weight of $L(\lambda)$, whence $\mu = \lambda$. This also forces $\mathrm{Stab}_W(\lambda) = \mathsf{S}_j \times \mathsf{S}_{n-j}$, and so $\lambda = a_j\varpi_j$. Now if $a \geq 2$, then $\lambda \succ \lambda - \alpha_j = \varpi_{j-1} + (a_j - 2)\varpi_j + \varpi_{j+1}$, and the latter is another dominant weight of $L(\lambda)$, a contradiction. So $\lambda = \varpi_4$ or ϖ_{r-3} in such a case.

We may therefore assume that

(7.3.1) $$a_j = 0 \text{ for every } s+1 \leq j \leq r-s.$$

Next suppose that $\sum_{i=1}^{s} ia_i \geq s+1$. By [**14**, Lemma 2.6] (applied with $m = s$), there is some weight $\mu = \sum_{i=1}^{r} b_i\varpi_i \in \Lambda^+$ with $\lambda \succ \mu$ and $b_{s+1} > a_{s+1}$. This situation is already considered by the preceding analysis. Using the symmetry under the graph automorphism τ of G,

(7.3.2) $$\sum_{i=1}^{s} ia_i \leq s,\ \sum_{i=1}^{s} ia_{r+1-i} \leq s.$$

(b) Suppose $a_4 > 0$ or $a_{r-3} > 0$. By symmetry, we may assume $a_4 > 0$. By (7.3.2) $n = 22$, $a_4 = 1$, and $a_1 = a_2 = a_3 = 0$. Now if $\lambda \neq \varpi_4$, then $a_j \geq 1$ for some $r-3 \leq j \leq r$ by (7.3.1). By Lemma 7.2,

$$\mathrm{Stab}_W(\lambda) \leq \langle \rho_1, \rho_2, \rho_3, \rho_5, \rho_6, \ldots, \rho_{j-1}, \rho_{j+1}, \ldots, \rho_r \rangle \cong \mathsf{S}_4 \times Y_1,$$

with Y_1 a proper Young subgroup of S_{n-4} and hence $|Y_1| \leq (n-5)!$. It follows that

$$\dim L(\lambda) \geq [W : \mathrm{Stab}_W(\lambda)] \geq (n-4)\binom{n}{4} > M,$$

a contradiction. Hence from now on we may assume that $a_4 = a_{r-3} = 0$.

(c) Suppose that $a_3 > 0$ or $a_{r-2} > 0$. By symmetry, we may assume $a_3 > 0$, and so $a_3 = 1$ by (7.3.2). Suppose $a_j \geq 1$ for some $r-3 \leq j \leq r$. By Lemma 7.2,

(7.3.3) $$\mathrm{Stab}_W(\lambda) \leq \langle \rho_1, \rho_2, \rho_4, \rho_5, \ldots, \rho_{j-1}, \rho_{j+1}, \ldots, \rho_r \rangle \cong \mathsf{S}_3 \times Y_2,$$

with Y_2 a proper Young subgroup of S_{n-3} and hence $|Y_2| \leq (n-4)!$. It follows that

$$\dim L(\lambda) \geq [W : \mathrm{Stab}_W(\lambda)] \geq (n-3)\binom{n}{3} > M$$

(as $n \geq 8$), a contradiction. Now, if $n \neq 22$, then using (7.3.1) and (7.3.2) we see that $\lambda = \varpi_3$. If $n = 22$ but $\lambda \neq \varpi_3$, then $\lambda = \varpi_1 + \varpi_3$. In this case, $\lambda - (\alpha_1 + \alpha_2 + \alpha_3) = \varpi_4$ is also a weight of $L(\lambda)$ by Theorem 7.1, and since

$$\mathrm{Stab}_W(\varpi_1 + \varpi_3) = \langle \rho_2, \rho_4, \rho_5, \ldots, \rho_r \rangle \cong \mathsf{S}_2 \times \mathsf{S}_{n-3},\ \mathrm{Stab}_W(\varpi_4) = \mathsf{S}_4 \times \mathsf{S}_{n-4}$$

by Lemma 7.2,

(7.3.4) $$|\mathcal{O}(\varpi_1 + \varpi_3)| + |\mathcal{O}(\varpi_4)| = \binom{22}{3}/2 + \binom{22}{4} = 11935 > M,$$

which contradicts $\dim L(\lambda) < M$.

(d) We may now assume that $a_i = 0$ for $3 \leq i \leq r-2$. Suppose that $a_2 > 0$ or $a_{r-1} > 0$. By symmetry, we may assume $a_2 > 0$. If $a_2 \geq 2$, then by (7.3.2) $n = 22$, $(a_2, a_1) = (2, 0)$. Note that $\lambda \succ \lambda - \alpha_2 = \varpi_1 + \varpi_3 + a_{r-1}\varpi_{r-1} + a_r\varpi_r$. It follows from Theorem 7.1 that $\Pi(\lambda)$ contains $\varpi_1 + \varpi_3 + a_{r-1}\varpi_{r-1} + a_r\varpi_r$ and $\varpi_4 + a_{r-1}\varpi_{r-1} + a_r\varpi_r$. The lengths of W-orbits of these two weights are at least 11395 by (7.3.4), and so $\dim L(\lambda) > M$.

If $a_2 = 1$ but $a_1 \geq 2$, then by (7.3.2) $n = 22$, $(a_2, a_1) = (1, 2)$. Note that

$$\lambda \succ \lambda - \alpha_1 = 2\varpi_2 + a_{r-1}\varpi_{r-1} + a_r\varpi_r.$$

The preceding arguments show that $\Pi(\lambda)$ contains the weights $\varpi_1+\varpi_3+a_{r-1}\varpi_{r-1}+a_r\varpi_r$ and $\varpi_4+a_{r-1}\varpi_{r-1}+a_r\varpi_r$, and again $\dim L(\lambda) > M$.

Hence $a_2 = 1$ and $a_1 \leq 1$. If $a_r = a_{r-1} = 0$, then $\lambda = \varpi_2$ or $\varpi_1+\varpi_2$. So assume that $a_r + a_{r-1} > 0$.

Suppose that $a_{r-1} > 0$. Then by Lemma 7.2

(7.3.5) $$\mathrm{Stab}_W(\lambda) \leq \langle \rho_1, \rho_3, \rho_4, \ldots, \rho_{r-2}, \rho_r\rangle = \mathsf{S}_2 \times \mathsf{S}_{n-4} \times \mathsf{S}_2.$$

It follows that

$$\dim L(\lambda) \geq [W : \mathrm{Stab}_W(\lambda)] \geq \frac{n!}{(n-4)!\cdot(2!)^2} > M,$$

a contradiction. Hence $a_{r-1} = 0$.

Suppose next that $a_r \geq 2$. Then

$$\lambda \succ \nu_1 := \lambda - \alpha_r = \lambda - (2\varpi_r - \varpi_{r-1}) = a_1\varpi_1 + \varpi_2 + (a_{r-1}+1)\varpi_{r-1} + (a_r-2)\varpi_r.$$

Thus $\nu_1 \in \Pi(\lambda)\cap\Lambda^+$, and (7.3.5) applied to ν_1 shows that

$$\dim L(\lambda) \geq |\mathcal{O}(\nu_1)| \geq [\mathsf{S}_n : (\mathsf{S}_2 \times \mathsf{S}_{n-4} \times \mathsf{S}_2)] > M.$$

We may therefore assume that $a_r = 1$. If $a_1 = 0$ then $\lambda = \varpi_2 + \varpi_r$. Otherwise $\lambda = \varpi_1+\varpi_2+\varpi_r$, in which case $\lambda \succ \nu_2 := \lambda - (\alpha_1+\alpha_2) = \lambda - (\varpi_1+\varpi_2-\varpi_3) = \varpi_3 + \varpi_r$. But in such a case, (7.3.3) applied to ν_2 shows that

$$\dim L(\lambda) \geq |\mathcal{O}(\nu_2)| \geq (n-3)\binom{n}{3} > M.$$

(e) We may now assume that $a_i = 0$ for $2 \leq i \leq r-1$, i.e., $\lambda = a\varpi_1 + b\varpi_r$ with $s \geq a \geq b \geq 0$ (by symmetry). Suppose $a \geq 4$, whence $n = 22$ and $a = 4$ by (7.3.2). Then $\lambda \succ \lambda - 2\alpha_1 = 2\varpi_2 + b\varpi_r$, and so $\dim L(\lambda) > M$ by the first paragraph of (d).

Suppose $a = 3$ but $b > 0$. Then $\lambda \succ \nu_3 := \lambda - (2\alpha_1+\alpha_2) = \lambda - (3\varpi_1 - \varpi_3) = \varpi_3 + b\varpi_r$, and (7.3.3) applied to ν_3 shows that $\dim L(\lambda) \geq |\mathcal{O}(\nu_3)| > M$. Hence $\lambda = \varpi_3$.

Suppose $a = b = 2$, i.e., $\lambda = 2\varpi_1 + 2\varpi_r$. Then $\lambda \succ \nu_4 := \lambda - (\alpha_1+\alpha_r) = \varpi_2 + \varpi_{r-1}$. In such a case,

$$\dim L(\lambda) \geq |\mathcal{O}(\nu_4)| \geq \frac{n!}{(n-4)!\cdot(2!)^2} > M,$$

by using (7.3.5) for ν_4. So $a+b \leq 3$, and thus $\lambda = 2\varpi_1+\varpi_r$, $2\varpi_1$, $\varpi_1+\varpi_r$, ϖ_1, or 0. □

To handle the other classical groups, we first consider a special case. Again, we use the weight labeling as in [**40**].

PROPOSITION 7.4. *Let G be a simply connected simple algebraic group over $\mathbb{F}$ of type B_r, C_r, or D_r, with $r \geq 7$. Then*

$$\dim L(\varpi_1+\varpi_2) \geq \begin{cases} 4r(r^2-1)/3, & \text{if } p = 3,\\ 4r(r-1)(2r-1)/3, & \text{if } p \neq 3.\end{cases}$$

PROOF. Note that $|\mathcal{O}(\varpi_3)| = 8\binom{r}{3}$ and $|\mathcal{O}(\varpi_1+\varpi_2)| = 4r(r-1)$. Since $\lambda := \varpi_1+\varpi_2$ is the highest weight of $L(\lambda)$, it suffices to show that $\mu := \varpi_3$ is a weight of $L(\lambda)$, with multiplicity $m_{L(\lambda)}(\mu) \geq 1$ if $p = 3$ and $m_{L(\lambda)}(\mu) \geq 2$ if $p \neq 3$. If $p = 3$, then $\lambda \succ \lambda - (\alpha_1+\alpha_2) = \mu$, whence $\mu \in \Pi(\lambda)$ by Theorem 7.1, and we are done.

In what follows we may assume $p \neq 3$. We realize the roots and the weights of G using an orthonormal basis $(e_i \mid 1 \leq i \leq r)$ of $\mathbb{R}^r$ (with scalar product $(\cdot,\cdot)$); in particular, $\varpi_1 = e_1$, $\varpi_2 = e_1 + e_2$, $\varpi_3 = e_1 + e_2 + e_3$. Consider the simple (Weyl) module $V(\lambda)$ of the corresponding algebraic group over $\mathbb{C}$. Then $\nu \in \Lambda^+$ is a weight of $V(\lambda)$ precisely when $\lambda \succ \nu$. Writing ϖ_i in terms of simple roots (see [**40**, Table 2]), it is straightforward to check that this is equivalent to

$$\nu \in \begin{cases} \{\lambda, \mu, \varpi_1\}, & \text{if } G = C_r \text{ or } D_r, \\ \{\lambda, \mu, \varpi_2, 2\varpi_1, 0\}, & \text{if } G = B_r. \end{cases} \tag{7.4.1}$$

Next we use Freudenthal's formula [**17**, p. 122] to find the multiplicity $m_{V(\lambda)}(\mu)$ of μ as a weight of $V(\lambda)$. Then we must find all multiples $l\alpha$ of positive roots α, with $l \in \mathbb{Z}_{\geq 1}$, such that $\mu + l\alpha$ is a weight of $V(\lambda)$, i.e., W-conjugate to one of the weights listed in (7.4.1). Again using [**40**, Table 2], we can check that this happens precisely when $l = 1$ and $\alpha = e_1 - e_2, e_1 - e_3, e_2 - e_3$. In all these cases, $\mu + l\alpha$ is W-equivalent to λ, and we readily obtain $m_{V(\lambda)}(\mu) = 2$.

Suppose now that

$$m_{L(\lambda)}(\mu) \leq 1. \tag{7.4.2}$$

This can happen only when μ is a weight of some composition factor $L(\nu_0)$ of a reduction modulo p of $V(\lambda)$, where ν_0 is listed in (7.4.1). Note that if ν is such a weight and $\nu \neq \mu$, then $\nu \not\succ \mu$. It follows that $\nu_0 = \mu$, i.e., $L(\mu)$ is a composition factor of a reduction modulo p of $V(\lambda)$. By the linkage principle, see [**21**], this implies that $w \circ \lambda = \mu$ for some $w \in W_p$. Here, the affine Weyl group W_p is generated by the map $\rho_{\alpha,l} \circ \lambda = \rho_\alpha \circ \lambda + lp\alpha$, where α is a simple root, $l \in \mathbb{Z}$ and ρ_α denotes the reflection corresponding to α; furthermore, $\rho_\alpha \circ \lambda = \rho_\alpha(\lambda + \delta) - \delta$, where $\delta := \sum_{i=1}^r \varpi_i$.

Write $|v|^2$ instead of (v, v) for $v \in \mathbb{R}^r$. Then, for any root α and any $l \in \mathbb{Z}$

$$|\rho_{\alpha,l} \circ \lambda + \delta|^2 = |\lambda + \delta|^2 + l^2p^2|\alpha|^2 + 2lp(\rho_\alpha(\lambda + \delta), \alpha).$$

Observe that $|\alpha|^2 \in \mathbb{Z}$ and

$$(\rho_\alpha(\lambda + \delta), \alpha) = \left((\lambda + \delta) - \frac{2(\lambda + \delta, \alpha)}{(\alpha, \alpha)}, \alpha\right) = -(\lambda + \rho, \alpha) \in \mathbb{Z}.$$

It follows that $|\rho_{\alpha,l} \circ \lambda + \delta|^2 \equiv |\lambda + \delta|^2 \pmod{\gcd(2, p)}$. Thus we have shown:

(7.4.3) If two weights λ, μ are linked, then $|\lambda + \delta|^2 \equiv |\mu + \delta|^2 \pmod{\gcd(2, p)p}$.

(Note that a slightly weaker result than (7.4.3), namely only modulo p, was obtained in [**46**, Lemma 2.1].) In our case, $|\lambda + \delta|^2 - |\mu + \delta|^2 = 6$. Applying (7.4.3), we conclude that (7.4.2) can happen only when $p = 3$. □

THEOREM 7.5. *Let G be a simply connected simple algebraic group over $\mathbb{F}$ of type B_r, C_r, or D_r, with $r \geq 12$. Let $L(\lambda)$ be an irreducible $\mathbb{F}G$-representation, which is restricted if $p := \mathrm{Char}(\mathbb{F}) > 0$. Suppose that $\dim L(\lambda) \leq M$, where $M := 2(r+1)^3$ if $r \geq 14$, and $M := r^3$ if $r = 12, 13$. Then $\lambda = a\varpi_1$ with $0 \leq a \leq 3$, ϖ_2, ϖ_3, or $\varpi_1 + \varpi_2$. Moreover, if $\lambda = \varpi_1 + \varpi_2$ and $r \geq 16$, then $p = 3$.*

PROOF. Write the highest weight λ as $\sum_{i=1}^r a_i\varpi_i$ with $a_i \in \mathbb{Z}_{\geq 0}$ (note that $L(\varpi_1)$ is the natural module for G).

(a) First suppose that $a_i > 0$ for some $r - 2 \leq i \leq r$. In this case, by Lemma 7.2, the length of the W-orbit $\mathcal{O}(\lambda)$ of λ is at least $[W : \mathrm{Stab}_W(\lambda)] \geq 2^{r-1} > M$.

Next suppose that $a_i > 0$ for some $4 \le i \le r-3$. In this case, if G is of type X_r so that $W = W(X_r)$, then $\mathrm{Stab}_W(\lambda)$ is contained in $W(A_{i-1}) \times W(X_{r-i})$ by Lemma 7.2; hence

$$\dim L(\lambda) \ge |\mathcal{O}(\lambda)| \ge 2^i \binom{r}{i} \ge \min\left(2^4\binom{r}{4}, 2^{r-3}\binom{r}{3}\right) > 2(r+1)^3,$$

since $r \ge 9$.

Applying this argument to $\mu = \sum_{i=1}^r b_i\varpi_i \in \Pi(\lambda) \cap \Lambda^+$, we deduce that

$$b_i = 0 \text{ for } 4 \le i \le r. \tag{7.5.1}$$

(b) Suppose $a_3 > 0$. If $a_1 > 0$ or $a_2 > 0$, then $\mathrm{Stab}_W(\lambda) \le W(A_1) \times W(X_{r-3})$, whence

$$|\mathcal{O}(\lambda)| \ge 4r(r-1)(r-2) > 2(r+1)^3 \tag{7.5.2}$$

(since $r \ge 9$), contradicting the bound on $\dim L(\lambda)$. So $\lambda = a_3\varpi_3$. Now, if $a_3 \ge 2$ (and so $p \ne 2$ as λ is restricted), then by Theorem 7.1

$$\lambda \succ \lambda - \alpha_3 = \varpi_2 + \varpi_4 \in \Pi(\lambda) \cap \Lambda^+,$$

violating (7.5.1). Hence $\lambda = \varpi_3$ in this case.

We have shown that $a_3 = 0$. Suppose $a_2 > 0$. Now, if $a_2 \ge 2$ (so again $p \ne 2$), then

$$\lambda \succ \lambda - \alpha_2 = (a_1+1)\varpi_1 + (a_2-2)\varpi_2 + \varpi_3 \in \Pi(\lambda),$$

leading to a contradiction by applying (7.5.2) to $\lambda - \alpha_2$. If $a_2 = 1$ but $a_1 \ge 2$, then $p \ne 2$ and

$$\lambda \succ \lambda - (\alpha_1 + \alpha_2) = (a_1 - 1)\varpi_1 + \varpi_3 \in \Pi(\lambda),$$

again yielding a contradiction by applying (7.5.2) to $\lambda - \alpha_1 - \alpha_2$. Hence $\lambda \in \{\varpi_2, \varpi_1 + \varpi_2\}$ in this case.

We are left with the case $\lambda = a\varpi_1$. If $a \ge 4$, then again $p \ne 2$ and

$$\lambda \succ \lambda - (2\alpha_1 + \alpha_2) = (a_1 - 3)\varpi_1 + \varpi_3 \in \Pi(\lambda),$$

leading to a contradiction by applying (7.5.2) to $\lambda - 2\alpha_1 - \alpha_2$. So $0 \le a \le 3$ as stated.

(c) We make some more comments about the cases $\lambda \in \{3\varpi_1, \varpi_3, \varpi_1 + \varpi_2\}$. Note that

$$3\varpi_1 - \alpha_1 = \varpi_1 + \varpi_2, \ (\varpi_1 + \varpi_2) - (\alpha_1 + \alpha_2) = \varpi_3.$$

So, assuming $p \ne 2$ when $\lambda = \varpi_1 + \varpi_2$, we may assume $\varpi_3 \in \Pi(\lambda)$ in all these three cases. It now follows from Lemma 7.2 that in these cases

$$\dim L(\lambda) \ge |\mathcal{O}(\varpi_3)| = 4r(r-1)(r-2)/3 > r^3.$$

Proposition 7.4 shows that if $p = 2$, then $\dim L(\varpi_1 + \varpi_2) > 8\binom{r}{3} > r^3$ (as $r \ge 12$), and if $r \ge 16$ and $p \ne 3$, then $\dim L(\varpi_1 + \varpi_2) > 4r(r-1)(2r-1)/3 > 2(r+1)^3$. Thus,

$$\text{If } r \ge 12 \text{ and } \dim L(\lambda) \le r^3, \text{ then } \lambda \in \{0, \varpi_1, 2\varpi_1, \varpi_2\}. \tag{7.5.3}$$

This statement (7.5.3) was recorded in [**33**, Theorem 5.1], but we note that the treatment of the weights $a_1\varpi_1 + a_2\varpi_2$ therein is incorrect. Also note that $\dim L(\varpi_1 + \varpi_2)$ may be smaller than $2(r+1)^3$ when $p = 3$, see [**33**] for examples. □

8. A dichotomy for monodromy groups

We will need some preliminary facts:

LEMMA 8.1. *Let $n \in \mathbb{Z}_{\geq 1}$ and let $D := 2^n(2^{2n+1} - 1)$. Then none of the following equations*

(i) $D = x^2 - 1$,
(ii) $D = x(x-1)/2$,
(iii) $D = x(x-1)/2 - 1$ *and* $n \geq 2$,

has a solution in the positive integers.

PROOF. (i) Suppose $x^2 - 1 = D$ for some $x \in \mathbb{Z}_{\geq 1}$. Checking the cases $1 \leq n \leq 7$ directly, we may assume $n \geq 8$. Now $x > 1$ is odd, and $\gcd(x-1, x+1) = 2$, but $2^n | (x^2 - 1)$. It follows that there is some $\epsilon = \pm 1$ such that $2^{n-1} | (x - \epsilon)$. Write $x - \epsilon = 2^{n-1}y$ for some $y \in \mathbb{Z}_{\geq 1}$. Then

$$2^{3n+1} - 2^n = D = x^2 - 1 = (2^{n-1}y + \epsilon)^2 - 1 = 2^{2n-2}y^2 + 2^n \epsilon y,$$

and so $\epsilon y + 1 = 2^{n-2}(2^{n+3} - y^2)$. This implies $y > 1$, and $y + \epsilon$ is divisible by 2^{n-2}. Hence $y + \epsilon = 2^{n-2}z$ for some $z \in \mathbb{Z}_{\geq 1}$. In this case, $y \geq 2^{n-2} - 1$, $x \geq 2^{2n-3} - 2^{n-1} - 1$, and so $x^2 - 1 > 2^{4n-7} \geq 2^{3n+1} > D$ (as $n \geq 8$), a contradiction.

(ii) Suppose $x(x-1)/2 = D$ for some $x \in \mathbb{Z}_{\geq 1}$. Then $\gcd(x-1, x) = 1$, but $2^{n+1} | x(x-1)$. It follows that there is some $\epsilon \in \{0, 1\}$ such that $2^{n+1} | (x - \epsilon)$. Write $x - \epsilon = 2^{n+1}y$ for some $y \in \mathbb{Z}_{\geq 1}$. If $\epsilon = 0$, then

$$2^{3n+2} - 2^{n+1} = 2D = x(x-1) = 2^{2n+2}y^2 - 2^{n+1}y,$$

and so $y - 1 = 2^{n+1}(y^2 - 2^n)$. This implies $y > 1$, and $y - 1$ is divisible by 2^{n+1}. Hence $y - 1 \geq 2^{n+1}$, $x > 2^{2n+2}$, and so $x(x-1) > 2^{4n+4} > 2^{3n+1} > D$, a contradiction. If $\epsilon = 1$, then

$$2^{3n+2} - 2^{n+1} = 2D = x(x-1) = 2^{2n+2}y^2 + 2^{n+1}y,$$

and so $y + 1 = 2^{n+1}(2^n - y^2)$. This implies $1 \leq y < 2^{n/2}$, and $y + 1$ is divisible by 2^{n+1}, which is impossible.

(iii) Suppose $n \geq 2$ and $D = x(x-1)/2 - 1 = (x+1)(x-2)/2$ for some $x \in \mathbb{Z}_{\geq 1}$. Then $\gcd(x+1, x-2) | 3$, but $2^{n+1} | (x+1)(x-2)$. It follows that there is some $\epsilon \in \{-1, 2\}$ such that $2^{n+1} | (x - \epsilon)$. Write $x - \epsilon = 2^{n+1}y$ for some $y \in \mathbb{Z}_{\geq 1}$. If $\epsilon = -1$, then

$$2^{3n+2} - 2^{n+1} = 2D = (x+1)(x-2) = 2^{2n+2}y^2 - 3 \cdot 2^{n+1}y,$$

and so $3y - 1 = 2^{n+1}(y^2 - 2^n)$. This implies that $3y - 1 \geq 2$ is divisible by 2^{n+1}. Hence $y \geq (2^{n+1} + 1)/3 > 2^{n-1}$, $x + 1 > 2^{2n}$, and so $(x+1)(x-2) > 2^{2n}(2^{2n} - 3) > 2^{3n+1} > D$ (as $n \geq 2$), a contradiction. If $\epsilon = 2$, then

$$2^{3n+2} - 2^{n+1} = 2D = (x+1)(x-2) = 2^{2n+2}y^2 + 3 \cdot 2^{n+1}y,$$

and so $3y + 1 = 2^{n+1}(2^n - y^2)$. This implies $1 \leq y < 2^{n/2} \leq 2^{n-1}$, and $3y + 1$ is divisible by 2^{n+1}, which is impossible when $n \geq 2$. $\square$

Note that if $(n, D) = (1, 14)$, then $D = \binom{6}{2} - 1 = \binom{6}{3} - 6 = (3^3 + 1)/2$.

THEOREM 8.2. *Suppose the sheaf $\mathcal{F}(q,f)$ in (4.1.1), of rank $D = 2^n(2^{2n+1}-1)$, has infinite geometric monodromy group $G = G_{\rm geom}$. Then $G^\circ = \mathrm{SL}_D$. Under the more restrictive condition (4.1.3), $G = \mathrm{SL}_D$; moreover, the sheaf $\mathcal{G}(q,f,t(q))$ has geometric monodromy group equal to G.*

PROOF. By Theorem 5.12, G satisfies (**S**+). Thus $G^\circ \le \mathrm{SL}_D$, and $\mathbf{Z}(G)$ is finite, but G° is infinite. It follows from [**30**, Lemma 1.4] that G° is irreducible on the underlying representation V, i.e., $\mathcal{F}(q,f)$ is Lie-irreducible. By Theorem 5.13, G° is a simple algebraic group of dimension $\ge D$.

(a) Suppose $n \ge 3$, so that $D \ge 1016$. Then G° must be a classical group of rank say r, where $r(2r+1) \ge \dim G^\circ \ge D \ge 1016$, whence $r \ge 23$. Also, if G° is of type A_r, then

$$D \le \dim G^\circ = r(r+2) < \binom{r+1}{4}, \tag{8.2.1}$$

and if G° is of type B_r, C_r, or D_r, then

$$D \le r(2r+1) < r^3. \tag{8.2.2}$$

In the case of A_r, we can apply Theorem 7.3 and, using the dimension formula for $L(\lambda)$ given in [**40**, Table 5] and the bound (8.2.1), we see that the highest weight λ of the G°-module V is, up to duality, $a\varpi_1$ with $a = 1, 2$, ϖ_2, or $\varpi_1 + \varpi_r$. If $\lambda = \varpi_1$, then $D = r+1$, $G^\circ = \mathrm{SL}_D$, and hence $G = \mathrm{SL}_D$ as stated. In the other cases, $D = \binom{r+1}{2}$, $\binom{r}{2}$, or $(r+1)^2 - 1$; all are impossible by Lemma 8.1.

In the case of types B_r, C_r, and D_r, we can apply Theorem 7.5 (more precisely, (7.5.3)), and, using the dimension formula for $L(\lambda)$ given in [**40**, Table 5] and the bound (8.2.2), we see that the highest weight λ of the G°-module V is $a\varpi_1$ with $a = 1, 2$, or ϖ_2. If $\lambda = \varpi_1$, then $G^\circ = \mathrm{Sp}(V)$ or $\mathrm{SO}(V)$, whence $\mathcal{F}(q,f)$ is Lie-self-dual, contrary to Corollary 5.8. In the other cases, $D = \binom{m}{2}$ or $\binom{m}{2} - 1$ for some integer $m \ge 2$, and this is impossible by Lemma 8.1.

(b) Suppose $n = 2$, so that $D = 124$. Then G° is of type A_r with $r \ge 11$, B_r, C_r, or D_r with $r \ge 8$, E_7, or E_8. Neither E_7 nor E_8 has irreducible representations of degree 124, see [**33**], so G° is classical of rank r. If G° is of type B_r, C_r, or D_r with $8 \le r \le 11$, then using [**33**] we check that G° has no irreducible representation of degree 124. So $r \ge 12$, and we apply Theorem 7.3, respectively (7.5.3), as above to conclude that $G = \mathrm{SL}_D$.

Finally, assume that $n = 1$, so that $D = 14$. The arguments in part (b) of the proof of Theorem 6.1 repeated verbatim show that either $G^\circ = \mathrm{SL}_D$, or $V|_{G^\circ}$ is self-dual. In the former case $G = \mathrm{SL}_D$ as in (a), and the latter case is ruled out by Corollary 5.8.

(c) Now assume (4.1.3). Then we can consider the descent $\mathcal{G}(q,f,t(q))$, for which the trace function takes values in $\mathbb{Q}(\zeta_4)$ and all slopes are less than 1. Since $p = 2$, Theorem 2.1(iii) implies that $\mathcal{G}(q,f,t(q))$ has trivial determinant, and thus $\mathcal{G}(q,f,t(q))$ has geometric monodromy group $H \le \mathrm{SL}_D$. But $H \ge G$, so we conclude $H = G = \mathrm{SL}_D$. □

THEOREM 8.3. *Suppose the sheaf $\mathcal{F}(q,f)$ in (4.1.3), of rank $D = 2^n(2^{2n+1}-1)$, has finite geometric monodromy group $G = G_{\rm geom}$. Then $G = {}^2B_2(q)$ with $q = 2^{2n+1}$.*

PROOF. (a) Let V denote the underlying representation. By Theorem 5.12, G satisfies $(\mathbf{S}+)$ on V. But the dimension $D = \dim(V) = q_0(q-1)$, with $q_0 := 2^n$, is not a prime power. Hence G is almost quasisimple by [**30**, Lemma 1.1]: $S \lhd G/\mathbf{Z}(G) \leq \mathrm{Aut}(S)$ for some finite, non-abelian simple group S. Then the quasisimple subgroup $L := G^{(\infty)}$ acts irreducibly on V by [**30**, Lemma 1.4].

The condition (4.1.3) allows us to consider the descent $\mathcal{G}(q, f, t(q))$ on $\mathbb{G}_m$, with geometric monodromy group H. Then $G \lhd H$ of finite index, whence H is finite and satisfies $(\mathbf{S}+)$, and $L = H^{(\infty)}$, as $H/G \hookrightarrow C_{t(q)}$. The representation of H on V has ∞-slopes $\sigma := (q_0+1)/D < 1$, and is not tame at ∞. Hence Theorem 2.5 applies to G and H. We collect some further facts about (G, V) and the character φ of H on V that we will use in the proof:

(i) $\mathbb{Q}(\varphi|_G) = \mathbb{Q}(\varphi) = \mathbb{Q}(i)$. Indeed, by Theorem 2.1(i), the arithmetic monodromy group H_{arith,k_0} of $\mathcal{G}(q, f, t(q))$ over k_0 has finite determinant. But it normalizes the finite irreducible subgroup H, so finite determinant implies that H_{arith,k_0} is finite. Now, by Chebotarev density, the finiteness of H_{arith,k_0} implies that all elements of it are Frobenii, and all Frobenii have traces in $\mathbb{Q}(i)$. But $G \leq H \leq H_{\mathrm{arith},k_0}$, so $\mathbb{Q}(\varphi|_G) \subseteq \mathbb{Q}(\varphi) \subseteq \mathbb{Q}(i)$. But $V|_G$ is not self-dual by Theorem 5.7, hence $\mathbb{Q}(\varphi|_G) = \mathbb{Q}(i) = \mathbb{Q}(\varphi)$. Since each element of $\mathbf{Z}(H)$ acts as a root of unity on V, and the only roots of unity in $\mathbb{Q}(i)$ are in μ_4, both $|\mathbf{Z}(G)|$ and $|\mathbf{Z}(H)|$ divide 4.

(ii) If $n \neq 2$, then G is perfect and hence $G = L$. Indeed, by [**1**, Proposition 6], $\pi_1(\mathbb{A}^1/\overline{\mathbb{F}_p})$ has no nontrivial finite p'-quotient. Since $\mathcal{F}(q, f)$ lives on $\mathbb{A}^1/\overline{\mathbb{F}_2}$, G has no nontrivial quotient of odd order. On the other hand, the proof of Theorem 4.5 shows that G has no quotient of order 2 when $n \neq 2$.

(iii) The image $J = QC$ of $I(\infty)$ has $Q = \mathbf{O}_2(J)$ and $C = \langle g_\infty \rangle$, where the central order $\bar{\mathsf{o}}(g_\infty)$ is divisible by $q-1$. Indeed, since the ∞-slope of $\mathcal{F}(q, f)$ is $1 + 1/D$, this implies, by [**23**, Proposition 1.14], that $I(\infty)$ acts irreducibly on V, of dimension $D = q_0(q-1)$. Since the image J of $I(\infty)$ is cyclic of p'-order modulo the image Q of $P(\infty)$, it follows that g_∞ permutes the pairwise non-isomorphic $q-1$ simple Q-summands on V, each of dimension q_0, transitively.

(b) Consider the case $n \geq 3$, so that $D \geq 1016$, and $G = L$ by (ii).

(b1) First suppose that $S = \mathsf{A}_m$ for some $m \geq 3$. As $G/\mathbf{Z}(G) = \mathsf{A}_m \hookrightarrow \mathrm{GL}_{m-1}(\mathbb{C})$, by Theorem 2.5, $m - 1 \geq 1/\sigma$. Note that $\lceil 1/\sigma \rceil = q - 2q_0 + 1$, so $m \geq q - 2q_0 + 2 \geq 114$. It follows that $D = q_0(q-1) < (m^2 - 5m + 2)/2$. In this case, by [**15**, Lemma 6.1] $m = D + 1$, $G = \mathsf{A}_{D+1}$, and $V = \mathbb{C}^D$ is the heart of the natural permutation module. But then V is self-dual, contrary to (i).

(b2) Next suppose that S is a sporadic simple group. By (iii), the maximum order $\mathsf{meo}(S)$ of elements in S is at least $q - 1 \geq 127$. On the other hand, $\mathsf{meo}(S) \leq 119$, as one can check using the [**6**] (see also [**30**, Table 2]), a contradiction.

(b3) Consider the case S is a simple group of Lie type in characteristic $r \neq 2$. By Theorem 2.5, the degree e of every nontrivial projective representation over $\overline{\mathbb{F}_r}$ of S satisfies $e^2 - 1 \geq 1/\sigma$; in particular, $e \geq 11$. Similarly, the degree d of every faithful linear representation over $\overline{\mathbb{F}_r}$ of S satisfies $d \geq 1/\sigma$; in particular, $d \geq 114$. This rules out all classical groups of types A_m or ${}^2A_m$ with $m \leq 9$, B_m with $m \leq 56$, and

C_m or D_m with $m \le 7$ (as $\mathrm{PSp}_{2m}(r^a)$ and $\mathrm{P}\Omega^{\pm}_{2m}(r^a)$ have faithful representations of degree $m(2m-1)$ over $\overline{\mathbb{F}_r}$). This also rules out exceptional groups of types G_2, ${}^2G_2$, and ${}^3D_4$. For the remaining exceptional groups of type $F_4(s)$, $E_6(s)$, ${}^2E_6(s)$, $E_7(s)$, and $E_8(s)$, with $s = r^a$, by [**22**, Table A7] we have the following upper bounds: $s(s+1)(s^2+1)$, $s(s^6-1)/((s-1)\gcd(3,s-1))$, $(s+1)(s^2+1)(s^3-1)/\gcd(3,s+1)$, $(s+1)(s^2+1)(s^4+1)/2$, and $(s+1)(s^2+s+1)(s^5-1)$ for $\mathsf{meo}(S)$, respectively. On the other hand, $q_0(q-1) = D$ is at least the smallest degree $\mathfrak{d}(S)$ of nontrivial projective complex representations of S, which in turn is at least $s^6(s^2-1)$, $s^9(s^2-1)$, $s^9(s^2-1)$, $s^{15}(s^2-1)$, $s^{27}(s^2-1)$, respectively, see e.g. [**47**, Table I]; and we arrive at a contradiction in all five cases, as $q - 1 \le \mathsf{meo}(S)$ by (iii) and $D < \sqrt{q^3/2}$.

If $S = \mathrm{PSL}_m(s)$ or $\mathrm{PSU}_m(s)$ with $m \ge 11$, then $m^2 - 1 \ge 1/\sigma$, and so $m^2 \ge q - 2q_0 + 2$, by Theorem 2.5. Since $m \ge 11$, by [**47**, Theorem 1.1]

$$D \ge \frac{s^m - s}{s+1} \ge \frac{3^m - 3}{4} > m^3 \ge (q - 2q_0 + 2)^{3/2} > q_0(q-1),$$

a contradiction. For S of type BCD_m with $m \ge 7$, we have $\min(m(2m-1), 2m+1) \ge 1/\sigma$, and so $2m^2 \ge q - 2q_0 + 1$, by Theorem2.5. Since $m \ge 7$, by [**47**, Theorem 1.1],

$$D \ge \frac{s^m - 1}{2} \ge \frac{3^m - 1}{2} > 3m^3 \ge (3/2^{3/2})(q - 2q_0 + 1)^{3/2} > q_0(q-1),$$

again a contradiction.

(b4) We may now assume that S is a simple group of Lie type in characteristic 2, defined over a field $\mathbb{F}_s$ with $s = 2^a$. Since $n \ge 3$, by Theorem 3.4, $t(q) = q - 2q_0 + 1$ admits a divisor $\ell = \mathrm{ppd}(2, 4(2n+1))$. Next, we use the fact that the image of $I(0)$ in H has order $q - 2q_0 + 1$, which implies that H has an element of prime order ℓ that normalizes $G = L$. But $\mathbf{C}_H(G) = \mathbf{Z}(H)$ has order dividing 4 by (i), so

$$\ell \text{ divides } |\mathrm{Aut}(L)|. \tag{8.3.1}$$

First suppose that $S = \mathrm{Sp}_{2m}(s)$ with $m \ge 2$ or $\mathrm{P}\Omega^{\pm}_{2m}(s)$ with $m \ge 3$. Then, (8.3.1) implies that $\ell \ge 4(2n+1) + 1 \ge 29$ divides $a|S|$. If moreover $\ell \nmid a$, then ℓ divides $\prod_{i=1}^{m}(s^{2i} - 1)$, which implies $2ma \ge 4(2n+1)$ by primitivity of ℓ. In either case,

$$s^m = 2^{ma} \ge 2^{2(2n+1)} = q^2 \ge 2^{14} \tag{8.3.2}$$

Now, applying [**47**, Theorem 1.1] for $m \ge 3$ we obtain

$$q^{3/2} > D \ge \mathfrak{d}(S) \ge \frac{(s^m - 1)(s^{m-1} - 1)}{s^2 - 1} > s^{2m-3}/2 \ge s^m/2 \ge q^2/2,$$

a contradiction. If $S = \mathrm{PSp}_4(s)$, then $s \ge q \ge 2^7$ by (8.3.2), and so

$$q^{3/2} > D \ge \mathfrak{d}(S) = s(s-1)^2/2 \ge q(q-1)^2/2 > q^2,$$

again a contradiction.

Next suppose that $S = \mathrm{PSL}_m(s)$ or $\mathrm{PSU}_m(s)$ with $m \ge 2$. The same arguments as above show that (8.3.2) still holds; in fact, $s^m \ge q^4$ for $S = \mathrm{PSL}_m(s)$. Assume that $S = \mathrm{PSL}_m(s)$ with $m \ge 2$; in particular, $s^{m-1} \ge s^{m/2} \ge q^2$. Then using [**47**, Theorem 1.1] we obtain that

$$q^{3/2} > D \ge \mathfrak{d}(S) \ge \frac{s^m - s}{s - 1} > s^{m-1} \ge q^2,$$

a contradiction. Next, assume that $S = \mathrm{PSU}_m(s)$ with $m \geq 5$; in particular, $q^{3/2} \leq s^{3m/4} \leq s^{m-5/4}$. Then using [**47**, Theorem 1.1] we obtain that

$$\begin{aligned}\frac{q^{3/2}}{\sqrt{2}} > D \geq \mathfrak{d}(S) \geq \frac{s^m - s}{s+1} \geq \frac{2}{3}(s^{m-1} - 1) > \frac{2^{5/4}}{3}(s^{m-5/4} - 1)\\ > \frac{2^{5/4}}{3}(q^{3/2} - 1) > \frac{q^{3/2}}{\sqrt{2}},\end{aligned}$$

again a contradiction. When $S = \mathrm{PSU}_4(s)$, $\ell | a$ or ℓ divides $(s+1)(s^2-1)(s^3+1)(s^4-1)$, so instead of (8.3.2) now $s^3 \geq q^2$. Again using [**47**, Theorem 1.1] we obtain

$$q^{3/2} > D \geq \mathfrak{d}(S) \geq \frac{s^4-1}{s+1} > s^3/2 \geq q^2/2,$$

a contradiction. Finally, if $S = \mathrm{PSU}_3(s)$, then $s^3 \geq q^2 \geq 2^{14}$ from (8.3.2). But every irreducible character of $\mathrm{SU}_3(s)$ of even degree has degree divisible by $s \geq q^{2/3}$, and hence cannot be equal to $D = q_0(q-1)$.

Let S be one of the exceptional groups of type $G_2(s)$ with $s > 2$, ${}^3D_4(s)$, ${}^2F_4(s)'$ with $s > 2$, $F_4(s)$ with $s > 2$, $E_6(s)$, ${}^2E_6(s)$, $E_7(s)$, and $E_8(s)$. That ℓ divides $|\mathrm{Aut}(S)|$ implies $q^4 \leq s^c$, where

$$c = 6,\ 12,\ 12,\ 12,\ 12,\ 18,\ 18,\ 30,$$

respectively. It follows that $D < q^{3/2} \leq s^{3c/8}$, with

$$3c/8 = 9/4,\ 9/2,\ 9/2,\ 9/2,\ 9/2,\ 27/4,\ 27/4,\ 45/4,$$

respectively. But this contradicts the lower bounds $D \geq \mathfrak{d}(S) \geq s(s^2-1)$, $s^3(s^2-1)$, $s^4(s-1)\sqrt{s/2}$, $s^6(s^2-1)$, $s^9(s^2-1)$, $s^9(s^2-1)$, $s^{15}(s^2-1)$, $s^{27}(s^2-1)$, respectively, see e.g. [**47**, Table I]; and we arrive at a contradiction in all eight cases. The cases ${}^2F_4(2)'$ and $F_4(2)$ are ruled out because $|\mathrm{Aut}(S)|$ is not divisible by the prime $\ell \geq 29$.

The only remaining case is that $S = {}^2B_2(s)$. That ℓ divides $|\mathrm{Aut}(S)|$ implies $q^4 \leq s^4$, i.e., $q \leq s$. But $D = q_0(q-1)$ is the degree of some irreducible character of G, so $q = s$, i.e., $S = {}^2B_2(q)$. In this case also $G = L = {}^2B_2(q)$, as stated.

(c) Now we consider the case $n = 2$, i.e., $D = 124$.

(c1) As the quasisimple subgroup $L = G^{(\infty)}$ acts irreducibly on $V = \mathbb{C}^D$, by [**16**] we have the following possibilities for S:

$$\mathrm{PSL}_2(125),\ \mathrm{SL}_3(5),\ \mathrm{SL}_5(2),\ G_2(5),\ \mathsf{A}_{125},\ \text{or } {}^2B_2(32).$$

Here, a generator g_0 of the image of $I(0)$ in H has order $t(q) = 25$ and normalizes G and L. Since $\mathbf{C}_H(L) = \mathbf{Z}(H) \leq C_4$ by (i), 25 divides $|\mathrm{Aut}(S)|$. This rules out the first three cases.

In the two cases $S = G_2(5)$ and A_{125}, $L = S$ and $V|_L$ is self-dual. Since $\mathbf{C}_H(L) \leq C_4$ and $|\mathrm{Out}(S)| \leq 2$, it follows that H/L is a 2-group. On the other hand, H has no quotient of order 2, as shown in part (i) of the proof of Theorem 4.5 (which also works for $n = 2$). Hence $H = L = G$, and so $\mathcal{F}(q, f)$ is self-dual. But this contradicts Theorem 5.7.

The only remaining case is $S = {}^2B_2(32)$, in which case $L = S$.

(c2) The rest of this paragraph applies to all $n \geq 1$, for which we know $L = S = {}^2B_2(q)$. Since $\mathrm{Out}(S) \cong C_{2n+1}$, but $G = \mathbf{O}^{2'}(G)$, we see that $G = \mathbf{Z}(G)S$, with $\mathbf{Z}(G) \leq C_4$. Recall that $H/G \hookrightarrow C_{t(q)}$ and $\mathbf{Z}(H) \leq C_4$. In particular,

$\mathbf{Z}(H)G/G \cong \mathbf{Z}(H)/\mathbf{Z}(G)$ has order dividing both 4 and $t(q)$, whence $\mathbf{Z}(H) = \mathbf{Z}(G)$. Next, H/G acts trivially on $G/S \cong \mathbf{Z}(G)$, i.e., $G/S \leq \mathbf{Z}(H/S)$, and the quotient $(H/S)/(G/S) \cong H/G$ is cyclic. It follows that H/S is an abelian group. As mentioned above, H has no quotient of order 2, so $2 \nmid |H/S|$, whence $\mathbf{Z}(G) = 1$; in particular, $G = S$ when $n = 2$. We have also shown that $\mathbf{C}_H(S) = \mathbf{Z}(H) = 1$, so

$$S \lhd H \leq \mathrm{Aut}(S) = S \rtimes C_{2n+1} \tag{8.3.3}$$

(d) Finally, we consider the case $n = 1$, i.e., $D = 14$. As the quasisimple group $G = L$ acts irreducibly on $V = \mathbb{C}^D$ and $\mathbb{Q}(\varphi) = \mathbb{Q}(i)$, by [**16**] the only possibility is that $G = {}^2B_2(8)$. □

Now we are ready to prove the second main result of the paper:

THEOREM 8.4. *For the geometric monodromy group $G = G_{\mathrm{geom}}$ of the sheaf $\mathcal{F}(q, f)$ in (4.1.3), of rank $D = 2^n(2^{2n+1} - 1)$, either $G = \mathrm{SL}_D$ or $G = {}^2B_2(q)$ with $q = 2^{2n+1}$. Furthermore, for any $r|t(q)$, the geometric monodromy group of the descent $\mathcal{G}(q, f, r)$ is also equal to G.*

PROOF. Suppose G is infinite. Then the statements follow from Theorem 8.2.

From now on, assume that G is finite. Then $G = S = {}^2B_2(q)$ by Theorem 8.3. Since for any $r|t(q)$, the geometric monodromy group of $\mathcal{G}(q, f, r)$ contains G and is contained in the geometric monodromy group H of $\mathcal{F}(q, f, t(q))$, it suffices to show that $H = S$.

First, note that the ∞-slope of $\mathcal{G}(q, f, t(q))$ is $\sigma = (q_0 + 1)/D$ and $D + 1 = (q_0 + 1)t(q)$, so $\gcd(D, q_0 + 1) = 1$. It follows from [**23**, Proposition 1.14] that $I(\infty)$ acts irreducibly on V, of dimension $D = q_0(q - 1)$. Since the image J of $I(\infty)$ is cyclic of p'-order modulo the image Q of $P(\infty)$, $Q = \mathbf{O}_2(J)$ and $J = \langle Q, g_\infty \rangle$, where the p'-element g_∞ transitively permutes the pairwise non-isomorphic $q - 1$ simple Q-summands on V, each of dimension q_0.

Since $q = 2^{2n+1}$, using [**50**] we can find a primitive prime divisor $\ell = \mathrm{ppd}(2, 2n + 1)$, and fix a power h of g_∞ that has order ℓ. Clearly, the prime ℓ is at least $2n + 3$, so it is coprime to $2n + 1 = |\mathrm{Out}(S)|$. On the other hand, $S \leq H \leq \mathrm{Aut}(S)$ by (8.3.3). Hence $h \in S$ and $Q < S$.

We can write $h = x^{(q-1)/\ell}$, where on the natural module $U = \mathbb{F}_q^4$ for $S = {}^2B_2(q) < \mathrm{Sp}(U)$, the spectrum of $x \in S$ consists of 4 eigenvalues ξ^{2^n}, ξ^{-2^n}, ξ^{1-2^n}, ξ^{2^n-1} where $\xi \in \overline{\mathbb{F}_2}^\times$ has order $q - 1$, see [**4**], [**45**]. We may write $\mathrm{Aut}(S) = \langle S, \theta \rangle$, where θ acts as the Galois automorphism $\lambda \mapsto \lambda^2$ of $\overline{\mathbb{F}_2}$. Suppose that for some $1 \leq a \leq 2n$ and for some $y \in S$, the element $y\theta^a$ centralizes h. Note that θ^a sends x to x^{2^a} and $h = x^{(q-1)/\ell}$ to $x^{2^a(q-1)/\ell}$. It follows that

$$x^{2^a(q-1)/\ell} = \theta^a h \theta^{-a} = y^{-1}(y\theta^a)h(y\theta^a)^{-1}y = y^{-1}hy$$

is S-conjugate to $h = x^{(q-1)/\ell}$. On the other hand, it is known [**4**] that if $b, c \in \mathbb{Z}$ then x^b and x^c are S-conjugate if and only if $c \equiv \pm b \pmod{(q-1)}$. It follows that ℓ divides $2^a \pm 1$; in particular, ℓ divides $2^{2a} - 1$. The primitivity of ℓ then implies that $2n + 1$ divides $2a$, a contradiction.

We have shown that $\mathbf{C}_{\mathrm{Aut}(S)}(h) \leq S$. As g_∞ centralizes h, it follows that $g_\infty \in S$. Hence $J = Q\langle g_\infty \rangle < S$. Now $S \lhd H$ and H is finite, so $H = S$ by Theorem 4.3. □

Theorems 6.1 and 8.4 imply the following.

COROLLARY 8.5. *The sheaves $\mathcal{F}_q$ and $\mathcal{G}_q$ in §1, of rank $D = 2^n(2^{2n+1}-1)$, have the same geometric monodromy group $G = G_{\rm geom}$. Furthermore, either $G = {}^2B_2(q)$ with $q = 2^{2n+1}$, or $n \geq 2$ and $G = \mathrm{SL}_D$.*

REMARK 8.6. It is plausible that for each $q = 2^{2n+1} \geq 32$, we can find a polynomial $f_1 \in \mathbb{F}_2[x]$ of degree $2^n + 1$ such that the sheaf $\mathcal{F}(q, f)$ with $f(x) = f_1(x^{t(q)})$ as in (4.1.3) has infinite geometric monodromy group, which then is SL_D by Theorem 8.4. Indeed, a MAGMA calculation shows for each $2 \leq n \leq 25$ that $\mathrm{Frob}_{1,\mathbb{F}_{2^{k(n)}}}$ has non-integral trace on $\mathcal{F}(q, x^{(1+2^n)t(q)})$ for some choice of $k(n) \in \mathbb{Z}_{\geq 1}$. For instance, $\mathrm{Frob}_{1,\mathbb{F}_{2^{k(n)}}}$ has trace $(2i-7)/2$ for $(n,(k(n)) = (2, 7)$, $(5i+3)/2$ for $(n, k(n)) = (3, 5)$, $(7-3i)/4$ for $(n, k(n)) = (4, 7)$, and $5/2$ for $(n, k(n)) = (5, 7)$. In fact, for infinitely many integers $n \geq 2$, we offer in Theorem 9.18 a construction of a sheaf $\mathcal{F}(q, f)$ with $G_{\rm geom} = \mathrm{SL}_D$.

9. Arithmetic vs. geometric monodromy groups

9A. Glauberman and Dade correspondences. Our next results depend on the Glauberman correspondence. Recall that if A is a solvable finite group acting by automorphisms on another finite group S with $(|A|, |S|) = 1$, then there exists a canonical bijection * : $\mathrm{Irr}_A(S) \to \mathrm{Irr}(C)$, where $C = \mathbf{C}_S(A)$ is the fixed-point subgroup and $\mathrm{Irr}_A(S)$ is the set of A-invariant irreducible characters of S. (See [**20**, Chapter 13].) Since the map is canonical, it is not difficult to see that $\mathbb{Q}(\theta) = \mathbb{Q}(\theta^*)$, where $\mathbb{Q}(\theta)$ is the field of values of θ. (See [**20**, Problem 3.1].)

Suppose now that $S = {}^2B_2(2^n)$, where n is odd. Then it is well known that S admits a field automorphism a of order n. Assume further that $(m, |S|) = 1$ for some divisor $m > 1$ of n. Consider $A = \langle a^{n/m} \rangle \cong C_m$. If $m = n$, then $C = \mathbf{C}_S(A) = {}^2B_2(2) \cong C_5 \rtimes C_4$ has 5 irreducible characters; two of its four linear characters are rational, the others have field of values $\mathbb{Q}(i)$. It also has a rational irreducible character of degree 4. In particular, it follows that $\mathrm{Irr}(S)$ has exactly 5 irreducible A-invariant characters, and exactly two of them have field of values $\mathbb{Q}(i)$. On the other hand, if $m < n$, then $C = \mathbf{C}_S(A) = {}^2B_2(2^{n/m})$ has exactly 2 irreducible A-invariant characters with field of values $\mathbb{Q}(i)$, namely the ones of degree $(r-1)\sqrt{r/2}$ with $r := 2^{n/m}$; see [**4**]. This proves the following.

LEMMA 9.1. *Suppose that $S = {}^2B_2(2^n)$, where n is odd. Assume further that $(m, |S|) = 1$ for a divisor $m > 1$ of n, and let A be the subgroup of field automorphisms of S of order m. Let $C = \mathbf{C}_S(A)$. If $\theta \in \mathrm{Irr}_A(S)$ has field of values $\mathbb{Q}(i)$, then θ^* has degree $(r-1)\sqrt{r/2}$ with $r := 2^{n/m}$.*

LEMMA 9.2. *Suppose that A is a cyclic group of order m acting faithfully and coprimely on S. Let $\theta \in \mathrm{Irr}(S)$ be A-invariant, $C = \mathbf{C}_S(A)$, and let $\eta \in \mathrm{Irr}(C)$ be the A-Glauberman correspondent of θ. Let $G = S \rtimes A$ be the semidirect product, and let $\psi \in \mathrm{Irr}(G)$ be an extension of θ. Consider $x \in G \smallsetminus S$ such that $\langle x, S \rangle = G$.*

(i) *If η is linear, then $|\psi(x)| = 1$.*
(ii) *In all cases, there exist a root of unity $\gamma \in \mathbb{C}^\times$ of order dividing $2m$ and $c \in C$ such that $\psi(x) = \gamma\eta(c)$.*

PROOF. Let $\psi \in \mathrm{Irr}(G)$ be the canonical extension of θ to G. (This is the unique extension ψ such that the determinantal order is coprime with $|A|$, see [**20**, Corollary 6.28].) Since every extension of θ to G is a multiple of ψ by a linear character λ of G/S, we may assume that $\psi = \lambda\psi_0$, where ψ_0 is the canonical extension.

Observe that m divides the order of x. Since m is coprime to $|S|$, we can write $x = cb = bc$, with b being a π-element and c being a π'-element, where π is the set of prime divisors of m. Also $|b| = m$, and $S \ni x^m = b^m c^m = c^m$, implying $c \in S$ since $\gcd(m, |S|) = 1$. Moreover, $G = S \rtimes \langle b \rangle$, so without any loss we may replace A by $\langle b \rangle$, and now $c \in C = \mathbf{C}_S(b) = \mathbf{C}_S(A)$. By [**20**, Theorem 13.6]

$$\psi_0(x) = \psi_0(cb) = \epsilon\eta(c)\,,$$

where $\epsilon = \pm 1$. Hence, $\psi(x) = \gamma\eta(c)$, where $\gamma = \epsilon\lambda(x)$ and $\lambda(x)^m = 1$. □

Since each of the two irreducible characters of $S := {}^2B_2(q)$ of degree $(q-1)\sqrt{q/2}$ has field of values $\mathbb{Q}(i)$ and is $\mathrm{Aut}(S)$-invariant, Lemmas 9.1 and 9.2 imply the following.

COROLLARY 9.3. *Let $q = 2^n$ with $2 \nmid n$ and let θ be either of the two irreducible characters of $S := {}^2B_2(q)$ of degree $(q-1)\sqrt{q/2}$. Then θ extends to $G = \mathrm{Aut}(S) \cong S \rtimes C_n$. If furthermore n is coprime to $|S|$, then $|\psi(x)| = 1$ for any extension ψ of θ to G and for any $x \in G \smallsetminus S$ with $G = \langle x, S \rangle$.*

We remark that the character θ in Corollary 9.3 has a canonical extension to G. Indeed, by [**38**, Theorem A], there exists a unique $\psi \in \mathrm{Irr}(G)$ such that the field of values of ψ is $\mathbb{Q}(i)$. In particular, notice that the restriction ψ to C_n is rational-valued.

In some non-coprime situations, we can use the following statement. This also follows from results in [**9**, §9], but in the situation under consideration, our approach is more straightforward.

LEMMA 9.4. *Suppose that $G = SA$, where $A = \langle a \rangle$, $S \lhd G$ and $A \cap S = 1$. Let $C = \mathbf{C}_S(A)$, and assume that $\chi \in \mathrm{Irr}_A(S)$ has an extension $\psi \in \mathrm{Irr}(G)$ such that ψ_A is rational-valued. Suppose that for every $x \in G \smallsetminus S$, there exists some $g \in G$ such that $x^g = cb$ for some $c \in C$ and $b \in A$. Then there exist a character $\theta \in \mathrm{Irr}(C)$ and a sign $\epsilon = \pm$ such that $\psi(cb) = \epsilon\theta(c)$ for every $c \in C$ and every generator b of A.*

PROOF. By the proof of [**20**, Lemma 13.5], we can write

$$\psi_{CA} = \sum_{\beta \in \mathrm{Irr}(C), [\chi_C, \beta] \neq 0} \beta \boxtimes \psi_\beta\,,$$

where ψ_β is a rational-valued character of S. Now, define $\theta(c) = \psi(ca)$ for $c \in C$. Then

$$\theta = \sum_\beta \psi_\beta(s)\beta\,.$$

Thus θ is a virtual character of C. We claim that $[\theta, \theta] = 1$. Let T be a set of representatives for the right cosets of C in S. In order to use the proof of [**20**, Theorem 13.6], we claim that

$$Sa = \bigcup_{t \in T} (Ca)^t$$

is a disjoint union. By hypothesis,

$$Sa = \bigcup_{s \in S} (Ca)^s = \bigcup_{t \in T} (Ca)^t\,.$$

Now,

$$|S| = |Sa| = \left|\bigcup_{t\in T}(Ca)^t\right| \leq \sum_{t\in T}|(Ca)^t| = |T||C| = |S|,$$

and the claim follows. The rest of the proof follows as in [**20**, Theorem 13.6]. □

Finally, to address the general situation in the Suzuki case, we must go much deeper into [**9**, §9].

COROLLARY 9.5. *Let $q = 2^n$ with $2 \nmid n$ and let θ be either of the two irreducible characters of $S := {}^2B_2(q)$ of degree $(q-1)\sqrt{q/2}$. Then θ extends to $G = \mathrm{Aut}(S) \cong S \rtimes C_n$ and $|\psi(x)| = 1$ for any extension ψ of θ to G and for any $x \in G \smallsetminus S$ with $G = \langle x, S\rangle$.*

PROOF. We adapt the notation of [**9**, §9] to ours. We know that $\mathbf{C}_S(A) = B \rtimes R$, where $B = \langle\beta\rangle$ has order 5, and R is a cyclic group of order 4. Now, let $G' = \mathbf{N}_G(B) = S' \rtimes A$, where $S' = \mathbf{N}_S(B)$. Let G_0 be the set of $x \in G$ such that $\langle xS\rangle = G/S$, and let G_0' be the set of $x \in G'$ such that $\langle xS'\rangle = G'/S'$. By [**9**, Lemma 9.3], $S' = C \rtimes R$, where C is a cyclic group described there. By [**9**, Proposition 9.7], G_0' is a trivial intersection subset of G with normalizer G'. In Dade's language, it satisfies (6.4) of [**9**]. In particular, by [**9**, Lemma 6.5],

$$G_0 = \bigcup_{\tau\in T}(G_0')^\tau$$

is a disjoint union, where $G = \bigcup_{\tau\in T} G'\tau$ is a disjoint union. (In that lemma right cosets are used.) In particular, if $x \in G_0'$ and η is a class function of G', then

$$\eta^G(x) = \eta(x).$$

By [**9**, Theorem 9.8], θ naturally corresponds to some irreducible character $\eta \in \mathrm{Irr}_A(S')$, which therefore has field of values $\mathbb{Q}(i)$. Now, S only has two A-invariant irreducible characters with field of values $\mathbb{Q}(i)$, so using the inverse of Dade's natural correspondence from that theorem, necessarily η is one of the two linear characters of $S'/C = R$.

By hypothesis $\langle x, S\rangle = G$, and so $x \in G_0$. Since

$$G_0 = \bigcup_{\tau\in T}(G_0')^\tau,$$

we may assume that $x \in G_0'$. By [**19**, Lemma G], $\psi(x) = \epsilon\gamma(x)$, where $\gamma \in \mathrm{Irr}(S'A)$ is an extension of η, and ϵ is a root of unity. In particular, γ is linear. We conclude that $|\psi(x)| = 1$, as desired. □

LEMMA 9.6. *Let $S \lhd G < \mathrm{GL}(V) \cong \mathrm{GL}_d(\mathbb{C})$, where G is finite, $S = {}^2B_2(q)$, $D = q_0(q-1)$ with $q_0 = 2^n$, $q = 2^{2n+1}$, and $V = \mathbb{C}^D$ is irreducible over S. Suppose that G contains two elements g_0, g_1 such that $\mathrm{Trace}(g_0) = \pm i$, $\mathrm{Trace}(g_1) = \pm 1$, and $g_1 \in g_0S$. Suppose in addition that the trace of every element in $\mathbf{Z}(G)$ belongs to $\mathbb{Q}(i)$. Then g_0 and g_1 both induce non-inner automorphisms of S.*

PROOF. Assume the contrary. As $g_1 \in g_0S$, g_0 induces an inner automorphism of S. Hence we can write $g_0 = zs$ for some $s \in S$ and some $z \in \mathbf{C}_G(S) = \mathbf{Z}(G)$. Since G is finite, z has finite order, whence $z = \zeta \cdot \mathrm{id}_V$ for some root of unity ζ. By assumption, $\zeta \in \mathbb{Q}(i)$, whence $\zeta^4 = 1$.

First suppose that $\zeta = \pm 1$. Then $\mathrm{Trace}(s) = \zeta^{-1}\,\mathrm{Trace}(g_0) = \pm i$, but this is impossible for any element in S, see [**45**]. Hence, $\zeta = \pm i$. Recalling that $g_1 \in g_0 S$, we can write $g_1 = g_0 t = zst$ for some $t \in S$. In such a case, $\mathrm{Trace}(st) = \zeta^{-1}\,\mathrm{Trace}(g_1) = \pm i$, and we again arrive at a contradiction since $st \in S$. □

9B. Traces of Frobenii and arithmetic monodromy groups. We now give a lemma on traces for those local systems $\mathcal{F}(q, f)$ in which the polynomial $f(x)$ of degree $(q_0 + 1)t(q)$ lies in $\mathbb{F}_2[x]$ and has $f(0) = 0$. Recall that $q_0 = 2^n$, $n \geq 1, q = 2q_0^2$, $t(q) = q + 1 - 2q_0$; for $k/\mathbb{F}_2$ a finite extension and $i = \zeta_4$, the trace function of $\mathcal{F}(q, f)$ is

$$t \in k \mapsto \frac{-1}{\left(1 - (-1)^n\, i\right)^{\deg(k/\mathbb{F}_2)}} \sum_{x \in k} \psi_2(\mathrm{Trace}_{W_2(k)/W_2(\mathbb{F}_2)}([x^{t(q)}, f(x) + tx])).$$

LEMMA 9.7. *For $\mathcal{F}(q, f)$ as above, i.e., with $f(x) \in \mathbb{F}_2[x]$ of degree $(q_0 + 1)t(q)$ and $f(0) = 0$, define*

$$A := \textit{ the number of nonzero monomials in } f(x),$$

so that $A \equiv f(1) \pmod 2$. Then the traces at points of $\mathbb{F}_2$ are given as follows, with $i = \zeta_4$:

$$\mathrm{Trace}(\mathrm{Frob}_{0,\mathbb{F}_2} | \mathcal{F}(q, f)) = \begin{cases} -1, & \textit{if } 2 \nmid (A - n), \\ -(-1)^n i, & \textit{if } 2 | (A - n), \end{cases}$$

$$\mathrm{Trace}(\mathrm{Frob}_{1,\mathbb{F}_2} | \mathcal{F}(q, f)) = \begin{cases} -(-1)^n i, & \textit{if } 2 \nmid (A - n), \\ -1, & \textit{if } 2 | (A - n). \end{cases}$$

Furthermore, for any finite extension $k/\mathbb{F}_2$ and any $t \in k$, $|\,\mathrm{Trace}(\mathrm{Frob}_{t,k} | \mathcal{F}(q, f))| \leq \sqrt{\#k}$.

PROOF. The trace at time $0 \in \mathbb{F}_2$ is $\frac{-1}{(1-(-1)^n\, i)}$ times the sum

$$\psi_2[0, 0] + \psi_2[1, A] = 1 + (-1)^A i,$$

while the trace at time $1 \in \mathbb{F}_2$ is $\frac{-1}{(1-(-1)^n\, i)}$ times the sum

$$\psi_2[0, 0] + \psi_2[1, A + 1] = 1 + (-1)^{A+1} i.$$

If $(-1)^A = -(-1)^n$, these traces are respectively -1 and $-(-1)^n\, i$. If $(-1)^A = (-1)^n$, these traces are respectively $-(-1)^n i$ and -1. For the second statement, note that $\mathrm{Trace}(\mathrm{Frob}_{t,k} | \mathcal{F}(q, f))$ is a sum of $\#k$ terms, each of absolute value 1, divided by a clearing factor of absolute value $\sqrt{\#k}$; hence it has absolute value $\leq \sqrt{\#k}$. □

THEOREM 9.8. *Let $n \in \mathbb{Z}_{\geq 1}$ and $q = 2^{2n+1}$. Suppose the Airy sheaf $\mathcal{F}(q, f)$ defined in (4.1.3) with $q = 2^{2n+1}$, $f(x) \in \mathbb{F}_2[x]$, and $f(0) = 0$, has finite geometric monodromy group S. Then the following statements hold for S and the arithmetic monodromy group $G := G_{\mathrm{arith},\mathbb{F}_2}$ of $\mathcal{F}(q, f)$ over $\mathbb{F}_2$:*

(i) *$S = {}^2B_2(q)$. Furthermore, G induces non-inner automorphisms of S, $G = \mathbf{Z}(G) \times R$ for some $S < R \leq \mathrm{Aut}(S) = S \rtimes C_{2n+1}$, and $\mathbf{Z}(G) \leq C_4$.*

(ii) *If $2n + 1$ is coprime to $|S|$, then $G \cong \mathbf{Z}(G) \times \mathrm{Aut}(S)$.*

(iii) *Suppose that $|\,\mathrm{Trace}(\mathrm{Frob}_{0,\mathbb{F}_{q^2}})| = q_0 := 2^n$. Then $G \cong \mathbf{Z}(G) \times \mathrm{Aut}(S)$. Moreover, $|\mathbf{Z}(G)| \leq 2$ if $\mathrm{Trace}(\mathrm{Frob}_{0,\mathbb{F}_{q^2}}) = -q_0$ or $\mathrm{Trace}(\mathrm{Frob}_{0,\mathbb{F}_q}) = \pm q_0 i$, and $\mathbf{Z}(G) = 1$ if $\mathrm{Frob}_{1,\mathbb{F}_2}$ has odd order.*

PROOF. (i) By Theorem 8.4, the geometric monodromy group of $\mathcal{F}(q, f)$ is $S = {}^2B_2(q)$ in its irreducible representation of degree $D = q_0(q-1)$. By Lemma 9.7, for the images g_0 of $\mathrm{Frob}_{0,\mathbb{F}_2}$ and g_1 of $\mathrm{Frob}_{1,\mathbb{F}_2}$ in G, one has trace $\pm i$ and the other has trace ± 1. Note that $g_1 \in g_0 S$, and since G is finite, the traces of all elements in G belong to $\mathbb{Q}(i)$ by Chebotarev density. By Lemma 9.6, g_0 induces a non-inner automorphism of S.

Recall that $\mathrm{Out}(S) \cong C_{2n+1}$. Then G projects onto a subgroup $S \rtimes C_m$ of $\mathrm{Aut}(S)$ with kernel $\mathbf{C}_G(S) = \mathbf{Z}(G)$, for some divisor $m > 1$ of $2n+1$. Again, each element in $\mathbf{Z}(G)$ acts as a scalar α on $\mathbb{C}^D$ and has trace a root of unity belonging to $\mathbb{Q}(i)$, whence $\alpha^4 = 1$, and thus $\mathbf{Z}(G) \leq C_4$. Now $(G/S)/(\mathbf{Z}(G)S/S) \cong G/\mathbf{Z}(G)S \cong C_m$ is cyclic, and $\mathbf{Z}(G)/S \leq \mathbf{Z}(G/S)$. It follows that G/S is abelian of order $m|\mathbf{Z}(G)|$, with cyclic quotient of odd order m and a cyclic 2-subgroup $\mathbf{Z}(G)S/S$ of order $|\mathbf{Z}(G)|$. Hence $G/S = R/S \times \mathbf{Z}(G)/S$, with $R \cong S \rtimes C_m$. The composition factors of R are S and cyclic groups of odd order (dividing m), so $R \cap \mathbf{Z}(G) = 1$ and $G = \mathbf{Z}(G) \times R$.

(ii) Now assume that $2n+1$ is coprime to $|S|$, but $m < 2n+1$. Let ψ be the character of $R = S \rtimes C_m$ acting on the sheaf, which extends the character θ of S. Let η denote the Glauberman correspondent of θ as in Lemma 9.2; in particular, it has degree $(r-1)\sqrt{r/2}$ for $r := 2^{(2n+1)/m} \geq 8$. We can write $g_0 = zh_0$ and $g_1 = zh_1$ for some $h_0, h_1 \in R$ and $z \in \mathbf{Z}(G)$ (recall $g_0 S = g_1 S$). Now z acts as a root of unity β with $\beta^4 = 1$, and $\psi(h_j) = \gamma_j \eta(c_j)$ with $c_j \in C = {}^2B_2(r)$ and $\gamma_j^{2m} = 1$ for $j = 0, 1$ by Lemma 9.2. Since $\eta(c_j), \beta, \psi(g_j) \in \mathbb{Q}(i)$, we see that $\gamma_j \in \mathbb{Q}(i)$. But $\gamma_j^{2m} = 1$ and $2 \nmid m$, so $\gamma_j = \pm 1$.

Note that, since $r \geq 8$, the character η does not take value $\pm i$, see [**45**]. Without loss of generality, we may assume $\psi(g_0) = \pm 1$ and $\psi(g_1) = \pm i$. Now, if $\beta = \pm i$, then $\eta(g_0) = \psi(g_0)/(\beta\gamma_0) = \pm i$, which is impossible. On the other hand, if $\beta = \pm 1$, then $\eta(g_1) = \psi(g_1)/(\beta\gamma_1) = \pm i$, which is again impossible. Hence $m = 2n+1$, as stated.

(iii) By assumption, $|\varphi(g_0^{4n+2})| = q_0$, where φ is the character of G acting on $\mathcal{F}(q, f)$. Using $G = \mathbf{Z}(G) \times R$, we again write $g_0 = zh_0$ with $z \in \mathbf{Z}(G)$ acting on $\mathcal{F}(q, f)$ as a root of unity β, and $h_0 \in R$. Then $\mathbf{Z}(G) = \langle z \rangle$ since $G = \langle g_0, S \rangle$, and $\beta^4 = 1$. Now $s := h_0^{2n+1} \in S$ as $R/S \hookrightarrow C_{2n+1}$, and $|\varphi(s^2)| = q_0$. Checking the character table of S [**4**], we see that $|s^2| = 2$ or 4 and thus s is a 2-element. But every 2-element of S has order dividing 4, so $|s| = 4$. Hence we can write $|h_0| = 4e$ for some $e|(2n+1)$.

Suppose that $e < 2n+1$, whence $e \leq (2n+1)/3$. Then g_0^e is the image of $\mathrm{Frob}_{0,2^e}$, and so

$$q_0 = 2^n > 2^{(2n+1)/6} \geq 2^{e/2} \geq |\varphi(g_0^e)| = |\varphi(h_0^e)|$$

by Lemma 9.7. On the other hand, since $R/S \hookrightarrow C_{2n+1}$ and h_0^e has order 4, $h_0^e \in S$ and hence $|\varphi(h_0^e)| = q_0$, a contradiction. Thus $|h_0| = 4(2n+1)$. Note that $|\mathbf{C}_S(s^2)| = q^2$, hence $|\mathbf{C}_G(s^2)|$ has order dividing $q^2|\mathbf{Z}(G)| \cdot |R/S|$. But h_0 belongs to $\mathbf{C}_G(s^2)$ and has order $4(2n+1)$; thus, $(2n+1)$ divides $|R/S|$ and so $R = \mathrm{Aut}(S)$.

Since $G = \langle g_1, S \rangle$, the assumption that the image g_1 of $\mathrm{Frob}_{1,\mathbb{F}_2}$ has odd order implies that $2 \nmid |G/S|$, and so $\mathbf{Z}(G) = 1$. Next suppose that $|\mathbf{Z}(G)| > 2$ but $\varphi(g_0^{4n+2}) = -q_0$ or $\varphi(g_0^{2n+1}) = \pm q_0 i$. Then $\beta = \pm i$. In the former case,

$$\varphi(g_0^{4n+2}) = \varphi(\beta^{4n+2} h_0^{4n+2}) = -\varphi(h_0^{4n+2}) = -\varphi(s^2),$$

and thus the involution $s^2 \in S$ has trace q_0, which is impossible, cf. [**4**]. In the latter case,

$$\varphi(g_0^{2n+1}) = \varphi(\beta^{2n+1} h_0^{4n+2}) = \pm i\varphi(h_0^{2n+1}) = \pm i\varphi(s),$$

and thus $s \in S$ has trace $\pm q_0$, which is again impossible, cf. [**4**]. □

PROPOSITION 9.9. *We consider the sheaf $\mathcal{F}_q$, $q = 2^{2n+1}$, as defined in the Introduction. Thus $\mathcal{F}_q := \mathcal{F}(q, f)$ with $f(x) := f_1(x^{t(q)})$, $f_1(x) := \sum_{i=1}^n x^{1+2^i}$ as in (4.1.3). For a finite extension $k/\mathbb{F}_2$, define*

$$\mathrm{Ker}(k) := \left\{ x \in k \mid \sum_{i=0}^{2n} x^{2^i} = 0 \right\}.$$

Then we have the following results:

(i) *For any subfield k of $\mathbb{F}_{q^2}$, $|\,\mathrm{Trace}(\mathrm{Frob}_{0,k}|\mathcal{F}_q)|^2$ is either 0 or $\#\,\mathrm{Ker}(k)$.*
(ii) $|\,\mathrm{Trace}(\mathrm{Frob}_{0,\mathbb{F}_{q^2}}|\mathcal{F}_q)|^2 = \#\,\mathrm{Ker}(\mathbb{F}_{q^2}) = \#\,\mathrm{Ker}(\mathbb{F}_q) = q/2.$

PROOF. We first observe that $\gcd(t(q), q^2 - 1) = 1$. To see this, note that $t(q) = q + 1 - 2q_0$ divides $q^2 + 1$ (indeed $(q + 1 - 2q_0)(q + 1 + 2q_0) = q^2 + 1$), while $\gcd(q^2 - 1, q^2 + 1) = \gcd(q^2 - 1, 2) = 1$. Thus for any subfield k of of $\mathbb{F}_{q^2}$, the map $x \mapsto x^{t(q)}$ is bijective on k.

The sheaf $\mathcal{F}_q$ was built out of the Witt vector

$$\left[x^{t(q)}, \sum_{i=1}^{n} x^{t(q)(1+2^i)} \right].$$

Let us denote by $\mathcal{H}_q$ the sheaf built by the same recipe, with same clearing factor, out of the Witt vector

$$\left[x, \sum_{i=1}^{n} x^{1+2^i} \right].$$

Then for any subfield k of of $\mathbb{F}_{q^2}$

$$\mathrm{Trace}(\mathrm{Frob}_{0,k}|\mathcal{F}_q) = \mathrm{Trace}(\mathrm{Frob}_{0,k}|\mathcal{H}_q), \tag{9.9.1}$$

precisely because the map $x \mapsto x^{t(q)}$ is bijective on k.

Let us rewrite the input Witt vector for $\mathcal{H}_q$ as

$$\left[x, \sum_{i=1}^{n} x^{1+2^i} \right] = [x, xR(x)] \text{ with } R(x) := \sum_{i=1}^{n} x^{2^i}. \tag{9.9.2}$$

With this rewriting, we apply the idea of van der Geer-van der Vlugt, cf. [**49**, §5], as follows. Let us define

$$V(x) := [x, xR(x)]. \tag{9.9.3}$$

In Witt vector addition in $\mathbb{F}_2$-algebras, using the fact that $R(x)$ is an additive polynomial, we get

$$\begin{aligned} V(x+y) &- V(x) - V(y) \\ &= [x+y, (x+y)(R(x)+R(y))] + [x, xR(x)+x^2] + [y, yR(y)+y^2] \\ &= [y, (x+y)x + (x+y)(R(x)+R(y)) + xR(x) + x^2] + [y, yR(y)+y^2] \\ &= [0, y^2 + (x+y)x + (x+y)(R(x)+R(y)) + xR(x) + x^2 + yR(y) + y^2] \\ &= [0, xy + xR(y) + yR(x)] \\ &= [0, \langle x, y\rangle] \end{aligned}$$

for

$$\langle x, y\rangle := xy + xR(y) + yR(x). \tag{9.9.4}$$

The key point is that $\langle x, y\rangle$ on $k \times k$ is a symmetric $\mathbb{F}_2$-bilinear map to k, and $\mathrm{Trace}_{k/\mathbb{F}_2}(\langle x, y\rangle)$ is a symmetric $\mathbb{F}_2$-bilinear form on $k \times k$ as $\mathbb{F}_2$ vector space.

Then

$$|\,\mathrm{Trace}(\mathrm{Frob}_{0,k}|\mathcal{H}_q)|^2 = (1/\#k) \sum_{x,y\in k} \psi_2\big(\mathrm{Trace}_{k/\mathbb{F}_2}(V(x) - V(y))\big)$$

(by the shearing transformation $(x, y) \mapsto (x + y, y)$)

$$= (1/\#k) \sum_{x,y\in k} \psi_2\big(\mathrm{Trace}_{k/\mathbb{F}_2}(V(x+y) - V(y))\big)$$

$$= (1/\#k) \sum_{x,y\in k} \psi_2\big(\mathrm{Trace}_{k/\mathbb{F}_2}(V(x) + [0, \langle x, y\rangle])\big)$$

$$= \sum_{x\in k} \psi_2\big(\mathrm{Trace}_{k/\mathbb{F}_2}(V(x))\big) \Big((1/\#k) \sum_{y\in k} \psi\big(\mathrm{Trace}_{k/\mathbb{F}_2}(\langle x, y\rangle)\big) \Big).$$

The second summand vanishes unless the given $x \in k$ has $\mathrm{Trace}_{k/\mathbb{F}_2}(\langle x, y\rangle) = 0$ for all $y \in k$, in which case it is 1. But $x \in k$ has $\mathrm{Trace}_{k/\mathbb{F}_2}(\langle x, y\rangle) = 0$ for all $y \in k$ if and only if $x \in \mathrm{Ker}(k)$. To see this, note that for $x, y \in k$,

$$\langle x, y\rangle = xy + xR(y) + yR(x) = xy + \sum_{i=1}^{n} xy^{2^i} + \sum_{i=1}^{n} yx^{2^i}$$

has the same $\mathrm{Trace}_{k/\mathbb{F}_2}$ as $\big(x + \sum_{i=1}^n x^{1/2^i} + \sum_{i=1}^n x^{2^i}\big)y$. So by nondegeneracy of the trace, $x \in k$ has $\mathrm{Trace}_{k/\mathbb{F}_2}(\langle x, y\rangle) = 0$ for all $y \in k$ if and only if

$$x + \sum_{i=1}^{n} x^{1/2^i} + \sum_{i=1}^{n} x^{2^i} = 0, \text{ i.e., if and only if } \sum_{i=0}^{2n} x^{2^i} = 0.$$

Thus for k a subfield of $\mathbb{F}_{q^2}$,

$$\big|\mathrm{Trace}(\mathrm{Frob}_{0,k}\mathcal{H}_q)\big|^2 = \sum_{x\in \mathrm{Ker}(k)} \psi_2\big(\mathrm{Trace}_{k/\mathbb{F}_2}(V(x))\big).$$

To show (i), notice that on $\mathrm{Ker}(k)$, $x \mapsto \mathrm{Trace}_{k/\mathbb{F}_2}(V(x))$ is additive, i.e., it is a linear form on $\mathrm{Ker}(k)$. If it is nontrivial, the sum giving $|\,\mathrm{Trace}(\mathrm{Frob}_{0,k}\mathcal{H}_q)|^2$ vanishes. If it is trivial, this sum is $\#\,\mathrm{Ker}(k)$.

To show (ii), notice that for any $k \subseteq \overline{\mathbb{F}_2}$, $\mathrm{Ker}(k)$ is precisely the set of elements $x \in k \cap \mathbb{F}_q$ with $\mathrm{Trace}_{\mathbb{F}_q/\mathbb{F}_2}(x) = 0$. (Indeed, if $x \in \mathrm{Ker}(k)$, then

$$0 = \sum_{i=0}^{2n} F^i(x) = F\Big(\sum_{i=0}^{2n} F^i(x)\Big) = \sum_{i=1}^{2n+1} F^i(x),$$

and so $x = F^{2n+1}(x)$, i.e., $x \in \mathbb{F}_q$.) In particular, $\mathrm{Ker}(\mathbb{F}_{q^2}) = \mathrm{Ker}(\mathbb{F}_q)$ and $\#\mathrm{Ker}(\mathbb{F}_q) = q/2$. Now for $x \in \mathrm{Ker}(\mathbb{F}_{q^2})$, we have

$$\begin{aligned}\mathrm{Trace}_{\mathbb{F}_{q^2}/\mathbb{F}_2}(V(x)) &= \mathrm{Trace}_{\mathbb{F}_q/\mathbb{F}_2}(\mathrm{Trace}_{\mathbb{F}_{q^2}/\mathbb{F}_q}(V(x))) \\ &= \mathrm{Trace}_{\mathbb{F}_q/\mathbb{F}_2}(V(x) + V(x)) \\ &= \mathrm{Trace}_{\mathbb{F}_q/\mathbb{F}_2}([0, x^2]) \\ &= [0, \mathrm{Trace}_{\mathbb{F}_q/\mathbb{F}_2}(x^2)] \\ &= [0, 0],\end{aligned}$$

precisely because every $x \in \mathrm{Ker}(\mathbb{F}_{q^2})$ is an element of $\mathbb{F}_q$ of trace zero. So we see directly that each of the summands in the sum giving $|\,\mathrm{Trace}(\mathrm{Frob}_{0,\mathbb{F}_{q^2}}|\mathcal{H}_q)|^2$ is simply 1. $\square$

PROPOSITION 9.10. *For $q = 2^{2n+1}$ and the Airy sheaf $\mathcal{F}_q$,*

$$\mathrm{Trace}(\mathrm{Frob}_{0,\mathbb{F}_q}|\mathcal{F}_q) = -\epsilon 2^n i,$$

where $\epsilon := (-1)^{n(n+1)/2}$ is the Jacobi symbol

$$\epsilon_{2n+1} = \left(\frac{2}{2n+1}\right) = \begin{cases} 1, & \text{if } 2n+1 \equiv \pm 1 \pmod 8, \\ -1, & \text{if } 2n+1 \equiv \pm 3 \pmod 8. \end{cases}$$

PROOF. (i) Proposition 9.9 implies that $\mathrm{Trace}(\mathrm{Frob}_{0,\mathbb{F}_q}|\mathcal{F}_q)$ lies in $\mathbb{Z}[i]$ (because the only possible non-integrality is at the unique place of $\mathbb{Q}(i)$ over 2, where this trace and its complex conjugate have the same 2-adic ord). Furthermore, we can work with traces over $\mathcal{H}_q$ instead of $\mathcal{F}_q$.

Let us denote by F the absolute Frobenius $x \mapsto x^2$, and define

$$R_n(x) := \sum_{i=1}^{n} F^i(x), \quad V_n(x) := [x, xR_n(x)] = \Big[x, x\Big(\sum_{i=1}^{n} F^i(x)\Big)\Big].$$

Consider the non-normalized sum

$$\mathrm{RawTrace}(\mathrm{Frob}_{0,\mathbb{F}_q}|\mathcal{F}_q) := -\sum_{x \in \mathbb{F}_q} \psi_2(\mathrm{Trace}_{\mathbb{F}_q/\mathbb{F}_2}(V_n(x))),$$

so that

$$\mathrm{RawTrace}(\mathrm{Frob}_{0,\mathbb{F}_q}|\mathcal{F}_q) = (1 - (-1)^n i)^{2n+1}\,\mathrm{Trace}(\mathrm{Frob}_{0,\mathbb{F}_q}|\mathcal{F}_q).$$

(ii) To explain the idea of the proof, consider first the case when $2n+1$ is an odd prime p. Then $\mathbb{F}_{2^p}/\mathbb{F}_2$ has degree p, and $\mathbb{F}_{2^p} \setminus \mathbb{F}_2$ is the disjoint union of F-orbits of length p. On each such F-orbit, the value of $\mathrm{Trace}_{\mathbb{F}_q/\mathbb{F}_2}(V_n(x))$ is constant, and this constant value is then repeated p times as we sum over this orbit. So we have a congruence modulo $p\mathbb{Z}[i]$:

$$\begin{aligned}\mathrm{RawTrace}(\mathrm{Frob}_{0,\mathbb{F}_q}|\mathcal{F}_q) &\equiv -\sum_{x\in\mathbb{F}_2}\psi_2\big(\mathrm{Trace}_{\mathbb{F}_q/\mathbb{F}_2}(V_n(x))\big)\\ &\equiv -\psi_2\big(\mathrm{Trace}_{\mathbb{F}_q/\mathbb{F}_2}(V_n(0))\big)-\psi_2\big(\mathrm{Trace}_{\mathbb{F}_q/\mathbb{F}_2}(V_n(1))\big)\\ &\equiv -\psi_2\big(\mathrm{Trace}_{\mathbb{F}_q/\mathbb{F}_2}[0,0]\big)-\psi_2\big(\mathrm{Trace}_{\mathbb{F}_q/\mathbb{F}_2}[1,n]\big)\end{aligned}$$

(remembering that $\mathbb{F}_q/\mathbb{F}_2$ has odd degree p, and both $V_n(0), V_n(1)$ are already $\mathbb{F}_2$-rational)

$$\begin{aligned}\equiv -\psi_2(p[0,0])-\psi_2(p[1,(p-1)/2]) &= -1-\psi_2([1,(p-1)/2])^p\\ &= -1-(i^{1+(p-1)})^p = -1-i^{p^2} = -1-i.\end{aligned}$$

So when $2n+1=p$, we have a congruence modulo $p\mathbb{Z}[i]$:

$$(1-(-1)^{(p-1)/2}i)^p\,\mathrm{Trace}(\mathrm{Frob}_{0,\mathbb{F}_q}|\mathcal{F}_q)\equiv -1-i.$$

Multiplying both sides by $(1+(-1)^{(p-1)/2}i)^p$, we get a congruence modulo $p\mathbb{Z}[i]$:

$$2^p\,\mathrm{Trace}(\mathrm{Frob}_{0,\mathbb{F}_q}|\mathcal{F}_q)\equiv -(1+i)(1+(-1)^{(p-1)/2}i)^p.$$

If $p\equiv 1 \pmod 4$, the right side is

$$-(1+i)^{p+1} = -(2i)(2i)^{(p-1)/2} = -(2i)2^{(p-1)/2}i^{(p-1)/2}.$$

If $p\equiv 3 \pmod 4$, the right side is

$$-(1+i)(1-i)^p = -2(1-i)^{p-1} = -2(-2i)^{(p-1)/2} = 2^{(p+1)/2}i^{(p-1)/2}.$$

Recalling that $2^p\equiv 2 \pmod p$, and that $(p-1)/2=n$, we get a congruence modulo $p\mathbb{Z}[i]$:

$$\mathrm{Trace}(\mathrm{Frob}_{0,\mathbb{F}_q}|\mathcal{F}_q)\equiv\begin{cases}-i2^ni^{(p-1)/2}, & \text{if } p\equiv 1 \pmod 4,\\ 2^ni^{(p-1)/2}, & \text{if } p\equiv 3 \pmod 4.\end{cases}$$

Thus

$$\mathrm{Trace}(\mathrm{Frob}_{0,\mathbb{F}_q}|\mathcal{F}_q)\equiv -\epsilon_p 2^n i, \tag{9.10.1}$$

where ϵ_p is given by $i^{(p-1)/2}=(-1)^{(p-1)/4}=(-1)^{(p^2-1)/8}$ when $p\equiv 1 \pmod 4$, and by $i^{(p+1)/2}=(-1)^{(p+1)/4}=(-1)^{(p^2-1)/8}$ when $p\equiv 3 \pmod 4$. Thus in both cases ϵ_p is the Legendre symbol $\left(\frac{2}{p}\right)=(-1)^{(p^2-1)/8}$.

In view of Proposition 9.9, $\mathrm{Trace}(\mathrm{Frob}_{0,\mathbb{F}_q}|\mathcal{F}_q)$ is either 0 or an element of $\mathbb{Z}[i]$ of absolute value 2^n. It cannot be 0 because of (9.10.1). So it must be one of $\pm 2^n$ or $\pm 2^n i$. Of these four possibilities, only $-\epsilon_p 2^n i$ is congruent modulo $p\mathbb{Z}[i]$ to $-\epsilon_p 2^n i$ (this is just the statement that for an odd prime p, the four powers of i are distinct modulo $p\mathbb{Z}[i]$).

(iii) We now turn to the general case, where we proceed by induction on the total number (counting multiplicity) of primes dividing $2n+1$. Thus we write

$$2n+1=ps,\ \ s=2a+1,\ \ p=2b+1,\ \text{ with } a,b\geq 1 \text{ and } p \text{ prime}.$$

We will need to deal with both $\mathcal{F}_{2^{ps}}$ and $\mathcal{F}_{2^s}$. To simplify notation, let us write

$$Q:=2^{ps},\ q':=2^s.$$

Here

$$ps=s(2b+1)=2bs+s=2(sb+a)+1.$$

The idea is that $\mathbb{F}_Q \setminus \mathbb{F}_{q'}$ is the disjoint union of F^s-orbits, each of length p, and on each of these orbits, the value of $\mathrm{Trace}_{\mathbb{F}_Q/\mathbb{F}_2}(V_{sb+a}(x))$ is constant, and this constant value is then repeated p times as we sum over this orbit. Thus we get a congruence modulo $p\mathbb{Z}[i]$:

$$\mathrm{RawTrace}(\mathrm{Frob}_{0,\mathbb{F}_Q}|\mathcal{F}_Q) \equiv -\sum_{x\in\mathbb{F}_{q'}} \psi_2\big(\mathrm{Trace}_{\mathbb{F}_Q/\mathbb{F}_2}(V_{sb+a}(x))\big).$$

But for $x \in \mathbb{F}_{q'}$, the trace from $\mathbb{F}_Q$ down to $\mathbb{F}_{q'}$ is just multiplication by p, so

$$\mathrm{Trace}_{\mathbb{F}_Q/\mathbb{F}_2}(V_{sb+a}(x)) = \mathrm{Trace}_{\mathbb{F}_{q'}/\mathbb{F}_2}(pV_{sb+a}(x)).$$

Suppose first that b is even, i.e., that $p \equiv 1 \pmod 4$. Recall $4[x,y] = 0$. So for $x \in \mathbb{F}_{q'}$,

$$\begin{aligned} pV_{sb+a}(x) = V_{sb+a}(x) &= \left[x, xR_a(x) + x\left(\sum_{i=a+1}^{sb+a} F^i(x)\right)\right] \\ &= [x, xR_a(x) + x\big(b\,\mathrm{Trace}_{\mathbb{F}_{q'}/\mathbb{F}_2}(x)\big)] = [x, xR_a(x)] = V_a(x), \end{aligned}$$

where the last equality holds because b is even. So in this even b case,

$$\mathrm{RawTrace}(\mathrm{Frob}_{0,\mathbb{F}_Q}|\mathcal{F}_Q) \equiv \mathrm{RawTrace}(\mathrm{Frob}_{0,\mathbb{F}_a}|\mathcal{F}_{q'}) \pmod{p\mathbb{Z}[i]}.$$

Suppose now that b is odd, i.e., that $p \equiv -1 \pmod 4$. Then for $x \in \mathbb{F}_{q'}$,

$$\begin{aligned} pV_{sb+a}(x) = -V_{sb+a}(x) &= -[x, xR_a(x) + x\left(\sum_{i=a+1}^{sb+a} F^i(x)\right)] \\ &= -\left[x, xR_a(x) + x\,\mathrm{Trace}_{\mathbb{F}_{q'}/\mathbb{F}_2}(x)\right] = [x, xR_a(x) + x^2 + x\,\mathrm{Trace}_{\mathbb{F}_{q'}/\mathbb{F}_2}(x)] \\ &= [x, xR_a(x)] + [0, x^2 + x\,\mathrm{Trace}_{\mathbb{F}_{q'}/\mathbb{F}_2}(x)] \\ &= V_a(x) + [0, x^2 + x\,\mathrm{Trace}_{\mathbb{F}_{q'}/\mathbb{F}_2}(x)]. \end{aligned}$$

But the term $[0, x^2 + x\,\mathrm{Trace}_{\mathbb{F}_{q'}/\mathbb{F}_2}(x)]$ has

$$\mathrm{Trace}_{\mathbb{F}_{q'}/\mathbb{F}_2}([0, x^2+x\,\mathrm{Trace}_{\mathbb{F}_{q'}/\mathbb{F}_2}(x)]) = [0, \mathrm{Trace}_{\mathbb{F}_{q'}/\mathbb{F}_2}(x^2+x\,\mathrm{Trace}_{\mathbb{F}_{q'}/\mathbb{F}_2}(x))] = [0,0],$$

where the last equality holds because

$$\mathrm{Trace}_{\mathbb{F}_{q'}/\mathbb{F}_2}(x^2) = \mathrm{Trace}_{\mathbb{F}_{q'}/\mathbb{F}_2}(x) = (\mathrm{Trace}_{\mathbb{F}_{q'}/\mathbb{F}_2}(x))^2.$$

So in this odd b case as well, also

$$\mathrm{RawTrace}(\mathrm{Frob}_{0,\mathbb{F}_Q}|\mathcal{F}_Q) \equiv \mathrm{RawTrace}(\mathrm{Frob}_{0,\mathbb{F}_a}|\mathcal{F}_{q'}) \pmod{p\mathbb{Z}[i]}.$$

We have shown that

$$\begin{aligned} (9.10.2)\qquad (1-(-1)^{sb+a}i)^{ps}\,&\mathrm{Trace}(\mathrm{Frob}_{0,\mathbb{F}_Q}|\mathcal{F}_Q) \\ &\equiv (1-(-1)^a i)^s\,\mathrm{Trace}(\mathrm{Frob}_{0,\mathbb{F}_{q'}}|\mathcal{F}_q) \pmod{p\mathbb{Z}[i]}. \end{aligned}$$

By the induction hypothesis,

$$(9.10.3)\qquad \mathrm{Trace}(\mathrm{Frob}_{0,\mathbb{F}_{q'}}|\mathcal{F}_{q'}) = -\epsilon_s 2^a i.$$

The clearing factors are invertible modulo $p\mathbb{Z}[i]$. We next show that the clearing factors are equal modulo $p\mathbb{Z}[i]$. Their ratio is

$$\frac{(1-(-1)^{sb+a}i)^{ps}}{(1-(-1)^a i)^s} = \frac{(1-(-1)^{sb+a}i)^{ps}(1+(-1)^a i)^s}{2^s}.$$

If b is odd, then $(p+1)/2$ is even, and this ratio is

$$\frac{(1+(-1)^a i)^{ps+s}}{2^s} = \frac{(2(-1)^a i)^{s(p+1)/2}}{2^s}$$
$$= 2^{s(p-1)/2}((-1)^a i)^{s(p+1)/2} = 2^{s(p-1)/2} i^{s(p+1)/2}.$$

If $b = (p-1)/2$ is even, this ratio is

$$(1-(-1)^a i)^{(p-1)s} = (-2(-1)^a i)^{s(p-1)/2} = 2^{s(p-1)/2} i^{s(p-1)/2}.$$

Let $\chi_{p,\mathrm{quad}}$ be the quadratic character of $\mathbb{F}_p^\times$, so $2^{(p-1)/2} \equiv \chi_{p,\mathrm{quad}}(2) \pmod{p\mathbb{Z}[i]}$. As s is odd, the ratio of clearing factors modulo $p\mathbb{Z}[i]$ is $\chi_{p,\mathrm{quad}}(2) i^{s(p-1)/2}$ when $p \equiv 1 \pmod 4$, and $\chi_{p,\mathrm{quad}}(2) i^{s(p+1)/2}$ when $p \equiv 3 \pmod 4$. Next, when $p \equiv 1 \pmod 4$,

$$i^{(p-1)/2} = \chi_{p,\mathrm{quad}}(2).$$

When $p \equiv 3 \pmod 4$,

$$i^{(p+1)/2} = \chi_{p,\mathrm{quad}}(2).$$

As s is odd, we find that in all cases the ratio of clearing factors is $1 \pmod{p\mathbb{Z}[i]}$, as stated. Hence (9.10.2) and (9.10.3) imply the congruence

$$\mathrm{Trace}(\mathrm{Frob}_{0,\mathbb{F}_Q}|\mathcal{F}_Q) \equiv \mathrm{Trace}(\mathrm{Frob}_{0,\mathbb{F}_{q'}}|\mathcal{F}_{q'}) = -\epsilon_s 2^a i \pmod{p\mathbb{Z}[i]}.$$

Note that

$$2^{sb} = (2^{(p-1)/2})^s \equiv \chi_{p,\mathrm{quad}}(2)^s = \chi_{p,\mathrm{quad}}(2) = \epsilon_p \pmod p.$$

Hence

$$\epsilon_{ps} 2^{sb+a} \equiv \epsilon_{ps}\epsilon_p 2^a = \epsilon_s 2^a \pmod p$$

by multiplicativity of the Jacobi symbol. It follows that

$$\mathrm{Trace}(\mathrm{Frob}_{0,\mathbb{F}_Q}|\mathcal{F}_Q) \equiv -\epsilon_{ps} 2^{sb+a} i \pmod{p\mathbb{Z}[i]}.$$

This congruence shows that $\mathrm{Trace}(\mathrm{Frob}_{0,\mathbb{F}_Q}|\mathcal{F}_Q)$ is nonzero, so by Proposition 9.9 it is one of $\pm 2^{sb+a}$ or $\pm 2^{sb+a} i$. Again, of these four possibilities, only $-\epsilon_{ps} 2^n i$ is congruent modulo $p\mathbb{Z}[i]$ to $-\epsilon_{ps} 2^n i$, and the induction step is complete. □

PROPOSITION 9.11. *For $q = 2^{2n+1}$, consider the Airy sheaf $\mathcal{F}_q$. Then for any subfield k of $\mathbb{F}_q$*

$$|\,\mathrm{Trace}(\mathrm{Frob}_{1,k}|\mathcal{F}_q)|^2 = 1.$$

PROOF. Let k be a subfield of $\mathbb{F}_{q^2}$. Note that $N := (q + 2q_0 + 1)q^2/2$ divides $(q^2+1)q^2$ and so is coprime to q^2-1. Hence the map $x \to x^N$ is a bijection on k, and it sends $x^{t(q)}$ to $x^{(q^2+1)q^2/2} = x$ for any $x \in k$. As in the proof of Proposition 9.10, let us denote by F the absolute Frobenius $x \mapsto x^2$. For each integer $j \geq 0$, we define

$$R_j(x) := \sum_{i=1}^{j} F^i(x).$$

Consider the non-normalized sum

$$\mathrm{RawTrace}(\mathrm{Frob}_{1,k}|\mathcal{F}_q) := -\sum_{x \in k} \psi_2\big(\mathrm{Trace}_{k/\mathbb{F}_2}(V(x))\big),$$

where $V(x) = [x, xR_n(x) + x^N]$. Then we have

$$\mathrm{RawTrace}(\mathrm{Frob}_{1,k}|\mathcal{F}_q) = (1-(-1)^n i)^{\deg(k/\mathbb{F}_2)}\,\mathrm{Trace}(\mathrm{Frob}_{1,k}|\mathcal{F}_q).$$

We now take k to be a subfield of $\mathbb{F}_q$. We examine the function $x \mapsto x^N$ on k. This function depends only on $N \pmod{(q-1)}$. Then

$$N \equiv (1+1+2q_0)q/2 = (2+2q_0)q/2 = q + qq_0 \equiv 1 + q_0 = 1 + 2^n \pmod{(q-1)}.$$

Thus if $x \in k$, then $x^N = xF^n(x)$, and hence

$$V(x) = [x, xR_n(x) + x^N] = [x, xR_n(x) + xF^n(x)] = [x, xR_{n-1}(x)].$$

At this point, we repeat the van der Geer-van der Vlugt argument of Theorem 9.9. We find that if k is a subfield of $\mathbb{F}_q$, then

$$|\operatorname{Trace}(\mathrm{Frob}_{1,k}|\mathcal{F}_q)|^2 = \sum_{x \in \mathrm{Ker}'(k)} \psi_2(\operatorname{Trace}_{k/\mathbb{F}_2}(V(x))),$$

with

$$\mathrm{Ker}'(k) := \left\{ x \in k | \sum_{i=0}^{2n-2} F^i(x) = 0 \right\}.$$

The key observation is that for k a subfield of $\mathbb{F}_q$, $\mathrm{Ker}'(k) = \{0\}$. Indeed, since

$$0 \in \mathrm{Ker}'(k) \subseteq \mathrm{Ker}'(\mathbb{F}_q),$$

it suffices to show that $\mathrm{Ker}'(\mathbb{F}_q) = \{0\}$. But for $x \in \mathrm{Ker}'(\mathbb{F}_q)$,

$$\operatorname{Trace}_{\mathbb{F}_q/\mathbb{F}_2}(x) = \left(\sum_{i=0}^{2n-2} F^i(x) \right) + F^{2n-1}(x) + F^{2n}(x) = F^{2n-1}(x) + F^{2n}(x).$$

Thus for $x \in \mathrm{Ker}'(\mathbb{F}_q)$, $\operatorname{Trace}_{\mathbb{F}_q/\mathbb{F}_2}(x) = F^{2n-1}(x + F(x))$. As $\operatorname{Trace}_{\mathbb{F}_q/\mathbb{F}_2}(x) \in \mathbb{F}_2$, this gives

$$\operatorname{Trace}_{\mathbb{F}_q/\mathbb{F}_2}(x) = x + x^2.$$

If $\operatorname{Trace}_{\mathbb{F}_q/\mathbb{F}_2}(x) = 0$, then $x + x^2 = 0$, i.e., $x \in \mathbb{F}_2$, so x is 0 or 1. Of these, only $x = 0$ has $\operatorname{Trace}_{\mathbb{F}_q/\mathbb{F}_2}(x) = 0$. If $\operatorname{Trace}_{\mathbb{F}_q/\mathbb{F}_2}(x) = 1$, then $x + x^2 = 1$ and so $\mathbb{F}_2(x) = \mathbb{F}_4$. But $\mathbb{F}_4$ is not a subfield of $\mathbb{F}_q$, which has odd degree $2n+1$ over $\mathbb{F}_2$. So in this second case, there are no possible x.

Thus

$$\begin{aligned} \left|\operatorname{Trace}(\mathrm{Frob}_{1,k}|\mathcal{F}_q)\right|^2 &= \sum_{x \in \mathrm{Ker}'(k)} \psi_2(\operatorname{Trace}_{k/\mathbb{F}_2}(V(x))) = \psi_2(\operatorname{Trace}_{k/\mathbb{F}_2}(V(0))) \\ &= \psi_2([0,0]) = 1. \end{aligned}$$

□

PROPOSITION 9.12. *For* $q = 2^{2n+1}$ *and the Airy sheaf* $\mathcal{F}_q$,

$$\operatorname{Trace}(\mathrm{Frob}_{1,\mathbb{F}_q}|\mathcal{F}_q) = -1.$$

PROOF. (i) The quantity $\operatorname{Trace}(\mathrm{Frob}_{1,k}|\mathcal{F}_q)$ lies in $\mathbb{Q}(i)$ and is integral outside the unique place over 2, so by Proposition 9.11, $\operatorname{Trace}(\mathrm{Frob}_{1,\mathbb{F}_q}|\mathcal{F}_q)$ is one of $\{1, -1, i, -i\}$. For any odd prime p, these four elements are distinct modulo $p\mathbb{Z}[i]$. We proceed by induction on the total number (counting multiplicity) of primes in the factorization of $2n+1$.

For the induction base, suppose that $k = \mathbb{F}_q = \mathbb{F}_{2^p}$. Then $\mathbb{F}_q \setminus \mathbb{F}_2$ is the disjoint union of F-orbits of length divisible by p. On each such F-orbit, the value of $\mathrm{Trace}_{\mathbb{F}_q/\mathbb{F}_2}(V(x))$ is constant, and this constant value is then repeated a multiple of p times as we sum over this orbit. So we have a congruence modulo $p\mathbb{Z}[i]$:

$$\begin{aligned}\mathrm{RawTrace}(\mathrm{Frob}_{1,\mathbb{F}_q}|\mathcal{F}_q) &\equiv -\sum_{x\in\mathbb{F}_2}\psi_2\big(\mathrm{Trace}_{\mathbb{F}_q/\mathbb{F}_2}(V(x))\big)\\ &\equiv -\psi_2\big(\mathrm{Trace}_{\mathbb{F}_q/\mathbb{F}_2}(V(0))\big)-\psi_2\big(\mathrm{Trace}_{\mathbb{F}_q/\mathbb{F}_2}(V(1))\big)\\ &\equiv -\psi_2\big(\mathrm{Trace}_{\mathbb{F}_q/\mathbb{F}_2}[0,0]\big)-\psi_2\big(\mathrm{Trace}_{\mathbb{F}_q/\mathbb{F}_2}[1,n+1]\big)\\ &\equiv -\psi_2(s[0,0])-\psi_2(p[1,(p+1)/2])\\ &= -1-\psi_2([1,(p+1)/2])^p\\ &= -1-(i^{1+(p+1)})^p=-1-i^{p^2+2p}=-1+i.\end{aligned}$$

So for $\mathrm{Frob}_{1,\mathbb{F}_q}$, we have a congruence modulo $p\mathbb{Z}[i]$:

$$(1-(-1)^{(p-1)/2}i)^p\,\mathrm{Trace}(\mathrm{Frob}_{1,\mathbb{F}_q}|\mathcal{F}_q)=\mathrm{RawTrace}(\mathrm{Frob}_{1,\mathbb{F}_q}|\mathcal{F}_q)\equiv -1+i.$$

Multiplying both sides by $(1+(-1)^{(p-1)/2}i)^p$, we get a congruence modulo $p\mathbb{Z}[i]$:

$$2^p\,\mathrm{Trace}(\mathrm{Frob}_{0,\mathbb{F}_q}|\mathcal{F}_q)\equiv -(1-i)(1+(-1)^{(p-1)/2}i)^p.$$

If $p\equiv 3 \pmod 4$, the right side is

$$\begin{aligned}-(1-i)^{p+1}=-(2i)^{(p+1)/2}=-2^{(p+1)/2}(-1)^{(p+1)/4}&\equiv -2(-1)^{(p^2-1)/8+(p+1)/4}\\ &= -2 \pmod p.\end{aligned}$$

If $p\equiv 1 \pmod 4$, the right side is $-(1-i)(1+i)^p$, which is

$$\begin{aligned}-2(1+i)^{s-1}&=-2(2i)^{(p-1)/2}=-2^{(p+1)/2}(-1)^{(p-1)/4}\equiv -2(-1)^{(p^2-1)/8+(p-1)/4}\\ &= -2 \pmod p.\end{aligned}$$

Thus in both cases $q\,\mathrm{Trace}(\mathrm{Frob}_{1,\mathbb{F}_q}|\mathcal{F}_q)\equiv -2 \pmod{p\mathbb{Z}[i]}$. Using Proposition 9.11, we conclude that $\mathrm{Trace}(\mathrm{Frob}_{1,\mathbb{F}_q}|\mathcal{F}_q)=-1$ in this case.

(ii) Suppose now that $2n+1=ps$ with an odd prime p and an odd integer s. Then just as in (iii) of the proof of Proposition 9.10 we write

$$s=2a+1,\ p=2b+1,\ ps=2(sb+a)+1.$$

We will need to deal with both $\mathcal{F}_{2^{ps}}$ and $\mathcal{F}_{2^s}$. To simplify notation, let us write

$$Q:=2^{ps},\ q':=2^s.$$

Then, just as in the proof of Proposition 9.11,

$$\mathrm{RawTrace}(\mathrm{Frob}_{1,\mathbb{F}_Q}|\mathcal{F}_Q)=-\sum_{x\in\mathbb{F}_Q}\psi_2(\mathrm{Trace}_{\mathbb{F}_Q/\mathbb{F}_2}([x,R_{sb+a-1}(x)])).$$

The elements of $\mathbb{F}_Q\smallsetminus\mathbb{F}_{q'}$ fall into F-orbits of length p, and the elements from each of these orbits have the same $\mathrm{Trace}_{\mathbb{F}_Q/\mathbb{F}_2}([x,R_{sb+a-1}(x)])$. So we get a congruence modulo $p\mathbb{Z}[i]$

$$\begin{aligned}\mathrm{RawTrace}(\mathrm{Frob}_{1,\mathbb{F}_Q}|\mathcal{F}_Q)&\equiv -\sum_{x\in\mathbb{F}_{q'}}\psi_2(\mathrm{Trace}_{\mathbb{F}_Q/\mathbb{F}_2}([x,R_{sb+a-1}(x)]))\\ &= -\sum_{x\in\mathbb{F}_{q'}}\psi_2(\mathrm{Trace}_{\mathbb{F}_{q'}/\mathbb{F}_2}(p[x,R_{sb+a-1}(x)])).\end{aligned}$$

Notice that for $x \in \mathbb{F}_{q'}$,

$$R_{sb+a-1}(x) = R_{a-1}(x) + b\,\mathrm{Trace}_{\mathbb{F}_{q'}/F_2}(x).$$

Suppose first that $p \equiv 1 \pmod 4$. Then b is even,

$$p[x, R_{sb+a-1}(x)] = [x, R_{sb+a-1}(x)] = [x, xR_{a-1}(x)],$$

and hence

$$\mathrm{RawTrace}(\mathrm{Frob}_{1,\mathbb{F}_Q}|\mathcal{F}_Q) \equiv \mathrm{RawTrace}(\mathrm{Frob}_{1,\mathbb{F}_{q'}}|\mathcal{F}_{q'}) \pmod{p\mathbb{Z}[i]}$$

when $p \equiv 1 \pmod 4$.

Suppose next that $p \equiv -1 \pmod 4$. Since b is odd,

$$\begin{aligned}p[x, R_{sb+a-1}(x)] &= -[x, R_{sb+a-1}(x)] = [x, x^2 + R_{sb+a-1}(x)]\\ &= [x, x^2 + x\,\mathrm{Trace}_{\mathbb{F}_{q'}/\mathbb{F}_2}(x) + xR_{a-1}(x)]\\ &= [x, xR_{a-1}(x)] + [0, x^2 + x\,\mathrm{Trace}_{\mathbb{F}_{q'}/\mathbb{F}_2}(x)].\end{aligned}$$

But for $x \in \mathbb{F}_{q'}$, $\mathrm{Trace}_{\mathbb{F}_{q'}/\mathbb{F}_2}$ annihilates $x^2 + x\,\mathrm{Trace}_{\mathbb{F}_{q'}/\mathbb{F}_2}(x)$, so again

$$\mathrm{RawTrace}(\mathrm{Frob}_{1,\mathbb{F}_Q}|\mathcal{F}_Q) \equiv \mathrm{RawTrace}(\mathrm{Frob}_{1,\mathbb{F}_{q'}}|\mathcal{F}_{q'}) \pmod{p\mathbb{Z}[i]}$$

when $p \equiv -1 \pmod 4$.

It remains only to show that the ratio of clearing factors is $1 \pmod{p\mathbb{Z}[i]}$ in both cases. When $p \equiv 1 \pmod 4$, $(p-1)/2$ is even, and this ratio is

$$\begin{aligned}(1-(-1)^a i)^{ps}/(1-(-1)^a i)^s &= (1-(-1)^a i)^{(p-1)s} = (-2(-1)^a i)^{s((p-1)/2)}\\ &= \left(2^{(p-1)/2}(-1)^{(p-1)/4}\right)^s \equiv 1 \pmod p,\end{aligned}$$

since $2^{(p-1)/2} \equiv (-1)^{(p^2-1)/8} \pmod p$. When $p \equiv -1 \pmod 4$, $(p+1)/2$ is even, and the ratio is

$$\begin{aligned}(1+(-1)^a i)^{ps}/(1-(-1)^a i)^s &= (1+(-1)^a i)^{ps}(1+(-1)^a i)^s/2^s\\ &= (1+(-1)^a i)^{(p+1)s}/2^s = (2(-1)^a i)^{s(p+1)/2}/2^s\\ &= \left(2^{(p-1)/2}(-1)^{(p+1)/4}\right)^s \equiv 1 \pmod p,\end{aligned}$$

again because $2^{(p-1)/2} \equiv (-1)^{(p^2-1)/8} \pmod p$. □

LEMMA 9.13. *For $q = 2^{2n+1}$ and $S := {}^2B_2(q)$, suppose that $s \in \mathrm{Aut}(S)$ has odd order and that $\langle s, S\rangle = \mathrm{Aut}(S)$. Then $|s|$ divides $5(2n+1)$.*

PROOF. Let $\mathcal{G} := \mathrm{Sp}_4(\overline{\mathbb{F}_2})$. It is well known that there is a Steinberg endomorphism $\sigma : \mathcal{G} \to \mathcal{G}$ such that σ^2 is the standard Frobenius map $(a_{ij}) \mapsto (a_{ij}^2)$ on $\mathcal{G}$, and then we can identify S with $\mathcal{G}^{\sigma^{2n+1}} := \{X \in \mathcal{G} \mid \sigma^{2n+1}(X) = X\}$, which is then σ-invariant. Letting σ also denote its action on S, we can write $\mathrm{Aut}(S) = \langle \sigma, S\rangle$. Without loss of generality, we may assume $Ss = S\sigma^{-1}$, and write $s = x\sigma^{-1}$. By the Lang-Steinberg theorem, there is some $a \in \mathcal{G}$ such that

$$x = a\sigma(a)^{-1}. \tag{9.13.1}$$

We also note that in $\mathrm{Aut}(S)$

$$\left(x\sigma^{-1}\right)^{2n+1} = x \cdot \sigma(x) \cdot \sigma^2(x) \cdot \ldots \cdot \sigma^{2n}(x). \tag{9.13.2}$$

As in the proof of [**13**, Theorem 2.16], define $t := a^{-1}s^{(2n+1)}a = \big(x\sigma^{-1}\big)^{2n+1}a$. Then using (9.13.1) and (9.13.2) we obtain

$$\begin{aligned}\sigma(t) &= \sigma(a)^{-1}\cdot\big(\sigma(x)\cdot\sigma^2(x)\cdot\ldots\cdot\sigma^{2n}(x)\cdot\sigma^{2n+1}(x)\big)\sigma(a)\\ &= \big(\sigma(a)^{-1}x^{-1}\big)\cdot\big(x\cdot\sigma(x)\cdot\ldots\cdot\sigma^{2n}(x)\big)\cdot\big(x\sigma(a)\big)\\ &= a^{-1}\big(x\sigma^{-1}\big)^{2n+1}a = t.\end{aligned}$$

Thus $t \in \mathcal{G}^\sigma \cong {}^2B_2(2)$, and note that $|{}^2B_2(2)| = 20$. Since s has odd order and t, s^{2n+1} are conjugate in $\mathcal{G}$, $|t|$ is odd, whence $|s^{2n+1}| = |t|$ divides 5. Thus $|s|$ divides $5(2n+1)$, as stated. □

THEOREM 9.14. *Suppose the Airy sheaf $\mathcal{F}_q$ for $q = 2^{2n+1}$ has finite geometric monodromy group $G_{\rm geom}$. Then the following statements hold:*

(i) *The arithmetic monodromy group over $\mathbb{F}_2$ of $\mathcal{F}_q$ is $G_{{\rm arith},\mathbb{F}_2} = C\times{\rm Aut}({}^2B_2(q))$, with $|C| \le 2$.*
(ii) *Moreover, $C = 1$ if $2n+1 \equiv \pm3 \pmod 8$, and $C \cong C_2$ if $2n+1 \equiv \pm1 \pmod 8$.*
(iii) *Suppose $2n+1 \equiv \pm3 \pmod 8$. For the arithmetic monodromy group $G_{{\rm arith},k}$ of $\mathcal{F}_q$ over a finite extension $k/\mathbb{F}_2$, we have $G_{{\rm arith},\mathbb{F}_2} = {\rm Aut}({}^2B_2(q))$, $G_{{\rm arith},k} = G_{\rm geom} = {}^2B_2(q)$ when $k \supseteq \mathbb{F}_q$, and $[G_{{\rm arith},k} : G_{\rm geom}] = \deg(\mathbb{F}_q/k)$ when $k \subseteq \mathbb{F}_q$.*
(iv) *Suppose $2n+1 \equiv \pm1 \pmod 8$.*
   (α) *For the arithmetic monodromy group $G_{{\rm arith},k}$ of $\mathcal{F}_q$ over a finite extension $k/\mathbb{F}_2$, we have $G_{{\rm arith},k} = G_{\rm geom} = {}^2B_2(q)$ when $k \supseteq \mathbb{F}_{q^2}$, and $[G_{{\rm arith},k} : G_{\rm geom}] = \deg(\mathbb{F}_{q^2}/k)$ when $k \subseteq \mathbb{F}_{q^2}$.*
   (β) *For the arithmetic monodromy group $G_{{\rm arith},\tilde{\mathcal{F}}_q,k}$ of the sheaf $\tilde{\mathcal{F}}_q := \mathcal{F}_q \otimes (-1)^{\deg/\mathbb{F}_2}$ over $k/\mathbb{F}_2$, we have $G_{{\rm arith},\tilde{\mathcal{F}}_q,\mathbb{F}_2} = {\rm Aut}({}^2B_2(q))$, $G_{{\rm arith},\tilde{\mathcal{F}}_q,k} = G_{{\rm geom},\tilde{\mathcal{F}}_q} = {}^2B_2(q)$ when $k \supseteq \mathbb{F}_q$, and $[G_{{\rm arith},\tilde{\mathcal{F}}_q,k} : G_{{\rm geom},\tilde{\mathcal{F}}_q}] = \deg(\mathbb{F}_q/k)$ when $k \subseteq \mathbb{F}_q$.*

PROOF. Part (i) follows from Propositions 9.9, 9.10, and Theorem 9.8(iii).

(ii) Let g denote the image of ${\rm Frob}_{1,\mathbb{F}_2}$ in $G_{{\rm arith},\mathbb{F}_2}$. As $\langle g, G_{\rm geom}\rangle = C\times{\rm Aut}(S)$ for $S := {}^2B_2(q)$, we can write $g = zs$ with $C = \langle z\rangle \le C_2$ and $s \in {\rm Aut}(S) \cong S \rtimes C_{2n+1}$. The central element z acts on $\mathcal{F}_q$ as the scalar ξ, where $\xi = 1$ if $|C| = 1$ and $\xi = -1$ if $|C| = 2$. The finiteness of $G_{\rm geom}$ implies that $G_{{\rm arith},\mathbb{F}_2}$ is finite, and so ${\rm Trace}(g^m)$ is a Gaussian integer for any $m \in \mathbb{Z}$. Now Proposition 9.12 implies that

$$(9.14.1)\qquad {\rm Trace}(s^{2n+1}) = \xi\,{\rm Trace}(g^{2n+1}) = -\xi,$$

and note that $t := s^{2n+1} \in S$.

Also note that since ${\rm Aut}(S) = \langle s, S\rangle$, s has order $2n+1$ modulo S, whence $|s| = e(2n+1)$ for some $e \in \mathbb{Z}_{\ge1}$, whence $|t| = e$. Inspecting the character table of S [**4**] and using (9.14.1), we see that $e = |t|$ is odd and greater than 1; in fact e divides $q - 2q_0 + 1$ if $\xi = 1$ and e divides $q + 2q_0 + 1$ if $\xi = -1$, where $q_0 := 2^n$. It follows that $|s| = e(2n+1)$ is odd. By Lemma 9.13, $e > 1$ divides 5, so $e = 5$; in particular, $|s| = 5(2n+1)$. An easy computation shows that $5|(q-2q_0+1)$ precisely when $n \equiv 1, 2 \pmod 4$ (equivalently, $2n+1 \equiv \pm3 \pmod 8$), and $5|(q+2q_0+1)$ precisely when $n \equiv 0, 3 \pmod 4$ (equivalently, $2n+1 \equiv \pm1 \pmod 8$). Hence the statement follows, and we have also proved that the image g of $\mathrm{Frob}_{1,\mathbb{F}_2}$ has order

$$(9.14.2)\qquad \begin{cases} 10(2n+1), & \text{if } 2n+1 \equiv \pm 1 \pmod 8, \\ 5(2n+1), & \text{if } 2n+1 \equiv \pm 3 \pmod 8. \end{cases}$$

Parts (iii) and (iv)(α) follow from (i), (ii), and the facts that $G_{\mathrm{geom}} = {}^2B_2(q)$ and $G_{\mathrm{arith},\mathbb{F}_2}/G_{\mathrm{geom}}$ is cyclic of order $|C|(2n+1)$.

To prove (iv)(β), note that when $k \supseteq \mathbb{F}_4$, the images of $\pi_1(\mathbb{A}^1/k)$ on $\mathcal{F}_q$ and $\tilde{\mathcal{F}}_q$ are the same. Hence $G_{\mathrm{arith},\tilde{\mathcal{F}}_q,k} = G_{\mathrm{arith},k} = {}^2B_2(q)$ whenever $k \supseteq \mathbb{F}_{q^2}$, whence $G_{\mathrm{geom},\tilde{\mathcal{F}}_q} = {}^2B_2(q) = S$. Now $G_{\mathrm{arith},\tilde{\mathcal{F}}_q,\mathbb{F}_2} = \langle \tilde{g}, S\rangle$, where $\tilde{g}$ is the image of $\mathrm{Frob}_{1,\mathbb{F}_2}$ on $\tilde{\mathcal{F}}_q$. By its definition, $\tilde{g} = -g$, where g is the image of $\mathrm{Frob}_{1,\mathbb{F}_2}$ on $\mathcal{F}_q$, and the proof of (ii) shows (recalling $2n+1 \equiv \pm 1 \pmod 8$) that $g = -s$ with $\mathrm{Aut}(S) = \langle s, S\rangle$. Hence $G_{\mathrm{arith},\tilde{\mathcal{F}}_q,\mathbb{F}_2} = \langle s, S\rangle = \mathrm{Aut}(S) = S \rtimes C_{2n+1}$, and the assertion follows. □

REMARK 9.15. Computations in MAGMA suggest that, for the Airy sheaf $\mathcal{F}_q$ with $q = 2^{2n+1}$, the Frobenii $\mathrm{Frob}_{a,\mathbb{F}_{2^j}}$ with $a \in \mathbb{F}_{2^j}$ and $\gcd(j, 2n+1) = 1$, all have traces of absolute value 1. If one knew that $\mathcal{F}_q$ has geometric monodromy group $G_{\mathrm{geom},\mathcal{F}_q} = {}^2B_2(q)$, then this "absolute value one" property agrees with Corollary 9.5. Also, for $n = 1, 2$, computations show that $\mathrm{Frob}_{1,\mathbb{F}_2}$ has order $5(2n+1)$, and this again agrees with (9.14.2).

On the other hand, the infinite case of the dichotomy, namely $G_{\mathrm{geom}} = \mathrm{SL}_D$ would imply by Deligne's equidistribution theorem [**29**, Theorem 9.7.13] that when j is large enough (compared to q), some (in fact most) Frobenii $\mathrm{Frob}_{a,\mathbb{F}_{2^j}}$ would have traces of absolute value $\neq 1$. This again gives some evidence in support of the geometric part of [**26**, Conjecture 2.2] asserting that $G_{\mathrm{geom},\mathcal{F}_q} = {}^2B_2(q)$. However, Theorem 9.14(iii) shows that the arithmetic part of [**26**, Conjecture 2.2] stating that $G_{\mathrm{arith},\mathcal{F}_q,\mathbb{F}_2} = \mathrm{Aut}({}^2B_2(q))$ is false when $q = 2^m$ with $m \equiv \pm 1 \pmod 8$; it should be corrected in that case by replacing $\mathcal{F}_q$ by $\tilde{\mathcal{F}}_q$ as in (iv)(β) of Theorem 9.14.

9C. Local systems with infinite monodromy groups. Theorem 9.8 allows us to prove the following criterion for infinite monodromy.

PROPOSITION 9.16. *Let $n \in \mathbb{Z}_{\geq 1}$, $q = 2^{2n+1}$, and consider the Airy sheaf $\mathcal{F}(q,f)$ defined in (4.1.3) with $q = 2^{2n+1}$, $f(x) \in \mathbb{F}_2[x]$, and $f(0) = 0$. Suppose that*

(i) *$2n+1$ is coprime to $|{}^2B_2(q)|$, and*
(ii) *there is some odd integer m coprime to $2n+1$ such that* $|\mathrm{Trace}(\mathrm{Frob}_{0,\mathbb{F}_{2^m}}|\mathcal{F}(q,f))| \neq 1$.

Then $\mathcal{F}(q,f)$ has infinite geometric monodromy group.

PROOF. Assume the contrary: $\mathcal{F}(q,f)$ has finite geometric monodromy group S. By Theorem 9.8, $S = {}^2B_2(q)$ and the arithmetic monodromy group of $\mathcal{F}(q,f)$ over $\mathbb{F}_2$ is $G = \mathbf{Z}(G) \times \mathrm{Aut}(S)$, where $\mathbf{Z}(G) \leq C_4$. In particular, $|G/S|$ divides $4(2n+1)$, and so S equals the arithmetic monodromy group of $\mathcal{F}(q,f)$ over $\mathbb{F}_{q^4}$. Moreover, $G = \langle g, S\rangle$ for the image g of $\mathrm{Frob}_{0,\mathbb{F}_2}$ in G, and we can write $g = zs$ with $z \in \mathbf{Z}(G)$, $s \in \mathrm{Aut}(S)$, and $\mathrm{Aut}(S) = \langle s, S\rangle$. Applying Corollary 9.5, we see that $|\mathrm{Trace}(g)| = |\mathrm{Trace}(s)| = 1$, a contradiction. □

PROPOSITION 9.17. *Let $n \in \mathbb{Z}_{\geq 2}$, $q := 2^{2n+1}$, $r := [(n-1)/2]$, and define*

$$f_1(x) := \sum_{i=0}^{r} x^{1+2^{n-2i}}.$$

Consider the sheaf $\mathcal{F}(q, f)$ with $f(x) := f_1(x^{t(q)})$ as in (4.1.3). Then for $m := 2[n/2]+1$, we have $|\operatorname{Trace}(\operatorname{Frob}_{0,\mathbb{F}_{2^m}}|\mathcal{F}(q,f))| \neq 1$.

PROOF. As in the proof of Proposition 9.9, the starting point is that $\gcd(t(q), 2^m - 1) = 1$. To see this, note that $t(q) = q + 1 - 2q_0$ divides $q^4 - 1$, and $\gcd(q^4 - 1, 2^m - 1) = 2^{\gcd(4(2n+1),m)} - 1 = 1$. Thus for the field $k := \mathbb{F}_{2^m}$, the map $x \mapsto x^{t(q)}$ is bijective on k.

The sheaf $\mathcal{F}(q, f)$ was built out of the Witt vector $\left[x^{t(q)}, f_1(x^{t(q)})\right]$. Let us denote by $\mathcal{H}(q, f_1)$ the sheaf built by the same recipe, with same clearing factor, out of the Witt vector $[x, f_1(x)]$. Then we have

$$\operatorname{Trace}(\operatorname{Frob}_{0,k}|\mathcal{F}(q,f)) = \operatorname{Trace}(\operatorname{Frob}_{0,k}|\mathcal{H}(q,f_1)),$$

precisely because the map $x \mapsto x^{t(q)}$ is bijective on k.

Next, as in (9.9.2), we write the input vector $[x, f_1(x)]$ as $[x, xR(x)]$, with

$$R(x) := \sum_{i=0}^{r} x^{2^{n-2i}} = F^n(x) + F^{n-2}(x) + \ldots + F^{n-2r}(x),$$

and F denotes the absolute Frobenius. Now we can repeat the arguments in the proof of Proposition 9.9, and compute the form $\operatorname{Trace}_{k/\mathbb{F}_2}(\langle x, y\rangle)$ on $k \times k$ with $\langle x, y\rangle := xy + xR(y) + yR(x)$ as in (9.9.4). For any $x, y \in k$ we have $F^m(x) = x$, $F^m(y) = y$. If $2|n$, then $n = 2r+2$, $m = 2r+3$, and

$$\langle x, y\rangle = xy + y\Big(\sum_{i=1}^{r+1} F^{2i}(x)\Big) + \sum_{i=1}^{r+1} xF^{2i}(y)$$

has the same trace over $\mathbb{F}_2$ as

$$\begin{aligned} &xy + y\Big(\sum_{i=1}^{r+1} F^{2i}(x)\Big) + \sum_{i=1}^{r+1} F^{2r+3-2i}(xF^{2i}(y)) \\ &= xy + y\Big(\sum_{i=1}^{r+1} F^{2i}(x)\Big) + \sum_{i=1}^{r+1} yF^{2r+3-2i}(x) \\ &= y\Big(\sum_{j=0}^{2r+2} F^j(x)\Big) = y \operatorname{Trace}_{k/\mathbb{F}_2}(x). \end{aligned}$$

If $2 \nmid n$, then $n = m = 2r+1$, then

$$\begin{aligned} \langle x, y\rangle &= xy + y\Big(\sum_{i=0}^{r} F^{2i+1}(x)\Big) + \sum_{i=0}^{r} xF^{2i+1}(y) \\ &= 3xy + y\Big(\sum_{i=0}^{r-1} F^{2i+1}(x)\Big) + \sum_{i=0}^{r-1} xF^{2i+1}(y) \end{aligned}$$

has the same trace over $\mathbb{F}_2$ as

$$xy + y\Big(\sum_{i=0}^{r-1} F^{2i+1}(x)\Big) + \sum_{i=0}^{r-1} F^{2r-2i}\big(xF^{2i+1}(y)\big)$$
$$= xy + y\Big(\sum_{i=0}^{r-1} F^{2i+1}(x)\Big) + \sum_{i=0}^{r-1} yF^{2r-2i}(x) = y\Big(\sum_{j=0}^{2r} F^{j}(x)\Big) = y\,\mathrm{Trace}_{k/\mathbb{F}_2}(x).$$

So in both cases, the symmetric bilinear form $\mathrm{Trace}_{k/\mathbb{F}_2}(\langle x,y\rangle)$ on $k\times k$ has kernel consisting of the elements $x\in k$ with $\mathrm{Trace}_{k/\mathbb{F}_2}(x)=0$, that is, of exactly 2^{m-1} elements. Then the proof of Proposition 9.9 shows that $|\,\mathrm{Trace}\big(\mathrm{Frob}_{0,\mathbb{F}_{2^m}}|\mathcal{F}(q,f)\big)|$ is either 0 or $2^{(m-1)/2}>1$, and hence can never be equal to 1 (since by hypothesis $n\geq 2$, and hence $m\geq 3$). □

THEOREM 9.18. *Let $n\in\mathbb{Z}_{\geq 2}$, $q:=2^{2n+1}$, and consider the sheaf $\mathcal{F}(q,f)$ of rank $D:=2^n(2^{2n+1}-1)$, with $f(x)=f_1(x^{t(q)})$ and $f_1(x)$ as defined in Proposition 9.17. Assume in addition that $2n+1$ is coprime to $|^2B_2(q)|$; for instance, take $2n+1=\ell^a$ for any odd prime $\ell\neq 5$ and any $a\in\mathbb{Z}_{\geq 1}$. Then the geometric monodromy group of $\mathcal{F}(q,f)$ is SL_D.*

PROOF. We first apply Proposition 9.9, and note that m is coprime to $2(2n+1)$. It then follows from Proposition 9.16 that $\mathcal{F}(q,f)$ has infinite geometric monodromy group G_{geom}. By Theorem 8.4, $G_{\mathrm{geom}}=\mathrm{SL}_D$.

Suppose that $2n+1=\ell^a$ for a prime ℓ, but ℓ divides $|^2B_2(q)|$. Then ℓ divides $q^4-1=2^{4\ell^a}-1$. Since $\ell|(2^{\ell-1}-1)$, ℓ divides $\gcd(2^{4\ell^a}-1,2^{\ell-1}-1)=2^{\gcd(4\ell^a,\ell-1)}-1$, and so ℓ divides $2^4-1=15$. But $3\nmid|^2B_2(q)|$, so $r=5$. □

More generally, to ensure that $2n+1$ is coprime to $|^2B_2(q)|$ for $q=2^{2n+1}$, we can take any n such that $2n+1=p_1^{a_1}p_2^{a_2}\dots p_t^{a_t}$, where $p_1<p_2<\dots<p_t$ are primes, $p_i\neq 5$, $a_i\in\mathbb{Z}_{\geq 1}$, and $p_i\nmid(p_j-1)$ whenever $i<j$. Indeed, suppose p_j divides $|^2B_2(q)|$ for some j. Then p_j divides both $2^{p_j-1}-1$ and $q^4-1=2^{4(2n+1)}-1$. Since $\gcd(p_j-1,4(2n+1))=\gcd(p_j-1,4\prod_{i=1}^t p_i^{a_i})$ divides 4, it follows that p_j divides $2^4-1=15$. But $p_j\neq 5$ by assumption, and $p_j\neq 3$ since $3\nmid|^2B_2(q)|$, a contradiction.

Added in Proof

After the proofs of the paper were produced and proofread, we were informed that a stronger version of Theorem 7.5 was obtained in [**36**].

References

[1] S. Abhyankar, *Coverings of algebraic curves*, Amer. J. Math. **79** (1957), 825–856, DOI 10.2307/2372438. MR94354

[2] A. Borel, *Linear algebraic groups*, 2nd ed., Graduate Texts in Mathematics, vol. 126, Springer-Verlag, New York, 1991, DOI 10.1007/978-1-4612-0941-6. MR1102012

[3] W. Bosma, J. Cannon, and C. Playoust, *The Magma algebra system. I. The user language*, J. Symbolic Comput. **24** (1997), no. 3-4, 235–265, DOI 10.1006/jsco.1996.0125. Computational algebra and number theory (London, 1993). MR1484478

[4] R. Burkhardt, *Die Zerlegungsmatrizen der Gruppen* PSL$(2,p^f)$ (German), J. Algebra **40** (1976), no. 1, 75–96, DOI 10.1016/0021-8693(76)90088-0. MR480710

[5] R. W. Carter, *Finite groups of Lie type: Conjugacy classes and complex characters*, Pure and Applied Mathematics (New York), John Wiley & Sons, Inc., New York, 1985. MR794307

[6] J. H. Conway, R. T. Curtis, S. P. Norton, R. A. Parker, and R. A. Wilson, *ATLAS of finite groups: Maximal subgroups and ordinary characters for simple groups*, Oxford University Press, Eynsham, 1985. MR827219

[7] C. W. Curtis and I. Reiner, *Representation theory of finite groups and associative algebras*, Pure and Applied Mathematics, Vol. XI, Interscience Publishers (a division of John Wiley & Sons, Inc.), New York-London, 1962. MR144979

[8] C. W. Curtis and I. Reiner, Methods of representation theory—with applications to finite groups and orders, vol. I, Wiley & Sons, New York et al, 1981.

[9] E. C. Dade, *A new approach to Glauberman's correspondence*, J. Algebra **270** (2003), no. 2, 583–628, DOI 10.1016/j.jalgebra.2002.11.002. MR2019631

[10] P. Deligne, La conjecture de Weil II, *Publ. Math. I.H.E.S.* **52** (1981), 313–428.

[11] P. Deligne, *Les constantes des équations fonctionnelles des fonctions L* (French), Modular functions of one variable, II (Proc. Internat. Summer School, Univ. Antwerp, Antwerp, 1972), Lecture Notes in Math., Vol. 349, Springer, Berlin-New York, 1973, pp. 501–597. MR349635

[12] B. H. Gross, *Rigid local systems on $\mathbb{G}_m$ with finite monodromy*, Adv. Math. **224** (2010), no. 6, 2531–2543, DOI 10.1016/j.aim.2010.02.008. MR2652215

[13] S. Guest, J. Morris, C. E. Praeger, and P. Spiga, *On the maximum orders of elements of finite almost simple groups and primitive permutation groups*, Trans. Amer. Math. Soc. **367** (2015), no. 11, 7665–7694, DOI 10.1090/S0002-9947-2015-06293-X. MR3391897

[14] R. M. Guralnick, M. Larsen, and P. H. Tiep, *Representation growth in positive characteristic and conjugacy classes of maximal subgroups*, Duke Math. J. **161** (2012), no. 1, 107–137, DOI 10.1215/00127094-1507300. MR2872555

[15] R. M. Guralnick and P. H. Tiep, *The non-coprime $k(GV)$ problem*, J. Algebra **293** (2005), no. 1, 185–242, DOI 10.1016/j.jalgebra.2005.02.001. MR2173972

[16] G. Hiss and G. Malle, *Low-dimensional representations of quasi-simple groups*, LMS J. Comput. Math. **4** (2001), 22–63, DOI 10.1112/S1461157000000796. MR1835851

[17] J. E. Humphreys, *Introduction to Lie algebras and representation theory*, Graduate Texts in Mathematics, Vol. 9, Springer-Verlag, New York-Berlin, 1972. MR323842

[18] I. M. Isaacs, *Characters of solvable and symplectic groups*, Amer. J. Math. **95** (1973), 594–635, DOI 10.2307/2373731. MR332945

[19] I. M. Isaacs, Glauberman correspondence Dade-method, preprint, July 2002.

[20] I. M. Isaacs, *Character theory of finite groups*, AMS Chelsea Publishing, Providence, RI, 2006. Corrected reprint of the 1976 original [Academic Press, New York; MR0460423], DOI 10.1090/chel/359. MR2270898

[21] J. C. Jantzen, *Representations of Chevalley groups in their own characteristic*, The Arcata Conference on Representations of Finite Groups (Arcata, Calif., 1986), Proc. Sympos. Pure Math., vol. 47, Amer. Math. Soc., Providence, RI, 1987, pp. 127–146, DOI 10.1090/pspum/047.1/933356. MR933356

[22] W. M. Kantor and Á. Seress, *Large element orders and the characteristic of Lie-type simple groups*, J. Algebra **322** (2009), no. 3, 802–832, DOI 10.1016/j.jalgebra.2009.05.004. MR2531224

[23] N. M. Katz, *Gauss sums, Kloosterman sums, and monodromy groups*, Annals of Mathematics Studies, vol. 116, Princeton University Press, Princeton, NJ, 1988, DOI 10.1515/9781400882120. MR955052

[24] N. M. Katz, *Exponential sums and differential equations*, Annals of Mathematics Studies, vol. 124, Princeton University Press, Princeton, NJ, 1990, DOI 10.1515/9781400882434. MR1081536

[25] N. M. Katz, *From Clausen to Carlitz: low-dimensional spin groups and identities among character sums* (English, with English and Russian summaries), Mosc. Math. J. **9** (2009), no. 1, 57–89, back matter, DOI 10.17323/1609-4514-2009-9-1-57-89. MR2567397

[26] N. M. Katz, *Exponential sums, Ree groups and Suzuki groups: conjectures*, Exp. Math. **28** (2019), no. 1, 49–56, DOI 10.1080/10586458.2017.1334246. MR3938577

[27] N. M. Katz, A. Rojas-León, and P. H. Tiep, *A rigid local system with monodromy group the big Conway group* $2.\mathsf{Co}_1$ *and two others with monodromy group the Suzuki group* $6.\mathsf{Suz}$, Trans. Amer. Math. Soc. **373** (2020), no. 3, 2007–2044, DOI 10.1090/tran/7967. MR4068288

[28] N. M. Katz, A. Rojas-León, and P. H. Tiep, Rigid local systems and sporadic simple groups, *Mem. Amer. Math. Soc.* (to appear).

[29] N. M. Katz and P. Sarnak, *Random matrices, Frobenius eigenvalues, and monodromy*, American Mathematical Society Colloquium Publications, vol. 45, American Mathematical Society, Providence, RI, 1999, DOI 10.1090/coll/045. MR1659828

[30] N. M. Katz and P. H. Tiep, *Monodromy groups of Kloosterman and hypergeometric sheaves*, Geom. Funct. Anal. **31** (2021), no. 3, 562–662, DOI 10.1007/s00039-021-00578-0. MR4311580

[31] N. M. Katz and P. H. Tiep, Airy sheaves and generalized Airy sheaves, (in preparation).

[32] P. Kleidman and M. Liebeck, *The subgroup structure of the finite classical groups*, London Mathematical Society Lecture Note Series, vol. 129, Cambridge University Press, Cambridge, 1990, DOI 10.1017/CBO9780511629235. MR1057341

[33] F. Lübeck, *Small degree representations of finite Chevalley groups in defining characteristic*, LMS J. Comput. Math. **4** (2001), 135–169, DOI 10.1112/S1461157000000838. MR1901354

[34] H. Lüneburg, Ein einfacher Beweis für den Satz von Zsigmondy über primitive Primteiler von $A^n - 1$, in: Geometries and Groups, (edited by M. Aigner and D. Jungnickel), *Lect. Notes in Math.* **983**, 219–222, Springer Verlag, New York, 1981.

[35] O. Manz and T. R. Wolf, *Representations of solvable groups*, London Mathematical Society Lecture Note Series, vol. 185, Cambridge University Press, Cambridge, 1993, DOI 10.1017/CBO9780511525971. MR1261638

[36] A. Martínez, *Low-dimensional irreducible rational representations of classical algebraic groups*, arXiv:1811.07019.v1.

[37] H. L. Montgomery, I. Niven, and H. S. Zuckerman, An Introduction to Number Theory, fifth ed., Wiley & Sons, 1991.

[38] G. Navarro, *Fields, values and character extensions in finite groups*, J. Group Theory **10** (2007), no. 3, 279–285, DOI 10.1515/JGT.2007.022. MR2320967

[39] G. Navarro, *Restriction of characters to Sylow normalizers*, Glasg. Math. J. **43** (2001), no. 2, 311–315, DOI 10.1017/S0017089501020146. MR1838634

[40] A. L. Onishchik and È. B. Vinberg, *Lie groups and algebraic groups*, Springer Series in Soviet Mathematics, Springer-Verlag, Berlin, 1990. Translated from the Russian and with a preface by D. A. Leites, DOI 10.1007/978-3-642-74334-4. MR1064110

[41] A. A. Premet, *Weights of infinitesimally irreducible representations of Chevalley groups over a field of prime characteristic* (Russian), Mat. Sb. (N.S.) **133(175)** (1987), no. 2, 167–183, 271, DOI 10.1070/SM1988v061n01ABEH003200; English transl., Math. USSR-Sb. **61** (1988), no. 1, 167–183. MR905003

[42] M. Roitman, *On Zsigmondy primes*, Proc. Amer. Math. Soc. **125** (1997), no. 7, 1913–1919, DOI 10.1090/S0002-9939-97-03981-6. MR1402885

[43] J.-P. Serre and J. Tate, *Good reduction of abelian varieties*, Ann. of Math. (2) **88** (1968), 492–517, DOI 10.2307/1970722. MR236190

[44] O. Šuch, *Monodromy of Airy and Kloosterman sheaves*, Duke Math. J. **103** (2000), no. 3, 397–444, DOI 10.1215/S0012-7094-00-10332-8. MR1763654

[45] M. Suzuki, On a class of doubly transitive groups, *Annals of Math.* **75** (1962), 105–145.

[46] Phạm Hữu Tiệp, *Globally irreducible representations of the finite symplectic group* $\mathrm{Sp}_4(q)$, Comm. Algebra **22** (1994), no. 15, 6439–6457, DOI 10.1080/00927879408825199. MR1303014

[47] P. H. Tiep and A. E. Zalesskii, *Minimal characters of the finite classical groups*, Comm. Algebra **24** (1996), no. 6, 2093–2167, DOI 10.1080/00927879608825690. MR1386030

[48] J. Tits, *Algebraic and abstract simple groups*, Ann. of Math. (2) **80** (1964), 313–329, DOI 10.2307/1970394. MR164968

[49] G. van der Geer and M. van der Vlugt, *Reed-Muller codes and supersingular curves. I*, Compositio Math. **84** (1992), no. 3, 333–367. MR1189892

[50] K. Zsigmondy, *Zur Theorie der Potenzreste* (German), Monatsh. Math. Phys. **3** (1892), no. 1, 265–284, DOI 10.1007/BF01692444. MR1546236

Department of Mathematics, Harvard University, Cambridge, Massachusetts 02138
Email address: alpoge@math.harvard.edu

Department of Mathematics, Princeton University, Princeton, New Jersey 08544
Email address: nmk@math.princeton.edu

Departament de Matemàtiques, Universitat de València, 46100 Burjassot, València, Spain
Email address: gabriel@uv.es

University of Auckland, Auckland, New Zealand
Email address: e.obrien@auckland.ac.nz

Department of Mathematics, Rutgers University, Piscataway, New Jersey 08854
Email address: tiep@math.rutgers.edu

Contemporary Mathematics
Volume **800**, 2024
https://doi.org/10.1090/conm/800/16051

On the generation of simple groups by Sylow subgroups

Timothy C. Burness and Robert M. Guralnick

ABSTRACT. Let G be a finite simple group of Lie type and let P be a Sylow 2-subgroup of G. In this paper, we prove that for any nontrivial element $x \in G$, there exists $g \in G$ such that $G = \langle P, x^g \rangle$. By combining this result with recent work of Breuer and Guralnick, we deduce that if G is a finite nonabelian simple group and r is any prime divisor of $|G|$, then G is generated by a Sylow 2-subgroup and a Sylow r-subgroup.

1. Introduction

Let G be a finite group, let π be a set of primes and let π' be the complementary set of primes. In [**6**], Breuer and Guralnick prove that G can be generated by a π-subgroup and a π'-subgroup of G (moreover, each generating subgroup H can be chosen to be intravariant, which means that H and $\varphi(H)$ are conjugate in G for all automorphisms φ of G). By taking $\pi = \{2\}$, it follows that G can be generated by a Sylow 2-subgroup and an intravariant (solvable) subgroup of odd order.

A key ingredient in the proof of this theorem is a stronger result for finite simple groups (see [**6**, Theorem 1.8]), which relies on the classification of finite simple groups. This asserts that if G is simple then there exists a prime r (depending on G) such that G can be generated by a Sylow r-subgroup and a Sylow s-subgroup for any prime divisor s of $|G|$. This builds on earlier work of Aschbacher and Guralnick [**1**], who proved that every simple group G can be generated by a pair of Sylow p-subgroups for some prime p. In turn, this was extended in [**16**], where Guralnick shows that G is generated by a Sylow 2-subgroup and an involution (in particular, G is generated by a pair of Sylow 2-subgroups).

In fact, Breuer and Guralnick conjecture that if G is simple and r, s are *any* prime divisors of $|G|$ (not necessarily distinct) then G can be generated by a Sylow r-subgroup and a Sylow s-subgroup. A proof for all alternating and sporadic simple groups is given in [**6**]. In addition, [**11**, Corollary 19] implies that if r and s are fixed primes then the conclusion of the conjecture holds for all but finitely many finite simple groups. The full version of the conjecture remains open.

2020 *Mathematics Subject Classification.* Primary 20E32, 20D20, 20P05.

Key words and phrases. Simple groups, Sylow subgroups, generation.

Both authors thank the Department of Mathematics at the California Institute of Technology for their generous hospitality during a research visit in spring 2022. They also thank an anonymous referee for their careful reading of the paper and for several helpful comments and suggestions. Guralnick was partially supported by the NSF grant DMS-1901595 and a Simons Foundation Fellowship 609771.

Our main result is the following.

THEOREM 1.1. *Let G be a finite simple group of Lie type over $\mathbb{F}_q$ and let P be a Sylow 2-subgroup of G. Then for all nontrivial $x \in G$, there exists $g \in G$ such that $G = \langle P, x^g \rangle$.*

By combining this with results in [**6**] for alternating and sporadic groups, we obtain the following corollary, which establishes a special case of the Breuer-Guralnick conjecture. It also shows that the prime in [**6**, Theorem 1.8], which a priori depends on the choice of the simple group, can in fact be chosen uniformly as 2.

COROLLARY 1.2. *Let G be a finite nonabelian simple group and let r be a prime divisor of $|G|$. Then G is generated by a Sylow 2-subgroup and a Sylow r-subgroup of G.*

REMARK 1.3. Let us record some comments on the statement of Theorem 1.1.

(a) By [**6**, Proposition 2.2], the conclusion to Theorem 1.1 holds if G is a sporadic simple group. In fact, in this setting Breuer and Guralnick have used a computational approach to verify that the same conclusion holds when P is a Sylow p-subgroup of G and p is an arbitrary prime divisor of $|G|$ (see [**5**, Section 8.19]).

(b) Let G be a finite simple group of Lie type in characteristic p. If P is a Sylow p-subgroup and $x \in G$ is nontrivial, then [**6**, Proposition 2.3] states that $G = \langle P, x^g \rangle$ for some $g \in G$. So in order to prove Theorem 1.1 we may assume q is odd.

(c) As noted in [**6**], Theorem 1.1 is false for alternating groups. For example, let $G = A_n$ and let $\mathrm{orb}(P)$ be the number of P-orbits on $[n] = \{1, \ldots, n\}$ of a Sylow p-subgroup P of G. Clearly, if $x \in G$ and $\mathrm{orb}(x) > n - \mathrm{orb}(P)$, where $\mathrm{orb}(x)$ is the number of orbits of $\langle x \rangle$ on $[n]$, then $\langle P, x^g \rangle$ is intransitive for all $g \in G$. For example, if $G = A_{15}$, P is a Sylow 2-subgroup and x is a 3-cycle, then $\mathrm{orb}(x) = 13$ and $\mathrm{orb}(P) = 4$, so $G \neq \langle P, x^g \rangle$ for all $g \in G$.

(d) It is also worth noting that the natural extension of Theorem 1.1 to odd primes is false in general. For example, let P be a Sylow r-subgroup of $G = \mathrm{L}_n(2) = \mathrm{SL}(V)$, where $n = 2k - 1$ and $r = 2^k - 1$ is a prime. Then $P = \langle y \rangle$ has order r and $\dim C_V(y) = k - 1$, so $G \neq \langle P, x \rangle$ for any element x with $\dim C_V(x) \geqslant k + 1$.

Our proof of Theorem 1.1 relies heavily on a probabilistic approach based on fixed point ratio estimates. Computational methods also play a key role and we will also use character theory and some elementary linear algebra to handle certain special cases that arise. In order to explain the general set-up, let G be a finite group, let $P \neq G$ be a Sylow 2-subgroup of G and let $\mathcal{M}$ be the set of maximal subgroups of G containing P. Fix a nontrivial element $x \in G$ and let $\mathcal{Q}(x)$ be the probability that $G \neq \langle P, x^g \rangle$, where x^g is a randomly chosen conjugate of x. Then $\mathcal{Q}(x) \leqslant \Sigma(x)$, where

$$\Sigma(x) = \sum_{H \in \mathcal{M}} \mathrm{fpr}(x, G/H) \tag{1}$$

and

$$\mathrm{fpr}(x, G/H) = \frac{|x^G \cap H|}{|x^G|}$$

is the fixed point ratio of x with respect to the action of G on G/H. In particular, if $\Sigma(x) < 1$ then $x^G \neq \bigcup_{H \in \mathcal{M}}(x^G \cap H)$ and so there exists a conjugate $y \in x^G$ that is not contained in any maximal overgroup of P. Therefore, $G = \langle P, y \rangle$ and so it suffices to show that $\Sigma(x) < 1$ for all nontrivial $x \in G$.

Since $\mathrm{fpr}(x, G/H) \leqslant \mathrm{fpr}(x^m, G/H)$ for all $m \in \mathbb{Z}$, we only need to verify the bound $\Sigma(x) < 1$ for elements of prime order. Therefore, in view of Remark 1.3(b), the following result reduces the proof of Theorem 1.1 to the groups in part (ii), which can be handled using a different approach (see Theorem 3.24).

THEOREM 1.4. *Let G be a finite simple group of Lie type in odd characteristic and let P be a Sylow 2-subgroup of G. Then either*

(i) $\Sigma(x) < 1$ *for all $x \in G$ of prime order; or*
(ii) $G = \mathrm{L}_n(3)$ *and $n \geqslant 7$ is odd.*

REMARK 1.5. We expect that the bound in (i) also holds for the groups in (ii), but this would require sharper fixed point ratio estimates for subspace actions and it is easier to verify Theorem 1.1 directly in this special case. For instance, if $G = \mathrm{L}_n(3)$ and $n \geqslant 7$ is a Mersenne prime then $\mathcal{M} = \{P_m \,:\, 1 \leqslant m < n\}$, where P_m is the stabilizer of an m-dimensional subspace of the natural module for G. By [**17**, Proposition 3.1], we have $\mathrm{fpr}(x, G/P_m) < 2 \cdot 3^{-m}$ if $m < n/2$ and thus

$$\Sigma(x) < 4 \sum_{m=1}^{(n-1)/2} 3^{-m},$$

which does not yield $\Sigma(x) < 1$. If $n = 7$ and $x = (-I_6, I_1)$, then with the aid of MAGMA we compute $\Sigma(x) = 1086/1093$.

In order to prove Theorem 1.4, we first need to determine the subgroups in $\mathcal{M}$ and then we need to apply upper bounds on fixed point ratios for primitive actions of simple groups of Lie type. Here $\mathcal{M}$ coincides with the set of odd-index maximal subgroups of G and the various possibilities have been determined by Liebeck and Saxl [**26**] (and independently by Kantor [**20**]). We will also appeal to more recent work of Maslova [**27**], which gives a precise description of the odd-index maximal subgroups of simple classical groups in odd characteristic. Given a certain type of subgroup H in $\mathcal{M}$ (for example, the stabilizer of an orthogonal decomposition of the natural module if G is classical), we need to know the number of conjugacy classes of such subgroups in G and we also need to estimate the number of distinct conjugates of H that contain our given Sylow 2-subgroup P. In each case, the number of conjugacy classes is readily available in the literature (for example, this can be read off from [**4, 21**] when G is classical). And if $H \in \mathcal{M}$ then it is straightforward to show that P is contained in exactly $|N_G(P) : N_H(P)|$ distinct conjugates of H (see Lemma 2.2). In particular, P is contained in at most $|N_G(P) : P|$ conjugates of H, and this allows us to apply work of Kondrat'ev and Mazurov [**22, 23**] on the normalizers of Sylow 2-subgroups in simple groups.

There is an extensive literature on fixed point ratios for groups of Lie type. This includes general bounds, such as the main results in [**13, 25**], as well as more specialized results in [**7–10, 17**] for classical groups and [**24**] for exceptional groups. It will also be convenient to use MAGMA [**3**] to handle certain low rank groups defined over small fields, which allows us to compute $\Sigma(x)$ precisely for all $x \in G$ of prime order.

REMARK 1.6. It is worth noting that the proof of [**6**, Theorem 1.8] for classical groups is more constructive than our proof of Theorem 1.4. In this paper, with some additional work, one could bound the probability $\mathcal{Q}(x)$ away from 0 for all nontrivial $x \in G$, but we do not pursue this here.

2. Exceptional groups

In this section we prove Theorem 1.4 for exceptional groups of Lie type, while the classical groups will be handled in Section 3. All logarithms in this paper are in base 2 and we adopt the standard notation for simple groups used in [**21**].

Let G be a finite simple group of exceptional Lie type over $\mathbb{F}_q$, where q is odd. Let P be a Sylow 2-subgroup of G. Let $\mathcal{M}$ be the set of maximal subgroups of G containing P and note that each $H \in \mathcal{M}$ has odd index in G, so the possibilities for H are described by Liebeck and Saxl in [**26**]. We will also need the following theorem of Kondrat'ev and Mazurov (see [**23**, Theorem 6]). Given a positive integer n, let $n_{2'}$ be the largest odd divisor of n.

THEOREM 2.1. *Let G be a finite simple group of exceptional Lie type over $\mathbb{F}_q$ with q odd and let P be a Sylow 2-subgroup of G. Then either $N_G(P) = P$, or one of the following holds:*

(i) *$G = E_6^{\epsilon}(q)$ and $N_G(P) = P \times C_m$, where $m = (q-\epsilon)_{2'}/(3, q-\epsilon)$.*
(ii) *$G = {}^2G_2(q)'$ and $N_G(P) = P{:}L$, where $L = C_7$ if $q = 3$ and $L = C_7{:}C_3$ if $q > 3$.*

Given a subgroup $H \in \mathcal{M}$, let $n(H,P)$ be the number of distinct conjugates of H containing P. The following elementary observation will be useful.

LEMMA 2.2. *Let G be a finite group, let P be a Sylow p-subgroup of G and let H be a subgroup of G containing P. Then $n(H,P) = |N_G(P) : N_H(P)|$.*

PROOF. Suppose $P \leqslant H^g$ for some $g \in G$. Then P and $P^{g^{-1}}$ are Sylow p-subgroups of H, so Sylow's theorem implies that $P^{g^{-1}} = P^h$ for some $h \in H$ and thus $g \in Hx$ for some $x \in N_G(P)$. We deduce that $n(H,P)$ is the number of distinct cosets of the form Hx with $x \in N_G(P)$, whence

$$n(H,P) = |HN_G(P) : H| = |N_G(P) : N_H(P)|,$$

as required. □

THEOREM 2.3. *The conclusion to Theorem 1.4 holds if G is an exceptional group.*

PROOF. Let G be a finite simple exceptional group of Lie type over $\mathbb{F}_q$, where $q = p^f$ and p is an odd prime. Fix a Sylow 2-subgroup P of G and define $\mathcal{M}$ and $n(H,P)$ as above, noting that $n(H,P) \leqslant |N_G(P) : P|$ by Lemma 2.2. Let $x \in G$ be an element of prime order and define $\Sigma(x)$ as in (1). Our aim is to verify the bound $\Sigma(x) < 1$.

First assume $G = E_8(q)$ and let $H \in \mathcal{M}$. By inspecting the main theorem of [**26**], we deduce that one of the following holds (up to conjugacy in G):

(a) $H = E_8(q_0)$ is a subfield subgroup, where $q = q_0^e$ and e is an odd prime.
(b) H is a maximal rank subgroup of type $D_8(q)$, $D_4(q)^2$, $A_1(q)^8$ or $(C_{q-\epsilon})^8$ with $\epsilon = \pm$.

By Theorem 2.1, we have $n(H,P) = 1$ and [**24**, Theorem 1] gives $\mathrm{fpr}(x, G/H) \leqslant q^{-8}(q^4-1)^{-1}$. Since there are at most $\log\log q$ odd prime divisors of f, it follows that

$$\Sigma(x) < (5 + \log\log q)q^{-8}(q^4-1)^{-1} < 1$$

and the result follows.

Next assume $G = E_6^{\epsilon}(q)$ and set $\widetilde{G} = \mathrm{Inndiag}(G)$ and $d = (3, q-\epsilon)$, so $\widetilde{G} = G.d$ is the subgroup of $\mathrm{Aut}(G)$ generated by the inner and diagonal automorphisms. By [**26**], each $H \in \mathcal{M}$ is either a subfield subgroup of type $E_6^{\epsilon}(q_0)$, where $q = q_0^e$ and $e \geqslant 3$ is an odd prime divisor of f (for each e there is a unique $\widetilde{G}$-class of such subgroups, so at most d distinct G-classes) or H is $\widetilde{G}$-conjugate to one of at most 4 maximal rank subgroups. Now Theorem 2.1 gives $n(H,P) \leqslant (q-\epsilon)/2d$ and [**24**, Theorem 1] implies that $\mathrm{fpr}(x, G/H) \leqslant (q^4-q^2+1)^{-1}$ for all nontrivial $x \in G$. Putting this together, we deduce that

$$\Sigma(x) < \frac{1}{2}(q+1)(4+\log\log q)(q^4-q^2+1)^{-1} < 1$$

for all $q \geqslant 3$.

The remaining cases are entirely similar and we omit the details since no special difficulties arise. Note that if $G = {}^2G_2(3)'$ then $\mathcal{M} = \{H\}$, where $H = N_G(P) = 2^3{:}7$, and we compute $\mathrm{fpr}(x, G/H) \leqslant 2/9$ for all nontrivial $x \in G$. □

3. Classical groups

Here we complete the proof of Theorem 1.1 by handling the classical groups. We adopt the same notation as in the previous section and we begin by recalling the following theorem of Kondrat'ev [**22**]. Given a positive integer n, let $t(n)$ be the number of nonzero digits in the binary expansion of n. In part (ii), note that $\mathrm{PSp}_2(q) = \mathrm{L}_2(q)$.

THEOREM 3.1. *Let G be a finite simple classical group over $\mathbb{F}_q$ with q odd and let P be a Sylow 2-subgroup of G. Then either $N_G(P) = P$, or one of the following holds:*

(i) $G = \mathrm{L}_n^{\epsilon}(q)$, $n \geqslant 3$, $t(n) \geqslant 2$ *and* $|N_G(P) : P| = ((q-\epsilon)_{2'})^{t(n)-1}/(n, q-\epsilon)_{2'}$.
(ii) $G = \mathrm{PSp}_n(q)$, $n \geqslant 2$, $q \equiv \pm 3 \pmod 8$ *and* $|N_G(P) : P| = 3^{t(n)}$.

REMARK 3.2. For certain low rank classical groups defined over small fields, we can compute $\Sigma(x)$ precisely with the aid of MAGMA [**3**] (version V2.26-6). To do this, we typically work in the quasisimple version of the group, namely $L = \mathrm{SL}_n^{\epsilon}(q)$, $\mathrm{Sp}_n(q)$ or $\Omega_n^{\epsilon}(q)$, and we use the functions `ClassicalClasses` and `ClassicalMaximals` to construct representatives of the conjugacy classes of elements and maximal subgroups of L. It is then straightforward to calculate the relevant fixed point ratios and we compute $\max\{\Sigma(x) : 1 \neq x \in G\}$ by running over a set of representatives of the conjugacy classes of non-central elements in L.

EXAMPLE 3.3. Suppose $G = \mathrm{P}\Omega_8^-(5)$. We can use the following MAGMA code to show that $\Sigma(x) \leqslant 17629/203763$ for all $1 \neq x \in G$. Here we are using the fact that $n(H,P) = 1$ for all $H \in \mathcal{M}$ (see Theorem 3.1). We also use the function `FixedPointRatio` defined in [**14**, Section 1.2.1].

```
G:=OmegaMinus(8,5);
cl:=ClassicalClasses(G);
M:=ClassicalMaximals("O-",8,5);
S:=SylowSubgroup(G,2);
A:=[i : i in [1..#M] | #M[i] mod #S eq 0];
B:=[];
a:=[i : i in [1..#cl] | cl[i][2] gt 1];
for i in a do
  z:=0;
  x:=cl[i][3];
  for j in A do
    z:= z+FixedPointRatio(G,M[j],x);
  end for;
  Append(~B,z);
end for;
Maximum(B);
```

We divide the proof of Theorem 1.4 for classical groups into several subsections and we begin by handling the orthogonal groups in Sections 3.1 and 3.2.

3.1. Odd dimensional orthogonal groups. Let $G = \Omega_n(q)$, where $n \geqslant 7$ and $q \geqslant 3$ are odd. Let P be a Sylow 2-subgroup of G and let $H \in \mathcal{M}$ be a maximal overgroup of P in G. By Theorem 3.1 we have $n(H,P) = 1$ and by appealing to [**26, 27**] we deduce that one of the following holds:

(a) H is a subfield subgroup of type $\mathrm{O}_n(q_0)$, where $q = q_0^e$ for some odd prime e.
(b) H is the stabilizer of an orthogonal decomposition $V = V_1 \perp \cdots \perp V_n$ of the natural module, where the V_i are isometric 1-spaces and $q = p \equiv \pm 3 \pmod 8$.
(c) $n = 7$, $q = p \equiv \pm 3 \pmod 8$ and $H = \mathrm{Sp}_6(2)$.
(d) $H = G_U$ is the stabilizer in G of an even-dimensional nondegenerate subspace U of V with square discriminant in $\mathbb{F}_q$.

As a consequence, it will be convenient to write $\mathcal{M} = \mathcal{M}_1 \cup \mathcal{M}_2$, where $\mathcal{M}_1$ comprises the subgroups in cases (a), (b) and (c). In turn, we will write $\Sigma(x) = \Sigma_1(x) + \Sigma_2(x)$ for all nontrivial $x \in G$, where $\Sigma(x)$ is defined as in (1) and $\Sigma_i(x)$ denotes the contribution from the subgroups in $\mathcal{M}_i$.

Using MAGMA, we can handle the cases $(n,q) = (7,3)$, $(7,5)$ and $(9,3)$ directly (see Remark 3.2 and Example 3.3). Indeed, we compute $\Sigma(x) \leqslant \alpha$, where α is defined as follows:

(n,q)	$(7,3)$	$(7,5)$	$(9,3)$
α	$322/351$	$298/1125$	$1090/3321$

So for the remainder, we will assume that

$$(n,q) \notin \{(7,3),(7,5),(9,3)\}. \tag{2}$$

LEMMA 3.4. *If $H \in \mathcal{M}_1$ then $\mathrm{fpr}(x,G/H) < q^{(4-n)/2}$ for all nontrivial $x \in G$.*

PROOF. We may assume x has prime order and thus the main theorem of [**7**] (which is proved in [**8–10**]) yields

$$\text{fpr}(x, G/H) < |x^G|^{-\frac{1}{2}+\frac{1}{n}} \tag{3}$$

since $(n,q) \neq (7,3)$. The conjugacy classes of elements of prime order in G are described in [**12**, Section 3.5] and it is straightforward to show that

$$|x^G| \geqslant \frac{1}{2}q^{\frac{1}{2}(n-1)}(q^{\frac{1}{2}(n-1)} - 1),$$

with equality if $x = (-I_{n-1}, I_1)$ is an involution with a minus-type (-1)-eigenspace on the natural module. The result now follows by combining this with the upper bound in (3). □

PROPOSITION 3.5. *We have* $\Sigma_1(x) < 1/12$ *for all nontrivial* $x \in G$.

PROOF. Fix a nontrivial element $x \in G$ and first assume $H \in \mathcal{M}$ is a subfield subgroup of type $\mathrm{O}_n(q_0)$, where $q = q_0^e$ and e is an odd prime. There are at most $\log\log q$ possibilities for e and for each choice there is a unique G-class of subgroups by [**21**, Proposition 4.5.8]. Since there are no subgroups in $\mathcal{M}$ of type (b) or (c), by applying Lemma 3.4 we deduce that

$$\Sigma_1(x) < q^{\frac{1}{2}(4-n)} \log\log q < \frac{1}{12},$$

as required.

For the remainder, we may assume $q = p \equiv \pm 3 \pmod 8$, where $q \geqslant 11$ if $n = 7$ (see (2)). First let $H \in \mathcal{M}_1$ be a maximal subgroup of type $\mathrm{O}_1(q) \wr S_n$. By [**21**, Proposition 4.2.15], there is a unique G-class of such subgroups and Lemma 3.4 implies that $\text{fpr}(x, G/H) < q^{(4-n)/2}$. If $n \geqslant 9$ then $\Sigma_1(x) = \text{fpr}(x, G/H)$ and the result follows. Finally suppose $n = 7$, $q \geqslant 11$ and $H = \mathrm{Sp}_6(2)$. Here there are two G-classes of these subgroups (see [**4**, Table 8.40], for example) and thus $\Sigma_1(x) < 3q^{(4-n)/2} < 1/12$. □

PROPOSITION 3.6. *We have* $\Sigma_2(x) < 11/12$ *for all nontrivial* $x \in G$.

PROOF. Let $H = G_U$ be the stabilizer of a nondegenerate m-space U, where m is even and the restriction of the defining quadratic form on V has square discriminant. The latter condition uniquely determines the Witt index of U and by applying [**21**, Proposition 4.1.6] we deduce that there is a unique G-class of subgroups for each m, hence a total of at most $(n-1)/2$ subgroups in $\mathcal{M}_2$.

Since H is a subspace subgroup, we cannot appeal to the upper bound in (3). In its place, we work with the bound in [**17**, Proposition 3.16], which gives

$$\text{fpr}(x, G/H) < (2q+1)q^{\frac{1}{2}(1-n)} + q^{\frac{1}{2}(m+1-n)} + q^{-m}$$

if $2 \leqslant m \leqslant n-5$ and

$$\text{fpr}(x, G/H) < (2q+1)q^{\frac{1}{2}(1-n)} + q^{\frac{1}{2}(2-m)} + q^{m-n}$$

for $m \in \{n-3, n-1\}$. Putting these bounds together, we deduce that

$$\Sigma_2(x) < \left(\frac{1}{2}(n-1)(2q+1) + q(q+1)\right) q^{\frac{1}{2}(1-n)} + q^{-3} + a + b,$$

where

$$a = q^{-1}\sum_{i=0}^{\infty} q^{-i} < \frac{1}{q-1}, \quad b = q^{-2}\sum_{i=0}^{\infty} q^{-2i} < \frac{1}{q^2-1}.$$

Since we may assume $(n, q) \neq (7,3), (9,3)$ (see (2)), it is straightforward to check that $\Sigma_2(x) < 11/12$, as required. □

By combining Propositions 3.5 and 3.6, we obtain the following result.

THEOREM 3.7. *The conclusion to Theorem 1.4 holds if $G = \Omega_n(q)$, where $n \geqslant 7$ is odd.*

3.2. Even dimensional orthogonal groups. In this section we assume $G = \mathrm{P}\Omega_n^{\epsilon}(q)$, where $n \geqslant 8$ is even and $q \geqslant 3$ is odd. As before, let P be a Sylow 2-subgroup of G and let $H \in \mathcal{M}$ be a maximal overgroup of P in G. By Theorem 3.1, we have $n(H, P) = 1$ and using [**26, 27**], we deduce that one of the following holds:

(a) H is a subfield subgroup of type $\mathrm{O}_n^{\epsilon}(q_0)$, where $q = q_0^e$ for some odd prime e.

(b) H is the stabilizer of an orthogonal decomposition $V = V_1 \perp \cdots \perp V_b$, where the V_i are isometric nondegenerate a-spaces with $a = 2^k$, $k \geqslant 0$.

(c) $G = \mathrm{P}\Omega_8^+(q)$, $q = p \equiv \pm 3 \pmod 8$ and $H = \Omega_8^+(2)$.

(d) $H = G_U$ is the stabilizer in G of a nondegenerate subspace U of V.

Write $\mathcal{M} = \mathcal{M}_1 \cup \mathcal{M}_2$, where $\mathcal{M}_1$ comprises the subgroups of type (a), (b) and (c). Similarly, write $\Sigma(x) = \Sigma_1(x) + \Sigma_2(x)$ for all nontrivial $x \in G$.

Using MAGMA, we compute $\Sigma(x) \leqslant 986/3731$ when $(n, q) = (8, 3)$ and $\Sigma(x) \leqslant 17629/203763$ for $(n, q) = (8, 5)$. Therefore, for the remainder we will assume

$$(n, q) \notin \{(8,3), (8,5)\}. \tag{4}$$

LEMMA 3.8. *If $H \in \mathcal{M}_1$ then $\mathrm{fpr}(x, G/H) < q^{5-n}$ for all nontrivial $x \in G$.*

PROOF. Suppose $x \in G$ has prime order and note that (3) holds via the main theorem of [**7**] (here we are using the fact that $(\epsilon, n, q) \neq (+, 8, 3)$). By inspecting [**12**, Section 3.5], it is straightforward to show that

$$|x^G| \geqslant \frac{(q^{n-2} - 1)(q^{n/2} + 1)(q^{n/2-2} - 1)}{q^2 - 1},$$

with equality if and only if $\epsilon = -$ and x is a long root element (that is, x is a unipotent element with Jordan form (J_2^2, J_1^{n-4}) on V). The result now follows by combining these two bounds. □

PROPOSITION 3.9. *We have $\Sigma_1(x) < 1/22$ for all nontrivial $x \in G$.*

PROOF. This is very similar to the proof of Proposition 3.5. First assume H is a subfield subgroup of type $\mathrm{O}_n^{\epsilon}(q_0)$, where $q = q_0^e$ and e is an odd prime. There are at most $\log\log q$ possibilities for e, with a unique G-class of subgroups for each e (see [**21**, Proposition 4.5.8]). Next suppose H is the stabilizer of an orthogonal decomposition as in (b) above. Now G has two classes of subgroups of type $\mathrm{O}_1(q) \wr S_n$ (see [**21**, Proposition 4.2.15]) and at most one class of subgroups of type $\mathrm{O}_a^{\epsilon'}(q) \wr S_{n/a}$ for each divisor $a = 2^k$ of n and each choice of sign $\epsilon' = \pm$, where $k \geqslant 1$. Since there are at most $\log n - 1$ possibilities for a, it follows that $\mathcal{M}_1$ contains at most $2\log n$ subgroups of type (b). Finally, if $q = p \equiv \pm 3 \pmod 8$ then [**4**, Table 8.50] indicates that $\mathrm{P}\Omega_8^+(q)$ has 4 classes of maximal subgroups isomorphic to $\Omega_8^+(2)$.

In view of Lemma 3.8 and (4), we conclude that

$$\Sigma_1(x) < q^{5-n}(\delta \log\log q + 2\log n + 4) \leqslant \frac{1}{22}$$

for all nontrivial $x \in G$, where $\delta = 1$ if $q \geqslant 27$, otherwise $\delta = 0$. □

We will need the following lemma in order to obtain an effective upper bound on $\Sigma_2(x)$.

LEMMA 3.10. *Let H be the stabilizer in G of a 1-dimensional nondegenerate subspace of the natural module. Then $\mathrm{fpr}(x, G/H) \leqslant (q^2-1)^{-1}$ for all nontrivial $x \in G$.*

PROOF. Let V be the natural module for G and set $\Omega = G/H$, where H is an almost simple group with socle $\Omega_{n-1}(q)$. Then

$$|\Omega| = \frac{1}{2}q^{\frac{1}{2}(n-2)}(q^{\frac{1}{2}n} - \epsilon)$$

and we may identify Ω with the set of 1-dimensional subspaces $\langle v \rangle$ such that $Q(v)$ is a square in $\mathbb{F}_q$, where Q is the defining quadratic form on V. It suffices to verify the bound for elements of prime order, so let us assume $x \in G$ has prime order r.

First assume $r = 2$. Here $|C_\Omega(x)|$ is maximal when $\epsilon = +$ and x is an involution of the form $(-I_2, I_{n-2})$, where both eigenspaces are nondegenerate minus-type spaces. By calculating the number of appropriate 1-spaces in each eigenspace, we deduce that

$$|C_\Omega(x)| \leqslant \frac{1}{2}(q+1) + \frac{1}{2}q^{\frac{1}{2}(n-4)}(q^{\frac{1}{2}(n-2)} + 1)$$

and the result follows. Next assume $r = p$, so x is unipotent. If x does not have Jordan form (J_3, J_1^{n-3}) on V, then one can check that the bounds on $|C_\Omega(x)|$ given in the proof of [**13**, Lemma 5.30] are sufficient. Now assume $x = (J_3, J_1^{n-3})$. There are two conjugacy classes of this form in G, and there are also two classes of elements in H with Jordan form (J_3, J_1^{n-4}) on the natural module for H. The two H-classes are not fused in G, so

$$|x^G \cap H| = |x^H| \leqslant \frac{|\,\mathrm{SO}_{n-1}(q)|}{2q^{n-3}|\,\mathrm{SO}^+_{n-4}(q)|}, \quad |x^G| = \frac{|\,\mathrm{SO}^\epsilon_n(q)|}{2q^{n-2}|\,\mathrm{SO}_{n-3}(q)|}$$

and thus

$$\mathrm{fpr}(x, G/H) \leqslant \frac{q^{(n-4)/2} + 1}{q^{n/2} - \epsilon} \leqslant \frac{1}{q^2 - 1}.$$

Finally, if $r \neq p$ is odd then it is easy to check that the bounds presented in the proof of [**13**, Lemma 5.30] are sufficient. □

PROPOSITION 3.11. *We have $\Sigma_2(x) < 21/22$ for all nontrivial $x \in G$.*

PROOF. Let H be the stabilizer of a nondegenerate m-space and let $x \in G$ be nontrivial. First assume $m = 2\ell + 1$ is odd, where $1 \leqslant \ell \leqslant \lfloor (n-2)/4 \rfloor$. By [**21**, Proposition 4.1.6], there are two conjugacy classes of subgroups of this type and [**17**, Proposition 3.16] gives

$$\mathrm{fpr}(x, G/H) < (2q+1)q^{\frac{1}{2}(2-n)} + q^{\frac{1}{2}(2\ell+2-n)} + q^{-2\ell-1}.$$

Now Lemma 3.10 implies that $\mathrm{fpr}(x, G/H) \leqslant (q^2-1)^{-1}$ if $m = 1$ or $n-1$, so the contribution to $\Sigma_2(x)$ from stabilizers of odd-dimensional spaces is less than

$$\alpha_1 = \sum_{\ell=1}^{\lfloor (n-2)/4 \rfloor} 2\left((2q+1)q^{\frac{1}{2}(2-n)} + q^{\frac{1}{2}(2\ell+2-n)} + q^{-2\ell-1}\right) + 2(q^2-1)^{-1}.$$

Similarly, the contribution from the stabilizers of even-dimensional spaces is less than

$$\alpha_2 = \sum_{\ell=1}^{\lfloor n/4 \rfloor} 2\left((2q+1)q^{\frac{1}{2}(2-n)} + q^{\frac{1}{2}(2\ell+2-n)} + q^{-2\ell}\right),$$

and we deduce that

$$\begin{aligned}\Sigma_2(x) < (n-1)(2q+1)q^{\frac{1}{2}(2-n)} + 2q^{-2}a + 2q^{\frac{1}{2}(2\lfloor n/4 \rfloor+2-n)}a \\ + 2q^{\frac{1}{2}(2\lfloor (n-2)/4 \rfloor+2-n)}a + 2(q^2-1)^{-1}\end{aligned}$$

where

$$a = \sum_{i=0}^{\infty} q^{-i} = \frac{q}{q-1}.$$

In view of (4), one can check that this upper bound is sufficient unless $(n,q) = (10,3)$ or $(12,3)$.

Suppose $(n,q) = (10,3)$. If $\epsilon = +$ then the main theorem of [**27**] implies that each $\mathcal{M}_2$ is of type $\mathrm{O}_1(3) \perp \mathrm{O}_9(3)$ and thus Lemma 3.10 yields $\Sigma_2(x) \leqslant 2/(3^2-1) = 1/4$. Similarly, if $\epsilon = -$ then $\Sigma_2(x) = \mathrm{fpr}(x, G/H)$ with H of type $\mathrm{O}_2^-(3) \perp \mathrm{O}_8^+(3)$, and the bound in [**17**, Proposition 3.16] is sufficient. Finally, suppose $(n,q) = (12,3)$. If $\epsilon = +$ then $\mathcal{M}_2 = \{H\}$ with H of type $\mathrm{O}_4^+(3) \perp \mathrm{O}_8^+(3)$ and the result quickly follows. On the other hand, if $\epsilon = -$ then there are 6 subgroups in $\mathcal{M}_2$: two each of type $\mathrm{O}_m(3) \perp \mathrm{O}_{12-m}(3)$ with $m \in \{1,3\}$, together with subgroups of type $\mathrm{O}_m^-(3) \perp \mathrm{O}_{12-m}^+(3)$ for $m \in \{2,4\}$. A routine calculation working with the bounds in Lemma 3.10 and [**17**, Proposition 3.16] now establishes the result. □

THEOREM 3.12. *The conclusion to Theorem 1.4 holds if $G = \mathrm{P\Omega}_n^{\epsilon}(q)$, where $n \geqslant 8$ is even.*

3.3. Symplectic groups. Let $G = \mathrm{PSp}_n(q)$, where $n \geqslant 4$ and $q \geqslant 3$ is odd. As usual, let P be a Sylow 2-subgroup of G and let $H \in \mathcal{M}$ be a maximal overgroup of P in G. By inspecting [**26, 27**] we see that one of the following holds:

(a) H is a subfield subgroup of type $\mathrm{Sp}_n(q_0)$, where $q = q_0^e$ for some odd prime e.
(b) H is the stabilizer of an orthogonal decomposition $V = V_1 \perp \cdots \perp V_b$, where the V_i are nondegenerate a-spaces with $a = 2^k$, $k \geqslant 1$.
(c) $H = G_U$ is the stabilizer in G of a nondegenerate subspace U of V with $\dim U < n/2$.
(d) $n = 4$, $q = p \equiv \pm 3 \pmod 8$ and $H = 2^4.A_5$.

By Theorem 3.1 we have $n(H,P) \leqslant |N_G(P):P| = 3^t$, where either $t = 0$, or $q \equiv \pm 3 \pmod 8$ and t is the number of nonzero digits in the binary expansion of n. In fact, we have the following result.

LEMMA 3.13. *Let $H \in \mathcal{M}$ be a maximal overgroup of P. Then $n(H,P) = 1$.*

PROOF. In view of Theorem 3.1, we may assume $q \equiv \pm 3 \pmod 8$. In (a) we have $H = \mathrm{PSp}_n(q_0)$, so Theorem 3.1 implies that $|N_G(P):P| = |N_H(P):P|$ and thus $n(H,P) = 1$. Similarly, $|P| = 2^6$ and $N_G(P) = P{:}3 < H$ in case (d), whence $n(H,P) = 1$.

Next consider (c). As a P-module we have $V = V_1 \oplus \cdots \oplus V_t$, where each V_i is an irreducible submodule of dimension 2^{a_i} with $a_1 > a_2 > \cdots > a_t \geqslant 0$. It follows that P fixes at most one subspace of any given dimension and thus $n(H,P) = 1$.

Finally, let us assume $n = ab$ and H is the stabilizer of an orthogonal decomposition $V = V_1 \perp \cdots \perp V_b$, where each V_i is a nondegenerate a-space and $a = 2^k$ with $k \geqslant 1$. Here it is convenient to work in the quasisimple group $G = \mathrm{Sp}_n(q)$, in which case $H = \mathrm{Sp}_a(q) \wr S_b$ and it suffices to show that $N_G(P) \leqslant H$. If we write $b = 2^{b_1} + \cdots + 2^{b_t}$ with $b_1 > b_2 > \cdots > b_t \geqslant 0$ then $P = P_1 \times \cdots \times P_t$, where P_i is a Sylow 2-subgroup of $H_i = \mathrm{Sp}_a(q) \wr S_{2^{b_i}}$ and P_i acts transitively on the set of 2^{b_i} a-spaces in the corresponding orthogonal decomposition preserved by H_i. Therefore, we may assume b is a 2-power and P acts transitively on $\{V_1, \ldots, V_b\}$. Set $P_0 = P \cap \mathrm{Sp}_a(q)^b$ and observe that there exists an element $z \in N_G(P_0) \cap \mathrm{Sp}_a(q)^b$ of order 3, which is centralized by a Sylow 2-subgroup of S_b. Therefore, $z \in N_G(P) \cap H$ and thus $N_G(P) = P{:}\langle z \rangle \leqslant H$, as required. □

For now we will assume $n \geqslant 6$, postponing the analysis of the case $n = 4$ to the end of the section. Write $\mathcal{M} = \mathcal{M}_1 \cup \mathcal{M}_2$ and $\Sigma(x) = \Sigma_1(x) + \Sigma_2(x)$, where $\mathcal{M}_1$ comprises the subgroups of type (a) and (b) as above.

With the aid of MAGMA, we compute $\Sigma(x) \leqslant \alpha$ in the following cases:

(n,q)	$(6,3)$	$(6,5)$	$(8,3)$	$(8,5)$	$(10,3)$
α	$2/13$	$10/217$	$6/205$	$14/4069$	$826/7381.$

Therefore, for the remainder we will assume

$$(n,q) \notin \{(6,3),(6,5),(8,3),(8,5),(10,3)\}. \tag{5}$$

LEMMA 3.14. *Suppose $n \geqslant 6$, $H \in \mathcal{M}_1$ and $x \in G$ is nontrivial and not a transvection. Then*

$$\mathrm{fpr}(x, G/H) < q^{\ell - n},$$

where $\ell = 6$ if H is of type $\mathrm{Sp}_{n/2}(q) \wr S_2$, otherwise $\ell = 4$.

PROOF. First assume H is not of type $\mathrm{Sp}_{n/2}(q) \wr S_2$. Here (3) holds by the main theorem of [**7**] and the result follows since

$$|x^G| \geqslant \frac{|\mathrm{Sp}_n(q)|}{|\mathrm{Sp}_{n-2}(q)||\mathrm{Sp}_2(q)|} = \frac{q^{n-2}(q^n-1)}{q^2-1},$$

with equality if $x = (-I_2, I_{n-2})$. Similarly, if H is of type $\mathrm{Sp}_{n/2}(q) \wr S_2$ then $n \geqslant 8$,

$$\mathrm{fpr}(x, G/H) < |x^G|^{-\frac{1}{2} + \frac{2}{n}}$$

by [**7**] and the result follows once again by applying the above lower bound on $|x^G|$. □

PROPOSITION 3.15. *If $n \geqslant 6$ then $\Sigma_1(x) < 1/43$ for all nontrivial $x \in G$.*

PROOF. Let $x \in G$ be nontrivial and write $\Sigma_1(x) = \alpha(x) + \beta(x)$, where $\alpha(x)$ is the contribution to $\Sigma_i(x)$ from subgroups of type (a). We will show that $\alpha(x) < 1/200$ and $\beta(x) < 1/45$, which implies that $\Sigma_1(x) < 1/43$, as required.

First assume H is a subfield subgroup of type $\mathrm{Sp}_n(q_0)$, where $q = q_0^e$ and $e \geqslant 3$ is a prime. Here there are at most $\log\log q$ possibilities for e and a unique G-class for each e (see [**21**, Proposition 4.5.4]). In addition, $n(H,P) = 1$ by Lemma 3.13.

Suppose x is a transvection. There are two conjugacy classes of transvections in both G and H, and the two H-classes are not fused in G (see the proof of [**12**, Proposition 5.5.8], for example). Therefore,

$$\mathrm{fpr}(x, G/H) = \frac{|x^G \cap H|}{|x^G|} = \frac{q_0^n - 1}{q^n - 1} < 2q^{-\frac{2}{3}n}$$

and thus $\alpha(x) < 2q^{-2n/3} \log\log q < 1/200$. On the other hand, if x is not a transvection then Lemma 3.14 yields $\alpha(x) < q^{4-n} \log\log q < 1/200$.

For the remainder, let us assume $H \in \mathcal{M}_1$ is a subgroup of type $\mathrm{Sp}_a(q) \wr S_{n/a}$, where $a = 2^k$ and $k \geqslant 1$. By [**21**, Proposition 4.2.10], there is a unique class of such subgroups for each k and Lemma 3.13 gives $n(H, P) = 1$. Also note that there are at most $\log(n/2)$ possibilities for a.

If x is a transvection then $|x^G| = (q^n - 1)/2$ and H contains $n(q^a - 1)/a$ transvections, whence

$$\mathrm{fpr}(x, G/H) \leqslant \frac{2n}{a}\left(\frac{q^a - 1}{q^n - 1}\right)$$

and in view of (5) we deduce that

$$\beta(x) < \left(\frac{4}{q^{n/2} + 1}\right)\log(n/2) < \frac{1}{45}.$$

Now assume x is not a transvection. If $n \equiv 2 \pmod 4$ then $a = 2$ is the only possibility, and thus Lemma 3.14 implies that $\beta(x) < q^{4-n} < 1/45$. On the other hand, if $n \equiv 0 \pmod 4$ then $\beta(x) < q^{4-n}(\log(n/2) - 1) + q^{6-n}$ and once again this bound is sufficient. □

PROPOSITION 3.16. *If $n \geqslant 6$ then $\Sigma_2(x) < 42/43$ for all nontrivial $x \in G$.*

PROOF. Let $H = G_U$ be the stabilizer of a nondegenerate m-space U, where $m < n/2$ is even. By [**21**, Proposition 4.1.3], there is a unique G-class of subgroups for each m. In addition, Lemma 3.13 gives $n(H, P) = 1$.

Set $m = 2\ell$. Then [**17**, Proposition 3.16] gives

$$\mathrm{fpr}(x, G/H) < 2q^{\frac{1}{2}(4-n)} + q^{-\frac{1}{2}n} + q^{-\ell} + q^{2\ell-n},$$

and thus

$$\begin{aligned}
\Sigma_2(x) &< \sum_{\ell=1}^{\lfloor n/4 \rfloor}\left(2q^{\frac{1}{2}(4-n)} + q^{-\frac{1}{2}n} + q^{-\ell} + q^{2\ell-n}\right) \\
&< \frac{1}{4}n(2q^2+1)q^{-\frac{1}{2}n} + q^{-1}\sum_{i=0}^{\infty} q^{-i} + q^{2\lfloor n/4\rfloor - n}\sum_{i=0}^{\infty} q^{-2i} \\
&< \frac{1}{4}n(2q^2+1)q^{-\frac{1}{2}n} + \frac{1}{q-1} + q^{2\lfloor n/4\rfloor - n}\left(\frac{q^2}{q^2-1}\right).
\end{aligned}$$

One can check that this bound is sufficient. □

By combining Propositions 3.15 and 3.16, we obtain the following result.

THEOREM 3.17. *The conclusion to Theorem 1.4 holds if $G = \mathrm{PSp}_n(q)$ and $n \geqslant 6$.*

Finally, we handle the 4-dimensional symplectic groups.

THEOREM 3.18. *The conclusion to Theorem 1.4 holds if $G = \mathrm{PSp}_4(q)$.*

PROOF. If $q = 3$ then using MAGMA [**3**] we compute $\Sigma(x) \leqslant 7/15$, so for the remainder we may assume $q \geqslant 5$. By the main theorem of [**25**], we have $\mathrm{fpr}(x, G/H) \leqslant 4/3q$ for all $H \in \mathcal{M}$ and by applying Lemma 3.13 and inspecting [**4**, Table 8.12], we observe that there are at most $2 + \log\log q$ subgroups in $\mathcal{M}$. Therefore,

$$\Sigma(x) \leqslant (2 + \log\log q) \cdot \frac{4}{3q} < 1$$

and the result follows. □

3.4. Linear and unitary groups. In this final section, we complete the proof of Theorem 1.1 by handling the linear and unitary groups. The low dimensional groups $G = \mathrm{L}_n^{\epsilon}(q)$ with $n \leqslant 5$ require special attention and they are treated separately at the end of the section.

THEOREM 3.19. *The conclusion to Theorem 1.4 holds if* $G = \mathrm{U}_n(q)$ *and* $n \geqslant 6$.

PROOF. Let $H \in \mathcal{M}$. By applying [**26, 27**] we deduce that one of the following holds:

(a) H is a subfield subgroup of type $\mathrm{GU}_n(q_0)$, where $q = q_0^e$ and $e \geqslant 3$ is a prime.
(b) H is the stabilizer of an orthogonal decomposition $V = V_1 \perp \cdots \perp V_b$, where the V_i are nondegenerate a-spaces with $a = 2^k$, $k \geqslant 0$.
(c) H is the stabilizer of a nondegenerate m-space with $m < n/2$.

Let $x \in G$ be nontrivial and write $\mathcal{M} = \mathcal{M}_1 \cup \mathcal{M}_2$ and $\Sigma(x) = \Sigma_1(x) + \Sigma_2(x)$, where $\mathcal{M}_1$ comprises the subgroups in (a) and (b). If $(n, q) = (6, 3)$ then a MAGMA computation yields $\Sigma(x) \leqslant 5/351$, so for the remainder we may assume $(n, q) \neq (6, 3)$. We claim that $\Sigma_i(x) < 1/2$ for $i = 1, 2$ and thus $\Sigma(x) < 1$, as required.

First assume $H \in \mathcal{M}_1$ and note that (3) holds (see [**7**]). Then

$$|x^G| > \frac{(q^{n-1} - 1)(q^n - 1)}{q + 1}$$

and by applying (3) we deduce that $\mathrm{fpr}(x, G/H) < q^{7/2-n}$. In addition, Theorem 3.1 yields

$$n(H, P) \leqslant |G : N_G(P)| = \frac{((q+1)_{2'})^{t-1}}{(n, q+1)_{2'}} \leqslant \left(\frac{1}{2}(q+1)\right)^{\log n - 1},$$

where t is the number of nonzero digits in the binary expansion of n. Note that $t < \log n$.

In (a), there are at most $\log\log q$ choices for e and there are $(n, (q+1)/(q_0+1))$ distinct G-classes for each e (see [**21**, Proposition 4.5.3]). Similarly, there are at most $\log n$ choices for a in (b) and by [**21**, Proposition 4.2.9] we see that there is a unique conjugacy class of subgroups for each given a. Since $(n, (q+1)/(q_0+1)) \leqslant (n, q+1)_{2'}$, it follows that

$$\Sigma_1(x) < (\delta \log\log q + \log n) \left(\frac{1}{2}(q+1)\right)^{\log n - 1} q^{7/2-n},$$

where $\delta = 1$ if $q \geqslant 27$, otherwise $\delta = 0$. One can check that this yields $\Sigma_1(x) < 1/2$.

Now assume $H \in \mathcal{M}_2$ is the stabilizer of a nondegenerate m-space, where $m < n/2$. Note that $n(H, P) = 1$ since P fixes at most one subspace of V of any given dimension. We divide the analysis into two cases according to the parity of n.

First assume $n \geqslant 6$ is even. Here [**17**, Proposition 3.16] gives

$$\mathrm{fpr}(x, G/H) < (2q^3+1)q^{1-n} + q^{m+d-n} + q^{-2m},$$

where $d = 1$ if $m = 1$, otherwise $d = 0$. Therefore,

$$\Sigma_2(x) < (n/2-1)(2q^3+1)q^{1-n} + q^{2-n} + q^{-2}a + q^{-n/2-1}b,$$

where

$$a = \sum_{i=0}^{\infty} q^{-2i} = \frac{q^2}{q^2-1}, \quad b = \sum_{i=0}^{\infty} q^{-i} = \frac{q}{q-1},$$

and it is straightforward to check that $\Sigma_2(x) < 1/2$, as required.

Finally, suppose $n \geqslant 7$ is odd. Here

$$\mathrm{fpr}(x, G/H) < (2q^3+1)q^{-n} + q^{m+d-n} + q^{-2m}$$

by [**17**, Proposition 3.16], where $d = 1$ if m is even, otherwise $d = 0$. Therefore,

$$\Sigma_2(x) < \frac{1}{2}(n-1)(2q^3+1)q^{-n} + q^{1-n} + (q^{-2} + 2q^{\frac{1}{2}(1-n)})a,$$

where a is defined as above. This yields $\Sigma_2(x) < 1/2$ and the result follows. □

THEOREM 3.20. *The conclusion to Theorem* 1.4 *holds if* $G = \mathrm{L}_n(q)$ *and* $n \geqslant 6$.

PROOF. We need to show that $\Sigma(x) < 1$ unless n is odd and $q = 3$. So for the remainder of the proof we will assume n is even if $q = 3$.

Let $H \in \mathcal{M}$. By inspecting [**26, 27**] we see that one of the following holds:

(a) H is a subfield subgroup of type $\mathrm{GL}_n(q_0)$, where $q = q_0^e$ and $e \geqslant 3$ is a prime.

(b) H is the stabilizer of a decomposition $V = V_1 \oplus \cdots \oplus V_b$, where the V_i are a-spaces with $a = 2^k$, $k \geqslant 0$.

(c) H is the stabilizer of an m-space with $1 \leqslant m < n$.

As before, write $\mathcal{M} = \mathcal{M}_1 \cup \mathcal{M}_2$ and $\Sigma(x) = \Sigma_1(x) + \Sigma_2(x)$, where $\mathcal{M}_1$ comprises the subgroups in (a) and (b). If $(n,q) = (6,3)$ then a direct computation gives $\Sigma(x) \leqslant 235/1001$, so for the remainder we may assume $(n,q) \neq (6,3)$.

By repeating the argument in the proof of Theorem 3.19, it is straightforward to show that

$$\Sigma_1(x) < (\delta \log\log q + \log n)\left(\frac{1}{2}(q-1)\right)^{\log n - 1} q^{3-n} < \frac{1}{3},$$

where $\delta = 1$ if $q \geqslant 27$, otherwise $\delta = 0$. Therefore, to complete the proof we just need to verify the bound $\Sigma_2(x) \leqslant 2/3$.

Write $n = 2^{a_1} + \cdots + 2^{a_t}$, where $a_1 > a_2 > \cdots > a_t \geqslant 0$. Then P fixes an m-space if and only if $m = b_1 2^{a_1} + \cdots + b_t 2^{a_t}$ for some $b_i \in \{0,1\}$. Moreover, if m is of this form then P fixes a unique m-space and thus $n(H,P) = 1$ for all $H \in \mathcal{M}_2$. Suppose $n \geqslant 6$ is even, so $m \neq 1, n-1$. If $m \leqslant n/2$ then [**17**, Proposition 3.1] gives $\mathrm{fpr}(x, G/H) < 2q^{-m}$, and thus

$$\Sigma_2(x) < 4q^{-2}\sum_{i=1}^{\infty} q^{-i} = \frac{4}{q(q-1)} \leqslant \frac{2}{3}.$$

Similarly, if $n \geqslant 7$ is odd then [**17**, Proposition 3.1] yields

$$\Sigma_2(x) < \frac{4}{q(q-1)} + 2\left(\frac{1}{q} + \frac{1}{q^{n-1}}\right),$$

and we deduce that $\Sigma_2(x) < 2/3$ for all $q \geqslant 5$. The result follows. □

To complete the proof of Theorem 1.1 for the groups $G = \mathrm{L}_n^{\epsilon}(q)$ with $n \geqslant 6$ we may assume $\epsilon = +$, n is odd and $q = 3$, in which case $G = \mathrm{SL}_n(3)$. The following lemma handles the case where $x \in G$ is semisimple, and the proof relies on a theorem of Gow [**15**], which is proved using character theory.

LEMMA 3.21. *Let $G = \mathrm{L}_n(3)$, where $n \geqslant 7$ is odd, and let P be a Sylow 2-subgroup of G. If $x \in G$ is non-central and semisimple, then $G = \langle P, x^g \rangle$ for some $g \in G$.*

PROOF. Let $y \in G$ be an element of order $(3^n - 1)/2$ and note that each maximal overgroup of P in G is a parabolic subgroup (see [**27**]). Since y acts irreducibly on the natural module, it follows that $G = \langle P, y^g \rangle$ for all $g \in G$. By [**18**, Lemma 7.5], P contains a regular semisisimple element z and by combining the main theorems of [**15, 19**] we deduce that $x \in y^G z^G$, say $x = y^a z^b$. Then $y^{ab^{-1}} = x^{b^{-1}} z^{-1}$ and the result now follows since $G = \langle P, x^{b^{-1}} z^{-1} \rangle = \langle P, x^{b^{-1}} \rangle$. □

REMARK 3.22. The method used in the proof of Lemma 3.21 can be applied more generally. For example, suppose $G = \mathrm{SL}_n(q)$, where q is odd and $n \geqslant 3$. Let P be a Sylow 2-subgroup of G and let $y \in G$ be an element of order $(q^n - 1)/(q-1)$. By [**2**], every maximal subgroup of G containing y is a field extension subgroup of type $\mathrm{GL}_{n/k}(q^k)$ for some prime divisor k of n. Since none of these subgroups contain P, it follows that $G = \langle P, y^g \rangle$ for all $g \in G$. And so we can repeat the argument in the proof of Lemma 3.21 to conclude that if $x \in G$ is non-central and semisimple, then $G = \langle P, x^g \rangle$ for some $g \in G$. But this still leaves us to deal with unipotent elements, which explains why we work with fixed point ratios in the proof of Theorem 3.20.

Next we turn to unipotent elements. In the following lemma, we use elementary linear algebra to handle transvections.

LEMMA 3.23. *Let $G = \mathrm{L}_n(3)$, where $n \geqslant 7$ is odd, and let P be a Sylow 2-subgroup of G. If $x \in G$ is a transvection, then $G = \langle P, x^g \rangle$ for some $g \in G$.*

PROOF. First observe that any two transvections in G are conjugate and recall that each maximal overgroup of P in G acts reducibly on the natural module V. Therefore, it suffices to construct a specific transvection y such that $\langle P, y \rangle$ acts irreducibly on V.

The space V decomposes as a direct sum $V = V_1 \oplus \cdots \oplus V_t$ of irreducible submodules for P, where $\dim V_i = 2^{a_i}$ and $a_1 > a_2 > \cdots > a_t \geqslant 0$. Without loss of generality, we may assume that each V_i has a basis contained in the standard basis $\{e_1, \ldots, e_n\}$ for V, where we view e_i as a column vector with 1 in the i-th position and 0 in all other positions.

Let N be an $n \times n$ matrix with columns $c_1, \ldots, c_n$ such that every entry in c_1 is equal to 1 and $c_i = \alpha_i c_1$ for $i \geqslant 2$, where the nonzero scalars $\alpha_i \in \mathbb{F}_3$ are chosen so that $1 + \alpha_2 + \cdots + \alpha_n = 0$. Then N is a rank one nilpotent matrix and thus $y = I_n + N$ is a transvection. Moreover, every non-diagonal entry of y is nonzero and thus $\langle P, y\rangle$ acts irreducibly on V. As noted above, this implies that $G = \langle P, y\rangle$ and the proof is complete. □

THEOREM 3.24. *The conclusion to Theorem 1.1 holds if $G = \mathrm{L}_n(3)$ and $n \geqslant 7$ is odd.*

PROOF. Let $x \in G$ be nontrivial. In view of Lemmas 3.21 and 3.23, we may assume x is a unipotent element of order 3 with $\dim C_V(x) \leqslant n-2$. As before, let $\mathcal{M}$ be the set of maximal overgroups of P in G and recall that each $H \in \mathcal{M}$ is the stabilizer of an m-space. Moreover, there is a unique G-class of such subgroups for each m and we have $n(H,P) = 1$ since P fixes at most one subspace of any given dimension. As before, it suffices to show that $\Sigma(x) < 1$.

By applying the upper bound in [**17**, Proposition 3.1], we see that the contribution to $\Sigma(x)$ from the stabilizers of m-spaces with $m \neq 1, n-1$ is less than

$$4\sum_{i=2}^{(n-1)/2} 3^{-i} < \frac{2}{3}.$$

Finally, suppose $H \in \mathcal{M}$ is the stabilizer of a 1-space (or an $(n-1)$-space) and set $\Omega = G/H$, so $|\Omega| = (3^n-1)/2$. Since $C_\Omega(x)$ coincides with the set of 1-spaces in $C_V(x)$, it follows that $|C_\Omega(x)| \leqslant (3^{n-2}-1)/2$ and $\mathrm{fpr}(x, G/H) \leqslant (3^{n-2}-1)/(3^n-1)$. Therefore,

$$\Sigma(x) < \frac{2}{3} + 2\left(\frac{3^{n-2}-1}{3^n-1}\right) < 1$$

and the result follows. □

To complete the proof of Theorem 1.1 we may assume $G = \mathrm{L}^\epsilon_n(q)$ and $n \leqslant 5$.

THEOREM 3.25. *The conclusion to Theorem 1.4 holds if $G = \mathrm{L}_2(q)$.*

PROOF. With the aid of MAGMA, one can check that $\Sigma(x) \leqslant 51/55$ for all nontrivial $x \in G$ when $q < 101$ (maximal when $q = 11$), so we may assume $q \geqslant 101$. By Theorem 3.1 we have $|N_G(P) : P| = 3$ if $q \equiv \pm 3 \pmod 8$, otherwise $N_G(P) = P$. Let $H \in \mathcal{M}$ be a maximal subgroup of G containing P. Then by inspecting [**27**], we deduce that one of the following holds:

(a) H is a subfield subgroup of type $\mathrm{GL}_2(q_0)$, where $q = q_0^2$.
(b) H is a subfield subgroup of type $\mathrm{GL}_2(q_0)$, where $q = q_0^e$ and $e \geqslant 3$ is a prime.
(c) $H = D_{q-\epsilon}$ and $q \equiv \epsilon \pmod 4$.
(d) $H = A_4$ and $q = p \equiv \pm 3, \pm 13 \pmod{40}$.
(e) $H = S_4$ and $q = p \equiv \pm 7 \pmod{16}$.
(f) $H = A_5$ and $q = p \equiv \pm 11, \pm 19 \pmod{40}$.

Write $\mathcal{M} = \mathcal{M}_1 \cup \mathcal{M}_2$, where $\mathcal{M}_1$ comprises the subfield subgroups in (a). In addition, write $\Sigma(x) = \Sigma_1(x) + \Sigma_2(x)$ for all nontrivial $x \in G$, where $\Sigma_i(x)$ is the contribution from the subgroups in $\mathcal{M}_i$.

Suppose $q = q_0^2$, so $q \equiv 1 \pmod 8$ and thus $N_G(P) = P$. There are two G-classes of subfield subgroups H of type $\mathrm{GL}_2(q_0)$ and the main theorem of [**25**] yields $\mathrm{fpr}(x, G/H) \leqslant 2/(q+1)$. Since $n(H,P) = 1$, it follows that

$$\Sigma_1(x) \leqslant \frac{4}{q+1}.$$

Similarly, if H is one of the subgroups labelled (b)-(f), then $\mathrm{fpr}(x, G/H) \leqslant 4/3q$ by [**25**] and we note that $n(H,P) \leqslant 3$. Since there are at most $\log\log q$ possibilities for e in (b), we deduce that

$$\Sigma_2(x) \leqslant 3(\log\log q + 3) \cdot \frac{4}{3q}$$

and it is straightforward to check that $\Sigma(x) < 1$, as required. □

We will need the following lemma for the 3-dimensional groups.

LEMMA 3.26. *Let $G = \mathrm{L}_3^{\epsilon}(q)$ and let H be a subfield subgroup of type $\mathrm{GL}_3^{\epsilon}(q_0)$, where $q = q_0^e$ and e is an odd prime. Then $\mathrm{fpr}(x, G/H) < q^{-2}$ for all nontrivial $x \in G$.*

PROOF. Let $x \in G$ be an element of prime order r. First assume $r = p$, so x has Jordan form (J_2, J_1) or (J_3) on the natural module. If $x = (J_2, J_1)$ then

$$|x^G \cap H| = \frac{|\mathrm{GL}_3^{\epsilon}(q_0)|}{q_0^3|\mathrm{GL}_1^{\epsilon}(q_0)|^2} = (q_0+\epsilon)(q_0^3-\epsilon),\;\; |x^G| = (q+\epsilon)(q^3-\epsilon)$$

and the result follows. Similarly, if $x = (J_3)$ then

$$|x^G \cap H| \leqslant \frac{|\mathrm{GL}_3^{\epsilon}(q_0)|}{q_0^2|\mathrm{GL}_1^{\epsilon}(q_0)|} = q_0(q_0^2-1)(q_0^3-\epsilon),\;\; |x^G| \geqslant \frac{|\mathrm{GL}_3^{\epsilon}(q)|}{3q^2|\mathrm{GL}_1^{\epsilon}(q)|} = \frac{1}{3}q(q^2-1)(q^3-\epsilon),$$

and once again it is straightforward to check that these bounds are sufficient.

Now assume $r \neq p$. If x is regular, then the trivial bound $|x^G \cap H| \leqslant |\mathrm{PGL}_3^{\epsilon}(q_0)|$ combined with the lower bound

$$|x^G| \geqslant \frac{|\mathrm{GU}_3(q)|}{3(q+1)^3} = \frac{1}{3}q^3(q-1)(q^2-q+1)$$

is good enough. On the other hand, if x is non-regular then $|x^G \cap H| = q_0^2(q_0^2 + \epsilon q_0 + 1)$, $|x^G| = q^2(q^2+\epsilon q+1)$ and the result follows. □

THEOREM 3.27. *The conclusion to Theorem 1.4 holds if $G = \mathrm{L}_n^{\epsilon}(q)$ and $3 \leqslant n \leqslant 5$.*

PROOF. First assume $n = 3$. If $q < 29$ then with the aid of MAGMA it is straightforward to check that $\Sigma(x) \leqslant 10/13$ for all nontrivial $x \in G$. Therefore, we may assume $q \geqslant 29$. Let $H \in \mathcal{M}$ and note that

$$n(H,P) \leqslant |N_G(P) : P| = \frac{(q-\epsilon)_{2'}}{(3, q-\epsilon)} \leqslant \frac{1}{2}(q+1)$$

by Theorem 3.1. By inspecting [**27**] we see that one of the following holds:

(a) H is a subfield subgroup of type $\mathrm{GL}_3^{\epsilon}(q_0)$, where $q = q_0^e$ and $e \geqslant 3$ is a prime.
(b) H is of type $\mathrm{GL}_1^{\epsilon}(q) \wr S_3$ and $q \equiv \epsilon \pmod 4$.
(c) $\epsilon = +$ and H is the stabilizer of a 1-space or a 2-space.
(d) $\epsilon = -$ and H is the stabilizer of a nondegenerate 1-space.

Write $\mathcal{M} = \mathcal{M}_1 \cup \mathcal{M}_2$ and $\Sigma(x) = \Sigma_1(x) + \Sigma_2(x)$, where $\mathcal{M}_1$ comprises the subfield subgroups in (a).

Let $H \in \mathcal{M}_1$ be a subfield subgroup as in (a). By inspecting [**4**, Tables 8.3, 8.5] we note that for each e there are m distinct G-classes of subgroups of this form, where

$$m = \left(3, \frac{q-\epsilon}{q_0-\epsilon}\right) = \begin{cases} 3 & \text{if } e = 3 \text{ and } q_0 \equiv \epsilon \pmod 3 \\ 1 & \text{otherwise.} \end{cases}$$

If $m = 3$ then $(3, q-\epsilon) = 3$ and thus $n(H,P) \leqslant (q+1)/6$. Therefore, since there are at most $\log\log q$ possibilities for e, by applying Lemma 3.26 we deduce that

$$\Sigma_1(x) < \frac{1}{2}(q+1)q^{-2}\log\log q.$$

Now let us consider the subgroups H in $\mathcal{M}_2$. There is a unique conjugacy class of subgroups of type (b) and (d), and there are two classes in (c) (one for each dimension). Moreover, P fixes a unique 1-space and a unique 2-space (in (d), the relevant 2-space is simply the orthogonal complement of the 1-space fixed by P), so we have $n(H,P) = 1$ in (c) and (d). Therefore, by applying the main theorem of [**25**] we deduce that

$$\Sigma_2(x) < (2 + (q+1)/2) \cdot \frac{4}{3q},$$

and it is now straightforward to check that $\Sigma(x) < 1$ for all nontrivial $x \in G$.

Next assume $n = 4$. If $q < 11$ then we compute $\Sigma(x) \leqslant 34/117$ for all nontrivial $x \in G$, so we may assume $q \geqslant 11$. Note that Theorem 3.1 yields $N_G(P) = P$ and thus $n(H,P) = 1$ for all $H \in \mathcal{M}$. Then in the usual way, using [**25, 27**], we deduce that

$$\Sigma(x) < (\delta\log\log q + 6) \cdot \frac{4}{3q} < 1$$

for all nontrivial $x \in G$, where $\delta = 1$ if $q \geqslant 27$, otherwise $\delta = 0$.

Finally, let us assume $n = 5$. If $q < 9$ then we compute $\Sigma(x) \leqslant 82/121$ for all nontrivial $x \in G$, so we may assume $q \geqslant 9$. Let $H \in \mathcal{M}$ and note that

$$n(H,P) \leqslant |G : N_G(P)| = \frac{(q-\epsilon)_{2'}}{(5, q-\epsilon)} \leqslant \frac{1}{2}(q+1)$$

by Theorem 3.1. Then by applying the main theorem of [**27**], we observe that one of the following holds:

(a) H is a subfield subgroup of type $\mathrm{GL}_5^{\epsilon}(q_0)$, where $q = q_0^e$ and $e \geqslant 3$ is a prime.
(b) H is of type $\mathrm{GL}_1^{\epsilon}(q) \wr S_5$ and $q \equiv \epsilon \pmod 4$.
(c) $\epsilon = +$ and H is the stabilizer of a 1-space or a 4-space.
(d) $\epsilon = -$ and H is the stabilizer of a nondegenerate 1-space.

Write $\mathcal{M} = \mathcal{M}_1 \cup \mathcal{M}_2$ and $\Sigma(x) = \Sigma_1(x) + \Sigma_2(x)$, where $\mathcal{M}_1$ comprises the subgroups in (a) and (b).

Let $H \in \mathcal{M}_1$. By combining the bound in (3) (see [**7**]) with the lower bound

$$|x^G| \geqslant \frac{|\mathrm{GL}_5^{\epsilon}(q)|}{q^7|\mathrm{GL}_3^{\epsilon}(q)||\mathrm{GL}_1^{\epsilon}(q)|} = \frac{(q^4-1)(q^5-\epsilon)}{q-\epsilon}$$

(equality if x is unipotent with Jordan form (J_2, J_1^3)), we deduce that $\mathrm{fpr}(x, G/H) < q^{-2}$ for all nontrivial $x \in G$. Now G has a unique class of subgroups as in (b). In addition, if $q = q_0^e$ then G has m classes of subfield subgroups of type $\mathrm{GL}_5^\epsilon(q_0)$, where

$$m = \left(5, \frac{q-\epsilon}{q_0-\epsilon}\right) = \begin{cases} 5 & \text{if } e = 5 \text{ and } q_0 \equiv \epsilon \pmod 5 \\ 1 & \text{otherwise.} \end{cases}$$

If $m = 5$ then $(5, q-\epsilon) = 5$ and thus $n(H,P) \leqslant (q+1)/10$. Since there are at most $\log\log q$ possibilities for e we deduce that

$$\Sigma_1(x) < \frac{1}{2}(q+1)q^{-2}(\delta \log\log q + 1),$$

where $\delta = 1$ if $q \geqslant 27$, otherwise $\delta = 0$.

Finally, suppose $H \in \mathcal{M}_2$. As before, P fixes a unique 1-space and a unique 4-space, so $n(H,P) = 1$ and by applying the main theorem of [**25**] we deduce that

$$\Sigma_2(x) < 2 \cdot \frac{4}{3q}.$$

Since $q \geqslant 9$, one can now check that $\Sigma(x) < 1$ for all nontrivial $x \in G$. □

By combining Theorems 2.3, 3.7, 3.12, 3.17, 3.18, 3.19, 3.20, 3.25 and 3.27, we conclude that the proof of Theorem 1.4 is complete. This reduces the proof of Theorem 1.1 to the linear groups $\mathrm{L}_n(3)$ with $n \geqslant 7$ odd and we handled this special case in Theorem 3.24. Therefore, the proof of Theorem 1.1 is complete.

References

[1] M. Aschbacher and R. Guralnick, *Solvable generation of groups and Sylow subgroups of the lower central series*, J. Algebra **77** (1982), no. 1, 189–201, DOI 10.1016/0021-8693(82)90286-1. MR665173

[2] Á. Bereczky, *Maximal overgroups of Singer elements in classical groups*, J. Algebra **234** (2000), no. 1, 187–206, DOI 10.1006/jabr.2000.8458. MR1799483

[3] W. Bosma, J. Cannon, and C. Playoust, *The Magma algebra system. I. The user language*, J. Symbolic Comput. **24** (1997), no. 3-4, 235–265, DOI 10.1006/jsco.1996.0125. MR1484478

[4] J. N. Bray, D. F. Holt, and C. M. Roney-Dougal, *The maximal subgroups of the low-dimensional finite classical groups*, London Mathematical Society Lecture Note Series, vol. 407, Cambridge University Press, Cambridge, 2013, DOI 10.1017/CBO9781139192576. MR3098485

[5] T. Breuer, *The GAP Character Table Library, Version 1.3.3*, GAP package, `http://www.math.rwth-aachen.de/~Thomas.Breuer/ctbllib`, 2022.

[6] T. Breuer and R. M. Guralnick, *Finite groups can be generated by a π-subgroup and a π'-subgroup*, Israel J. Math. **257** (2023), no. 1, 56–70.

[7] T. C. Burness, *Fixed point ratios in actions of finite classical groups. I*, J. Algebra **309** (2007), no. 1, 69–79, DOI 10.1016/j.jalgebra.2006.05.024. MR2301233

[8] T. C. Burness, *Fixed point ratios in actions in finite classical groups. II*, J. Algebra **309** (2007), no. 1, 80–138, DOI 10.1016/j.jalgebra.2006.05.025. MR2301234

[9] T. C. Burness, *Fixed point ratios in actions of finite classical groups. III*, J. Algebra **314** (2007), no. 2, 693–748, DOI 10.1016/j.jalgebra.2007.01.011. MR2344583

[10] T. C. Burness, *Fixed point ratios in actions of finite classical groups. IV*, J. Algebra **314** (2007), no. 2, 749–788, DOI 10.1016/j.jalgebra.2007.01.012. MR2344584

[11] T. C. Burness, S. Gerhardt, and R. M. Guralnick, *Topological generation of exceptional algebraic groups*, Adv. Math. **369** (2020), 107177, 50, DOI 10.1016/j.aim.2020.107177. MR4095751

[12] T. C. Burness and M. Giudici, *Classical groups, derangements and primes*, Australian Mathematical Society Lecture Series, vol. 25, Cambridge University Press, Cambridge, 2016, DOI 10.1017/CBO9781139059060. MR3443032

[13] T. C. Burness and R. M. Guralnick, *Fixed point ratios for finite primitive groups and applications*, Adv. Math. **411** (2022), part A, Paper No. 108778, 90 pp., DOI 10.1016/j.aim.2022.108778. MR4512400

[14] T. C. Burness and S. Harper, *Computations concerning the uniform domination number of a finite simple group*, available at `http://seis.bristol.ac.uk/~tb13602/udncomp.pdf`

[15] R. Gow, *Commutators in finite simple groups of Lie type*, Bull. London Math. Soc. **32** (2000), no. 3, 311–315, DOI 10.1112/S0024609300007141. MR1750469

[16] R. M. Guralnick, *Generation of simple groups*, J. Algebra **103** (1986), no. 1, 381–401, DOI 10.1016/0021-8693(86)90194-8. MR860714

[17] R. M. Guralnick and W. M. Kantor, *Probabilistic generation of finite simple groups*, J. Algebra **234** (2000), no. 2, 743–792, DOI 10.1006/jabr.2000.8357. MR1800754

[18] R. M. Guralnick, M. W. Liebeck, E. A. O'Brien, A. Shalev, and P. H. Tiep, *Surjective word maps and Burnside's $p^a q^b$ theorem*, Invent. Math. **213** (2018), no. 2, 589–695, DOI 10.1007/s00222-018-0795-z. MR3827208

[19] R. M. Guralnick and P. H. Tiep, *Lifting in Frattini covers and a characterization of finite solvable groups*, J. Reine Angew. Math. **708** (2015), 49–72, DOI 10.1515/crelle-2013-0085. MR3420328

[20] W. M. Kantor, *Primitive permutation groups of odd degree, and an application to finite projective planes*, J. Algebra **106** (1987), no. 1, 15–45, DOI 10.1016/0021-8693(87)90019-6. MR878466

[21] P. Kleidman and M. Liebeck, *The subgroup structure of the finite classical groups*, London Mathematical Society Lecture Note Series, vol. 129, Cambridge University Press, Cambridge, 1990, DOI 10.1017/CBO9780511629235. MR1057341

[22] A. S. Kondrat′ev, *Normalizers of Sylow 2-subgroups in finite simple groups* (Russian, with Russian summary), Mat. Zametki **78** (2005), no. 3, 368–376, DOI 10.1007/s11006-005-0133-9; English transl., Math. Notes **78** (2005), no. 3-4, 338–346. MR2227510

[23] A. S. Kondrat′ev and V. D. Mazurov, *2-signalizers of finite simple groups* (Russian, with Russian summary), Algebra Logika **42** (2003), no. 5, 594–623, 638, DOI 10.1023/A:1025923522954; English transl., Algebra Logic **42** (2003), no. 5, 333–348. MR2025717

[24] R. Lawther, M. W. Liebeck, and G. M. Seitz, *Fixed point ratios in actions of finite exceptional groups of Lie type*, Pacific J. Math. **205** (2002), no. 2, 393–464, DOI 10.2140/pjm.2002.205.393. MR1922740

[25] M. W. Liebeck and J. Saxl, *Minimal degrees of primitive permutation groups, with an application to monodromy groups of covers of Riemann surfaces*, Proc. London Math. Soc. (3) **63** (1991), no. 2, 266–314, DOI 10.1112/plms/s3-63.2.266. MR1114511

[26] M. W. Liebeck and J. Saxl, *The primitive permutation groups of odd degree*, J. London Math. Soc. (2) **31** (1985), no. 2, 250–264, DOI 10.1112/jlms/s2-31.2.250. MR809946

[27] N. V. Maslova, *Classification of maximal subgroups of odd index in finite simple classical groups: addendum*, Sib. Èlektron. Mat. Izv. **15** (2018), 707–718, DOI 10.1007/s11633-015-0942-6. MR3827971

School of Mathematics, University of Bristol, Bristol BS8 1UG, United Kingdom
Email address: `t.burness@bristol.ac.uk`

Department of Mathematics, University of Southern California, Los Angeles, California 90089-2532
Email address: `guralnic@usc.edu`

Contemporary Mathematics
Volume **800**, 2024
https://doi.org/10.1090/conm/800/16052

Representations of the quantum double of quasitriangular Hopf algebras

Miriam Cohen and Sara Westreich

Dedicated to the memory of Prof. Amitsur, a teacher and guide, with great admiration

Abstract. Conjugacy classes for groups were generalized in previous works of the authors to semisimple Hopf algebras over algebraically closed fields of characteristic zero. An essential property of the set of conjugacy classes $\{\mathfrak{C}_j\}$ is that they are irreducible representations of the quantum double $D(H)$ of H. We show that more is true. When (H, R) is quasitriangular, then any irreducible representations of $D(H)$ is a direct summand of a tensor product of the form of $\mathfrak{C}_j \otimes V_i$, where V_i is an irreducible representation of H. The proof follows by applying a partition defined in our 2011 paper on the indices of the irreducible characters of $D(H)$ and identifying the partition with the equivalence classes obtained from Nichols-Richmond equivalence relations on the set of the simple subcoalgebra of $D(H)^*$. For any block of this partition, there is a unique $\mathfrak{C}_j$ whose character as a $D(H)$-representation belongs to it.

When V is a 1-dimensional representation of H then $\mathfrak{C}_j \otimes V$ is an irreducible representation of $D(H)$, isomorphic to $\mathfrak{C}_j$ as a vector space endowed with a twisted action of $D(H)$.

We apply the above results to $H = kG$, where G are the groups S_3 and D_4, and decompose all $\mathfrak{C}_j \otimes V_i$ into their irreducible components.

Introduction

Semisimple Hopf algebras H over algebraically closed fields of characteristic 0 have an elusive nature. On the one hand, they resemble finite groups (or group algebras) in many aspects. As a result, many of the techniques in dealing with Hopf algebras try to use methods of group theory and their representation theory, sometimes with a great deal of success. On the other hand, many elementary facts about finite groups are either hard to translate or even false.

A particular example of this resemblance is the concept of *Conjugacy Classes* that was first defined by us in [**1**], extending the notion of conjugacy classes of groups. Their properties and their contribution to the structure theory of H were further investigated in various works. An essential property of the set of conjugacy classes $\{\mathfrak{C}_j\}$, (see [**2**]), is that they are irreducible representations of $D(H)$, the quantum (Drinfeld) double of H. In this paper, we demonstrate how conjugacy

2020 *Mathematics Subject Classification.* Primary 16W22, 16W99.

Key words and phrases. Quantum double; Drinfrld double, Quasitriangular Hopf algebras, Hopf conjugacy classes.

Miriam Cohen passed away in November 2023.

classes of a semisimple quasitriangular Hopf algebra H are the key to capture all irreducible representations of $D(H)$.

To avoid ambiguity, we denote by $\{\mathfrak{C}^j\}$ the conjugacy classes of H. Objects in $D(H)$ and $D(H)^*$ will usually appear with $\widehat{\ }$.

For this purpose, we use a partition $I = \cup_s I_s$ defined in [**2**] on the indices of primitive idempotents $\{F_i\}$ of $R(D(H))$, the algebra generated by the irreducible characters of $D(H)$. Note that the primitive idempotents of $R(D(H))$ and the irreducible characters of $D(H)$ can be organized with respect to the same set of indices I. In Proposition 1.2, we identify this partition with the equivalence classes of irreducible characters obtained from the Nichols-Richmond equivalence relation with respect to a certain Hopf subalgebra $\widehat{B}$ of $D(H)$, as follows:

Proposition 1.2. *For any two irreducible characters $\widehat{\chi}_j, \widehat{\chi}_k$ of $D(H)$,*

$$j, k \in I_s \iff \widehat{\chi}_j \cong_{\widehat{B}} \widehat{\chi}_k.$$

We then focus on the characters $\{\widehat{\chi}_{j_{\mathfrak{C}^i}}\}$ of the irreducible $D(H)$-representations $\{\mathfrak{C}^i\}$. We denote by i^* the index of the primitive idempotent $s(F_i)$, and obtain:

Theorem 1.4. *Let (H, R) be a semisimple quasitriangular Hopf algebra, $\mathfrak{C}^i$ a conjugacy class and $\widehat{\chi}_{j_{\mathfrak{C}^i}} \in R(D(H))$ its corresponding irreducible character. Then $j_{\mathfrak{C}^i} \in I_{i^*}$.*

Recall that when (H, R) is quasitriangular we have a Hopf algebra projection $\Phi_R : D(H) \to H$ and thus a Hopf injection $\Phi_R^* : H^* \to D(H)^*$. Moreover, for $T = (R^\tau)^{-1}$, (H, T) is another quasitriangular structure on H. Based on the above, we can prove our main theorems.

Theorem 1.6. *Let (H, R) be a semisimple quasitriangular Hopf algebra and let $\widehat{B}_R = \Phi_T^*(H^*)$ where $T = (R^\tau)^{-1}$. Let $\mathfrak{C}^i$ be a conjugacy class, $\widehat{\chi}_{j_{\mathfrak{C}^i}} \in R(D(H))$ its corresponding irreducible character and $\widehat{B}_{j_{\mathfrak{C}^i}}$ the simple subcoalgebra of $D(H)^*$ generated by $\widehat{\chi}_{j_{\mathfrak{C}^i}}$. Then*

$$D(H)^* = \bigoplus_i \widehat{B}_{j_{\mathfrak{C}^i}} \widehat{B}_R.$$

In particular, each irreducible character of $D(H)$ is a constituent of

$$\widehat{\chi}_{j_{\mathfrak{C}^i}} \Phi_T^*(\chi_l)$$

for some conjugacy class $\mathfrak{C}^i$ and some irreducible H-character χ_l.

As a corollary, we obtain the main theorem of this paper:

Theorem 1.7. *Let (H, R) be a semisimple quasitriangular Hopf algebra. Then each irreducible $D(H)$-representation is isomorphic to a direct summand of $\mathfrak{C}^i \otimes V_l$, where V_l is an irreducible representation of H, $\mathfrak{C}^i$ some conjugacy class and $D(H)$ acts on $\mathfrak{C}^i$ by $\bullet$ and on V_l by the action induced from Φ_T.*

Of special interest are $\mathfrak{C}^i \otimes V$, where $V = k$ and τ is its character, that is, $h_{\dot{\tau}} 1 = \langle \tau, h \rangle 1$. In fact $\tau \in G(H^*)$, the group of grouplike elements of H^*. We can then modify the action of $D(H)$ on $\mathfrak{C}^i$ via $\underset{\Phi^*\tau}{\bullet}$ as follows:

$$(p \bowtie h) \underset{\Phi^*\tau}{\bullet} c = ((\tau \rightharpoonup h)_{\dot{ad}} c) \leftharpoonup s(p \leftharpoonup f(\tau^{-1})),$$

where $f : H^* \to H$ is given by the quasitriangular structure defined on H.

Based on the above, we obtain a family of $D(H)$-representations that are, on the one hand, twistings of $\mathfrak{C}^i$, and thus irreducible, and on the other hand have the form $\mathfrak{C}^i \otimes (k, \dot{\tau})$.

Proposition 2.1 *Let* $\tau \in G(H^*)$, *then*

$$\mathfrak{C}^i \otimes (k, \dot{\tau}) \cong (\mathfrak{C}^i, \underset{\Phi^*\tau}{\bullet})$$

as $D(H)$*-representations.*

We wish to emphasize that $(\mathfrak{C}^i, \underset{\Phi^*\tau}{\bullet})$ and $(\mathfrak{C}^i, \underset{\Phi^*\tau'}{\bullet})$ may be isomorphic as $D(H)$-representations even if $\tau \neq \tau'$. For the case $H = kG$, G a finite group, we obtain

Proposition 2.3. *Let* $H = kG$, $\sigma \in G$, $\tau, \tau' \in G(kG^*)$. *Let* $Z_G(\sigma)$ *denote the centralizer of* σ *in* G. *Let* $\mathfrak{C}^i = k\mathfrak{C}_\sigma$ *where* $\mathfrak{C}_\sigma$ *is the (group) conjugacy class of* σ. *Then*

$$\tau_{|Z_G(\sigma)} = \tau'_{|Z_G(\sigma)} \Longleftrightarrow (k\mathfrak{C}_\sigma, \underset{\tau}{\bullet}) \cong (k\mathfrak{C}_\sigma, \underset{\tau'}{\bullet})$$

as $D(H)$*-modules.*

In the last section, we apply the above results to $H = kG$, where G are the groups S_3 and D_4. We decompose each $\mathfrak{C}_j \otimes V_i$ into its irreducible components.

Preliminaries

Throughout this paper, H is a semisimple Hopf algebra over an algebraically closed field k of characteristic 0. We denote by S and s the antipodes of H and H^*, respectively. Denote by $Z(H)$ the center of H.

The Hopf algebra H^* becomes a right and left representation of H by the *hit* actions $\leftharpoonup$ and $\rightharpoonup$ defined for all $a \in H$, $p \in H^*$,

$$\langle p \leftharpoonup a, a' \rangle = \langle p, aa' \rangle \qquad \langle a \rightharpoonup p, a' \rangle = \langle p, a'a \rangle.$$

H becomes a left and right H^*-representation analogously.

Denote by $\underset{ad}{\cdot}$ the left adjoint action of H on itself, that is, for all $a, h \in H$,

$$h \underset{ad}{\cdot} a = \sum h_1 a S(h_2).$$

A left coideal subalgebra of H is called *normal* if it is stable under the left adjoint action of H.

Let $D(H)$ denote the quantum double of the Hopf algebra H. We have the action $\bullet$ of $D(H)$ on H given by

$$(p \bowtie a) \bullet h = (a \underset{ad}{\cdot} h) \leftharpoonup s(p) \tag{1}$$

for all $p \in H^*$, $a, h \in H$.

Let (H, R) be any finite dimensional quasitriangular Hopf algebra. Recall the Hopf algebra maps $f_R : H^{*cop} \to H$, defined by $f_R(p) = \langle p, R^1 \rangle R^2$ and $f_{R^\tau} : H^{*op} \mapsto H$, defined by $f_{R^\tau}(p) = \langle p, R^2 \rangle R^1$. We define also the Drinfeld map $f_Q = f_{R^\tau} * f_R$ and its transpose $f_Q^* = f_R * f_{R^\tau}$. (H, R) is **factorizable** if the map f_Q is a monomorphism. In particular, $D(H)$ is always factorizable, where

$$f_{Q_{D(H)}}(a \otimes p) = (\varepsilon \bowtie a)(p \bowtie 1). \tag{2}$$

When (H, R) is quasitriangular, we have a Hopf projection $\Phi : D(H) \to H$, given by:

$$\Phi(p \bowtie h) = f_R(p)h.$$

Its dual Hopf injection $\Phi^* : H^* \longrightarrow D(H)^*$ is given explicitly by:

$$\Phi^*(p) = \sum f_{R^\tau}(p_1) \otimes p_2. \tag{3}$$

Let H be a semisimple Hopf algebra and $\{V_0, \ldots V_{n-1}\}$ be a complete set of non-isomorphic irreducible representation of H with $\{E_0, \ldots E_{n-1}\}$ and $\mathrm{Irr}(H) = \{\chi_0, \ldots, \chi_{n-1}\}$ the associated central primitive idempotents and irreducible H-characters respectively. Let $\dim V_i = d_i = \langle \chi_i, 1 \rangle$ where $E_0 = \Lambda$, is the idempotent integral of H and $\chi_0 = \varepsilon$. We denote by $R(H) = sp_k\{\chi_i\}$ the subalgebra of H^* generated by the characters of H. Recall that when H is semisimple, then so is $R(H)$. Moreover,

$$\lambda = \chi_H = \sum_{i=0}^{n-1} d_i \chi_i$$

is an integral for H^* satisfying $\langle \lambda, \Lambda \rangle = 1$. In particular,

$$\lambda \leftharpoonup E_i = d_i \chi_i \quad \text{hence} \quad \Lambda \leftharpoonup \chi_i = \frac{1}{d_i} S E_i$$

and $\{\frac{\chi_i}{d_i}\}$, $\{E_j\}$ are dual bases for $R(H)$ and $Z(H)$, respectively.

Note that when (H, R) is quasitriangular, then the Hopf projection Φ implies in particular that if V_i is an irreducible representation of H with associated character χ_i, then it is also an irreducible $D(H)$-representation with associated character $\Phi^*(\chi_i) \in R(D(H)^*$.

For any semisimple Hopf algebra H, let $\{\frac{1}{d}\lambda = F_0, \ldots F_{m-1}\}$ be a complete set of central primitive idempotents of $R(H)$, and $\{f_0, \ldots, f_{m-1}\}$ be primitive orthogonal idempotents in $R(H)$ so that $f_i F_j = \delta_{ij} f_i$. We generalize from groups to semismple Hopf algebras the notions of conjugacy classes and conjugacy sums. We define the **conjugacy class** $\mathfrak{C}_i$ as:

$$\mathfrak{C}_i = \Lambda \leftharpoonup f_i H^*.$$

Then we have shown in [**2**, Th.1.4] that:

THEOREM. *Let H be a semisimple Hopf algebra and let $\{f_1, \ldots f_n\}$ be idempotents in $R(H)$ so that $\{f_i R(H)\}$ is the complete set of non-isomorphic irreducible $R(H)$-representations. Then $\mathcal{C}_i$ is an irreducible $D(H)$-representation. Moreover, let $\dim(f_i R(H)) = m_i$, then*

$$H \cong \oplus_{i=1}^n \mathcal{C}_i^{\oplus m_i}$$

as $D(H)$-representations.

We generalize also the notion of **Class sum** as follows:

$$C_i = \Lambda \leftharpoonup d f_i.$$

A representative of a conjugacy class is replaced by

$$\eta_i = \frac{C_i}{\dim(f_i H^*)} \in Z(H).$$

We refer to η_i as a **normalized class sum**.

If $R(H)$ is commutative (as is the case for quasitriangular Hopf algebras), then $\{F_i\}$ forms a basis of $R(H)$ and $f_i = F_i$ for all i. Hence

$$\mathfrak{C}_i = \Lambda \leftharpoonup F_i H^*, \qquad \eta_i = \frac{\Lambda \leftharpoonup d F_i}{\dim(F_i H^*)}. \tag{4}$$

In this case, $\{F_i\}$ and $\{\eta_j\}$ form another set of dual bases for $R(H)$ and $Z(H)$, respectively.

Recall that when (H, R) is factorizable and E_i is the central idempotent related to the irreducible character χ_i, then the set $\{F_i = f_Q^{-1}(E_i)\}$ is the set of primitive idempotents of $R(H)$ and then [**1**, Proof of Th. 4.3]

$$f_Q(\frac{\chi_i}{d_i}) = \eta_i. \tag{5}$$

It follows that $\{\chi_i,\, E_i,\, F_i,\, \eta_i\}$ can all be organized with respect to the same set of indices I.

For any normal left coideal subalgebra N of H, we denote by $H//N$ the Hopf quotient $\overline{H} = H/HN^+$, where $N^+ = N \cap \ker \varepsilon$. Recall that for any Hopf surjection $\pi : H \to \overline{H}$, we define

$$H^{co\pi} = \{h \in H | \sum h_1 \otimes \pi(h_2) = h \otimes \pi(1)\}.$$

Then by [**12**], $H^{co\pi}$ is a normal left coideal subalgebra of H and

$$H//H^{co\pi} \cong \overline{H}.$$

Related concepts are invariants. For any subalgebra T of H^*, we denote by H^T the set of T-invariants of H under the left *hit* action. That is,

$$H^T = \{h \in H\, |b \rightharpoonup h = \langle b, 1\rangle h,\ \forall b \in T\}. \tag{6}$$

It was shown in [**3**, Cor.2.5] that B is a Hopf subalgebra of H^* if and only if $N = H^B$ is a normal left coideal subalgebra of H. In this case, one also has $\dim B = \frac{d}{\dim N}$ and $B = (H^*)^N$.

Fix $\pi : H \to \overline{H}$, $B = \overline{H}^* \subset H^*$ and N a left coideal subalgebra. Then summing up:

$$N = H^{co\pi} \Leftrightarrow \overline{H} = B^* \cong H//N \Leftrightarrow \overline{H}^* = B = (H^*)^N \Leftrightarrow N = H^B. \tag{7}$$

1. Interrelations between $D(H)$ and H

In order to distinguish between H and $D(H)$, we use the following. Elements of $D(H)$ and $D(H)^*$ will usually appear with $\widehat{\ }$. In particular, we denote the irreducible characters of $D(H)$ by $\widehat{\chi}_j$ with corresponding irreducible modules of dimensions $\widehat{d}_j$ and write $f_{\widehat{Q}}$ for $f_{Q(D(H))}$. Moreover, let $\{F^i\}$ denote the set of primitive orthogonal idempotents of $R(H)$ and $\{\eta^i\}$ the corresponding normalized class sums in H. Let $\widehat{F^i} = \Phi^*(F^i)$, an idempotent of $D(H)^*$. Denote $s(F^i)$, which is another primitive idempotent of $R(H)$, by F^{i^*}. Then $S\eta^i = \eta^{i^*}$.

Recall, [**2**], we have a partition of the set $I = \{j|\widehat{F}_j$ is a primitive idempotents of $R(D(H))\}$, given by $I = \cup I_s$, where

$$I_s = \{j|\widehat{F}_j\widehat{F}^s = \widehat{F}_j\}. \tag{8}$$

We obtain

LEMMA 1.1. *Let $\{\widehat{\eta}_j\}$, $\{\eta^i\}$ be the sets of normalized class sums in $D(H)$ and H respectively. Then*

$$\Phi(\widehat{\eta}_j) = \eta^i \Longleftrightarrow j \in I_i.$$

PROOF. Let $\widehat{\eta}_j$ be a normalized class sum of $D(H)$. Recall, [**2**, Lemma 2.4], if $j \in I_i$ then $\Phi(\widehat{\eta}_j) = \eta^i$. As for the reverse direction, assume $\Phi(\widehat{\eta}_j) = \eta^i$. Since $\{I_s\}$ is a partition of the set I, it follows that $j \in I_k$ for some $k \in I$. By $\Leftarrow$, $\Phi(\widehat{\eta}_j) = \eta^k$; hence $k = i$ and so $j \in I_i$. □

The partition and the Nichols-Richmond equivalence relation. We wish to relate the partition defined in (8) with the Nichols-Richmond equivalence relation with respect to a certain Hopf subalgebra B of H^*. Recall [**9**, Proposition 18], the equivalence relation defined on the simple subcoalgebra of H^* via B by:

$$B_i \cong B_k \Longleftrightarrow B_i \subset B_k B.$$

Equivalently,

$$\chi_j \cong_B \chi_k \Longleftrightarrow \lambda_B \frac{\chi_j}{d_j} = \lambda_B \frac{\chi_k}{d_k} \tag{9}$$

where λ_B is an integral of B. Note that the Nichols-Richmond equivalence relation yields a decomposition of H^*

$$H^* = \bigoplus_s B_s B$$

where $\{B_s\}$ is a set of representatives of each equivalence class.

In what follows, we apply the Nichols-Richmond equivalence relation to $D(H)$ with respect to a specific Hopf subalgebra $\widehat{B}$ of $D(H)^*$.

Set $\widehat{N} = D(H)^{co\Phi}$, that is,

$$\widehat{N} = \{x \in D(H) | \sum x_1 \otimes \Phi(x_2) = x \otimes 1\}.$$

Recall that $\widehat{N}$ is the maximal normal left coideal subalgebra of $D(H)$ mapped to k under Φ. By (6) and (7) where $D(H)$ here stands for H there and H here stands for $\overline{H}$ there, we obtain that

$$\dim \widehat{N} = d. \tag{10}$$

Set

$$\widehat{B} = f_{\widehat{Q}}^{-1}(\widehat{N}).$$

By [**4**, Lemma 4.2(iii)], $\widehat{B}$ is a Hopf subalgebra of $D(H)^*$. Let $\lambda_{\widehat{B}} \neq 0$ denote an integral of $\widehat{B}$ and $\Lambda_{\widehat{N}}$ the idempotent integral of $\widehat{N}$. We have that for all $\widehat{n} \in \widehat{N}$, $\widehat{n} = f_{\widehat{Q}}(\widehat{b})$, some $\widehat{b} \in \widehat{B}$, (and so $\langle \widehat{b}, 1 \rangle = \langle \varepsilon, \widehat{n} \rangle$),

$$f_{\widehat{Q}}(\lambda_{\widehat{B}})\widehat{n} = f_{\widehat{Q}}(\lambda_{\widehat{B}}\widehat{b}) = \langle \widehat{b}, 1 \rangle f_{\widehat{Q}}(\lambda_{\widehat{B}}) = \langle \varepsilon, \widehat{n} \rangle f_{\widehat{Q}}(\lambda_{\widehat{B}})$$

where the first equality follows from [**6**] since $\lambda_{\widehat{B}}$ is cocommutative. This implies that $f_{\widehat{Q}}(\lambda_{\widehat{B}})$ is a non-zero integral for $\widehat{N}$.

PROPOSITION 1.2. *For any $\widehat{\chi}_j, \widehat{\chi}_k \in R(D(H))$,*

$$j, k \in I_s \Longleftrightarrow \widehat{\chi}_j \cong_{\widehat{B}} \widehat{\chi}_k.$$

PROOF. We have

$$\begin{aligned}
&\widehat{\chi}_j \cong_{\widehat{B}} \widehat{\chi}_k \Leftrightarrow\\
&\Leftrightarrow \lambda_{\widehat{B}}\frac{\widehat{\chi}_j}{d_j} = \lambda_{\widehat{B}}\frac{\widehat{\chi}_k}{d_k}\\
&\Leftrightarrow \Phi f_{\widehat{Q}}(\lambda_{\widehat{B}}\frac{\widehat{\chi}_j}{d_j}) = \Phi f_{\widehat{Q}}(\lambda_{\widehat{B}}\frac{\widehat{\chi}_k}{d_k})\\
&\Leftrightarrow \Phi f_{\widehat{Q}}(\frac{\widehat{\chi}_j}{d_j}) = \Phi f_{\widehat{Q}}(\frac{\widehat{\chi}_k}{d_k}) \text{ (since } f_{\widehat{Q}}(\lambda_{\widehat{B}}) = \langle \lambda_{\widehat{B}}, 1\rangle \Lambda_{\widehat{N}} \text{ and } \Phi(\Lambda_{\widehat{N}}) = 1)\\
&\Leftrightarrow \Phi(\widehat{\eta}_j) = \Phi(\widehat{\eta}_k) = \eta^s
\end{aligned}$$

for some s. By Lemma 1.1, $\Phi(\widehat{\eta}_j) = \Phi(\widehat{\eta}_k) = \eta^s \Longleftrightarrow j, k \in I_s$. □

A distinguished family of irreducible characters of $D(H)$. Motivated by [**13**, Lemma. 5.7], we focus on the following distinguished family of irreducible $D(H)$-representations.

Let $\widehat{\chi}_{j_{\mathfrak{C}^i}}$ denote the character of the irreducible $D(H)$-representation $\mathfrak{C}^i$. That is,

$$\langle \widehat{\chi}_{j_{\mathfrak{C}^i}}, p \bowtie h\rangle = Trace_{(\mathfrak{C}^i,\bullet)}(p \bowtie h) \tag{11}$$

where the action $\bullet$ is defined in (1). By (5),

$$f_{\widehat{Q}}\left(\frac{\widehat{\chi}_{j_{\mathfrak{C}^i}}}{\dim \mathfrak{C}^i}\right) = \widehat{\eta}_{j_{\mathfrak{C}^i}} \in Z(D(H)).$$

We have

PROPOSITION 1.3. *Let $\mathfrak{C}^i$ be a conjugacy class of H and η^i its normalized class sum. Let $\widehat{\chi}_{j_{\mathfrak{C}^i}} \in R(D(H))$ be the character of $\mathfrak{C}^i$ as an irreducible $D(H)$ module. Then*

$$\Phi(\widehat{\eta}_{j_{\mathfrak{C}^i}}) = S\eta^i.$$

PROOF. Let $\chi_k \in R(H)$, then $\Phi^*(\chi_k) \in R(D(H))$ which implies by (5) that $f_{\widehat{Q}}\Phi^*(\chi_k) \in Z(D(H)$, hence it acts on the $D(H)$-irreducible module $\mathfrak{C}^i$ as a scalar that equals

$$\left\langle \frac{\widehat{\chi}_{j_{\mathfrak{C}^i}}}{\dim \mathfrak{C}^i}, f_{\widehat{Q}}\Phi^*(\chi_k)\right\rangle.$$

Now, by (3)

$$f_{\widehat{Q}}\Phi^*(\chi_k) = \sum (\varepsilon \bowtie f_{R^\tau}(\chi_{k_1}))\,((\chi_{k_2} \bowtie 1)\,.$$

In particular, by the definition of the action $\bullet$,

$$f_{\widehat{Q}}\Phi^*(\chi_k) \bullet \eta^i = \left(\sum (f_{R^\tau}(\chi_{k_1})\right)_{ad} (\eta^i \leftharpoonup s(\chi_{k_2})) = \left\langle \frac{\widehat{\chi}_{j_{\mathfrak{C}^i}}}{\dim \mathfrak{C}^i}, f_{\widehat{Q}}\Phi^*(\chi_k)\right\rangle \eta^i.$$

Taking ε on the middle term yields

$$\langle \varepsilon, f_{R^\tau}(\chi_{k_1})\rangle \langle \eta^i, s(\chi_{k_2})\rangle = \langle \chi_k, S\eta^i\rangle\,,$$

while taking ε on the right hand side yields by [**1**, Lemma 4.1]

$$\left\langle \frac{\widehat{\chi}_{j_{\mathfrak{C}^i}}}{\dim \mathfrak{C}^i}, f_{\widehat{Q}}\Phi^*(\chi_k)\right\rangle = \left\langle \Phi f^*_{\widehat{Q}}(\frac{\widehat{\chi}_{j_{\mathfrak{C}^i}}}{\dim \mathfrak{C}^i}), \chi_k\right\rangle = \left\langle \Phi f_{\widehat{Q}}(\frac{\widehat{\chi}_{j_{\mathfrak{C}^i}}}{\dim \mathfrak{C}^i}), \chi_k\right\rangle.$$

Since $\Phi f_{\widehat{Q}}(\frac{\widehat{\chi}_{j_{\mathfrak{C}^i}}}{\dim \mathfrak{C}^i}) \in Z(H)$ and $\{\chi_k\}$ is a basis for $R(H)$, a dual space of $Z(H)$, it follows that $\Phi f_{\widehat{Q}}(\frac{\widehat{\chi}_{j_{\mathfrak{C}^i}}}{\dim \mathfrak{C}^i}) = S\eta^i$. □

Recall the index notation for $S\eta^i$ is η^{i^*}. As a consequence of Proposition 1.3 and Lemma 1.1, we obtain:

THEOREM 1.4. *Let (H, R) be a semisimple quasitriangular Hopf algebra, $\mathfrak{C}^i$ a conjugacy class and $\widehat{\chi}_{j_{\mathfrak{C}^i}} \in R(D(H))$ its corresponding irreducible character as defined in* (11). *Then $j_{\mathfrak{C}^i} \in I_{i^*}$.*

Recall that if (H, R) is a quasitriangular where $R = \sum R^1 \otimes R^2$, then (H, T) is quasitriangular as well, where $T = (R^\tau)^{-1} = \sum R^2 \otimes SR^1$. Note that $f_T(p) = \sum \langle p, R^2 \rangle SR^1(p) = Sf_{R^\tau}(p)$. We denote $T^\tau T$ by Q_T and the corresponding Hopf projection from $D(H)$ to H by Φ_T. To avoid confusion, we will refer to the original map Φ as Φ_R.

Recall (6) the definition of $(D(H)^*)^{\widehat{N}}$. We can now prove

PROPOSITION 1.5. *Let $\widehat{N}_R = D(H)^{co\Phi_R}$, $\widehat{N}_T = D(H)^{co\Phi_T}$ and let $\widehat{B}_R = f_{\widehat{Q}}^{-1}(\widehat{N}_R)$, $\widehat{B}_T = f_{\widehat{Q}}^{-1}(\widehat{N}_T)$ be the respective Hopf subalgebras of $D(H)^*$. Then*

$$\widehat{B}_R = \Phi_T^*(H^*) = (D(H)^*)^{\widehat{N}_T}.$$

Switching R and T, we have also

$$\widehat{B}_T = \Phi_R^*(H^*) = (D(H)^*)^{\widehat{N}_R}.$$

PROOF. We prove the equalities in the first line; the proof of the second part is identical.

Observe that for all $p \in H^*$ we have

$$\begin{aligned}
\Phi_R f_{\widehat{Q}} \Phi_T^*(p) &= \\
&= \Phi_R f_{\widehat{Q}} \left(\sum f_{R^{-1}}(p_1) \otimes p_2 \right) \qquad \text{(by (3))} \\
&= \Phi_R \left(\sum (\varepsilon \bowtie f_{R^{-1}}(p_1))(p_2 \bowtie 1) \right) \qquad \text{(by (2))} \\
&= \sum f_{R^{-1}}(p_1) f_R(p_2) \qquad \text{since } \Phi_R \text{ is multiplicative} \\
&= \langle p, 1 \rangle 1.
\end{aligned}$$

Now, $\Phi_T^*(H^*)$ is a Hopf subalgebra of $D(H)^*$; hence by [**4**, Lemma 4.1], $f_{\widehat{Q}} \Phi_T^*(H^*)$ is a normal left coideal subalgebra of $D(H)$, mapped to k by Φ_R. By definition of $D(H)^{co\Phi_R}$, we obtain $f_{\widehat{Q}} \Phi_T^*(H^*) \subset D(H)^{co\Phi_R}$. However, $f_{\widehat{Q}} \Phi_T^*(H^*)$ is d-dimensional since $f_{\widehat{Q}} \Phi_T^*$ is injective, while $D(H)^{co\Phi_R} = \widehat{N}_R$ is d-dimensional by (10). Hence equality holds, that is,

$$f_{\widehat{Q}} \Phi_T^*(H^*) = \widehat{N}_R.$$

Applying $f_{\widehat{Q}}^{-1}$ to both sides yields

$$\Phi_T^*(H^*) = \widehat{B}_R.$$

This proves the first equality.

We show now the second one. Observe that for all $p \in H^*$, $\widehat{n} \in \widehat{N}_T = D(H)^{co\Phi_T}$, we have

$$\langle \Phi_T^*(p), \widehat{n} \rangle = \langle p, \Phi_T(\widehat{n}) \rangle = \langle p, 1 \rangle \langle \varepsilon, \widehat{n} \rangle .$$

Since $\Phi_T^*(H^*)$ is a Hopf subalgebra and $\widehat{N}_T$ is a left coideal, it is straightforward to see that $\Phi_T^*(H^*) \subset (D(H)^*)^{\widehat{N}_T}$.

Calculating dimensions again, we have that $\dim(\Phi_T^*(H^*)) = d$, while by (7) $\dim(D(H)^*)^{\widehat{N}_T} = d$ as well. Hence equality holds. This concludes the proof of the second equality. □

We can prove now our main theorem.

THEOREM 1.6. *Let (H, R) be a semisimple quasitriangular Hopf algebra and let $\widehat{B}_R = \Phi_T^*(H^*)$ where $T = (R^\tau)^{-1}$. Let $\mathfrak{C}^i$ be a conjugacy class, $\widehat{\chi}_{j_{\mathfrak{C}^i}} \in R(D(H))$ its corresponding irreducible character and $\widehat{B}_{j_{\mathfrak{C}^i}}$ the simple subcoalgebra of $D(H)^*$ generated by $\widehat{\chi}_{j_{\mathfrak{C}^i}}$. Then*

$$D(H)^* = \bigoplus_i \widehat{B}_{j_{\mathfrak{C}^i}} \widehat{B}_R.$$

In particular, each irreducible character of $D(H)$ is a constituent of

$$\widehat{\chi}_{j_{\mathfrak{C}^i}} \Phi_T^*(\chi_l)$$

for some conjugacy class $\mathfrak{C}^i$ and some irreducible H-character χ_l

PROOF. By Proposition 1.5, for $\widehat{N} = \widehat{N}_R$, we have $\widehat{B}_R = \Phi_T^*(H^*)$. Proposition 1.2 and Theorem 1.4 imply now that $\{\widehat{B}_{j_{\mathfrak{C}^i}}\}$ form a set of representatives for the Nichols-Richmond equivalence classes with respect to $\widehat{B}_R$. Thus we obtain the desired composition.

In particular,

$$R(D(H)) = \bigoplus_i \widehat{\chi}_{j_{\mathfrak{C}^i}} \Phi_T^*(R(H)).$$

□

As a corollary we obtain:

THEOREM 1.7. *Let (H, R) be a semisimple quasitriangular Hopf algebra. Then each irreducible $D(H)$-representation is isomorphic to a direct summand of $\mathfrak{C}^i \otimes V_l$, where V_l is an irreducible representation of H, $\mathfrak{C}^i$ some conjugacy class and $D(H)$ acts on $\mathfrak{C}^i$ by $\bullet$ and on V_l by the action induced from Φ_T. Explicitly, for $v \in V_l$, $c \in \mathfrak{C}^i$, $p \in H^*$, $h \in H$,*

$$(p \bowtie h) \underset{\bullet \otimes \Phi_T}{\cdot} (c \otimes v) = \sum (h_1 \dot{_{ad}} c) \leftharpoondown s(p_2) \otimes \langle p_1, R^2 \rangle SR^1 h_2 \cdot v. \tag{12}$$

REMARK 1.8. When $H = kG$, G a finite group with $R = 1 \otimes 1$, then (12) boils down to

$$(p \bowtie h) \underset{\bullet \otimes \Phi_T}{\cdot} (c \otimes v) = \sum ((h_1 \dot{_{ad}} c) \leftharpoondown Sp) \otimes (h_2 \cdot v).$$

2. Twisted actions of $D(H)$

In what follows, we describe various twists of the action of $D(H)$ on its irreducible representations. Unless there is a matter of ambiguity, we denote by Φ either Φ_R or Φ_T and by f either f_R or f_T.

Of particular interest are $D(H)$-representations obtained from $\mathfrak{C}^i$ by tensoring it with 1-dimensional representations of H. Note that if H is any Hopf algebra and τ is a character of a 1-dimensional representation of H, then τ is multiplicative. In fact, $\tau \in G(H^*)$, the group of grouplike elements of H^*. Let $(M, \cdot)$ be any representation of H, then we have a modified action of H on M given by

$$h \dot{_\tau} m = (\tau \rightharpoonup h) \cdot m. \tag{13}$$

Moreover, if $(M, \cdot)$ is an irreducible representation of H, so is $(M, \dot{_\eta})$.

Back to $D(H)$. By [**10**, Prop.10], grouplike elements of $D(H)^*$ are of the form

$$g \otimes \tau,\ g \in G(H),\ \tau \in G(H^*),\ \tau \bowtie g \in Z(D(H)). \tag{14}$$

Observe in particular that if $g \in Z(H)$ and $\tau \in Z(H^*)$, then necessarily $g \otimes \tau \in G(D(H^*))$.

Twisted actions of $D(H)$ on conjugacy classes. Consider the action $\bullet$ of $D(H)$ on $\mathfrak{C}^i$. Given $g \otimes \tau \in G(D(H)^*)$, we have the modified $g \otimes \tau$-action of $D(H)$ on $\mathfrak{C}^i$, $(\mathfrak{C}^i, \underset{g\otimes\tau}{\bullet})$ given by:

$$(p \bowtie h) \underset{g\otimes\tau}{\bullet} c = ((\tau \rightharpoonup h) \dot{_{ad}} c) \leftharpoonup s(p \leftharpoonup g) \tag{15}$$

for all $p \in H^*,\ h \in H,\ c \in \mathfrak{C}^i$. When H is quasitriangular, then since Φ^* is a Hopf algebra injection, it follows that for $\tau \in G(H^*)$,

$$\Phi^*(\tau) = f(\tau^{-1}) \otimes \tau \in G(D(H)^*).$$

Hence we have a twisted $\Phi^*(\tau)$-action of $D(H)$ on $\mathfrak{C}^i$. We denote this action by $\underset{\tau}{\bullet}$. That is,

$$(p \bowtie h) \underset{\tau}{\bullet} c = ((\tau \rightharpoonup h) \dot{_{ad}} c) \leftharpoonup s(p \leftharpoonup (\tau^{-1})). \tag{16}$$

Recall that any 1-dimensional representation of H is of the form $(k, \dot{_\tau})$, where $\tau \in G(H^*)$ and $\dot{_\tau}$ is defined as:

$$h \dot{_\tau} 1 = \langle \tau, h \rangle\, 1.$$

Based on the above, we obtain for a quasitriangular Hopf algebra (H, R) a family of $D(H)$-representations that on the one hand are twistings of $\mathfrak{C}^i$ and on the other hand have the form $\mathfrak{C}^i \otimes (k, \dot{_\tau})$.

PROPOSITION 2.1. *Let $\tau \in G(H^*)$, then*

$$\mathfrak{C}^i \otimes (k, \dot{_\tau}) \cong (\mathfrak{C}^i, \underset{\tau}{\bullet})$$

as $D(H)$-representations. In particular,

$$\mathfrak{C}^i \otimes (k, \dot{_\varepsilon}) \cong \mathfrak{C}^i.$$

PROOF. For $c \in \mathfrak{C}^i$, let

$$\iota(c \otimes 1) = c.$$

Then

$$\begin{aligned}
&\iota\Big((p \bowtie h) \underset{\bullet \otimes \Phi}{\cdot} (c \otimes 1)\Big) = \\
&= \iota\left(\sum (h_1 \,\dot{}_{ad} c) \leftharpoonup s(p_2) \otimes \langle p_1, SR^2 \rangle \, R^1 h_2 \dot{}_{\tau} 1\right) \qquad \text{(by (12))} \\
&= \iota\left(\sum (h_1 \,\dot{}_{ad} c) \leftharpoonup s(p_2) \otimes \langle p_1, SR^2 \rangle \langle R^1 h_2, \tau \rangle \, 1\right) \\
&= \iota\left(\sum (h_1 \,\dot{}_{ad} c) \leftharpoonup s(p_2) \langle p_1, f_R(\tau^{-1}) \rangle \langle h_2, \tau \rangle \otimes 1\right) \quad (\text{since } Sf_R(\tau) = f_R(\tau^{-1})) \\
&= \iota\left((\tau \rightharpoonup h) \,\dot{}_{ad} c) \leftharpoonup s(p \leftharpoonup f_R(\tau^{-1})) \otimes 1\right) \\
&= (\tau \rightharpoonup h) \,\dot{}_{ad} c) \leftharpoonup s(p \leftharpoonup f_R(\tau^{-1})) \\
&= (p \bowtie h) \underset{\tau}{\bullet} c \\
&= (p \bowtie h) \underset{\tau}{\bullet} \iota(c \otimes 1).
\end{aligned}$$

□

REMARK 2.2. If $\tau \neq \tau'$, then $(k, \dot{}_{\tau}) \neq (k, \dot{}_{\tau'})$ as a 1-dimensional representation of H. However, $\mathfrak{C}^i \otimes (k, \dot{}_{\tau}) \cong (\mathfrak{C}^i, \underset{\tau}{\bullet})$ and $\mathfrak{C}^i \otimes (k, \dot{}_{\tau'}) \cong (\mathfrak{C}^i, \underset{\tau'}{\bullet})$ may be isomorphic as $D(H)$-representations. See below the case $H = kG$. Specific examples for $G = S_3$ and $G = D_4$ are given in the next paragraph.

When $H = kG$ with $R = 1 \otimes 1$, we can say more. Let $\{p_g\}_{g \in G}$ be a dual basis in $(kG)^*$ for $\{g\}_{g \in G}$. Then $\Phi^*(\tau) = 1 \otimes \tau$ and we have

$$(17) \qquad (p_g \bowtie g') \underset{\tau}{\bullet} \sigma = ((\tau \rightharpoonup g') \,\dot{}_{ad} \sigma) \leftharpoonup s(p_g) = \delta_{g^{-1}, g' \dot{}_{ad} \sigma} \langle \tau, g' \rangle \, (g' \,\dot{}_{ad} \sigma)$$

for all $\sigma \in \mathfrak{C}^i$, $g, g' \in G$. We then have

PROPOSITION 2.3. *Let $H = kG$, $\sigma \in G$, $\tau, \tau' \in G(kG^*)$. Let $Z_G(\sigma)$ denote the centralizer of σ in G. Let $\mathfrak{C}^i = k\mathfrak{C}_\sigma$ where $\mathfrak{C}_\sigma$ is the (group) conjugacy class of σ. Then*

$$\tau_{|Z_G(\sigma)} = \tau'_{|Z_G(\sigma)} \iff (k\mathfrak{C}_\sigma, \underset{\tau}{\bullet}) \cong (k\mathfrak{C}_\sigma, \underset{\tau'}{\bullet})$$

as $D(H)$-modules.

PROOF. It is enough to show that $(k\mathfrak{C}_\sigma, \underset{\tau}{\bullet}) \cong k\mathfrak{C}_\sigma \iff \tau_{|Z_G(\sigma)} = \varepsilon$. To show it, we calculate the characters of $k\mathfrak{C}_\sigma$ and $(k\mathfrak{C}_\sigma, \underset{\tau}{\bullet})$.

Let $\mathfrak{C}_\sigma = \{\sigma = \sigma_1, \sigma_2, \ldots \sigma_k\}$. Then

$$\widehat{\chi}_{\mathfrak{C}_\sigma}(p_g \bowtie g') = \sum_i \langle (p_g \bowtie g') \bullet \sigma_i, p_{\sigma_i} \rangle = \sum_i \left\langle (g' \,\dot{}_{ad} \sigma_i) \leftharpoonup p_{g^{-1}}), p_{\sigma_i} \right\rangle.$$

For each summand, $\left\langle (g' \,\dot{}_{ad} \sigma_i) \leftharpoonup p_{g^{-1}}), p_{\sigma_i} \right\rangle = (\delta_{g^{-1}, g' \dot{}_{ad} \sigma_i})(\delta_{g^{-1}, \sigma_i})$. In particular, $g' \in Z_G(\sigma_i)$.

Now, let $\{g_1, \dots, g_l\}$ be so that $g_i \cdot_{ad} \sigma = \sigma_i$, and let $\{a_1, \dots, a_k\} = Z_G(\sigma)$. Then $Z_G(\sigma_i) = \{g_i \cdot_{ad} a_1, \dots, g_i \cdot_{ad} a_k\}$. Thus $g' = g_i \cdot_{ad} a_j$ for some j. Thus we obtain

$$\begin{aligned}
\widehat{\chi}_{\mathfrak{C}_\sigma} &= \\
&= \sum_{g,g'} \langle \widehat{\chi}_{\mathfrak{C}_\sigma}, p_g \bowtie g' \rangle\, g \otimes p_{g'} \\
&= \sum_{g',i} \left\langle \widehat{\chi}_{\mathfrak{C}_\sigma}, p_{\sigma_i^{-1}} \bowtie g' \right\rangle \sigma_i^{-1} \otimes p_{g'} \\
&= \sum_{j,i} \left\langle \widehat{\chi}_{\mathfrak{C}_\sigma}, p_{\sigma_i^{-1}} \bowtie g_i \cdot_{ad} a_j \right\rangle \sigma_i^{-1} \otimes p_{g_i \cdot_{ad} a_j} \\
&= \sum_{j,i} \sigma_i^{-1} \otimes p_{g_i \cdot_{ad} a_j}.
\end{aligned}$$

To calculate $\widehat{\chi}_{(\mathfrak{C}_\sigma, \underset{\tau}{\bullet})}$, we have

$$\widehat{\chi}_{(\mathfrak{C}_\sigma, \underset{\tau}{\bullet})}(p_g \bowtie g') = \sum_i \langle ((\tau \rightharpoonup g') \cdot_{ad} \sigma_i) \leftharpoonup p_{g^{-1}}), p_{\sigma_i} \rangle.$$

Again, for each summand $\langle ((\tau \rightharpoonup g') \cdot_{ad} \sigma_i) \leftharpoonup p_{g^{-1}}), p_{\sigma_i} \rangle = (\delta_{g^{-1}, (\tau \rightharpoonup g') \cdot_{ad} \sigma_i})(\delta_{g^{-1}, \sigma_i})$. In particular, $\tau \rightharpoonup g' \in Z_G(\sigma_i)$. Hence $g' = g_i \cdot_{ad} (\tau^{-1} \rightharpoonup a_j)$ and so, similar calculations yield

$$\widehat{\chi}_{(\mathfrak{C}_\sigma, \underset{\tau}{\bullet})} = \sum_{i,j} \sigma_i^{-1} \otimes p_{g_i \cdot_{ad} (\tau^{-1} \rightharpoonup a_j)}.$$

Hence $\widehat{\chi}_{(\mathfrak{C}_\sigma, \underset{\tau}{\bullet})} = \widehat{\chi}_{\mathfrak{C}_\sigma}$ if and only if $\tau \rightharpoonup a_j = a_j$ for all $a_j \in Z_G(\sigma)$. Since we assume characteristic zero, it follows that in this case the two representations are isomorphic. □

Twisted actions of $D(H)$ on representations of (H, R). Let $\sigma \in G(H)$ and let $(V, \cdot)$ be an irreducible H-module. If $\sigma \in Z(H)$, then $\varepsilon \bowtie \sigma \in Z(D(H))$; hence $\sigma \otimes \varepsilon \in G(D(H)^*)$ by (14). Let $(V, \cdot)$ be an irreducible H-module, then V is an irreducible $D(H)$-module via Φ. We can now twist the action $\underset{\Phi}{\cdot}$ of $D(H)$ on V via $\sigma \otimes \varepsilon$. That is,

$$(p \bowtie h) \cdot_{\Phi, \sigma} v = (f(\sigma \rightharpoonup p))h \cdot v. \tag{18}$$

In particular, for $(kG, 1 \otimes 1)$, we have

$$(p_g \bowtie g') \cdot_{\Phi, \sigma} v = \delta_{\sigma, g}(g' \cdot v). \tag{19}$$

3. Examples

A method to construct all representations of $D(kG)$ was introduced in [**5**] and is in common use in the literature, see e.g., [**7**]. In this section, we present two examples based on the analysis developed in this paper.

Representations of the quantum double of kS_3. Set $x = (1,2)$ and $y = (1,2,3)$. Recall that $x^2 = 1$, $y^3 = 1$ and $xy = y^2x$. Then conjugacy classes of S_3 are given by:

$$\mathfrak{C}_1 = \{1\} \quad \mathfrak{C}_x = \{x, yx, y^2x\} \quad \mathfrak{C}_y = \{y, y^2\}.$$

The irreducible representations of kS_3 are the following:

$$V_1 = (k,_{\dot{\varepsilon}}), \qquad V_2 = (k,_{\dot{\text{sgn}}}), \qquad V_3 = Sp_k\{v_1, v_2\}$$

where for ω a third root of unity,

$$x \cdot v_1 = v_2, \quad x \cdot v_2 = v_1, \quad y \cdot v_1 = \omega v_1, \quad y \cdot v_2 = \omega^2 v_2.$$

EXAMPLE 3.1. The decompositions of $D(kS_3)$-modules of the form $k\mathfrak{C}_\sigma \otimes V_l$ is as follows:

Item 1. $k\mathfrak{C}_1 \otimes V_l \cong (V_l, \underset{\Phi_T}{\cdot})$ by definition.

Item 2. $k\mathfrak{C}_x \otimes V_1 \cong (k\mathfrak{C}_x, \underset{\varepsilon}{\bullet})$ by Proposition 2.1.

Item 3. $k\mathfrak{C}_x \otimes V_2 \cong (k\mathfrak{C}_x, \underset{\text{sgn}}{\bullet})$ by Proposition 2.1.

Item 4. $k\mathfrak{C}_y \otimes V_1 \cong k\mathfrak{C}_y \otimes V_2 \cong (k\mathfrak{C}_y, \underset{\varepsilon}{\bullet})$ by Proposition 2.1.

Item 5. $k\mathfrak{C}_x \otimes V_3 \cong (k\mathfrak{C}_x, \underset{\varepsilon}{\bullet}) \oplus (k\mathfrak{C}_x, \underset{\text{sgn}}{\bullet})$. Indeed, let

$$\{y^i x \otimes v_j,\ i = 0,1,2,\ j = 1,2\}$$

be a basis for $k\mathfrak{C}_x \otimes V_3$. Then

$$(k\mathfrak{C}_x, \underset{\varepsilon}{\bullet}) \cong Sp_k\{x \otimes (v_1 + v_2), yx \otimes (\omega v_1 + \omega^2 v_2), y^2x \otimes (\omega^2 v_1 + \omega v_2)\}$$

via

$$\begin{aligned} x \otimes (v_1 + v_2) &\mapsto x, \\ y^2x \otimes (\omega v_1 + \omega^2 v_2) &\mapsto y^2x, \\ yx \otimes (\omega^2 v_1 + \omega v_2) &\mapsto yx \end{aligned}$$

and

$$(k\mathfrak{C}_x, \underset{\text{sgn}}{\bullet}) \cong Sp_k\{x \otimes (v_1 - v_2), yx \otimes (\omega v_1 - \omega^2 v_2), y^2x \otimes (\omega^2 v_1 - \omega v_2)\}$$

via

$$\begin{aligned} x \otimes (v_1 - v_2) &\mapsto x, \\ yx \otimes (\omega v_1 - \omega^2 v_2) &\mapsto yx, \\ y^2x \otimes (\omega^2 v_1 - \omega v_2) &\mapsto y^2x. \end{aligned}$$

Item 6. $k\mathfrak{C}_y \otimes V_3 \cong W_1^y \oplus W_2^y$ as follows. Let

$$\{y^i \otimes v_j,\ i = 1,2,\ j = 1,2\}$$

be a basis for $k\mathfrak{C}_y \otimes V_3$. Then

$$W_1^y = sp_k\{y \otimes v_1,\ y^2 \otimes v_2\}$$

and

$$W_2^y = sp_k\{y^2 \otimes v_1,\ y \otimes v_2\}.$$

Observe that W_i^y, $i = 1, 2$ are irreducible $D(kS_3)$-representations. This follows since there are only two grouplike elements in $D(kS_3)^*$, $1 \otimes \varepsilon$ and $1 \otimes \mathrm{sgn}$, implying there are only two 1-dimensional representations of $D(kS_3)$, that is, $(k, \underset{1\otimes\varepsilon}{\cdot})$ and $(k, \underset{1\otimes\mathrm{sgn}}{\cdot})$. For both $\widehat{\chi}(p_y \bowtie 1) = 0$, while $\widehat{\chi}_{W_i^y}(p_y \bowtie 1) = 1$. Hence W_i^y cannot be a direct sum of two 1-dimensional representations.

REMARK 3.2. Note that Item 4 demonstrates Remark 2.2 and Proposition 2.3 as $\mathrm{sgn}_{|Z_G(y)} = \varepsilon$.

Representations of the quantum double of kD_4. Set $x = (1, 2)$ and $y = (1, 2, 3, 4)$ so that

$$x^2 = 1,\ y^4 = 1,\ xy = y^3x.$$

Conjugacy classes of D_4 are given by:

$$\mathfrak{C}_1 = \{1\} \quad \mathfrak{C}_{y^2} = \{y^2\} \quad \mathfrak{C}_y = \{y, y^3\} \quad \mathfrak{C}_x = \{x, y^2x\} \quad \mathfrak{C}_{yx} = \{yx, y^3x\}$$

$G(kD_4^*) = \{\varepsilon, \mathrm{sgn}, \tau, \tau'\}$, where

$$\tau(x) = 1,\ \tau(y) = -1,\ \tau'(x) = -1,\ \tau'(y) = -1.$$

Hence we have four 1-dimensional kD_4-representations given by:

$$V_1 = (k, \underset{\varepsilon}{\cdot}), \quad V_2 = (k, \underset{\mathrm{sgn}}{\cdot}), \quad V_3 = (k, \underset{\tau}{\cdot}), \quad V_4 = (k, \underset{\tau'}{\cdot})$$

and a unique 2-dimensional representation

$$V_5 = Sp_k\{v, v'\}$$

where

$$x \cdot v = v',\ x \cdot v' = v,\ y \cdot v = iv,\ y \cdot v' = -iv'. \tag{20}$$

EXAMPLE 3.3. The decompositions of $D(kD_4)$-modules of the form $k\mathfrak{C}_\sigma \otimes V_l$ is as follows:

V_l is 1-dimensional. Based on Proposition 2.1 and Proposition 2.3, the following twists of the actions on conjugacy classes are obtained from tensoring with 1-dimensional representations of kD_4.

$$\begin{array}{llll}
\mathfrak{C}_1, & (\mathfrak{C}_1, \underset{\mathrm{sgn}}{\bullet}), & (\mathfrak{C}_1, \underset{\tau}{\bullet}) & (\mathfrak{C}_1, \underset{\tau'}{\bullet}) \\
\mathfrak{C}_{y^2}, & (\mathfrak{C}_{y^2}, \underset{\mathrm{sgn}}{\bullet}), & (\mathfrak{C}_{y^2}, \underset{\tau}{\bullet}) & (\mathfrak{C}_{y^2}, \underset{\tau'}{\bullet}) \\
\mathfrak{C}_x, & (\mathfrak{C}_x, \underset{\mathrm{sgn}}{\bullet}) & & \\
\mathfrak{C}_y, & (\mathfrak{C}_y, \underset{\mathrm{sgn}}{\bullet}) & & \\
\mathfrak{C}_{yx}, & (\mathfrak{C}_{yx}, \underset{\tau}{\bullet}). & &
\end{array}$$

Decompositions of $k\mathfrak{C}_\sigma \otimes V_5$. We have:

Item 1. $k\mathfrak{C}_1 \otimes V_5 \cong (V_5, \underset{\Phi}{\cdot})$ by definition.

Item 2. $k\mathfrak{C}_{y^2} \otimes V_5 \cong (V_5, \underset{\Phi, y^2}{\cdot})$, as described in (19). This follows since $y^2 \in Z(D_4)$; hence we have the twisted action via $y^2 \otimes \varepsilon$ on V_5.

Item 3. $k\mathfrak{C}_y \otimes V_5 \cong W_1^y \oplus W_2^y$, as follows. Let
$$\{y \otimes v,\ y^3 \otimes v,\ y \otimes v',\ y^3 \otimes v'\}$$
be a basis for $\mathfrak{C}_y \otimes V_5$. Then
$$W_1^y = Sp_k\{y \otimes v, y^3 \otimes v'\}$$
and
$$W_2^y = Sp_k\{y^3 \otimes v, y \otimes v'\}.$$

Item 4. $k\mathfrak{C}_x \otimes V_5 \cong W_1^x \oplus W_2^x$, as follows. Let
$$\{x \otimes v,\ y^2x \otimes v,\ x \otimes v',\ y^2x \otimes v'\}$$
be a basis for $\mathfrak{C}_x \otimes V_5$. Then
$$W_1^x = Sp_k\{x \otimes (v - v'), y^2x \otimes (v + v')\}$$
and
$$W_2^x = Sp_k\{x \otimes (v + v'), y^2x \otimes (v - v')\}.$$

Item 5. $k\mathfrak{C}_{yx} \otimes V_5 \cong W_1^{yx} \oplus W_2^{yx}$, as follows. Let
$$\{yx \otimes v,\ y^3x \otimes v,\ yx \otimes v',\ y^3x \otimes v'\}$$
be a basis for $\mathfrak{C}_{yx} \otimes V_5$. Then
$$W_1^{yx} = Sp_k\{yx \otimes (v - v'), y^3x \otimes (v + v')\}$$
and
$$W_2^{yx} = Sp_k\{yx \otimes (v + v'), y^3x \otimes (v - v')\}.$$

Similar to Example 3.1, it can be checked that all $D(kD_4)$-modules W_i^y, W_i^x, W_i^{yx}, $i = 1, 2$ are irreducible. Otherwise, they would have been decomposed into 1-dimensional representations. However, there are eight 1-dimensional representations V with corresponding characters $\widehat{\chi}_V$,
$$\{1 \otimes \varepsilon, 1 \otimes \mathrm{sgn}, 1 \otimes \tau, 1 \otimes \tau',\ \sigma^2 \otimes \varepsilon, \sigma^2 \otimes \mathrm{sgn}, \sigma^2 \otimes \tau, \sigma^2 \otimes \tau'\}.$$
By Remark 1.8, we see for example that $\widehat{\chi}_{W_i^{yx}}(p_{yx} \bowtie 1) = 1$ while for each 1-dimensional representation V, $\widehat{\chi}_V(p_{yx} \bowtie 1) = 0$.

References

[1] M. Cohen and S. Westreich, *Higman ideals and Verlinde-type formulas for Hopf algebras*, Ring and module theory, Trends Math., Birkhäuser/Springer Basel AG, Basel, 2010, pp. 91–114, DOI 10.1007/978-3-0346-0007-1_5. MR2744044

[2] M. Cohen and S. Westreich, *Conjugacy classes, class sums and character tables for Hopf algebras*, Comm. Algebra **39** (2011), no. 12, 4618–4633, DOI 10.1080/00927872.2011.617265. MR2863455

[3] M. Cohen and S. Westreich, *Hit-invariants and commutators for Hopf algebras*, Bull. Math. Soc. Sci. Math. Roumanie (N.S.) **56(104)** (2013), no. 3, 299–313. MR3114476

[4] M. Cohen and S. Westreich, *Character tables and normal left coideal subalgebras*, J. Pure Appl. Algebra **218** (2014), no. 10, 1845–1866, DOI 10.1016/j.jpaa.2014.02.010. MR3195413

[5] R. Dijkgraaf, V. Pasquier, and P. Roche, *Quasi Hopf algebras, group cohomology and orbifold models*, Nuclear Phys. B Proc. Suppl. **18B** (1990), 60–72 (1991), DOI 10.1016/0920-5632(91)90123-V. MR1128130

[6] V. G. Drinfel′d, *Almost cocommutative Hopf algebras* (Russian), Algebra i Analiz **1** (1989), no. 2, 30–46; English transl., Leningrad Math. J. **1** (1990), no. 2, 321–342. MR1025154

[7] M. D. Gould, *Quantum double finite group algebras and their representations*, Bull. Austral. Math. Soc. **48** (1993), no. 2, 275–301, DOI 10.1017/S0004972700015707. MR1238802

[8] W. D. Nichols and M. B. Richmond, *The Grothendieck group of a Hopf algebra*, J. Pure Appl. Algebra **106** (1996), no. 3, 297–306, DOI 10.1016/0022-4049(95)00023-2. MR1375826

[9] W. D. Nichols and M. B. Richmond, *The Grothendieck algebra of a Hopf algebra. I*, Comm. Algebra **26** (1998), no. 4, 1081–1095, DOI 10.1080/00927879808826185. MR1612188

[10] D. E. Radford, *Minimal quasitriangular Hopf algebras*, J. Algebra **157** (1993), no. 2, 285–315, DOI 10.1006/jabr.1993.1102. MR1220770

[11] S. Skryabin, *Projectivity and freeness over comodule algebras*, Trans. Amer. Math. Soc. **359** (2007), no. 6, 2597–2623, DOI 10.1090/S0002-9947-07-03979-7. MR2286047

[12] M. Takeuchi, *Quotient spaces for Hopf algebras*, Comm. Algebra **22** (1994), no. 7, 2503–2523, DOI 10.1080/00927879408824973. MR1271619

[13] J. Zhou and S. Zhu, *The applications of probability groups on Hopf algebras*, Comm. Algebra **49** (2021), no. 1, 14–32, DOI 10.1080/00927872.2020.1791897. MR4193611

Department of Mathematics, Ben Gurion University, Beer Sheva, Israel

Department of Management, Bar-Ilan University, Ramat-Gan, Israel
Email address: swestric@biu.ac.il

Contemporary Mathematics
Volume **800**, 2024
https://doi.org/10.1090/conm/800/16053

Minimal quadratic forms for the function field of a conic in characteristic 2

Adam Chapman and Anne Quéguiner-Mathieu

Abstract. In this note, we construct explicit examples of F_Q-minimal quadratic forms of dimension 5 and 7, where F_Q is the function field of a conic over a field F of characteristic 2. The construction uses the fact that any set of n cyclic p algebras over a field of characteristic p can be described using only $n+1$ elements of the base field. It also uses a general result that provides an upper bound on the Witt index of an orthogonal sum of two regular anisotropic quadratic forms over a henselian valued field.

1. Introduction

Let Q be a quaternion division algebra over a field F. Its Severi-Brauer variety is a conic, which is a codimension 1 subform of the norm form n_Q of Q. We denote by F_Q the function field of this conic. One may try to describe anisotropic quadratic forms that become either hyperbolic or isotropic over this function field F_Q. For hyperbolicity, there is a well-known answer: since n_Q is a 2-fold Pfister form, the subform theorem [**6**, Cor. 23.6] asserts that an anisotropic quadratic form becomes hyperbolic over F_Q if and only if it is a multiple of n_Q. The question is much harder for isotropy. Clearly, any quadratic form having the conic as a subform becomes isotropic over F_Q. But the converse does not hold. This was initially noticed by Wadsworth, and led to the notion of minimal form, introduced by Hoffmann: a quadratic form φ over F is called F_Q-minimal if φ is anisotropic, φ_{F_Q} is isotropic, and φ does not contain any strict subform which becomes isotropic over F_Q. Minimal forms in characteristic different from 2 were studied in the 90's by Hoffmann, Lewis and Van Geel, see [**11**], [**18**], [**12**] and [**13**]. In particular, they described a field and a conic for which there exist minimal forms of any odd dimension ≥ 3, see [**13**, Thm. 4.6].

Large parts of this theory extend to the characteristic 2 case, as was shown by Faivre in [**8**]. In particular, he defined an invariant, called the minimal splitting

---

2020 *Mathematics Subject Classification.* Primary 16K20; Secondary 11E04, 16K20, 16W60.

Key words and phrases. Quaternion algebras, quadratic forms, function fields of conics, algebras with involution.

The first author acknowledges the receipt of the Chateaubriand Fellowship (969845L) offered by the French Embassy in Israel, and thanks LAGA, Université Sorbonne Paris Nord, for the hospitality in Fall 2020.

Both authors thank Jean-Pierre Tignol for useful conversations on v-value functions and norms.

index $\bar{s}_{F_Q}(\varphi)$, which can be used to characterize F_Q-minimal forms. For forms of dimension 5, he gave another characterization using the Clifford algebra. Nevertheless, his thesis does not contain any explicit example of F_Q-minimal forms. In this note, we construct such examples in dimension 5 and 7, see Proposition 4.1 and Proposition 4.6, which are the first to appear in the literature in characteristic 2.

Recall that the even Clifford algebra of a quadratic space (V, q) is naturally endowed with an involution, called the canonical involution, which is of symplectic type if the base field has characteristic 2 and V is odd dimensional, see [**15**, §8.A]. Moreover, a central simple algebra with symplectic involution (A, σ) is called hyperbolic if A contains an isotropic ideal of maximal dimension, that is a right ideal $I \subset A$ such that $\sigma(I)I = \{0\}$ and $\dim_F(I) = \frac{1}{2}\dim_F(A)$ [**15**, (6.8)]. Starting from an F_Q-minimal quadratic form of dimension 5, the even Clifford algebra functor produces an example of a degree 4 algebra with symplectic involution which is non hyperbolic, becomes hyperbolic over F_Q, and does not contain any subalgebra with involution isomorphic to $(Q, \bar{\ })$, where $\bar{\ }$ denotes the canonical involution. This was already observed in [**20**] in characteristic different from 2 and in [**4**] in characteristic 2. So our paper also provides explicit examples of such algebras with involution in characteristic 2.

It is true in general that the even Clifford algebra of a quadratic form decomposes as a tensor product of quaternion algebras with involution. Over a field of characteristic 2, any algebra which decomposes as a tensor product of n quaternion algebras may be described in a specific way, using only $n+1$ elements of the base field. This fact, which actually holds over any field of positive characteristic for tensor product of cyclic p algebras, see Theorem 2.1, has been a crucial ingredient in designing our examples of minimal forms.

The base fields in those examples are iterated Laurent series fields. Over such a field, any vector space endowed with a regular quadratic form admits a v-value function which is a norm. To prove the F_Q-minimality of the 7-dimensional example, we use a general result of independant interest and valid in arbitrary characteristic, which provides an upper bound for the Witt index of an orthogonal sum of two regular quadratic forms over a henselian valued field, see Proposition 3.1.

The content of the paper is as follows. In Section 2, we provide a general description of any set of n cyclic p algebras over a field of characteristic p, where p is an arbitrary prime number. The bound on the Witt index of an orthogonal sum of two regular quadratic forms over a henselian valued field of arbitrary characteristic is provided in Section 3. Finally, the examples of F_Q-minimal forms over a base field of characteristic 2 are described in Section 4.

Notations. Throughout the paper, F is a field of positive characteristic p, except in Section 3 where the characteristic is arbitrary; in Section 4 on minimal quadratic forms, we will assume $p = 2$.

For $\alpha \in F$ and $\beta \in F^\times$, we let $[\alpha, \beta)_{p,F}$ be the cyclic p-algebra generated over F by two elements i and j subject to the relations $i^p - i = \alpha$, $j^p = \beta$ and $jij^{-1} = i - 1$. It follows from the definition that $[\alpha + x^p - x, \beta y^p)_{p,F} \simeq [\alpha, \beta)_{p,F}$ for all $x \in F$ and $y \in F^\times$. Recall that $[\alpha, \beta)_{p,F}$ is either division or isomorphic to $M_p(F)$. In particular, for all $\alpha \in F$ and $\beta \in F^\times$, we have $[\alpha, 1)_{p,F} \simeq [0, \beta)_{p,F} \simeq [\beta, \beta)_{p,F} \simeq M_p(F)$. In the sequel, we will use the following, which is proved in

[**1**, Lemma 2.3(a)]:

$$[\alpha,\beta)_{p,F}\otimes_F[\gamma,\delta)_{p,F}\simeq[\alpha+\gamma,\beta)_{p,F}\otimes[\gamma,\beta^{-1}\delta)_{p,F},\tag{1}$$

for all $\alpha,\gamma\in F$ and $\beta,\delta\in F^\times$.

We denote by Ω_F^* the ring of differential forms. It contains an element $d\beta$ for all $\beta\in F$, and we let $\operatorname{dlog}(\beta)=\beta^{-1}\,d\beta$ for all $\beta\neq 0$. The subspace spanned by n-fold differentials Ω_F^n admits a well-defined Artin-Schreier map $\wp:\Omega_F^n\to\Omega_F^n/d\Omega_F^{n-1}$ given by

$$\alpha\operatorname{dlog}(\beta_1)\wedge\cdots\wedge\operatorname{dlog}(\beta_n)\mapsto(\alpha^p-\alpha)\operatorname{dlog}(\beta_1)\wedge\cdots\wedge\operatorname{dlog}(\beta_n).$$

The cokernel of this map is the Kato-Milne cohomology group $H_p^{n+1}(F)$. For $n=0$ one recovers the Artin-Schreier group $H_p^1(F)=F/\wp(F)$ which describes cyclic degree p extensions of F. For $n=1$, there is an isomorphism to the p-torsion of the Brauer group $H_p^2(F)\simeq {}_p\mathrm{Br}(F)$ (see [**10**, Section 9]) given by $\alpha\operatorname{dlog}(\beta)\mapsto[\alpha,\beta)_{p,F}$, where we use the same notation for the Brauer class of the cyclic algebra $[\alpha,\beta)_{p,F}$ as for the algebra itself. So, when viewed as an element of the Brauer group, the symbol $[\alpha,\beta)_{p,F}$ is additive in α and multiplicative in β, that is,

$$[\alpha_1+\alpha_2,\beta)_{p,F}=[\alpha_1,\beta)_{p,F}+[\alpha_2,\beta)_{p,F},\ \text{and}\ [\alpha,\beta_1\beta_2)_{p,F}=[\alpha,\beta_1)_{p,F}+[\alpha,\beta_2)_{p,F},$$

for all $\alpha,\alpha_1,\alpha_2\in F$ and $\beta,\beta_1,\beta_2\in F^\times$. Using these relations, one may also check the following identity:

$$[\alpha,\beta)_{p,F}\simeq[(\beta+x^p)\alpha\beta^{-1},\beta+x^p)_{p,F},\tag{2}$$

for all $\alpha\in F$, $\beta\in F^\times$ and $x\in F$ such that $\beta+x^p\neq 0$. Indeed, by definition, we have $(\beta+x^p)\operatorname{dlog}(\beta+x^p)=d(\beta+x^p)$. It follows that $(\beta+x^p)\alpha\beta^{-1}\operatorname{dlog}(\beta+x^p)=\alpha\beta^{-1}d(\beta+x^p)=\alpha\beta^{-1}d\beta=\alpha\operatorname{dlog}(\beta)$, as required. See also [**1**, Lemma 2.2(d)] for a direct argument.

Recall that when $p=2$, there is an isomorphism $H_2^{n+1}(F)=I_q^{n+1}F/I_q^{n+2}F$ where $I_q(F)$ is the fundamental ideal of the Witt group of nonsingular quadratic forms over F, as proved by Kato in [**14**]. The symbol $\alpha\operatorname{dlog}(\beta_1)\wedge\cdots\wedge\operatorname{dlog}(\beta_n)$ corresponds under this isomorphism to the n-fold Pfister form $\langle\!\langle\beta_n,\dots,\beta_1,\alpha]\!]$, which is, by definition, the tensor product of the $(n-1)$-fold bilinear Pfister form $\langle\!\langle\beta_n,\dots,\beta_1\rangle\!\rangle$ with the rank 2 non singular quadratic form $[1,\alpha]$. If $n=2$, we get the quadratic form $\langle\!\langle\beta,\alpha]\!]=[1,\alpha]\perp\beta[1,\alpha]$ which is the norm form n_Q of the quaternion algebra $Q=[\alpha,\beta)_{2,F}$.

Let (A,σ) be a central simple algebra, endowed with a symplectic involution. A right ideal $I\subset A$ is called isotropic (with respect to the involution σ) if $\sigma(x)y=0$ for all $x,y\in I$. An isotropic ideal has dimension at most $\frac{1}{2}\dim_F(A)$. The involution σ is called isotropic (respectively hyperbolic) if there exists a non trivial isotropic ideal $I\subset A$ (respectively an isotropic ideal of dimension $\frac{1}{2}\dim_F(A)$). Over a field of characteristic 2, these definitions apply in particular to the Clifford algebra and the even Clifford algebra of an odd dimensional quadratic space, since its canonical involution is symplectic, see [**15**, 6.A, 6.B, 8.A].

2. On sets and tensor products of cyclic p-algebras over a field of characteristic p

The main result in this section is the following:

THEOREM 2.1. *Let F be a field of positive characteristic p. Any n cyclic p-algebras can be written as*

$$[a_0, a_1)_{p,F}, [a_1, a_2)_{p,F}, \dots, [a_{n-1}, a_n)_{p,F},$$

for some non-zero elements $a_i \in F^\times$.

The proof uses the following lemma:

LEMMA 2.2. *Let $A_1 = [\alpha, \beta)_{p,F}$ and $A_2 = [\gamma, \delta)_{p,F}$ be two cyclic division p-algebras. Then, there exist $a, b \in F^\times$ such that $A_1 = [a, b)_{p,F}$ and $A_2 = [b, \delta)_{p,F}$.*

PROOF OF LEMMA2.2. Let $x = \gamma - \beta$, so that $\gamma + x^p - x = \beta + x^p$ and denote by b this element. Since A_1 is division, $\beta \notin F^{\times p}$, so that $b = \beta + x^p \neq 0$; hence, by (2), the algebra A_1 is given by

$$A_1 = [(\beta + x^p)\alpha\beta^{-1}, \beta + x^p)_{p,F} = [a, b)_{p,F},$$

where $a = (\beta + x^p)\alpha\beta^{-1}$. Again since A_1 is division we have $a \in F^\times$; moreover, we have

$$A_2 = [\gamma + x^p - x, \delta)_{p,F} = [b, \delta)_{p,F}$$

and this finishes the proof. □

PROOF OF THEOREM 2.1. We first prove the result when all algebras are division, by induction on their number n. It clearly holds if $n = 1$. Given n cyclic p-algebras $A_1, \dots, A_n$, the induction hypothesis provides $a'_1, a_2, \dots, a_n \in F^\times$ such that $A_2 = [a'_1, a_2)_{p,F}$ and $A_i = [a_{i-1}, a_i)_{p,F}$ for $i \in [\![3; n]\!]$. Lemma 2.2 applied to A_1 and A_2 now provides a_0 and $a_1 \in F^\times$ such that $A_1 = [a_0, a_1)_{p,F}$ and $A_2 = [a_1, a_2)_{p,F}$, and this concludes the proof in this case.
If the algebras are not all division, one may first write the division algebras of the considered set as $[a_0, a_1)_{p,F}, \dots, [a_{k-1}, a_k)_{p,F}$. Since $[a_k, 1)_{p,F} = [1, 1)_{p,F} = M_p(F)$, the result holds with $a_i = 1$ for $i \in [\![k + 1; n]\!]$. □

As an immediate corollary, we get

COROLLARY 2.3. *Let A be a tensor product of n cyclic p-algebras. There exists $a_0, \dots, a_n \in F^\times$ such that*

$$A = [a_0, a_1)_{p,F} \otimes [a_1, a_2)_{p,F} \otimes \cdots \otimes [a_{n-1}, a_n)_{p,F}.$$

REMARK 2.4. Theorem 2.1 provides an easy proof of the fact that the essential dimension of decomposable algebras of degree p^n and exponent p is at most $n + 1$. This follows from the essential dimension of the group $(\mathbb{Z}/p\mathbb{Z})^n$ over a field of characteristic p, which is 1 (see [**17**]), but the argument there is more elaborate. See [**19**] for a detailed discussion on the essential dimension of such algebras.

3. Witt index of an orthogonal sum of two anisotropic quadratic forms over a henselian valued field

The main result in this section provides an upper bound for the Witt index of an orthogonal sum of two anisotropic quadratic forms over a henselian valued field, under some condition on the value function associated to one of them. We refer the reader to [**22**, §3.1.1] for general information on value functions and norms.

Let F be a field of arbitrary characteristic, and V a finite-dimensional vector space over F. We assume F is endowed with a valuation $v : F \to \Gamma \cup \{\infty\}$, where Γ is a totally ordered group. Replacing Γ by its divisible hull if necessary, we may assume Γ is divisible, and we denote the value group of F by $\Gamma_F = v(F^\times) \subset \Gamma$. Recall that a v-value function on V is a map $\alpha : V \to \Gamma \cup \{\infty\}$ such that

(1) $\alpha(x) = 0$ if and only if $x = 0$;
(2) $\alpha(x + y) \geq \min(\alpha(x), \alpha(y))$ for all $x, y \in V$;
(3) $\alpha(x\lambda) = \alpha(x) + v(\lambda)$ for all $x \in V$ and $\lambda \in F$.

The value set of α, denoted by $\Gamma_\alpha = \{\alpha(x),\ x \in V,\ x \neq 0\} \subset \Gamma$, need not be a subgroup of Γ. Nevertheless, condition (3) above shows that Γ_α consists of a union of cosets of Γ_F, and we denote by $|\Gamma_\alpha : \Gamma_F|$ the number of cosets. The v-value function α is called a v-norm if V admits a splitting base for α, that is, a base $(e_i)_{1\leq i\leq n}$ such that for all $\lambda_1, \dots, \lambda_n \in F$, we have

$$\alpha\Big(\sum_{i=1}^{n} e_i\lambda_i\Big) = \min\{\alpha(e_i) + v(\lambda_i),\ 1 \leq i \leq n\}.$$

The value set Γ_α then consists of the cosets $\alpha(e_i) + \Gamma_F$ for $1 \leq i \leq n$. Hence, we have $|\Gamma_\alpha : \Gamma_F| \leq \dim_F(V)$, with equality if and only if these cosets are pairwise distinct.

Let us assume now that V is endowed with an anisotropic quadratic form φ. As explained in [**7**, §4], see also [**21**], the function $\alpha : V \to \Gamma \cup \{\infty\}$ defined by $\alpha(x) = \frac{1}{2}v(\varphi(x))$ for all $x \in V$ is a v-value function on V. We call it the v-value function associated to φ. Our construction of examples of 7-dimensional F_Q-minimal forms uses the following :

PROPOSITION 3.1. *Let (V, φ) and (W, ψ) be two anisotropic quadratic spaces over a henselian valued field (F, v), such that the orthogonal sum $\varphi \perp \psi$ is regular. Denote by α and β the v-value functions on V and W, respectively, associated to φ and ψ. If α is a norm and $|\Gamma_\alpha : \Gamma_F| = \dim_F(V)$, then the Witt index of the orthogonal sum $\varphi \perp \psi$ satisfies*

$$i_w(\varphi \perp \psi) \leq |(\Gamma_\alpha \cap \Gamma_\beta) : \Gamma_F|.$$

REMARK 3.2. Since φ and ψ are anisotropic, both are regular. In characteristic not 2, it follows their orthogonal sum also is regular, but this is not true in general in characteristic 2. We need this additional assumption in the statement.

PROOF. Let Z be a maximal totally isotropic subspace of $(V \oplus W, \varphi \perp \psi)$. Since $\varphi \perp \psi$ is regular, the dimension of Z is equal to the Witt index of this form, and we need to prove that $\dim_F(Z) \leq |(\Gamma_\alpha \cap \Gamma_\beta) : \Gamma_F|$. Denote by p_1 and p_2 the projections of $V \oplus W$ on V and W, respectively. Since Z is totally isotropic, for all $z \in Z$, we have

$$(\varphi \perp \psi)(z) = \varphi(p_1(z)) + \psi(p_2(z)) = 0. \tag{3}$$

On the other hand, φ and ψ are both anisotropic. Hence, the projection p_1 induces an isomorphism between Z and $Z_1 = p_1(Z) \subset V$. Similarly, p_2 induces an isomorphism between Z and $Z_2 = p_2(Z) \subset W$. Combining those isomorphisms and using (3), we get an isometry f between the quadratic spaces $(Z_1, \varphi_{|Z_1})$ and $(Z_2, -\psi_{|Z_2})$. By definition of the v-value functions α and β associated to φ and ψ, respectively, it follows that for all $z_1 \in Z_1$ we have

$$\alpha(z_1) = \frac{1}{2}v(\varphi(z_1)) = \frac{1}{2}v\big(-\psi(f(z_1))\big) = \frac{1}{2}v\big(\psi(f(z_1))\big) = \beta(f(z_1)).$$

Therefore, since f is bijective, the v-value functions α and β induce v-value functions on Z_1 and Z_2 with the same value group, which we denote by Γ_Z. It satisfies

$$\Gamma_Z = \{\alpha(z_1),\ z_1 \in Z_1,\ z_1 \neq 0\} = \{\beta(z_2),\ z_2 \in Z_2,\ z_2 \neq 0\} \subset \Gamma_\alpha \cap \Gamma_\beta,$$

and it follows that

$$|\Gamma_Z : \Gamma_F| \leq |(\Gamma_\alpha \cap \Gamma_\beta) : \Gamma_F|.$$

By [**22**, Prop. 3.14], since α is a norm, its restriction to Z_1 also is a norm; hence, in particular, we have $|\Gamma_Z : \Gamma_F| \leq \dim_F(Z_1)$. We claim they actually are equal. Indeed, assume for the sake of contradiction that $|\Gamma_Z : \Gamma_F| < \dim_F(Z_1)$. Again by [**22**, Prop. .14], Z_1 admits a splitting complement Y, that is $V = Z_1 \oplus Y$ and $\alpha(v) = \min\big(\alpha(z_1), \alpha(y)\big)$ for all $v = z_1 + y \in V$ with $z_1 \in Z_1$ and $y \in Y$. Therefore, we have $\Gamma_\alpha = \Gamma_{\alpha|Z_1} \cup \Gamma_{\alpha|Y} = \Gamma_Z \cup \Gamma_{\alpha|Y}$. Moreover, the restriction of α to Y also is a norm, so that its value set satisfies $|\Gamma_{\alpha|Y} : \Gamma_F| \leq \dim_F(Y)$. So, we get

$$|\Gamma_\alpha : \Gamma_F| \leq |\Gamma_Z : \Gamma_F| + |\Gamma_{\alpha|Y} : \Gamma_F| < \dim_F(Z_1) + \dim_F(Y) = \dim_F(V),$$

which contradicts the assumption we made on α in the statement. So we have $|\Gamma_Z : \Gamma_F| = \dim_F(Z_1) = \dim_F(Z)$, and this concludes the proof. □

4. Explicit examples of F_Q-minimal forms

From now on, we assume $p = 2$, so that F has characteristic 2, and we use the notation $[\alpha, \beta) = [\alpha, \beta)_{2,F}$ for all $\alpha \in F$ and $\beta \in F^\times$. The purpose of this section is to construct explicit examples of F_Q-minimal forms of dimension 5 and 7. In both cases, the results are proved using characterisations of F_Q-minimal forms in characteristic 2 which are due to Faivre [**8**, §5.2]. As explained below, the method is slightly different in dimension 5 and in dimension 7.

We first prove the following:

PROPOSITION 4.1. *Assume a, b, c are elements of F such that $a\,\mathrm{dlog}(b) \wedge \mathrm{dlog}(c)$ is a non trivial symbol in $H_2^3(F)$, and let Q be the quaternion algebra $Q = [a, c)$. The quadratic form $\varphi_1 = c[1, a+b] \perp b[1, a] \perp \langle 1 \rangle$ is anisotropic, and becomes isotropic over F_Q. If in addition $[a, b) \otimes_F [b, c)$ is division over F, then φ_1 is F_Q-minimal.*

PROOF. Let F, a, b, c and φ_1 be as in the statement. Since the quaternion algebras $[a+b, b)$ and $[a, b)$ are isomorphic, they have isometric norm form $\langle\langle b, a+b]] \simeq \langle\langle b, a]]$. It follows that

$$\varphi_1 \perp bc[1, a+b] \simeq c\langle\langle b, a]] \perp b[1, a] \perp \langle 1 \rangle.$$

Therefore, one may easily check that φ_1 is isometric to a subform of $\langle\langle b, c, a]]$. This form has non trivial Arason invariant $a\,\mathrm{dlog}(b) \wedge \mathrm{dlog}(c)$; hence it is anisotropic and so is φ_1. Moreover, $\langle\langle b, c, a]] = \langle\langle b\rangle\rangle \otimes n_Q$ is hyperbolic over the field F_Q, and it follows that the 5-dimensional subform φ_1 is isotropic over F_Q. We now use [**8**, Prop. 5.2.12] to check φ_1 is F_Q-minimal. The first condition has already been checked: φ_1 is a Pfister neighbour of $\langle\langle b, c, a]] = \langle\langle b\rangle\rangle \otimes n_Q$. Moreover, its Clifford algebra is Brauer equivalent to $[a+b, c) \otimes [a, b) \simeq [a+b, bc)$, and the tensor product with Q is

$$[a+b, bc) \otimes [a, c) = [a, b) \otimes [b, c),$$

which is division as required. This finishes the proof. □

To get an explicit example, it remains to exhibit a field over which the assumptions of the theorem are fulfilled. This can be done as follows:

EXAMPLE 4.2. *Let F_0 be a field of characteristic* 2 *and consider the field of iterated Laurent series $F = F_0((c^{-1}))((b^{-1}))((a^{-1}))$. By* [**2**, Lem. 3.4], *the biquaternion algebra $[a, b) \otimes [b, c)$ is division over F. Moreover, the symbol $a\,\mathrm{dlog}\,b \wedge \mathrm{dlog}\,c$ is non trivial in $H_2^3(F)$. Therefore, φ_1 and Q defined as in Proposition* 4.1 *provide an explicit example of an F_Q-minimal form.*

REMARK 4.3.

(1) By contrast, in characteristic 0, there exists no 5-dimensional F_Q-minimal form over the field of iterated Laurent series $\mathbb{C}((x))((y))((z))$. Indeed, there is no biquaternion division algebra over this field, see [**16**, XIII, Rmk. 2.5]; hence the conditions in [**12**, Prop. 4.1] are not satisfied.

(2) Since $b[1, b] \perp \langle 1\rangle \simeq [1, b] \perp \langle 1\rangle \simeq \mathbb{H} + \langle 1\rangle$, the form φ_1 is isometric to

$$c[1, a+b] \perp b[1, a+b] \perp \langle 1\rangle \simeq c\big(\langle\langle bc, a+b]] \perp \langle c\rangle\big).$$

In characteristic p, we may still consider the homogeneous degree p form $\langle c\rangle \perp n_T$, where n_T now stands for the norm form of $[a+b, bc)_{p,F}$. It is still true in this case that this form is anisotropic and becomes isotropic over the function field of the Severi-Brauer variety of $[a, c)_{p,F}$. This can be checked using [**9**, Thm. 6].

The even Clifford algebra of the quadratic form φ_1, endowed with its canonical involution, may be computed as in [**5**, Ex 2.1]. Applying [**4**, Prop. 3.6], we get the following :

COROLLARY 4.4. *Over the field $F = F_0((c^{-1}))((b^{-1}))((a^{-1}))$, with F_0 of characteristic* 2, *the biquaternion algebra with symplectic involution*

$$(C, \gamma) = \big([a+b, c), ^-\big) \otimes \big([a, b), ^-\big)$$

satisfies the following : it is anisotropic, becomes hyperbolic over F_Q, and does not contain any γ-stable subalgebra isomorphic to $(Q, ^-)$, where $Q = [a, c)$.

REMARK 4.5. Alternately, using some explicit base change, and the characterization of orthogonal involutions on a split quaternion algebra given in [**3**, Lem. 7.1], one may check that the even Clifford algebra of φ_1 is isomorphic to

$$(C, \gamma) = \big([a+b, bc), ^-\big) \otimes \big(M_2(F), \mathrm{ad}_{\langle\langle b\rangle\rangle}\big).$$

The corollary then follows from [**4**, Thm 3.4].

The previous example of a 5-dimensional quadratic form extends to dimension 7 as follows :

PROPOSITION 4.6. *Consider the quaternion algebra $Q = [a, d)$ over the field of iterated Laurent series $F = F_0((d^{-1}))((c^{-1}))((b^{-1}))((a^{-1}))$. The quadratic form*

$$\varphi_2 = d[1, a + c] \perp c[1, a + b] \perp bcd[1, a] \perp \langle 1 \rangle$$

over F is F_Q-minimal.

PROOF. The first part of the argument follows the same line as in dimension 5. Using the following isometries

$$d[1, a + c] \perp cd[1, a + c] \simeq d\langle\langle c, a + c]] \simeq d\langle\langle c, a]] \simeq d[1, a] \perp cd[1, a],$$

and similarly

$$c[1, a + b] \perp bc[1, a + b] \simeq c[1, a] \perp bc[1, a],$$

one may check that φ_2 is a subform of $\langle 1, c, bc\rangle \otimes \langle\langle d, a]] = \langle 1, c, bc\rangle \otimes n_Q$, which in turn is a subform of $\langle\langle b, c\rangle\rangle \otimes n_Q = \langle\langle b, c, d, a]]$. Since $a\,\mathrm{dlog}\, b \wedge \mathrm{dlog}\, c \wedge \mathrm{dlog}\, d$ is a non-trivial symbol in $H_2^4(F)$, this form is anisotropic, and so is φ_2. Moreover, over the function field F_Q, the forms n_Q and $\langle 1, c, bc\rangle \otimes n_Q$ are hyperbolic; hence φ_2 is isotropic.

Denote by χ the quadratic form $cd[1, a + c] \perp bc[1, a + b]$. Since $[1, a] \perp \langle 1\rangle \simeq \mathbb{H} \perp \langle 1\rangle$, the computation above actually shows that $\varphi_2 \perp \chi$ is Witt equivalent to $\langle 1, c, bc\rangle \otimes n_Q \perp \langle 1\rangle$. In the language of [**8**, Def. 5.2.3], this shows that the invariant $\bar{s}_{F_Q}(\varphi_2)$ is at most 3. In view of [**8**, Prop. 5.2.7], to prove φ_2 is F_Q-minimal, it is enough to prove that $\bar{s}_{F_Q}(\varphi_2) = 3$, that is, there is no 4-dimensional regular quadratic form χ over F such that $\varphi_2 \perp \chi$ is Witt equivalent to $\tau \otimes n_Q \perp \langle 1\rangle$ with $\tau \otimes n_Q$ anisotropic, and τ of dimension 1 or 2.

Let us assume, for the sake of contradiction, that such forms exist. Using again $[1, \lambda] \perp \langle 1\rangle \simeq \mathbb{H} \perp \langle 1\rangle$ for all $\lambda \in F$, we get that

$$\tilde{\varphi}_2 \perp \tilde{\chi} \perp \langle 1\rangle \sim \tau \otimes n_Q \perp \langle 1\rangle,$$

where

$$\tilde{\varphi}_2 = d[1, a + c] \perp c[1, a + b] \perp bcd[1, a] + [1, a + b + c]$$

and $\tilde{\chi}$ is the Albert form $\chi \perp [1, \delta]$, where δ is the discriminant of χ. By [**6**, Prop. 13.6], it follows that

$$\tilde{\varphi}_2 \perp \tilde{\chi} \sim \tau \otimes n_Q, \text{ with } \tau \text{ of dimension 1 or 2.}$$

We may now use Clifford algebras to prove τ has dimension 1. Indeed, since both forms have trivial discriminant, by additivity of the Clifford invariant [**6**, Lem. 14.2], the Witt equivalence above implies that $\mathcal{C}(\tilde{\varphi}_2) \otimes \mathcal{C}(\tilde{\chi})$ is split if τ has dimension 2 and Brauer equivalent to Q if τ has dimension 1. The Clifford algebra of $\tilde{\varphi}_2$ may be computed as in [**5**, Appendix], and we get

$$\mathcal{C}(\tilde{\varphi}_2) \simeq [a + c, d) \otimes [a + b, c) \otimes [a, bcd) \simeq [a, b) \otimes [b, c) \otimes [c, d).$$

By [**2**, Lem. 3.4], it is a division algebra. On the other hand, $\tilde{\chi}$ is an Albert form, so $\mathcal{C}(\tilde{\chi})$ is a biquaternion algebra. Hence the tensor product is not split, and it follows that τ has dimension 1. Moreover, the Clifford algebra of $\tilde{\chi}$ is Brauer-equivalent to $\mathcal{C}(\tilde{\varphi}_2) \otimes Q$ which is

$$[a + c, d) \otimes [a + b, c) \otimes [a, bc) \sim [a + c, cd) \otimes [a + b, bc).$$

Hence, we get $\tilde{\chi} \simeq \mu\big(cd[1, a + c] + bc[1, a + b] + [1, b + c]\big)$ for some $\mu \in F^\times$.

To finish the proof, we use Proposition 3.1. Denote by v the $(d^{-1}, c^{-1}, b^{-1}, a^{-1})$-adic valuation of F. Its value group is $\Gamma_F = \mathbb{Z}^4 \subset \Gamma = \mathbb{Q}^4$. By [**22**, Prop 3.8 and

Ex. 3.11(ii)], every v-value function on a finite dimensional F-vector space is a norm. In particular, the v-value functions α and β respectively associated to the quadratic forms $\tilde{\varphi}_2$ and $\tilde{\chi}$ are norms. Hence, the value set of α is the union of the cosets of values of vectors of a splitting base of V. Specifically, one may check that the base of V corresponding to the following presentation of the quadratic form,

$$\tilde{\varphi}_2 \simeq d[1, a+c] \perp c[1, a+b] \perp bcd[1, a] + [1, a+b+c],$$

is a splitting base. It follows that the value set Γ_α of the norm α consists of the cosets of the following elements of Γ:

$$\{\alpha_1; \alpha_1 + \alpha'; \alpha_2; \alpha_2 + \alpha'; \alpha_3; \alpha_3 + \alpha'; \alpha_4; \alpha_4 + \alpha'\},$$

where $\alpha_1 = \big(\frac{1}{2}, 0, 0, 0\big)$, $\alpha_2 = \big(0, \frac{1}{2}, 0, 0\big)$, $\alpha_3 = \big(\frac{1}{2}, \frac{1}{2}, \frac{1}{2}, 0\big)$, $\alpha_4 = \big(0, 0, 0, 0\big)$, and $\alpha' = \big(0, 0, 0, \frac{1}{2}\big)$.

Any two of those clearly give rise to distinct cosets, so that $|\Gamma_\alpha : \Gamma_F| = 8$, which is the dimension of the underlying vector space. Hence the assumptions in Proposition 3.1 are satisfied and we get that the Witt index of $\tilde{\varphi}_2 \perp \tilde{\chi}$ is at most $|(\Gamma_\alpha \cap \Gamma_\beta) : \Gamma_F|$, where Γ_β denotes the value set of the norm β. On the other hand, since $\tilde{\varphi}_2 \perp \tilde{\chi} \sim \tau \otimes n_Q$ with τ of dimension 1, this orthogonal sum has Witt index 5, so that

$$|(\Gamma_\alpha \cap \Gamma_\beta) : \Gamma_F| \geq 5.$$

Besides, we have

$$\tilde{\chi} \simeq \mu\big(cd[1, a+c] + bc[1, a+b] + [1, b+c]\big).$$

Hence, the value set Γ_β consists of the cosets of the elements

$$\{\beta_5 + \beta''; \beta_5 + \beta'' + \alpha'; \beta_6 + \beta''; \beta_6 + \beta'' + \alpha', \beta'', \beta_7 + \beta''\}$$

where $\beta_5 = \big(\frac{1}{2}, \frac{1}{2}, 0, 0\big)$, $\beta_6 = \big(0, \frac{1}{2}, \frac{1}{2}, 0\big)$, $\beta_7 = \Big(0, 0, \frac{1}{2}, 0\Big)$ and $\beta'' = \frac{1}{2}\, v(\mu)$, and α' is as defined above.

In order to get a contradiction, we use the following :

Lemma 4.7. *The element $\frac{1}{2}v(\mu) + \Gamma_F$ is the coset of one of the following :*

$$\{\alpha_4,\ \alpha_4 + \alpha', \alpha_2, \alpha_2 + \alpha'\}.$$

It follows that β'' in the description of Γ_2 may only take one of those four values, and in each case, one may check that $\Gamma_\alpha \cap \Gamma_\beta$ consists of at most four distinct cosets, which contradicts the inequality above.

Hence, to complete the argument, it only remains to prove Lemma 4.7. Consider the quadratic extension $F' = F[t]/(t^2 + t + a)$. There is an isomorphism

$$F' \simeq F_0((d^{-1}))((c^{-1}))((b^{-1}))((z^{-1})), \text{where } z \in F' \text{ satisfies } z^2 + z = a.$$

Over F', n_Q is hyperbolic, so $\tilde{\varphi}_2$ and $\tilde{\chi}$ are Witt equivalent. Computing the Witt classes of both forms over this field, we get

$$d[1, c] \perp c[1, b] \perp [1, b+c] \sim \mu\big(d[1, c] \perp c[1, b] \perp [1, b+c]\big).$$

Hence, μ is a similarity factor of the Albert form $d[1, c] \perp c[1, b] \perp [1, b+c]$ over the field F'. By [**15**, (16.6)], it follows that μ belongs to $F'^{\times 2}\,\mathrm{Nrd}_B\, B^\times$, where B is the biquaternion algebra $\big([b, c) \otimes [c, d)\big)_{F'}$. Since F is henselian, the value v extends uniquely to the field F' and to the division algebra B, with value groups respectively denoted by $\Gamma_{F'}$ and Γ_B. Moreover, we have $v_B(x) = \frac{1}{4}v'\big(\mathrm{Nrd}_B(x)\big)$ for all x in B. Therefore, $\frac{1}{2}v(\mu)$ belongs to $\Gamma_{F'} + 2\Gamma_B$. Clearly, we have

$$\Gamma_{F'} = \mathbb{Z}^3 \times \frac{1}{2}\mathbb{Z} \subset \Gamma.$$

To compute the value group of Γ_B, we proceed as follows. Consider a generating pair (i,j) of $[b,c)_{F'}$ and a generating pair (k,ℓ) of $[c,d)_{F'}$ with $i^2+i=b$, $j^2=c$, $ij=j(i+1)$, and similarly $k^2+k=c$, $\ell^2=d$ and $k\ell=\ell(k+1)$. Using those relations, one may easily check that $v_B(i)=(0,0,-\frac{1}{2},0)$, $v_B(j)=(0,-\frac{1}{2},0,0)=v_B(k)$ and $v_B(\ell)=(-\frac{1}{2},0,0,0)$. Moreover, since j and k commute in B, we have $(j+k)^2=k$; hence $v_B(j+k)=(0,-\frac{1}{4},0,0)$. Since B has degree 4, it follows that B is unramified and

$$\Gamma_B \simeq \frac{1}{2}\mathbb{Z} \times \frac{1}{4}\mathbb{Z} \times \frac{1}{2}\mathbb{Z} \times \mathbb{Z} \subset \Gamma.$$

The assertion on $v(\mu)$ now follows by a direct computation. □

References

[1] A. Chapman, *Symbol length of p-algebras of prime exponent*, J. Algebra Appl. **16** (2017), no. 7, 1750136, 9, DOI 10.1142/S0219498817501365. MR3660419

[2] A. Chapman, *Field of iterated Laurent series and its Brauer group*, Bull. Belg. Math. Soc. Simon Stevin **27** (2020), no. 1, 1–6, DOI 10.36045/bbms/1590199296. MR4102695

[3] A. Dolphin, *Orthogonal Pfister involutions in characteristic two*, J. Pure Appl. Algebra **218** (2014), no. 10, 1900–1915, DOI 10.1016/j.jpaa.2014.02.013. MR3195416

[4] A. Dolphin and A. Quéguiner-Mathieu, *Symplectic involutions, quadratic pairs and function fields of conics*, J. Pure Appl. Algebra **221** (2017), no. 8, 1966–1980, DOI 10.1016/j.jpaa.2016.10.016. MR3623178

[5] A. Dolphin and A. Quéguiner-Mathieu, *The canonical quadratic pair on a Clifford algebra and triality*, Israel J. Math. **242** (2021), no. 1, 171–213, DOI 10.1007/s11856-021-2128-y. MR4282080

[6] R. Elman, N. Karpenko, and A. Merkurjev, *The algebraic and geometric theory of quadratic forms*, American Mathematical Society Colloquium Publications, vol. 56, American Mathematical Society, Providence, RI, 2008, DOI 10.1090/coll/056. MR2427530

[7] M. A. Elomary and J.-P. Tignol, *Springer's theorem for tame quadratic forms over Henselian fields*, Math. Z. **269** (2011), no. 1-2, 309–323, DOI 10.1007/s00209-010-0729-y. MR2836073

[8] F. Faivre, *Liaison des formes de Pfister et corps de fonctions de quadriques en caractéristique* 2, Thesis (Ph.D.), Université de Franche-Comté, 2006.

[9] P. Gille, *Invariants cohomologiques de Rost en caractéristique positive* (French, with English and French summaries), K-Theory **21** (2000), no. 1, 57–100, DOI 10.1023/A:1007839108933. MR1802626

[10] P. Gille and T. Szamuely, *Central simple algebras and Galois cohomology*, Cambridge Studies in Advanced Mathematics, vol. 101, Cambridge University Press, Cambridge, 2006, DOI 10.1017/CBO9780511607219. MR2266528

[11] D. W. Hoffmann, *Function fields of quadratic forms*, ProQuest LLC, Ann Arbor, MI, 1992. Thesis (Ph.D.)–University of California, Berkeley. MR2688477

[12] D. W. Hoffmann, D. W. Lewis, and J. Van Geel, *Minimal forms for function fields of conics*, K-theory and algebraic geometry: connections with quadratic forms and division algebras (Santa Barbara, CA, 1992), Proc. Sympos. Pure Math., vol. 58, Amer. Math. Soc., Providence, RI, 1995, pp. 227–237, DOI 10.1090/pspum/058.2/1327300. MR1327300

[13] D. W. Hoffmann and J. Van Geel, *Minimal forms with respect to function fields of conics*, Manuscripta Math. **86** (1995), no. 1, 23–48, DOI 10.1007/BF02567976. MR1314147

[14] K. Kato, *Symmetric bilinear forms, quadratic forms and Milnor K-theory in characteristic two*, Invent. Math. **66** (1982), no. 3, 493–510, DOI 10.1007/BF01389226. MR662605

[15] M.-A. Knus, A. Merkurjev, M. Rost, and J.-P. Tignol, *The book of involutions*, American Mathematical Society Colloquium Publications, vol. 44, American Mathematical Society, Providence, RI, 1998, DOI 10.1090/coll/044. MR1632779

[16] T. Y. Lam, *Introduction to quadratic forms over fields*, Graduate Studies in Mathematics, vol. 67, American Mathematical Society, Providence, RI, 2005, DOI 10.1090/gsm/067. MR2104929

[17] A. Ledet, *On the essential dimension of p-groups*, Galois theory and modular forms, Dev. Math., vol. 11, Kluwer Acad. Publ., Boston, MA, 2004, pp. 159–172, DOI 10.1007/978-1-4613-0249-0_8. MR2059762

[18] D. W. Lewis and J. Van Geel, *Quadratic forms isotropic over the function field of a conic*, Indag. Math. (N.S.) **5** (1994), no. 3, 325–339, DOI 10.1016/0019-3577(94)90008-6. MR1298779

[19] K. McKinnie, *Essential dimension of generic symbols in characteristic p*, Forum Math. Sigma **5** (2017), Paper No. e14, 30, DOI 10.1017/fms.2017.11. MR3668468

[20] A. Quéguiner-Mathieu and J.-P. Tignol, *Algebras with involution that become hyperbolic over the function field of a conic*, Israel J. Math. **180** (2010), 317–344, DOI 10.1007/s11856-010-0106-x. MR2735068

[21] T. A. Springer, *Quadratic forms over fields with a discrete valuation. I. Equivalence classes of definite forms*, Indag. Math. **17** (1955), 352–362. Nederl. Akad. Wetensch. Proc. Ser. A **58**. MR70664

[22] J.-P. Tignol and A. R. Wadsworth, *Value functions on simple algebras, and associated graded rings*, Springer Monographs in Mathematics, Springer, Cham, 2015. MR3328410

School of Computer Science, Academic College of Tel-Aviv-Yaffo, Rabenu Yeruham St., P.O.B 8401 Yaffo, 6818211, Israel

Email address: adam1chapman@yahoo.com

Université Sorbonne Paris Nord, Institut Galilée, LAGA - UMR 7539 du CNRS, 93430 Villetaneuse, France

Email address: queguin@math.univ-paris13.fr

Contemporary Mathematics
Volume **800**, 2024
https://doi.org/10.1090/conm/800/16054

Highly versal torsors

Uriya A. First

ABSTRACT. Let G be a linear algebraic group over an infinite field k. Loosely speaking, a G-torsor over a k-variety is said to be versal if it specializes to every G-torsor over any k-field. The existence of versal torsors is well-known. We show that there exist G-torsors that admit even stronger versality properties. For example, for every $d \in \mathbb{N}$, there exists a G-torsor over a smooth quasi-projective k-scheme that specializes to every torsor over a quasi-projective k-scheme after removing some codimension-d closed subset from the latter. Moreover, such specializations are abundant in a well-defined sense. Similar results hold if we replace k with an arbitrary base-scheme. In the course of the proof we show that every globally generated rank-n vector bundle over a d-dimensional k-scheme of finite type can be generated by $n+d$ global sections.

When G can be embedded in a group scheme of unipotent upper-triangular matrices, we further show that there exist G-torsors specializing to every G-torsor over any affine k-scheme. We show that the converse holds when $\operatorname{char} k = 0$.

We apply our highly versal torsors to show that, for fixed $m, n \in \mathbb{N}$, the symbol length of any degree-m period-n Azumaya algebra over any local $\mathbb{Z}[\frac{1}{n}, e^{2\pi i/n}]$-ring is uniformly bounded. A similar statement holds in the semilocal case, but under mild restrictions on the base ring.

CONTENTS

2020 *Mathematics Subject Classification.* Primary 14L15, 14L30.

Key words and phrases. Group scheme, linear algebraic group, torsor, principal homogeneous space, versal torsor, vector bundle, Azumaya algebra, Galois extension, symbol length.

1. Introduction

Let k be an infinite field and let G be a linear algebraic group over k, i.e., an affine group scheme of finite type over k. Informally, a G-torsor $E \to X$, where X is a k-variety, is said to be *versal* if $E \to X$ specializes to every G-torsor over a k-field. The question of how abundant are the specializations leads to several variations of this notion, e.g., see [**7**]. In particular, one says that:

(i) $E \to X$ is *weakly versal* if for every k-field K and G-torsor $E' \to \operatorname{Spec} K$, there is a morphism $f : \operatorname{Spec} K \to X$ such that $E' \cong \operatorname{Spec} K \times_{f,X} E$ as K-schemes with a G-action;

(ii) $E \to X$ is *versal* if for every k-field K and G-torsor $E' \to \operatorname{Spec} K$, there is a morphism f as in (i), and the joint image of all such f is dense in X; and

(iii) $E \to X$ is *strongly versal*[1] if for every k-field K and G-torsor $E' \to \operatorname{Spec} K$, there is a dominant G-equivariant rational map $\mathbb{A}^n_k \times_k E' \dashrightarrow E$.

Restricting K to be a finitely generated k-field does not affect the validity of (i)–(iii). The torsor $E \to X$ is also called *generic* if it is versal and X is a rational k-variety.

The study of versal torsors has a rich history and a large range of applications. It goes back to the universal bundles introduced by topologists in the first half of the 20th century, through Amitsur's work on the construction of central-simple algebras that are not crossed products [**1**], to the theory of essential dimension initiated by Z. Reichstein and collaborators, and the theory of cohomological invariants of algebraic groups [**29**]. We refer the reader to the introduction and appendix of [**7**] for a detailed historical survey.

We further note that torsors over particular algebraic groups G are sometimes equivalent to algebraic objects of a certain type, e.g., for a constant finite algebraic group Γ, the Γ-torsors are equivalent to Γ-Galois extensions, $\mathbf{PGL}_n(k)$-torsors are equivalent to Azumaya algebras of degree n, etcetera. (This can be formally stated as an equivalence between two categories fibered over k-schemes — the category of G-torsors and the [opposite] category of objects of the given type with specializations as its morphisms.) One can therefore discuss versal Γ-Galois extensions, versal Azumaya algebras — e.g., see [**27**, p. 73] — and so forth; in fact, much of the theory of versal torsors originated from such special cases.

In recent years, several examples of G-torsors, or equivalent algebraic objects, which admit versality properties going beyond k-fields have been exhibited. By this, we mean that these G-torsors specialize to every G-torsor $E' \to X'$ with X' ranging over a class of k-schemes that goes beyond the class of k-fields. We shall informally refer to such objects as "highly versal". The earliest such example that we are aware of goes back to Saltman [**26**, Theorem 2.3], who constructed for every p-group G a G-Galois extension of rings specializing to any other G-Galois extension of p-torsion rings. Another such construction was given by Fleischmann and Woodcock [**11**]. Saltman also constructed a highly versal cyclic Galois extension in arbitrary and mixed characteristic (but with more restricted specialization properties) in the recent paper [**28**]. Highly versal Azumaya algebras with involution were constructed by Auel, the author and B. Williams in [**3**, §5.1] (and applied in [**8**]). Taking a step further, in [**10**], the author, Z. Reichstein and B. Williams showed

[1] This is called *very versal* in [**7**].

that for every linear algebraic group G over k and every $d \in \mathbb{N} \cup \{0\}$, there exists a G-torsor $E \to X$, where X is a k-algebraic space (or a k-scheme if G is reductive), which specializes to every G-torsor over a finite type affine k-scheme of dimension $\leq d$; such torsors were called weakly d-versal in [op. cit.]. In fact, predating all of these works, a theorem of O. Förster [**14**] on the number of generators of a projective module over a noetherian (commutative) ring can be restated as saying that the generic rank-n projective module generated by $(n+d)$-elements studied by Raynaud [**24**] (equiv., its corresponding $\mathbf{GL}_n(\mathbb{Z})$-torsor) is highly versal in the sense that it specializes to any rank-n projective module over a noetherian ring of Krull dimension $\leq d$; see Section 7.

The present work is a systematic study of the higher versality properties that were glimpsed at in the works just mentioned. In particular, we introduce notions of weak, ordinary and strong higher-versality, analogous to (i)–(iii) above, construct examples of torsors having these properties, and apply them to prove finiteness results on the *symbol length* of local and semilocal rings.

While our definition of higher versality applies to group schemes over arbitrary base schemes, for the sake of simplicity and easier comparability with the classical definitions, *we give here a slightly weaker form of the definitions, and also restrict them to linear algebraic groups G over a field k*; the general definition is given in Section 5.

Let $d \in \mathbb{N} \cup \{0\}$ and let $\mathscr{C}$ be a class of k-schemes, e.g., nAff/k — the class of affine noetherian k-schemes —, or qProj/k — the class of quasi-projective k-schemes. Given a G-torsor $E \to X$, with X a k-algebraic space, we say that:

(i′) $E \to X$ is *weakly d-versal* for $\mathscr{C}$ if for every G-torsor $E' \to X'$ with $X' \in \mathscr{C}$, there is an open subscheme $U \subseteq X'$ and a k-morphism $f : U \to X$ such that $X' - U$ is of codimension $> d$ in X' and there is a G-torsor isomorphism $E'|_U \cong f^*E$;[2]

(ii′) $E \to X$ is *d-versal* for $\mathscr{C}$ if for every G-torsor $E' \to X'$ with $\emptyset \neq X' \in \mathscr{C}$, there is an open subscheme $U \subseteq X'$ and k-morphisms $\{f_i : U \to X\}_{i \in I}$ such that $X' - U$ is of codimension $> d$ in X', the induced map $\bigsqcup_{i\in I} U \to X$ is schematically dominant, and for every $i \in I$, there is a G-torsor isomorphism $E'|_U \cong f_i^*E$; and

(iii′) $E \to X$ is *strongly d-versal* for $\mathscr{C}$ if for every G-torsor $E' \to X'$ with $\emptyset \neq X' \in \mathscr{C}$, there is an open subscheme $U \subseteq X'$, $n \in \mathbb{N}$ and a schematically dominant morphism $f : \mathbb{A}^n_k \times_k U \to X$ such that $X' - U$ is of codimension $> d$ in X' and there is a G-torsor isomorphism $\mathbb{A}^n_k \times_k E'|_U \cong f^*E$.

Here, we have written $E'|_U$ for the G-torsor $U \times_{X'} E'$ and f^*E for the G-torsor which is the pullback of $E \to X$ along $f : U \to X$. A k-morphism $f : Y \to X$ is schematically dominant if the induced map $\mathcal{O}_X \to f_*\mathcal{O}_Y$ is a monomorphism (see [**16**, §11.10] or Section 4); when X and Y are k-varieties, this is the same as being dominant. By convention, the dimension of the empty scheme is $-\infty$, so we must have $U = X'$ if $\dim X' \leq d$. In particular, we must take $U = X'$ if $\mathscr{C}$ consists of schemes of dimension $\leq d$.

EXAMPLE 1.1. Let $\mathscr{C}$ be the class of affine irreducible k-varieties. Then a G-torsor is weakly 0-versal for $\mathscr{C}$ if and only if it is weakly versal in the sense of (i)

[2]This definition is more restrictive than the weak d-versality discussed in [**10**].

above. To see this, take the k-field K from (i) to be the fraction field of $X' \in \mathscr{C}$ and spread out.

The same reasoning shows that a G-torsor which is 0-versal for $\mathscr{C}$ is versal in the sense of (ii), but *a priori*, 0-versality for $\mathscr{C}$ is more restrictive than versality. The reason is that if $K = \mathrm{Frac}(X')$ and $\{f_i : \operatorname{Spec} K \to X\}_{i \in I}$ are morphisms such that their joint image is dense in X, then, while we can extend each f_i to some dense open $U_i \subseteq X'$, it is not clear why we can take all the U_i to be the same dense open subscheme.

Being strongly 0-versal for $\mathscr{C}$ is also *a priori* stronger than being strongly versal in the sense of (iii) above, because in (iii′) we insist on having a morphism $\mathbb{A}^n_k \times_k U \to X$ for some dense open $U \subseteq X'$, rather than a rational map $\mathbb{A}^n_k \times_k X' \dashrightarrow X$.

Let S be a scheme, let G be a linear group scheme over S that is flat and locally of finite presentation, and let nAff/S denote the class of affine noetherian schemes equipped with a morphism to S.[3] In Section 8, we give explicit constructions of G-torsors $E \to X$ which are strongly d-versal for nAff/S, for any prescribed $d \in \mathbb{N}$. For example, if G is a subgroup of $\mathbf{GL}_n(S)$, then the quotient morphism

$$E = \mathbf{GL}_{n+d}(S) / \begin{bmatrix} 1_{n\times n} & \mathbb{A}_S^{n\times d} \\ 0_{d\times n} & \mathbf{GL}_d(S) \end{bmatrix} \to \mathbf{GL}_{n+d}(S) / \begin{bmatrix} G & \mathbb{A}_S^{n\times d} \\ 0_{d\times n} & \mathbf{GL}_d(S) \end{bmatrix} = X$$

is such a G-torsor (Theorem 8.2), which also enjoys the following *extension property* (Theorem 8.3): If $E' \to X'$ is a G-torsor with X' affine of dimension $\leq d$, X'' is a closed subscheme of X' and $E'' \to X''$ is the pullback of $E' \to X'$ along $X'' \to X'$, then every specialization of E to E'' extends to a specialization of E to E'.

In general, the base X in our constructions is an S-algebraic space. However, if G is reductive or finite and S is noetherian, then X is an S-scheme. In fact, in the reductive and finite cases, we can even make X affine at the cost of replacing nAff/S with its subclass of schemes of dimension $\leq d$ (Theorem 8.1). When S is the spectrum of an infinite field k, things become even better in the sense that there exist G-torsors over quasi-projective k-schemes which are strongly d-versal for the class of quasi-projective k-schemes (Theorem 8.4).

REMARK 1.2. Our Theorem 8.2 implies in particular that for every linear algebraic group over a field k (possibly finite), and every $d \in \mathbb{N}$, there exists a finite-type smooth k-scheme X and a G-torsor $E \to X$ which specializes to every G-torsor $E' \to X'$ such that X' is an affine k-variety of dimension $\leq d$. However, when $d > 0$, the promised specialization $X' \to X$ is generally far from being an immersion. Indeed, we can choose X' to be a curve admitting a closed point x' such that $\dim T_{x'}X' > \dim T_x X$ for every $x \in X$, which means that there is no immersion $X' \to X$. Also, if k is finite, then we can arrange that $|X'(k)| > |X(k)|$, again making it impossible to define an immersion of X' in X. With this in mind, the existence of weakly d-versal G-torsors whose base scheme is of finite type of k may seem surprising. Informally, it also suggests that there should be limitations on the "complexity" of G-torsors over d-dimensional affine k-schemes.

We prove the existence of strongly d-versal G-torsors by reducing it to the existence of *weakly* d-versal rank-n vector bundles which admit some symmetry properties (Corollary 5.9). We then show that some constructions of "generic"

[3]Warning: In general, this is *not* the class of noetherian S-schemes Y whose structure morphism $Y \to S$ is affine.

rank-n vector bundles generated by $n+d$ global sections, e.g., those considered in [**24**] and (implicitly) in [**10**], satisfy the required conditions. The heart of the argument is a theorem of O. Förster [**14**] stating that a rank-n vector bundle over a d-dimensional noetherian affine scheme is generated by $n+d$ global sections, and its improvements in [**32**], [**9**], [**10**], which we need to strengthen even further for our purposes (Theorems 7.2, 7.3, 7.7). For example, letting k denote an infinite field, we show that every globally generated rank-n vector bundle over a finite type k-scheme is generated by $d+n$ global sections away from some closed subscheme of codimension $> d$.

Returning to group schemes over a general base S, we also consider the question of whether there exist G-torsors over finite type S-schemes which are ∞-versal for nAff/S. We show that the answer is "yes" if G can be embedded in a group of upper-triangular unipotent matrices over S (Theorem 10.1), and that the converse holds when S is the spectrum of a field of characteristic 0 (Theorem 10.6). This gives a conceptual explanation of why, for a field k of characteristic $p>0$, every finite p-group G admits a G-Galois extension of k-rings which specializes to any other G-Galois extension of k-rings [**26**], while such G-Galois extensions do not exist when $\operatorname{char} k = 0$ and G is nontrivial.

Finally, we apply one of our constructions of highly versal torsors together with Hoobler's extension of the Merkurjev–Suslin Theorem to semilocal rings [**17**] to prove finiteness results about the symbol length of semilocal (commutative) rings. In more detail, let $n,m \in \mathbb{N}$, and let R be a ring that is a $\mathbb{Z}[\frac{1}{n}, e^{2\pi i/n}]$-algebra. Recall that the (n,m)-symbol length of R is the minimal $L \in \mathbb{N}\cup\{0,\infty\}$ such that every degree-m period-n Azumaya algebra over R is Brauer equivalent to the tensor product of at most L degree-n *symbol Azumaya algebras* over R. We show in Theorem 11.2 that for every $n,m,s \in \mathbb{N}$, there is $L\in\mathbb{N}$ such that for every semilocal $\mathbb{Z}[\frac{1}{n}, e^{2\pi i/n}]$-ring R having at most s maximal ideals, the (n,m)-symbol length of R is at most L. (In particular, the (n,m)-symbol length of any semilocal $\mathbb{Z}[\frac{1}{n}, e^{2\pi i/n}]$-ring is finite.) A similar uniform upper bound on the symbol length also exists if, instead of bounding the number of maximal ideals of the semilocal ring R, we assume that R has infinite residue fields and is an algebra over some fixed semilocal ring R_0; see Theorem 11.1. Our results imply in particular that the (n,m)-symbol length of all fields (regardless of the characteristic) is uniformly bounded — a result known to experts.

The structure of the paper is as follows: Sections 2 and 3 are preliminary and collect facts about torsors and vector bundles. In Section 4, we establish some facts about schematically dominant morphisms. In Section 5, we introduce the definition of weakly, ordinary and strongly d-versal torsors and study some of their properties. In particular, we reduce the existence of *strongly* d-versal torsors for linear group schemes to the existence of *weakly* d-versal vector bundles admitting some symmetry properties. The subject matter of Sections 6 and 7 is to show existence of such d-versal vector bundles; their construction is given in Section 6 and their desired properties are established in Section 7. This is used in Section 8 to prove the existence of torsors that are strongly d-versal for some classes of schemes. We apply this to particular group schemes in Section 9 to deduce the existence of strongly d-versal Galois extensions, Azumaya algebras and so on. Section 10 is concerned with ∞-versal torsors. Finally, in Section 11, we apply some of our earlier results to bounding the symbol length of Azumaya algebras over semilocal rings.

Notation. Throughout, a ring means a commutative (unital, associative) ring. Algebras are not necessarily commutative.

As is common in algebraic geometry, the dimension of a topological space X is the supremum of the set of all $d \in \mathbb{N} \cup \{0\}$ for which there is chain of closed irreducible subsets $X_0 \subsetneq \cdots \subsetneq X_d$. By convention, the supremum of the empty set is $-\infty$, and so the dimension of the empty space is $-\infty$. The space X is noetherian if every ascending chain of open subsets eventually stabilizes. Recall that if Y is a subspace of X, then $\dim Y \leq \dim X$ and Y is noetherian if X is. The set of closed points of X is denoted X_{cl} and given the subspace topology. When $X = \operatorname{Spec} R$ for a ring R, the subspace X_{cl} is just the subspace of maximal ideals of R, denoted $\operatorname{Max} R$.

If X is irreducible and Y is a closed irreducible subset of X, then the codimension of Y in X — denoted $\operatorname{codim}(Y, X)$ — is defined to be the supremum of the set of $d \in \mathbb{N} \cup \{0\}$ for which there is chain of closed irreducible subsets $Y = Y_0 \subsetneq \cdots \subsetneq Y_d \subseteq X$. We extend this definition to general X and Y by letting $\operatorname{codim}(Y, X)$ be the infimum of $\operatorname{codim}(Y', X')$ when X' ranges over the irreducible components of X and Y' ranges over the irreducible components of $Y \cap X'$.[4] When X is catenary (see [**31**, Tag 02I1]; e.g., a finite type scheme over a field), $\operatorname{codim}(Y, X) \geq d$ if and only if $\dim X' \geq \dim(X' \cap Y) + d$ for every irreducible component X' of X.[5] By convention, the infimum of the empty set is ∞, and so $\operatorname{codim}(\emptyset, X) = \infty$ (even if $X = \emptyset$).

Let X be a scheme. The residue field of X at $x \in X$ is denoted $k(x)$. If $\mathcal{M}$ is an $\mathcal{O}_X$-module, we write $\mathcal{M}_x$ for the stalk of $\mathcal{M}$ at x and $\mathcal{M}(x) = \mathcal{M}_x \otimes_{\mathcal{O}_{X,x}} k(x)$. Given an open neighborhood U of x and $m \in \Gamma(U, \mathcal{M})$, we also write $m(x)$ for the image of m in $\mathcal{M}(x)$. As usual, the rank of $\mathcal{M}$ is the function $\operatorname{rank} \mathcal{M} : X \to \mathsf{cardinals}$ sending $x \in X$ to $\dim_{k(x)} \mathcal{M}(x)$. A vector bundle over X is a locally free $\mathcal{O}_X$-module of (point-wise) finite rank. A line bundle is a vector bundle of rank 1.

If not otherwise said, irreducible components of X are given the reduced induced subscheme structure.

Let S be a base scheme. Given S-schemes X and S_1, we write X_{S_1} for $X \times_S S_1$ viewed as an S_1-scheme. If $f : X \to Y$ is a morphism of S-schemes, then $f_{S_1} : X_{S_1} \to Y_{S_1}$ denotes the pullback of f along the structure morphism $S_1 \to S$. Given a class $\mathscr{C}$ of S-schemes, we write $\mathscr{C}/S_1$ for the class of S_1-schemes with underlying S-scheme in $\mathscr{C}$.

Recall that a collection of S-morphisms $\{f_i : Y_i \to X\}_{i \in I}$ is an fppf covering if each f_i is flat and locally of finite presentation (abbreviated: fppf) and $X = \bigcup_{i \in I} f_i(Y_i)$ as sets. The category Sch/S of S-schemes together with the Grothendieck topology generated by fppf coverings is the *large fppf site* of S, denoted $(\mathsf{Sch}/S)_{\mathrm{fppf}}$. We will freely identify schemes and algebraic spaces over S with the sheaves that they represent on the site $(\mathsf{Sch}/S)_{\mathrm{fppf}}$. In particular, given

[4]Caution: This definition differs from the definition in [**31**, Tag 02I3] even when Y is irreducible.

[5]The catenarity assumption is needed to guarantee that for every pair of irreducible closed subsets $Y' \subseteq X'$ of X, we have $\operatorname{codim}(Y', X') = \dim X' - \dim Y'$ and the three quantities used in the equality are finite.

an S-algebraic space X and an S-scheme U, we shall write $X(U)$ for the set of S-morphisms from U to X and call this set the U-sections of X (over S). Morphisms between S-algebraic spaces will usually be defined by specifying their action on U-sections for every S-scheme U.

An S-group means a group scheme over S. An algebraic group over a field k is a k-group of finite type. Let G be an S-group. A subgroup of G is a subgroup S-scheme of G. If G acts from the right on an S-algebraic space X, we let X/G denote the quotient of X by G in category of sheaves over $(\mathsf{Sch}/S)_{\mathrm{fppf}}$, i.e., X/G is the fppf-sheafification of the functor $U \mapsto X(U)/G(U)$ from S-schemes to sets. The quotient X/G is not always representable by an S-scheme, but we will treat it as an S-scheme when it is. Likewise for S-algebraic spaces.

An S-group G with structure morphism $p : G \to S$ is said to be *finite and locally free* if p is finite and $p_*\mathcal{O}_G$ is a locally free $\mathcal{O}_S$-module.

We write $\mathbf{GL}_n(S)$ for the S-group representing the sheaf of invertible $n \times n$ matrices over S, which is sometimes denoted $\mathbf{GL}_n(\mathcal{O}_S)$; for an S-scheme U, we have $\mathbf{GL}_n(S)(U) = \mathrm{GL}_n(\Gamma(U, \mathcal{O}_U))$. By default, sections of $\mathcal{O}_S^n$ are viewed as column vectors, so that $\mathbf{GL}_n(S)$ acts on $\mathcal{O}_S^n$ from the left by matrix multiplication. The subgroup of $\mathbf{GL}_n(S)$ consisting of matrices of determinant 1 is denoted $\mathbf{SL}_n(S)$. When $S = \operatorname{Spec} R$ for a ring R, we write $\mathbf{GL}_n(R)$ for $\mathbf{GL}_n(S)$ and $\mathbf{SL}_n(R)$ for $\mathbf{SL}_n(S)$.

An S-group G is said to be *linear* if there is an S-group monomorphism $G \to \mathbf{GL}_n(S)$ for some $n \in \mathbb{N}$.[6] It is *strictly linear* if there is an S-group morphism $G \to \mathbf{GL}_n(S)$ that is also a closed immersion.

We write $\mathbb{A}_S^{n\times m}$ for the affine space of $n \times m$ matrices over S, which is just $\mathbb{A}_S^{nm}$ but with the coordinates reorganized into an $n \times m$ matrix. Given $n, d \in \mathbb{N} \cup \{0\}$ and S-schemes A, B, C, D equipped with monomorphisms $A \to \mathbb{A}_S^{n\times n}$, $B \to \mathbb{A}_S^{n\times d}$, $C \to \mathbb{A}_S^{d\times n}$, $D \to \mathbb{A}_S^{d\times d}$, we write $\left[\begin{smallmatrix} A & B \\ C & D \end{smallmatrix}\right]$ for the scheme $A \times B \times C \times D$ endowed with the monomorphism $(a, b, c, d) \mapsto \left[\begin{smallmatrix} a & b \\ c & d \end{smallmatrix}\right]$ into $\mathbb{A}_S^{(n+d)\times(n+d)}$. When $A = \operatorname{Spec} S$ and $A \to \mathbb{A}_S^{n\times n}$ is the zero matrix (resp. unit matrix) section, we shall write $\left[\begin{smallmatrix} 0_{n\times n} & B \\ C & D \end{smallmatrix}\right]$ (resp. $\left[\begin{smallmatrix} 1_{n\times n} & B \\ C & D \end{smallmatrix}\right]$) instead of $\left[\begin{smallmatrix} A & B \\ C & D \end{smallmatrix}\right]$, and likewise with B, C, D. The subscript $n \times n$ will be dropped from $0_{n\times n}$ when it is clear from the context. (For example, such notation was used on page 132.)

2. Preliminaries on Torsors

Let S be a scheme. Throughout this section, all schemes and algebraic spaces are over S, and morphisms are S-morphisms.

Let G be an S-group and let X be an S-algebraic space. For our purposes, a G-torsor over X consists of a *sheaf* E on $(\mathsf{Sch}/S)_{\mathrm{fppf}}$ together with a right G-action $E \times_S G \to E$ and a morphism $p : E \to X$ such that

(1) $p(e \cdot g) = p(e)$ for every S-scheme U, $g \in G(U)$ and $e \in E(U)$, and
(2) there is an fppf covering $X' \to X$ such that $X' \times_X E \cong X' \times_S G$ as sheaves carrying a right G-action and a morphism to X'.

(Otherwise said, a G-torsor over X is a right sheaf G-torsor for the fppf topology.) When these conditions hold we will simply say that $p : E \to X$, or just

[6] Caution: Some texts call an S-group linear if there is an S-vector bundle V and a monomorphism $G \to \mathcal{E}nd_{\mathcal{O}_S}(V)$. This is equivalent to our definition when S is affine.

E, is a G-torsor over the base X. In this case, the induced sheaf morphism $[e] \mapsto p(e) : E/G \to X$ is an isomorphism. We sidestep the question of which G-torsors are representable by S-schemes or S-algebraic spaces and satisfy with noting that if $G \to S$ is affine and X is a scheme, then all G-torsors are schemes [**23**, III, Theorem 4.3].

A morphism from a G-torsor $p_1 : E_1 \to X_1$ to another G-torsor $p_2 : E_2 \to X_2$ is a pair $(\hat{f}, f)$ consisting of a G-equivariant S-morphism $\hat{f} : E_1 \to E_2$ and an S-morphism $f : X_1 \to X_2$ such that $p_2 \circ \hat{f} = f \circ p_1$. Since $X_i \cong E_i/G$, every G-equivariant morphism $\hat{f} : E_1 \to E_2$ determines and is determined by a morphism from $E_1 \to X_1$ to $E_2 \to X_2$. The collection of G-torsors and their morphisms form a category denoted $\mathsf{Tors}(G)$. The subcategory of $\mathsf{Tors}(G)$ consisting of G-torsors over a fixed S-algebraic space X and morphisms of the form $(\hat{f}, \mathrm{id}_X)$ is denoted $\mathsf{Tors}(G, X)$ and called the category of G-torsors over X. We write $E_1 \cong E_2$ when $E_1 \to X_1$ and $E_2 \to X_2$ are isomorphic in $\mathsf{Tors}(G)$. Isomorphism in the subcategory $\mathsf{Tors}(G, X)$ is denoted by $\cong_X$.

Let $p : E \to X$ be a G-torsor and let $f : Y \to X$ be an S-morphism. Then the pullback $p_f : Y \times_X E \to Y$ of p along f is a G-torsor (G acts on $Y \times_X E$ via E). We often abbreviate $Y \times_X E$ to f^*E. If $f : U \to X$ is an open immersion, we also write $E|_U$ instead of f^*E.

REMARK 2.1. Let $p : E \to X$ and $p' : E' \to X'$ be G-torsors. There is a one-to-one correspondence between morphisms $(\hat{f}, f)$ from $E' \to X'$ to $E \to X$ and pairs $(\hat{u}, f)$ consisting of an S-morphism $f : X' \to X$ and an isomorphism $\hat{u} = (\hat{u}, \mathrm{id}_X) : f^*E \to E'$ in $\mathsf{Tors}(G, X)$; the pairs $(\hat{u}, f)$ are called *specializations* of E to E'. Explicitly, given $(\hat{f}, f)$, define $\hat{u}$ to be the inverse of $e' \mapsto (p'(e'), \hat{f}(e')) : E' \to f^*X$, and given $(\hat{u}, f)$, let $\hat{f}$ be the composition of $\hat{u}^{-1}$ and the second projection $f^*X = X' \times_X E \to E$.

Let $\varphi : H \to G$ be a morphism of S-groups, and let $p : E \to X$ be an H-torsor. Then p and φ give rise to a G-torsor $p^G : E \times^{\varphi, H} G \to X$. Here, $E \times^{\varphi, H} G$ denotes the quotient of the sheaf $E \times G$ by the equivalence relation $E \times H \times G \to (E \times G) \times (E \times G)$ given by $(e, h, g) \mapsto ((eh, g), (e, \varphi(h)g))$ on sections, and p^G is given by $p^G([e, g]) = p(e)$ on sections. There is a natural H-equivariant morphism $E \to E \times^{\varphi, H} G$ given by $e \mapsto [e, 1_G]$ on sections. When φ is clear from the context, we will write $E \times^H G$ instead of $E \times^{\varphi, H} G$.

A left action of an S-group H on a G-torsor $p : E \to X$ consists of left actions of H on E and X such that p is H-equivariant and the actions of H and G on E commute.

Let G be an S-group acting from the right on an S-algebraic space E. We say that G acts freely on E if $G(U)$ acts freely on $E(U)$ for every S-scheme U. In general, the sheaf E/G on $(\mathsf{Sch}/S)_{\mathrm{fppf}}$ may not be an S-scheme even when E is, but under mild assumptions on G, it is an S-algebraic space and $E \to E/G$ is a G-torsor. (In fact, this is the reason why we need to consider algebraic spaces in this work.)

LEMMA 2.2. *Let G be a faithfully flat locally of finite presentation S-group acting freely from the right on an S-algebraic space E. Then the quotient sheaf E/G is an S-algebraic space and $E \to E/G$ is a G-torsor.*

PROOF. Consider the groupoid object (E, R, s, t, c) in S-algebraic spaces given by $R(U) = \{(x, xg) \mid x \in E(U), g \in G(U)\}$ (note that $R \cong E \times G$ because G acts freely on E), $s_U, t_U : R(U) \to E(U)$ being the first and second projections and $c_U = [x \mapsto (x, x)] : E(U) \to R(U)$. Since G acts freely on E, both s and t are isomorphic to the first projection $E \times_S G \to E$, so s and t are flat and locally of finite presentation (because $G \to S$ is). Thus, the stack $[E/R]$ is algebraic [**31**, Tag 06FI]. The freeness of the action further implies that $[E/R]$ is an S-algebraic space [**31**, Tag 04SZ], which must coincide with the quotient sheaf E/G. By [**31**, Tag 06FH], $E \to E/G$ is an fppf covering. Since the morphism $(e, g) \mapsto (e, eg) : E \times_S G \to E \times_{E/G} E$ is a G-equivariant isomorphism over E, it follows that $E \to E/G$ is a G-torsor. □

The question of whether E/G is a scheme when G acts freely on E is difficult in general, even if we impose the necessary condition that $(g, e) \mapsto (ge, e) : G \times_S E \to E \times_S E$ is an immersion. We refer the reader to Seshadri [**30**], Thomason [**33**, §3] and related works for an extensive discussion of the case where G is reductive[7] (see also Anantharaman [**2**, Appendix, Theorem 6] for the relation between the geometric quotients of [**30**] and the sheaf quotients we consider here). In this work, we will only be interested in the case where E is a strictly linear S-group and G is a subgroup of E, where we have the following positive result.

THEOREM 2.3. *Let S be a locally noetherian scheme, let G be a closed subgroup of $\mathbf{GL}_n(S)$ and let H be a subgroup of G. Suppose that there is an exact sequence $1 \to H' \to H \to H'' \to 1$ of group sheaves on $(\mathsf{Sch}/S)_{\mathrm{fppf}}$ with H' a reductive S-group and H'' a finite and locally free S-group. Then G/H is (represented by) an S-scheme of finite type and $G/H \to S$ is affine. If $H' = 1$, then the assumption that S is locally noetherian can be dropped.*

PROOF. It is enough to show this after passing to an affine covering of S, so we may assume that S is affine and noetherian.

We first show that G/H' is an affine S-scheme and $G/H' \to S$ is of finite type. Since H' is reductive, it is closed in G [**6**, XVI, Corollaire 1.5]. By embedding $\mathbf{GL}_n(S)$ in $\mathbf{SL}_{n+1}(S)$ via $x \mapsto \left[\begin{smallmatrix} x & 0 \\ 0 & (\det x)^{-1} \end{smallmatrix}\right]$, we may assume that G is closed inside the affine space $V = \mathbb{A}_S^{n \times n}$. Observe that H' acts linearly on V because G does. Write $G = \operatorname{Spec} B$ and define the fixed ring $B^{H'}$ as in [**30**, p. 229]. (Briefly, writing $S = \operatorname{Spec} A$, the ring $B^{H'}$ consists of those $b \in B$ whose image in $B \otimes_A A'$ is fixed under $H'(A')$ for any A-ring A'.) Since H' is closed in G, for every algebraically closed field k over S, the orbits of the right action of H'_k on G_k are closed. Now, [**30**, Theorem 3] tells us that $\operatorname{Spec} B^{H'}$ is a *geometric quotient* of G by H'. Consequently, $\operatorname{Spec} B^{H'}$ is the quotient of $\operatorname{Spec} B$ by G in the category of ringed spaces. The fact that H' is closed in G also tells us that the map $(h, g) \mapsto (gh, g) : H' \times G \to G \times G$ is a closed immersion. Since H' is flat of finite presentation over S (because it is reductive), we may apply [**2**, Appendix I, Theorem 6], which says that $\operatorname{Spec} B^{H'}$ represents the sheaf G/H' on $(\mathsf{Sch}/S)_{\mathrm{fppf}}$. This means that G/H' is an affine scheme. By Lemma 2.2, $G \to G/H'$ is also an fppf covering. Since $G \to S$ is of finite type, this means that $G/H' \to S$ is also of finite type [**31**, Tag 0367].

[7] An S-group G is called reductive if $G \to S$ is smooth and the geometric fibers of $G \to S$ are *connected* reductive algebraic groups.

Next, observe that H'' acts freely on the affine scheme $X := G/H'$, and $X/H'' = G/H$. Thus, by [**31**, Tag 07S7] and [**22**, Theorem B.18] (for instance, see also p. 604 in [op. cit.]), X/H'' is an affine scheme and $X \to X/H''$ is an H''-torsor. As in the previous paragraph, $G/H = X/H'' \to S$ is also of finite type. $\square$

As for the question of which affine S-groups G can be realized as closed subgroups of $\mathbf{GL}_n(S)$, we have:

THEOREM 2.4 (Thomason [**33**, Corollary 3.2]). *Let S be an affine noetherian scheme, and let G be an affine S-group of finite type. Then there is $n \in \mathbb{N}$ such that G is isomorphic to a closed subgroup of $\mathbf{GL}_n(S)$ if one of the following holds:*

(1) *S is regular and $\dim S \leq 1$ (e.g., if $S = \operatorname{Spec} \mathbb{Z}$),*
(2) *S is regular, $\dim S \leq 2$ and $G \to S$ is smooth with connected fibers,*
(3) *G is semisimple,*
(4) *S is normal and G is reductive.*

See [**15**] for more information about the reductive case.

When S is the spectrum of a field, things simplify significantly, and we have the following well-known stronger versions of Theorems 2.3 and 2.4.

THEOREM 2.5. *Let k be a field, let G be an algebraic group over k and let H be an algebraic subgroup of G. Then:*

(i) *G/H is a finite type k-scheme. When G is affine, G/H is quasi-projective over k, and if the neutral connected component of H is moreover reductive, then G/H is affine.*
(ii) *If G is affine, then it is strictly linear.*

PROOF. See [**22**, Theorem 5.28] (for instance) for the first assertion of (i), [op. cit., §7e, proof of Theorem 7.18] for the second assertion of (i), and [op. cit., Corollaries 4.10, 3.35] for (ii). The last assertion of (i) is part of Matsushima's Theorem, see, for instance, [**25**]. $\square$

3. Preliminaries on Vector Bundles

Let X be a scheme. Recall that a vector bundle V on X is a locally free $\mathcal{O}_X$-module of finite rank. A *basis* for V is a set of global sections $\{v_1, \dots, v_n\} \subseteq \Gamma(X, V)$ such that the map $(\alpha_1, \dots, \alpha_n) \mapsto \sum_i \alpha_i v_i : \mathcal{O}_X^n \to V$ is an isomorphism.

As usual, if $f : X' \to X$ is a morphism, then we let f^*V denote $f^{-1}V \otimes_{f^{-1}\mathcal{O}_X} \mathcal{O}_{X'}$, which is a vector bundle over X'. We write $V_{X'}$ instead of f^*V when f is clear from the context. When f is an open immersion, we also write $V|_{X'}$ for f^*V. The sheaf V extends to a representable sheaf on $(\mathsf{Sch}/X)_{\mathrm{fppf}}$ by setting $V(X') = \Gamma(X, V_{X'})$ for every X-scheme X', and we shall occasionally use V denote the scheme representing this sheaf. From this point of view, $V_{X'}$ is just $V \times_X X'$ as schemes over X'.

The dual of a vector bundle V over X is the $\mathcal{O}_X$-sheaf $V^\vee := \mathcal{H}om_{\mathcal{O}_X}(V, \mathcal{O}_X)$ on $(\mathsf{Sch}/X)_{\mathrm{fppf}}$ sending an X-scheme X' to $\operatorname{Hom}_{\mathcal{O}_{X'}}(V_{X'}, \mathcal{O}_{X'})$.

Fix $n \in \mathbb{N}$. We let $\mathsf{VB}_n(X)$ denote the category of rank-n vector bundles over X with $\mathcal{O}_X$-module isomorphisms as morphisms. It is well-known that $\mathsf{VB}_n(X)$ is equivalent to $\mathsf{Tors}(\mathbf{GL}_n(X), X)$, the category of $\mathbf{GL}_n(X)$-torsors over X. The equivalence is given by sending a rank-n vector bundle V to $\mathcal{I}so(\mathcal{O}_X^n, V)$, the sheaf mapping an X-scheme X' to the set of $\mathcal{O}_{X'}$-isomorphisms from $\mathcal{O}_{X'}^n$ to $V_{X'}$. Since $\mathbf{GL}_n(X) = \mathcal{I}so(\mathcal{O}_X^n, \mathcal{O}_X^n)$, composition defines a right action of $\mathbf{GL}_n(X)$ on $\mathcal{I}so(\mathcal{O}_X^n, V)$, making it into a torsor over this X-group. The equivalence $\mathsf{VB}_n(X) \to \mathsf{Tors}(\mathbf{GL}_n(X), X)$ is compatible with pullbacks. That is, if $f : X' \to X$ is a morphism, V is a vector bundle over X and E is its corresponding $\mathrm{GL}_n(X)$-torsor, then there is a natural isomorphism between f^*E and the $\mathbf{GL}_n(X')$-torsor corresponding to f^*V.

Suppose now that S is a scheme and X is an S-scheme. Then $\mathsf{Tors}(\mathbf{GL}_n(X), X)$ is equivalent to $\mathsf{Tors}(\mathbf{GL}_n(S), X)$. The following lemma describes the functor of points of the $\mathbf{GL}_n(S)$-torsor (viewed as an S-scheme) corresponding to a rank-n vector bundle over X.

LEMMA 3.1. *Let S be a scheme, let X be an S-scheme, let V be a rank-n vector bundle over X, and let $p : E \to X$ be the corresponding $\mathbf{GL}_n(S)$-torsor. Then the functor of points of E is isomorphic to the functor sending an S-scheme Y to the set of tuples $(f, v_1, \dots, v_n)$ consisting of a S-morphism $f : Y \to X$ and an ordered basis $v_1, \dots, v_n$ for f^*V. Under this isomorphism, the (right) action of $\mathbf{GL}_n(S)$ on E is given by $(f, v_1, \dots, v_n) \cdot (g_{ij})_{i,j} = (f, \sum_{i=1}^n g_{i1}v_i, \dots, \sum_{i=1}^n g_{in}v_i)$ on sections.*

PROOF. We just noted that the sheaf that E represents on $(\mathsf{Sch}/X)_{\mathrm{fppf}}$ is $\mathcal{I}so(\mathcal{O}_X^n, E)$ and the right action of $\mathbf{GL}_n(X) = \mathcal{I}so(\mathcal{O}_X^n, \mathcal{O}_X^n)$ is given section-wise by composition. Let $e_1, \dots, e_n$ denote the standard basis of $\mathcal{O}_X^n$. Then, for every X-scheme Y, the map $\psi \mapsto (\psi(e_1|_Y), \dots, \psi(e_n|_Y))$ defines a natural bijection between $E(Y) = \mathcal{I}so(\mathcal{O}_Y^n, V_Y)$ (E is viewed as an X-scheme) and the set of ordered bases of V_Y.

Now let Y be an S-scheme. The data of an S-morphism $\tilde{f} : Y \to E$ is equivalent to the data of a morphism $f : Y \to X$ and, viewing Y as an X-scheme via f, an X-morphism $f' : Y \to E$; the correspondence is given by $\tilde{f} \mapsto (f, f') := (p \circ \tilde{f}, \tilde{f})$. Combining this with the previous paragraph gives the lemma. □

LEMMA 3.2. *Let S be a scheme, let X be an S-scheme, let V be a rank-n vector bundle over X, and let E be its corresponding $\mathbf{GL}_n(S)$-torsor. Then, up to isomorphism in $\mathsf{Tors}(\mathbf{GL}_n(S), X)$, the $\mathbf{GL}_n(S)$-torsor corresponding to the dual bundle $V^\vee$ is E endowed with the right $\mathbf{GL}_n(S)$-action given by $(x, g) \mapsto x(g^{\mathrm{t}})^{-1}$ on sections.*

PROOF. Let Y be an S-scheme. By Lemma 3.1, $E(Y)$ is naturally isomorphic to the set of tuples $(f, v_1, \dots, v_n)$ where $f : Y \to X$ is an S-morphism and $v_1, \dots, v_n$ is a basis of f^*V. Likewise, if E' is the $\mathbf{GL}_n(S)$-torsor corresponding to $V^\vee$, then $E'(Y)$ is the set of tuples $(f, u_1, \dots, u_n)$ consisting of an S-morphism $f : Y \to X$ and a basis $u_1, \dots, u_n$ of $f^*(V^\vee) \cong (f^*V)^\vee$. There is a natural bijection between $E(Y)$ and $E'(Y)$ mapping $(f, v_1, \dots, v_n)$ to $(f, u_1, \dots, u_n)$, where $u_1, \dots, u_n$ is the dual basis of $v_1, \dots, v_n$. Thus, $E \cong E'$ as S-schemes. One readily checks that pulling the $\mathbf{GL}_n(S)$-action on E' via this isomorphism gives the $\mathbf{GL}_n(S)$-action on E in the lemma. □

PROPOSITION 3.3. *Let S be a scheme, and let G be a subgroup of $\mathbf{GL}_n(S)$ that is flat and locally of finite presentation over S, and such that $\mathbf{GL}_n(S)/G$ is an S-scheme. Let E be a $\mathbf{GL}_n(S)$-torsor over an S-scheme X. Then the S-algebraic space E/G (see Lemma 2.2) is a scheme. Moreover, if X and $\mathbf{GL}_n(S)/G$ are of finite type (resp. locally of finite type) over S, then so is E/G.*

PROOF. Let V be the vector bundle corresponding to E. Then there is an open covering $\{U_i\}_{i\in I}$ of X such that $V|_{U_i} \cong \mathcal{O}_{U_i}^n$ for all $i \in I$. Thus, if U is one of the U_i, or an intersection $U_i \cap U_j$, we have a $\mathbf{GL}_n(S)$-equivariant isomorphism $E|_U \cong U \times_S \mathbf{GL}_n(S)$. This means that $E|_U/G \cong U \times_S (\mathbf{GL}_n(S)/G)$, which is a scheme by our assumption. Observe that the pullback of $E|_U/G \to E/G$ along $E \to E/G$ is the open immersion $E|_U \to E$. Since $E \to E/G$ is an fppf covering (Lemma 2.2), this means that $E|_U/G \to E/G$ is an open immersion of algebraic spaces [**31**, Tag 041X]. Similarly, $E|_{U_i \cap U_j}/G$ is open in $E|_{U_i}/G$ for all $i, j \in I$. This shows that E/G can be obtained by gluing the schemes $E|_{U_i}/G$ ($i \in I$) along their open subschemes $E|_{U_i\cap U_j}/G$, so E/G is a scheme.

If X and $\mathbf{GL}_n(S)/G$ are locally of finite type over S, then so are the schemes $E|_U/G \cong U \times (\mathbf{GL}_n(S)/E)$ and it follows that E/G is also locally of finite type.

Suppose now that X and $\mathbf{GL}_n(S)/G$ are of finite type over S, i.e., they are locally of finite type and quasi-compact over S. We need to show that $E/G \to S$ is quasi-compact. To that end, it is enough to consider the case where S is affine. Then X is quasi-compact, and thus we may choose the covering $\{U_i\}_{i\in I}$ to be a *finite* covering. Since $\mathbf{GL}_n(S)/G$ is quasi-compact, each of the $E|_{U_i}/G$ is quasi-compact; and since I is finite, E/G is quasi-compact as well. □

4. Schematically Dominant Morphisms

In order to adapt the theory of versal torsors over fields to a general base scheme, we need to replace dominant morphisms with *schematically dominant* morphisms and *universally schematically dominant* morphisms. This section recalls these notions and establishes some related facts that will be needed in the sequel.

Throughout, S is a fixed base scheme.

Schematically dominant morphisms of schemes are discussed in detail in $\mathrm{EGA_{IV}}$ [**16**, §11.10]. The definition extends naturally to S-algebraic spaces as follows.

A morphism of S-algebraic spaces $f : Y \to X$ is said to be *schematically dominant* if the induced morphism $\mathcal{O}_X \to f_*\mathcal{O}_Y$ is a monomorphism of sheaves on the (small) *étale* site of X. Equivalently, for every étale morphism $U \to X$ of S-algebraic spaces, the induced map $f^* : \Gamma(U, \mathcal{O}_X) \to \Gamma(Y \times_X U, \mathcal{O}_Y)$ is injective. More generally, a family of morphisms $\{f_i : Y_i \to X\}_{i\in I}$ is said to be (jointly) schematically dominant if the morphism $\bigsqcup_{i\in I} Y_i \to X$ restricting to f_i on Y_i is schematically dominant.

The morphism $f : Y \to X$ is said to be *S-universally schematically dominant* if for every S-algebraic space S', the pullback $f_{S'} : Y_{S'} \to X_{S'}$ is schematically dominant. It is said to be *universally schematically dominant* if for every morphism of S-algebraic spaces $h : X' \to X$, the pullback $f_h : Y \times_X X' \to X'$ is schematically dominant. Similarly, a family of morphisms $\{f_i : Y_i \to X\}_{i\in I}$ is said to be (S-)universally schematically dominant if the induced morphism $\bigsqcup_{i\in I} Y_i \to X$ is (S-)universally schematically dominant.

One readily sees that the composition of schematically dominant morphisms is schematically dominant. If $g : Z \to Y$ and $f : Y \to X$ are two morphisms of S-algebraic spaces and $f \circ g$ is schematically dominant, then so is f. Similar statements hold for (S-)universally schematically dominant morphisms.

EXAMPLE 4.1. (i) Suppose $S = \operatorname{Spec} R$ for a ring R and $\varphi : A \to B$ is a morphism of R-rings. Then the induced morphism $f : \operatorname{Spec} B \to \operatorname{Spec} A$ is schematically dominant if and only if φ is injective. (Indeed, if $U := \operatorname{Spec} A' \to \operatorname{Spec} A$ is étale, then $\Gamma(U, \mathcal{O}_{\operatorname{Spec} A}) \to \Gamma(\operatorname{Spec} B \times_{\operatorname{Spec} A} U, \mathcal{O}_{\operatorname{Spec} B})$ is just the map $a' \mapsto 1_B \otimes a' : A' \to B \otimes_A A'$, and it is injective whenever φ is, because A' is flat over A.)

(ii) Any fpqc covering of S-algebraic spaces $f : Y \to X$ is universally schematically dominant. This follows from fpqc descent for the structure sheaf of X on the large fpqc site of X ([**31**, Tag 04P2]; the structure sheaf is represented by $\mathbb{A}^1_X$).

REMARK 4.2. In $\mathrm{EGA_{IV}}$ [**16**, §11.10], a morphism of S-schemes $f : Y \to X$ is called schematically dominant when $\mathcal{O}_X \to f_*\mathcal{O}_Y$ is a monomorphism of sheaves on the (small) *Zariski* site of X, and S-universally schematically dominant if $f_{S'} : Y_{S'} \to X_{S'}$ is schematically dominant for every S-*scheme* S'. This agrees with our definition by the next proposition. Consequently, any S-morphism of S-schemes $f : Y \to X$ that is schematically dominant is dominant, and the converse holds if X is reduced.

PROPOSITION 4.3. *Let S be a scheme and let $f : Y \to X$ be a morphism of S-schemes. Write $\mathcal{O}_X^{\mathrm{Zar}}$ (resp. $\mathcal{O}_X^{\text{ét}}$) for the structure sheaf of X on the small Zariski (resp. étale) site of X. Then:*

(i) *The map $\mathcal{O}_X^{\mathrm{Zar}} \to f_*\mathcal{O}_Y^{\mathrm{Zar}}$ induced by f is a monomorphism if and only if the map $\mathcal{O}_X^{\text{ét}} \to f_*\mathcal{O}_Y^{\text{ét}}$ induced by f is a monomorphism.*

(ii) *If $f_{S'} : Y_{S'} \to X_{S'}$ is schematically dominant for every S-scheme S', then this is also true for every S-algebraic space S'.*

PROOF.

(i) The "if" part is clear. As for the "only if" part, by [**16**, Theorem 11.10.5(ii)], for every morphism $g : X' \to X$ that is flat and locally of finite presentation, the pullback $f_g : Y \times_X X' \to X'$ also satisfies the condition that $\mathcal{O}_{X'}^{\mathrm{Zar}} \to (f_g)_*\mathcal{O}_{Y \times X'}^{\mathrm{Zar}}$ is a mononmorphism. Thus, $\mathcal{O}_X(X') \to \mathcal{O}_Y(Y \times_X X')$ is injective. Letting $g : X' \to X$ range over the étale morphisms to X shows that $\mathcal{O}_X^{\text{ét}} \to f_*\mathcal{O}_Y^{\text{ét}}$ is a monomorphism.

(ii) Let $g : S' \to S$ be a morphism of S-algebraic spaces. Then there is an S-scheme S'' and an étale covering $h : S'' \to S'$. By assumption, $f_{S''} : Y_{S''} \to X_{S''}$ is schematically dominant. Furthermore, since $h_X : X \times_S S'' \to X \times_S S'$ is an étale covering of S-algebraic spaces, it is schematically dominant (Example 4.1(ii)). Thus, the composition $h_X \circ f_{S''} : Y \times_S S'' \to X \times_S S'$ is schematically dominant. The morphism $h_X \circ f_{S''}$ also factors as $f_{S'} \circ h_Y$, so $f_{S'} : Y_{S'} \to X_{S'}$ is also schematically dominant. □

THEOREM 4.4 ([**16**, Theorem 11.10.5]). *Let X be a locally noetherian S-scheme and let $\{f_i : Z_i \to X\}_{i \in I}$ be morphisms of S-schemes such that each Z_i is flat over S. For $s \in S$, put $X_s = X \times_S \operatorname{Spec} k(s)$ and define $(Z_i)_s$, $(f_i)_s$ similarly. Then the family $\{f_i : Z_i \to X\}_{i \in I}$ is S-universally schematically dominant if and only if, for every $s \in S$, the family $\{(f_i)_s : (Z_i)_s \to X_s\}$ is schematically dominant.*

This theorem has a number of corollaries which we record.

COROLLARY 4.5. *Let k be a field, let X be a reduced locally noetherian k-scheme, and let $f : Y \to X$ be a morphism of k-schemes. Then f is dominant if and only if it is k-universally schematically dominant.*

COROLLARY 4.6. *Suppose that S is an affine noetherian scheme with infinite residue fields and let $n \in \mathbb{N} \cup \{0\}$. Let $\{f_i : S \to \mathbb{A}^n_S\}_{i\in I}$ be the S-sections of $\mathbb{A}^n$. Then the family $\{f_i : S \to \mathbb{A}^n_S\}_{i\in I}$ is S-universally schematically dominant.*

PROOF. Write $S = \operatorname{Spec} R$. The S-sections of $\mathbb{A}^n_S$ can be naturally identified with R^n. Let $s \in S$, $k = k(s)$ and put $\overline{R} = \operatorname{im}(R \to k)$. Then, once viewing k^n as the set of k-sections of $\mathbb{A}^n_k$, the family $\{(f_i)_s : \operatorname{Spec} k \to \mathbb{A}^n_k\}_{i\in I}$ is the subset $\overline{R}^n$. Since k is the fraction field of $\overline{R}$, the ring $\overline{R}$ is infinite. It is now routine to check that $\{(f_i)_s : \operatorname{Spec} k \to \mathbb{A}^n_k\}_{i\in I}$ is dominant, and thus schematically dominant because $\mathbb{A}^n_k$ is reduced. We finish by applying Theorem 4.4. □

COROLLARY 4.7. *Let k be an infinite field, let $n \in \mathbb{N}$ and let $\{f_i : \operatorname{Spec} k \to \mathbf{GL}_n(k)\}_{i\in I}$ denote the k-sections of the k-scheme $\mathbf{GL}_n(k)$. Then $\{f_i : \operatorname{Spec} k \to \mathbf{GL}_n(k)\}_{i\in I}$ is k-universally schematically dominant.*

We say that a morphism of S-algebraic spaces $f : Y \to X$ is *S-fppf surjective* if it is surjective when viewed as a morphism of sheaves on $(\mathsf{Sch}/S)_{\mathrm{fppf}}$. This is equivalent to saying that for every S-scheme V and $x \in X(V)$, there is an fppf covering $V' \to V$ and $y \in Y(V')$ such that $f(y) = x|_{V'}$.

LEMMA 4.8. *Let $f : Y \to X$ be an S-fppf surjective morphism of S-algebraic spaces. Then f is universally schematically dominant.*

PROOF. Let $g : X' \to X$ be a morphism of S-algebraic spaces. Let $\mathscr{S}$ denote the category of sheaves on $(\mathsf{Sch}/S)_{\mathrm{fppf}}$. By viewing X, Y and X' as objects of $\mathscr{S}$, one readily checks that $f_g : Y \times_X X' \to X'$ is also S-fppf surjective. It is therefore enough to show that f is schematically dominant. Let $U \to X$ be an étale morphism. We need to show that $\Gamma(U, \mathcal{O}_X) \to \Gamma(Y \times_X U, \mathcal{O}_Y)$ is injective. This is equivalent to saying that $\operatorname{Mor}_{\mathscr{S}}(U, \mathbb{A}^1_S) \to \operatorname{Mor}_{\mathscr{S}}(Y \times_X U, \mathbb{A}^1_S)$ is injective, and this holds because $Y \times_X U \to U$ is S-fppf surjective, i.e., surjective in $\mathscr{S}$. □

The following results give several examples of S-fppf surjective morphisms. By Lemma 4.8, these are also examples of universally schematically dominant morphisms.

Given a ring A, an element $a \in A$ and distinct $i, j \in \{1, \ldots, n\}$, we write $e_{ij}(a)$ for the matrix in $\mathrm{SL}_n(A)$ having 1-s on the diagonal, a in the (i, j)-entry, and 0 in all other entries.

PROPOSITION 4.9. *For every $m \in \mathbb{N}$, there exists $r \in \mathbb{N} \cup \{0\}$ and a sequence of pairs of distinct integers $(i_1, j_1), \ldots, (i_r, j_r) \in \{1, \ldots, m\}^2$ such that the morphism $f : \mathbb{A}^r_{\mathbb{Z}} \to \mathbf{SL}_m(\mathbb{Z})$ given by*

$$f(a_1, \ldots, a_r) = e_{i_1 j_1}(a_1) \cdots e_{i_r j_r}(a_r)$$

on sections is $\mathbb{Z}$-fppf surjective, and therefore universally schematically dominant.

In fact, the proof shows that $f : \mathbb{A}^r_{\mathbb{Z}} \to \mathbf{SL}_m(\mathbb{Z})$ is even surjective when regarded as a morphism of sheaves on the large Zariski site of $\mathbb{Z}$.

PROOF. *Step 1.* We first claim that there are $(i_1, j_1), \ldots, (i_r, j_r) \in \{1, \ldots, m\}^2$ such that for every semilocal ring A, the induced map $f_A : A^r \to \mathrm{SL}_m(A)$ is surjective. This is the same as showing that every matrix in $\mathrm{SL}_m(A)$ can be reduced to the identity matrix by using r *predetermined* (and possibly vacuous) row and column operations of the form $R_i + \alpha R_j \to R_i$ or $C_i + \alpha C_j \to C_i$. This is known, but we include a proof for the sake of completeness.

We prove this by induction on m. When $m = 1$, we can take $r = 0$. Suppose that $m > 1$, and let $(i_\ell, j_\ell)_{\ell=1}^{s}$ be the sequence of pairs in $\{1,\dots,m-1\}^2$ supplied by the induction hypothesis. Let A be a semilocal ring and let $a = (a_{ij}) \in \mathrm{SL}_m(A)$. Since $\det(a) = 1$, we have $\sum_{i=1}^{m} a_{im}A = A$. As A is semilocal, it has *stable range* 1 (see [**4**, §4, Corollary 6.5]); hence there are $\alpha_1,\dots,\alpha_{m-1} \in A$ such that

$$\sum_{i=1}^{m-1} (a_{im} + \alpha_i a_{mm})A = A.$$

We may therefore replace a with $e_{1m}(\alpha_1)e_{2m}(\alpha_2)\cdots\ e_{(m-1)m}(\alpha_{m-1})a$, and assume that $\sum_{i=1}^{m-1} a_{im}A = A$. (This is allowed because the pairs $(1,m),\dots,$ $(m-1,m)$ do not depend on A or a.) Now, if $m-1 > 1$, then again there are $\alpha_1,\dots,\alpha_{m-2} \in A$ with $\sum_{i=1}^{m-2}(a_{im} + \alpha_i a_{(m-1)m})A = A$, and we may replace a with $e_{1m}(\alpha_1)e_{2m}(\alpha_2)\cdots e_{(m-2)m}(\alpha_{m-2})a$ to assume $\sum_{i=1}^{m-2} a_{im}A = A$. Continuing in this manner, we reduce to the case where a_{1m} is invertible. Now, by replacing a with $e_{m1}(a_{1m}^{-1}(1 - a_{mm}))a$, we may assume that $a_{mm} = 1$. Replacing a once more by

$$e_{1m}(-a_{1m})\cdots e_{(m-1)m}(-a_{(m-1)m}) \cdot a \cdot e_{m(m-1)}(-a_{m(m-1)})\cdots e_{m1}(-a_{m1})$$

reduces us to the case where $a = \left[\begin{smallmatrix} a' & 0 \\ 0 & 1 \end{smallmatrix}\right]$ for some $a' \in \mathrm{SL}_{m-1}(A)$. By the induction hypothesis, there are $\alpha_1,\dots,\alpha_s \in A$ such that $a = e_{i_1j_1}(\alpha_1)\cdots e_{i_sj_s}(\alpha_s)$ and we are done.

Step 2. We now prove that $f : \mathbb{A}^r_{\mathbb{Z}} \to \mathbf{SL}_m(\mathbb{Z})$ is $\mathbb{Z}$-fppf surjective for the pairs $(i_1,j_1),\dots,(i_r,j_r)$ supplied by Step 1. Let V be a $\mathbb{Z}$-scheme and let $a \in \mathrm{SL}_m(V)$. By the previous paragraph, for every $x \in V$, there are $a_1^{(x)},\dots,a_r^{(x)} \in \mathcal{O}_{V,x}$ such that $e_{i_1j_1}(a_1^{(x)})\cdots e_{i_rj_r}(a_r^{(x)}) = a_x$, where a_x is the restriction of a to $\operatorname{Spec}\mathcal{O}_{X,x}$. By spreading out, we may assume that $a_1^{(x)},\dots,a_r^{(x)}$ are defined on an open affine neighborhood V_x of x and that $e_{i_1j_1}(a_1^{(x)})\cdots e_{i_rj_r}(a_r^{(x)}) = a|_{V_x}$. Put $V' = \bigsqcup_{x\in V} V_x$. Then $V' \to V$ is a Zariski covering of V such that $a|_{V'}$ is in the image of $\mathbb{A}^n_S(V') \to \mathrm{SL}_m(V')$. □

Let H be an S-group acting from the left on an S-algebraic space X. We say that the action of H on X is fppf transitive if the map $H \times_S X \to X \times_S X$ given section-wise by $(h,x) \mapsto (hx,x)$ is S-fppf surjective. This is equivalent to saying that for every X-scheme V and $x,x' \in X(V)$, there exist an fppf covering $V' \to V$ and $h \in H(V')$ such that $x'|_{V'} = h \cdot x|_{V'}$. In fact, it is enough to verify this only when V is affine.

LEMMA 4.10. *Let H be an S-group acting from the left on an S-algebraic space X, and let $f : Y \to X$ be an S-morphism of algebraic spaces. If the action of H on X is fppf transitive, then the morphism $H \times_S Y \to X \times_S Y$ given section-wise by $(h,y) \mapsto (h\cdot f(y), y)$ is S-fppf surjective, and thus universally schematically dominant.*

PROOF. Let V be an S-scheme and let $(x,y) \in X(V)\times Y(V)$. Since the action of H on X is fppf transitive, we may replace V with a suitable fppf covering to assume that there is $h \in H(V)$ such that $h\cdot f(y) = x$. Then $(h\cdot f(y), y) = (x,y)$ in $X(V)\times Y(V)$. □

5. d-Versal Torsors: Definition and Properties

Throughout this section, S is a fixed base scheme. If not indicated otherwise, all schemes and algebraic spaces are over S, all morphisms between them are S-morphisms, and G denotes an S-group. Given S-algebraic spaces X, Y, Z and morphisms $f : Z \to X$, $g : Z \to Y$, we write (f, g) to denote the morphism $Z \to X \times_S Y$ given by $z \mapsto (f(z), g(z))$ on sections.

DEFINITION 5.1. Let X be an S-algebraic space, let $E \to X$ be a G-torsor, let $\mathscr{C}$ be a class of S-schemes, and let $d \in \mathbb{N} \cup \{0\}$.

(i) We say that $E \to X$ is *weakly d-versal for $\mathscr{C}$* if for every G-torsor $E' \to X'$ with $X' \in \mathscr{C}$, there are:
  (1) an open $U \subseteq X'$ such that $\operatorname{codim}(X' - U, X') > d$ and
  (2) an S-morphism $f : U \to X$
  such that $E'|_U \cong_U f^*E$ (the isomorphism is in $\mathsf{Tors}(G, U)$; see Section 2).

(ii) We say that $E \to X$ is *d-versal for $\mathscr{C}$* if for every G-torsor $E' \to X'$ with $X' \in \mathscr{C}$, there are:
  (1) an open $U \subseteq X'$ such that $\operatorname{codim}(X' - U, X') > d$ and
  (2) a collection of S-morphisms $\{f_i : U \to X\}_{i \in I}$
  such that $E'|_U \cong_U f_i^*E$ for all $i \in I$, and the family $\{(f_i, \mathrm{id}_U) : U \to X \times_S U\}_{i \in I}$ is U-universally schematically dominant.

(iii) We say that $E \to X$ is *strongly d-versal for $\mathscr{C}$* if for every G-torsor $E' \to X'$ with $X' \in \mathscr{C}$, there are:
  (1) an open $U \subseteq X'$ such that $\operatorname{codim}(X' - U, X') > d$ and
  (2) an S-morphism $f : \mathbb{A}^n_S \times_S U \to X$ (for some $n \in \mathbb{N}$)
  such that $\mathbb{A}^n_S \times_S E'|_U \cong_{\mathbb{A}^n \times U} f^*E$ and the morphism $(f, p_2) : \mathbb{A}^n_S \times_S U \to X \times_S U$, where $p_2 : \mathbb{A}^n_S \times_S U \to U$ is the second projection, is universally schematically dominant.

We say that $E \to X$ is (weakly, strongly) ∞-versal for $\mathscr{C}$ if it is (weakly, strongly) d-versal for every $d \in \mathbb{N}$.

We say that $E \to X$ is (weakly, strongly) versal for $\mathscr{C}$ if we can always take $U = X'$ in (1).

Admittedly, there is some room for variation in the definition, e.g., we can replace "(U-)universally schematically dominant" with "schematically dominant" in (ii) and (iii). The present formulation allows for both strong and clean statements. Before we proceed, a few remarks are in order.

REMARK 5.2. (i) Suppose that $S = \operatorname{Spec} k$ for an infinite field k. As noted in the introduction, a G-torsor is weakly 0-versal for the class of irreducible k-varieties if and only if it is weakly versal in the sense of Duncan and Reichstein [7]. However, being (strongly) 0-versal for this class is *a priori* stronger than being (strongly) versal in the sense of [op. cit.].

(ii) Suppose that $S = \operatorname{Spec} k$ for a field k. Conditions (ii) and (iii) of Definition 5.1 differ from the simplified versions (ii′) and (iii′) of the introduction in two ways: First, in Definition 5.1, we require (S-)universal schematic dominance and not just schematic dominance. Second, in Definition 5.1, we require dominance for the family $\{(f_i, \mathrm{id}_U) : U \to X \times_S U\}_{i \in I}$ (resp. the morphism $(f, p_2) : \mathbb{A}^n_S \times_S U \to X \times_S U$) instead of the family $\{f_i : U \to X\}_{i \in I}$ (resp. the morphism $f : \mathbb{A}^n_S \times_S U \to X$). This is more restrictive than (ii′) (resp. (iii′)), because $X \times_k U \to X$ is schematically dominant whenever $U \neq \emptyset$ (Corollary 4.5). The reason for making this change is that when S is general, the morphism $X \to S$ is typically surjective (see Proposition 5.6 below), and thus the families considered in (ii′) and (iii′) can be dominant only when $X' \to S$ is dominant. In addition, by making this change, we get that (weak, strong) versality persists under base-change of S (Proposition 5.5).

(iii) When the class $\mathscr{C}$ consists of finite-dimensional schemes, being (weakly, strongly) ∞-versal for $\mathscr{C}$ is the same as being (weakly, strongly) versal for $\mathscr{C}$.

REMARK 5.3. Similarly to Definition 5.1, one can define rank-n vector bundles that are (weakly, strongly) d-versal for a class of S-schemes $\mathscr{C}$. Since rank-n vector bundles over S-schemes are equivalent to $\mathbf{GL}_n(S)$-torsors, a rank-n bundle would be (weakly, strongly) d-versal for $\mathscr{C}$ if and only if the same holds for its corresponding $\mathbf{GL}_n(S)$-torsor.

More generally, if F is a fibered category over S-algebraic spaces, then we can talk about objects in F that are (weakly, strongly) d-versal for $\mathscr{C}$. Propositions 5.4, 5.5 and 5.6 below actually hold in this more general setting. Examples of fibered categories that will be considered later include the category of degree-n Azumaya algebras over S-algebraic spaces, which is equivalent to $\mathsf{Tors}(\mathbf{PGL}_n(S))$, and, for a fixed finite group Γ, the category of Γ-Galois extensions of S-algebraic spaces, which is equivalent to $\mathsf{Tors}(\underline{\Gamma})$. (Here, $\underline{\Gamma}$ is the constant S-group associated to Γ.) In both examples, the morphisms from A to B are the specializations of B to A. Another example that is not equivalent to a category of torsors under an S-group is the category of degree-m Azumaya algebras over S-algebraic spaces whose Brauer class has period dividing n (m and n are fixed).

PROPOSITION 5.4. *Let $\mathscr{C}$ be a class of S-schemes, let X be an S-algebraic space and let $E \to X$ be a G-torsor.*

(i) *If $E \to X$ is d-versal for $\mathscr{C}$ then it is weakly d-versal for $\mathscr{C}$.*
(ii) *If $E \to X$ is strongly d-versal for $\mathscr{C}$, then it is weakly d-versal for $\mathscr{C}$. If S is moreover a scheme over an affine noetherian scheme with infinite residue fields, then $E \to X$ is also d-versal for $\mathscr{C}$.*

PROOF. (i) is immediate from the definition.

(ii) Let $E' \to X'$ be a G-torsor with $X' \in \mathscr{C}$, and let U and $f : \mathbb{A}^n_S \times U \to X$ be as in Definition 5.1(iii). Let $u : \mathbb{A}^n_U \to U$ be the structure morphism and let $z : U \to \mathbb{A}^n_U$ be the zero section. Then $u \circ z = \mathrm{id}_U$ and thus

$$z^* f^* E \cong z^*(\mathbb{A}^n_S \times_S E'|_U) = z^* u^*(E'|_U) = E'|_U.$$

This shows that $E \to X$ is weakly d-versal for $\mathscr{C}$.

Keeping the previous notation, suppose now that there is a morphism of schemes $S \to S_0$, where S_0 is affine, noetherian and has infinite residue fields. Let $\{g_i : S \to \mathbb{A}^n_S\}_{i \in I}$ be the family of sections of $\mathbb{A}^n_S \to S$. We claim that the family $\{f_i := f \circ (g_i)_U : U \to X\}_{i \in I}$ satisfies the requirements listed in Definition 5.1(ii). Indeed, for every $i \in I$, we have $u \circ (g_i)_U = \mathrm{id}_U$ and thus, as in the previous paragraph, $f_i^* E \cong_U E'|_U$. Next, by Corollary 4.6, the family $\{(g_i)_U : U \to \mathbb{A}^n_U\}_{i \in I}$ is S-universally schematically dominant. By assumption, $(f, p_2) : \mathbb{A}^n_S \times_S U \to X \times_S U$ is universally schematically dominant, so the family $\{(f_i, \mathrm{id}_U) = (f, p_2) \circ (g_i)_U\}_{i \in I}$ is S-universally schematically dominant. □

PROPOSITION 5.5. *Let $\mathscr{C}$ be a class of S-schemes, let X be an S-algebraic space and let $E \to X$ be a G-torsor. Let S' be an S-scheme and let $\mathscr{C}/S'$ denote the class of S'-schemes with underlying S-scheme in $\mathscr{C}$. If $E \to X$ is (weakly, strongly) d-versal for $\mathscr{C}$, then the $G_{S'}$-torsor $E_{S'} \to X_{S'}$ is (weakly, strongly) d-versal for $\mathscr{C}/S'$.*

PROOF. Let $E' \to X'$ be a $G_{S'}$-torsor with $X' \in \mathscr{C}/S'$. Then $E' \to X'$ is also a G-torsor. Let s be the structure morphism $X' \to S'$. The proposition follows from the following three observations (applied with $X' = U$ or $X' = \mathbb{A}^n_S \times_S U$ for U as in Definition 5.1), which can be readily verified on the level of sections: First, if $f : X' \to X$ is an S-morphism such that $f^*E \cong_{X'} E'$, then $g = (f, s) : X' \to X_{S'}$ satisfies $g^*E_{S'} \cong_{X'} E'$ as $G_{S'}$-torsors. Second, for every $f : X' \to X$, the morphism $(f, \mathrm{id}_{X'}) : X' \to X \times_S X'$ coincides with $(f_{S'}, \mathrm{id}_{X'}) : X' \to X_{S'} \times_{S'} X'$ under the evident identification $X \times_S X' \cong X_{S'} \times_{S'} X'$. Third, for every $f : \mathbb{A}^n_S \times_S X' \to X$, the morphism $(f, p_2) : \mathbb{A}^n_S \times_S X' \to X \times_S X'$ coincides with $(f_{S'}, p_2) : \mathbb{A}^n_{S'} \times_{S'} X' \to X_{S'} \times_{S'} X'$ under the evident identifications. □

PROPOSITION 5.6. *Let $E \to X$ be a G-torsor, where X is an S-scheme.*

(i) *If $E \to X$ is weakly versal for $\{S\}$, then $X \to S$ is surjective (i.e., set-theoretically surjective).*
(ii) *If $E \to X$ is versal for $\{S\}$ and S is reduced, then X is reduced.*
(iii) *If $E \to X$ is strongly versal for $\{S\}$ and S is integral, then X is integral.*
(iv) *If $E \to X$ is strongly versal for the S-schemes $\{\operatorname{Spec} k(s) \mid s \in S\}$, then $X \to S$ has integral fibers.*

PROOF. (i) View $G \to S$ as a G-torsor by letting G act on itself from the right. Since $E \to X$ is weakly versal for $\{S\}$, there is an S-morphism $S \to X$. The composition $S \to X \to S$ is id_S, so $X \to S$ is surjective.

(ii) Applying the versality to the G-torsor $G \to S$, we get a schematically dominant family of morphisms $\{f_i : S \to X \times_S S = X\}_{i \in I}$. The induced map $\mathcal{O}_X^{\mathrm{Zar}} \to \prod_i (f_i)_* \mathcal{O}_S^{\mathrm{Zar}}$ is therefore injective (notation as in Proposition 4.3). Since S is reduced, this means that $\mathcal{O}_X^{\mathrm{Zar}}$ has no nilpotent sections; hence X is reduced.

(iii) Applying strong versality to the G-torsor $G \to S$, we get a morphism $f : \mathbb{A}^n_S \to X$ such that induced map $\mathcal{O}_X^{\mathrm{Zar}} \to f_* \mathcal{O}_{\mathbb{A}^n_S}^{\mathrm{Zar}}$ is injective. Since S is integral, so is $\mathbb{A}^n_S$, and thus $f_* \mathcal{O}_{\mathbb{A}^n_S}^{\mathrm{Zar}}$ is a sheaf of integral domains. It follows that $\mathcal{O}_X^{\mathrm{Zar}}$ is also a sheaf of integral domains, hence X is integral.

(iv) Let $s \in S$, put $X_s = X \times_S \operatorname{Spec} k(s)$, and define E_s and G_s similarly. By Proposition 5.5, the G_s-torsor $E_s \to X_s$ is strongly versal for $\{\operatorname{Spec} k(s)\}$, so X_s is integral by (iii). □

PROPOSITION 5.7. *Let G be an S-group and let H be a subgroup of G that is flat and locally of finite presentation over S. Let $E \to X$ be a G-torsor that is weakly d-versal for a class of S-schemes $\mathscr{C}$. Then $E \to E/H$ is an H-torsor that is weakly d-versal for $\mathscr{C}$.*

PROOF. Lemma 2.2 and our assumptions on H imply that $E \to E/H$ is an H-torsor.

Let $E' \to X'$ be an H-torsor with $X' \in \mathscr{C}$, put $P = E' \times^H G$ and let $i : E' \to P$ denote the morphism given by $i(e) = [e, 1]$ on sections. Then $P \to X'$ is a G-torsor and i is H-equivariant. Since $E \to X$ is weakly d-versal for $\mathscr{C}$, there exists an open subscheme $U \subseteq X'$ with $\operatorname{codim}(X' - U, X') > d$ and a G-equivariant morphism

$P|_U \to E$. The composition $E'|_U \to P|_U \to E$ is H-equivariant, and thus induces a morphism of H-torsors from $E'|_U \to U$ to $E \to E/H$. We finish by Remark 2.1. □

Our next proposition shows that weakly versal G-torsors which admit an action by another S-group H (in the sense of Section 2) that is fppf transitive on the base (see Section 4) are sometimes versal, or even strongly versal.

PROPOSITION 5.8. *Let $\mathscr{C}$ be a class of S-schemes, let X be an S-algebraic space and let $E \to X$ be a G-torsor that is weakly d-versal for $\mathscr{C}$. Suppose that there is an S-group H acting on $E \to X$ (from the left) such that the action of H on X is fppf transitive. Then:*

(i) *If the family of S-sections of H is S-universally schematically dominant, then $E \to X$ is d-versal for $\mathscr{C}$.*
(ii) *If there is an S-fppf surjective morphism $\mathbb{A}^n_S \to H$ for some $n \in \mathbb{N} \cup \{0\}$, then $E \to X$ is strongly d-versal for $\mathscr{C}$.*

PROOF. (i) Let $\{h_i : S \to H\}_{i\in I}$ be a collection of schematically dominant S-morphisms; we view the h_i as elements of the group $H(S)$.

Let $E' \to X'$ be a G-torsor with $X' \in \mathscr{C}$. Then there are an open $U \subseteq X'$ with $\operatorname{codim}(X' - U, X') > d$ and a G-torsor morphism $(\hat{f}, f)$ from $E'|_U$ to E. For every $i \in I$, define $f_i : U \to X$ by $f_i(u) = h_i \cdot f(u)$ and $\hat{f}_i : E'|_U \to E$ by $\hat{f}_i(e) = h_i \cdot \hat{f}(e)$ on sections. One readily checks that each $(\hat{f}_i, f_i)$ is a G-torsor morphism from E'_U to E. Thus, $f_i^* E \cong_U E'|_U$ for all $i \in I$ (Remark 2.1).

By Lemma 4.10, the map $g : H \times_S U \to X \times_S U$ given section-wise by $g(h, u) = (h \cdot f(u), u)$ is universally schematically dominant. Our assumption on H implies that $\{(h_i)_U : U \to H \times_S U\}_{i\in I}$ is U-universally schematically dominant, so the family $\{g \circ (h_i)_U : U \to X \times_S U\}_{i\in I}$ is U-universally schematically dominant. On the level of sections, we have $(g \circ (h_i)_U)(u) = g(h_i, u) = (h_i \cdot f(u), u) = (f_i(u), u)$, so $\{(f_i, \mathrm{id}_U) : U \to X \times_S U\}_{i\in I}$ is S-universally schematically dominant. We conclude that $E \to X$ is d-versal for $\mathscr{C}$.

(ii) Let $h : \mathbb{A}^n_S \to H$ be an S-fppf surjective morphism, and let $E' \to X'$, U, $(\hat{f}, f)$ and g be as in the proof of (i). Define $w : \mathbb{A}^n_S \times_S U \to X$ by $w(v, u) = h(v) \cdot f(u)$ and $\hat{w} : \mathbb{A}^n_S \times E'|_U \to E$ by $\hat{w}(v, e) = h(v) \cdot \hat{f}(e)$ on sections. One readily checks that $(\hat{w}, w)$ is a G-torsor morphism from $\mathbb{A}^n_S \times_S E'|_U$ to E, hence $w^* E \cong_{\mathbb{A}^n \times U} \mathbb{A}^n_S \times E'|_U$. Moreover, $(w, p_2) : \mathbb{A}^n_S \times_S U \to X \times_S U$ coincides with $g \circ h_U$, which is universally schematically dominant because g and h are. □

Putting everything together, we arrive at the following corollary, which reduces the existence of strongly d-versal torsors for linear group schemes to the existence of certain weakly d-versal vector bundles.

COROLLARY 5.9. *Let S_0 be a scheme, let $\mathscr{C}$ be a class of S_0-schemes and let $n \in \mathbb{N}$. Suppose that there exists an S_0-scheme X_0 and a rank-n vector bundle V_0 over X_0 that is* weakly *d-versal for $\mathscr{C}$. Let $E_0 \to X_0$ be the $\mathbf{GL}_n(S_0)$-torsor corresponding to V_0, and suppose moreover that there are $m_1, \dots, m_r \in \mathbb{N}$ such that $H_0 := \mathbf{SL}_{m_1}(S_0) \times \cdots \times \mathbf{SL}_{m_r}(S_0)$ acts on $E_0 \to X_0$ (see Section 2) and the action of H_0 on E_0 is fppf transitive[8]. Then, for every S_0-scheme S and every flat locally of finite presentation S-group G that is isomorphic to a subgroup of $\mathbf{GL}_n(S)$, there exists a G-torsor over an S-algebraic space that is* strongly *d-versal for $\mathscr{C}/S$.*

[8]Caution: It is not enough to assume that H_0 acts transitively on X_0.

Specifically, if $\rho : G \to \mathbf{GL}_n(S)$ is a monomorphism of S-groups, then we can take this torsor to be $(E_0 \times_{S_0} S) \to (E_0 \times_{S_0} S)/G$, where G acts on $E_0 \times_{S_0} S$ via ρ.

PROOF. Let $\rho : G \to \mathbf{GL}_n(S)$ be a monomorphism; we view G as a subgroup of $\mathbf{GL}_n(S)$ via ρ. Put $E = E_0 \times_{S_0} S$, $X = X_0 \times_{S_0} S$ and $H = H_0 \times_{S_0} S$. By Proposition 5.5, $E \to X$ is a $\mathbf{GL}_n(S)$-torsor that is weakly d-versal for $\mathscr{C}/S$. Thus, by Proposition 5.7, $E \to E/G$ is a G-torsor that is weakly d-versal for $\mathscr{C}/S$. The action of H on $E \to X$ gives rise to an action of H on $E \to E/G$, and H acts fppf transitively on E/G because it acts fppf transitively on E and $E \to E/G$ is S-fppf surjective. By Proposition 4.9, there exists a universally schematically dominant morphism $\mathbb{A}^n_S \to H$, so Proposition 5.8 tells us that $E \to E/G$ is strongly d-versal for $\mathscr{C}/S$. □

REMARK 5.10. In Corollary 5.9, if we assume that S_0 is a scheme over an infinite field and replace $\mathbf{SL}_{m_1}(S_0) \times \cdots \times \mathbf{SL}_{m_r}(S_0)$ with $\mathbf{GL}_{m_1}(S_0) \times \cdots \times \mathbf{GL}_{m_r}(S_0)$, then we can conclude that the G-torsor $(E_0 \times_{S_0} S) \to (E_0 \times_{S_0} S)/G$ is d-versal for $\mathscr{C}/S$. To see this, replace Proposition 4.9 with Corollary 4.7 in the proof.

In the next two sections, we will construct vector bundles that can be fed into Corollary 5.9.

6. Generic Vector Bundles with a Fixed Number of Generators

Let $m \geq n \geq 0$ be integers. In this section, we recall two constructions of "generic" m-generated rank-n locally free modules and describe their corresponding $\mathbf{GL}_n(\mathbb{Z})$-torsors. Here, being generic means that these vector bundles specialize to every rank-n vector bundle that is generated by m global sections, provided the base scheme is in some prescribed class of schemes, e.g., the affine schemes.

Throughout this section, all schemes are over $\mathbb{Z}$. The functor of points of a scheme means its functor of points when considered as a $\mathbb{Z}$-scheme.

Recall that if X is a scheme, $\mathcal{M}$ an $\mathcal{O}_X$-module, and $m_1, \dots, m_t \in \Gamma(X, \mathcal{M})$, then $\mathcal{M}$ is said to be generated by $m_1, \dots, m_t$ if the $\mathcal{O}_X$-module map $\mathcal{O}_X^t \to \mathcal{M}$ given section-wise by $(a_1, \dots, a_t) \mapsto \sum_{i=1}^t a_i m_i$ is surjective. When $X = \operatorname{Spec} A$ for a ring A and $\mathcal{M}$ is quasi-coherent, $m_1, \dots, m_t$ generate $\mathcal{M}$ if and only if they generate the A-module $M := \Gamma(X, \mathcal{M})$ (but this is false in the non-affine case, e.g., see Example 7.8 below).

We begin with recalling the generic m-generated rank-n projective module studied by M. Raynaud in [**24**]. The idea is that every rank-n projective module P over a ring R that is generated by m elements is isomorphic to a summand of R^m, and can therefore be expressed as the image of a rank-n idempotent matrix $e \in \mathrm{M}_m(R)$, or equivalently, the cokernel of $1-e$. Thus, the image of such a universal idempotent matrix is a rank-n projective module which specializes to every rank-n m-generated projective module over a commutative ring; an analogous statement holds for the corresponding vector bundle.

Formally, let $X_{n,m}$ be the closed subscheme of $\mathbb{A}^{m \times m}_{\mathbb{Z}}$ (the affine space of $m \times m$ matrices over $\mathbb{Z}$) consisting of idempotent matrices whose characteristic polynomial is $t^{m-n}(t-1)^n$. The functor of points of $X_{n,m}$ maps a ring R to the set of rank-n idempotent matrices in $\mathrm{M}_m(R)$.[9] Let $A_{n,m}$ denote the coordinate ring of $X_{n,m}$, and

[9]The rank of an idempotent $e \in \mathrm{M}_m(R)$ is defined to be the rank of the projective module $\mathrm{im}(e)$. Equivalently, this is the function $\operatorname{Spec} R \to \mathbb{N} \cup \{0\}$ mapping $\mathfrak{p} \in \operatorname{Spec} R$ to the rank of the image of e in $\mathrm{M}_m(k(\mathfrak{p}))$.

let $x_{ij} \in A_{n,m}$ be the (i,j)-coordinate function. Then $e_{n,m} := (x_{ij}) \in \mathrm{M}_m(A_{n,m})$ is an idempotent matrix of rank n. Set $P_{n,m}$ to be $\mathrm{coker}(1-e_{n,m} : A_{n,m}^m \to A_{n,m}^m)$. Then $P_{n,m}$ is a projective module of rank n over $A_{n,m}$ that is generated by m elements. The rank-n vector bundle over $X_{n,m}$ corresponding to $P_{n,m}$ is denoted $V_{n,m}$.

PROPOSITION 6.1. *Let X be an affine scheme and let V be a rank-n vector bundle over X that is generated by m global sections. Then there is a morhpism $f : X \to X_{n,m}$ such that $V \cong f^*V_{n,m}$.*

PROOF. Write $X = \mathrm{Spec}\, A$ and let $P = \Gamma(X, V)$. Then P is a rank-n projective A-module that is generated by m elements. Thus, there is a rank-n idempotent $e \in \mathrm{M}_m(A)$ and an A-module isomorphism $P \cong \mathrm{coker}(1-e : A^m \to A^m)$. The ring homomorphism $\varphi : A_{n,m} \to A$ which specializes $e_{n,m}$ to e satisfies $P_{n,m} \otimes_{A_{n,m}} A = \mathrm{coker}(1-e_{n,m}) \otimes_{A_{n,m}} A \cong \mathrm{coker}(1-e) \cong P$ and the proposition follows. □

We now describe the $\mathbf{GL}_n(\mathbb{Z})$-torsor $E_{n,m} \to X_{n,m}$ corresponding to $V_{n,m}$.

PROPOSITION 6.2. *Let $m \geq n \geq 0$ be integers and put $d = m-n$. With notation as above, the $\mathbf{GL}_n(\mathbb{Z})$-torsor corresponding to the vector bundle $V_{n,m}$ over $X_{n,m}$ is isomorphic to the quotient morphism*

$$\mathbf{GL}_m(\mathbb{Z})/\begin{bmatrix} 1_{n\times n} & 0 \\ 0 & \mathbf{GL}_d(\mathbb{Z}) \end{bmatrix} \to \mathbf{GL}_m(\mathbb{Z})/\begin{bmatrix} \mathbf{GL}_n(\mathbb{Z}) & 0 \\ 0 & \mathbf{GL}_d(\mathbb{Z}) \end{bmatrix}.$$

Here, $\mathbf{GL}_n(\mathbb{Z})$ acts on $\mathbf{GL}_m(\mathbb{Z})/[\begin{smallmatrix} 1 & 0 \\ 0 & \mathbf{GL}_d(\mathbb{Z}) \end{smallmatrix}]$ from the right via $x \cdot g = x[\begin{smallmatrix} g & 0 \\ 0 & 1_{d\times d} \end{smallmatrix}]$.

PROOF. *Step 1.* We first introduce an auxiliary morphism of affine schemes $p : T_{n,m} \to X_{n,m}$. The functor of points of $T_{n,m}$ (as a $\mathbb{Z}$-scheme) maps a ring A to

$$\begin{aligned} T_{n,m}(A) =&\{(a,x,b) \in \mathrm{M}_{n\times m}(A) \times \mathrm{M}_m(A) \times \mathrm{M}_{m\times n}(A) : \\ & x \in X_{n,m}(A), ba = x, ab = 1_{n\times n}\}. \end{aligned}$$

and $p : T_{n,m} \to X_{n,m}$ is given section-wise by $p(a,x,b) = x$. The $\mathbb{Z}$-group $\mathbf{GL}_n(\mathbb{Z})$ acts on $T_{n,m}$ from the right by $(a,x,b)\cdot g = (g^{-1}a, x, bg)$ on sections.

Step 2. Let $E \to X_{n,m}$ be the $\mathbf{GL}_n(\mathbb{Z})$-torsor corresponding to $V_{n,m}$. We claim that there is a $\mathbf{GL}_n(\mathbb{Z})$-equivariant $X_{n,m}$-isomorphism from $T_{n,m}$ to E. In particular, $p : T_{n,m} \to X_{n,m}$ is a $\mathbf{GL}_n(\mathbb{Z})$-torsor.

Indeed, by Lemma 3.1, the functor of points of E sends an $A_{n,m}$-ring R to the set of tuples $(f, v_1, \dots, v_n)$, where $f : \mathrm{Spec}\, R \to X_{n,m}$ is a morphism and $v_1, \dots, v_n$ form a basis to $f^*V_{n,m}$. The datum of f is equivalent to specifying a rank-n idempotent $e \in \mathrm{M}_m(R)$, and then $f^*V_{n,m}$ is just eR^m. Thus, $E(R)$ is in a natural bijection with the set of tuples $(e, v_1, \dots, v_n)$ where $e \in \mathrm{M}_m(R)$ is an idempotent and $v_1, \dots, v_n \in R^m$ form an R-basis to eR^m.

Let $e \in \mathrm{M}_m(R)$ be an idempotent of rank n, let $v_1, \dots, v_n \in R^m$ and let b be the $m \times n$ matrix whose columns are $v_1, \dots, v_n$. Then $v_1, \dots, v_n$ form an R-basis of eR^m if and only if the map $b : R^n \to R^m$ is injective with image eR^m. Since $R^m = eR^m \oplus (1-e)R^m$, the latter is equivalent to the existence of an R-module homomorphism $a : R^m \to R^n$ (viewed as an $m \times n$ matrix) which vanishes on $(1-e)R^m$ and satisfies $ab = 1_{n\times n}$ and $ba = \mathrm{id}_{eR^m} \oplus 0_{(1-e)R^m} = e$. Such a is determined uniquely by b and satisfies $ae = a$. We have therefore shown that each $(e, v_1, \dots, v_n) \in E(R)$ determines a triple $(a, e, b) \in T_{n,m}(R)$. Conversely, every

$(a, e, b) \in T_{n,m}(R)$ determines a tuple $(e, v_1, \ldots, v_n) \in E(R)$, by taking $v_1, \ldots, v_n$ to be the columns of b. It is routine to check that this one-to-one correspondence is compatible with base-change of R and is $\mathrm{GL}_n(R)$-equivariant, so it determines a $\mathbf{GL}_n(\mathbb{Z})$-equivariant isomorphism $E \cong T_{n,m}$, which is also easily seen to be an isomorphism over $X_{n,m}$.

Step 3. We finish by showing that $T_{n,m} \to X_{n,m}$ is isomorphic to the torsor defined in the proposition. Observe that $\mathbf{GL}_m(\mathbb{Z})$ acts on $T_{n,m}$ and $X_{n,m}$ from the left by

$$h \cdot (a, x, b) := (ah^{-1}, hxh^{-1}, hb) \qquad \text{and} \qquad h \cdot x = hxh^{-1},$$

respectively, and the map $p : T_{n,m} \to X_{n,m}$ is $\mathbf{GL}_m(\mathbb{Z})$-equivariant. We claim that $\mathbf{GL}_m(\mathbb{Z})$ acts fppf transitively on $T_{n,m}$ and $X_{n,m}$. Provided this holds, consider the $\mathbb{Z}$-point

$$t_0 = (a_0, x_0, b_0) := \left(\begin{bmatrix} 1_{n\times n} & 0_{n\times d} \end{bmatrix}, \begin{bmatrix} 1_{n\times n} & 0_{n\times d} \\ 0_{d\times n} & 0_{d\times d} \end{bmatrix}, \begin{bmatrix} 1_{n\times n} \\ 0_{d\times n} \end{bmatrix} \right) \in T_{n,m}(\mathbb{Z})$$

and its image $x_0 \in X_{n,m}(\mathbb{Z})$. One readily checks that the stablizer of t_0 in $\mathbf{GL}_m(\mathbb{Z})$ is $\left[\begin{smallmatrix} 1 & 0 \\ 0 & \mathbf{GL}_d(\mathbb{Z}) \end{smallmatrix}\right]$, so we have an isomorphism $\mathbf{GL}_m(\mathbb{Z})/\left[\begin{smallmatrix} 1 & 0 \\ 0 & \mathbf{GL}_d(\mathbb{Z}) \end{smallmatrix}\right] \to T_{n,m}$ given section-wise by sending $[h]$ to ht_0. Similarly, $[h] \mapsto h \cdot x_0 : \mathbf{GL}_m(\mathbb{Z})/\left[\begin{smallmatrix} \mathbf{GL}_n(\mathbb{Z}) & 0 \\ 0 & \mathbf{GL}_d(\mathbb{Z}) \end{smallmatrix}\right] \to X_{n,m}$ defines an isomorphism and the proposition follows.

Step 4. It remains to show that $\mathbf{GL}_m(\mathbb{Z})$ acts fppf transitively on $T_{n,m}$, and *a fortiori* on $X_{n,m}$. To see this, let R be a ring and let $(a, x, b) \in T_{n,m}(R)$. Then b defines an isomorphism from R^n to xR^m. By replacing R with a Zariski covering (i.e., a product of localizations $R_{s_1} \times \cdots \times R_{s_t}$ with $s_1, \ldots, s_t \in R$ generating the unit ideal), we may assume that $(1-x)R^m$ is a free R-module. Let $c : R^d \to (1-x)R^m$ be an isomorphism, viewed as an $m \times d$ matrix. Then $xc = 0$ and $(1-x)c = c$. Let $h = \begin{bmatrix} b & c \end{bmatrix} \in \mathrm{GL}_n(R)$. We have

$$\begin{aligned} hb_0 &= b, \\ xh &= \begin{bmatrix} xb & xc \end{bmatrix} = \begin{bmatrix} b & 0 \end{bmatrix} = hx_0, \\ ah &= \begin{bmatrix} ab & ac \end{bmatrix} = \begin{bmatrix} ab & abac \end{bmatrix} = \begin{bmatrix} ab & ax(1-x)c \end{bmatrix} = \begin{bmatrix} 1_{n\times n} & 0_{n\times d} \end{bmatrix} = a_0. \end{aligned}$$

This means that $h \cdot t_0 = (a, x, b)$, completing the proof. □

We proceed with showing that the dual of the tautological rank-n vector bundle over the Grassmannian of n-planes inside an m-plane also specializes to every rank-n vector bundle V that is generated by m global sections. However, in contrast with $V_{n,m}$, this is true even if the base scheme of V is not affine, and there is a natural bijection between the set of specializations and the set of m-tuples generating V. That said, $V_{n,m}$ has the advantage that its base scheme $X_{n,m}$ is affine, and therefore — as we shall see in Section 8 — it will ultimately give rise to d-versal torsors whose base is affine.

For the sake of simplicity, given a morphism of schemes $f : Y \to X$ and a vector bundle V over X, we shall identify V with $V^{\vee\vee}$ and $f^*(V^\vee)$ with $(f^*V)^\vee$ via the standard natural isomorphisms. We also identify $\mathcal{O}_X^\vee$ with $\mathcal{O}_X$ and $f^*\mathcal{O}_X$ with $\mathcal{O}_Y$ in the usual way.

Let $m \geq n \geq 0$ be integers. We let $\mathbf{Gr}(n, m)$ denote the Grassmannian $\mathbb{Z}$-scheme of n-planes inside an m-plane; see [**31**, Tag 089R], for instance. The functor of points of $\mathbf{Gr}(n, m)$ sends a scheme X to the set of rank-n subbundles of

$\mathcal{O}_X^m$, i.e., the set of rank-n locally free $\mathcal{O}_X$-submodules $\mathcal{V}$ of $\mathcal{O}_X^m$ for which $\mathcal{O}_X^m/\mathcal{V}$ is also locally free. The tautological rank-n vector bundle on $\mathbf{Gr}(n,m)$ is denoted $V(n,m)$ and its functor of points (as a $\mathbb{Z}$-scheme) is given by

$$V(n,m)(X) = \{(v,\mathcal{V}) \mid \mathcal{V} \in \mathbf{Gr}(n,m)(X),\ v \in \Gamma(X,\mathcal{V})\}.$$

The morphism $(v,\mathcal{V}) \mapsto v : V(n,m) \to \mathcal{O}^m_{\mathbf{Gr}(n,m)}$ makes $V(n,m)$ into a subbundle of $\mathcal{O}^m_{\mathbf{Gr}(n,m)}$. (That $\mathcal{O}^m_{\mathbf{Gr}(n,m)}/V(n,m)$ is locally free can be checked using the standard affine covering of $\mathbf{Gr}(n,m)$, see [**31**, Tag 089R].) Dualizing the embedding $V(n,m) \to \mathcal{O}^m_{\mathbf{Gr}(n,m)}$ gives an $\mathcal{O}_X$-epimorphism

$$p_0 : \mathcal{O}^m_{\mathbf{Gr}(n,m)} \to V(n,m)^\vee.$$

Let $e_1,\dots,e_m$ denote the standard basis of $\mathcal{O}^m_{\mathbf{Gr}(n,m)}$ and let $v_i^{(n,m)}$ be the image of e_i in $\Gamma(\mathbf{Gr}(n,m), V(n,m)^\vee)$. Then the m-tuple $(v_1^{(n,m)},\dots,v_m^{(n,m)})$ generates $V(n,m)^\vee$; we call it the canonical tuple of generators of $V(n,m)^\vee$.

LEMMA 6.3. *Let X be a scheme and let $V \in \mathbf{Gr}(n,m)(X)$, namely, V is a rank-n subbundle of $\mathcal{O}_X^m$. Let $f : X \to \mathbf{Gr}(n,m)$ be the morphism corresponding to V. Then there is canonical isomorphism $a_V : f^*V(n,m) \to V$ that is characterized by the commutativity of the following diagram.*

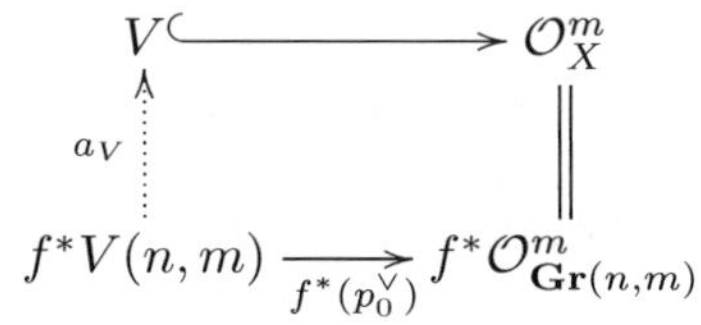

In particular, $V = \operatorname{im} f^*(p_0^\vee)$.

PROOF. This is known, but we include a proof since we were unable to find a satisfactory source.

Since $V \to \mathcal{O}_X^m$ is a monomorphism, a_V is uniquely determined if it exists, and it is necessarily $\mathcal{O}_X$-linear. Let T be a $\mathbb{Z}$-scheme. By the Yoneda Lemma, the map $f_T : X(T) \to \mathbf{Gr}(n,m)(T)$ induced by f is given by $f_T(x) = x^*V$. (Here, $x : T \to X$ is a morphism. Note that $x^*V \subseteq x^*\mathcal{O}_X^m = \mathcal{O}_T^m$.) We have $f^*V(n,m) = X \times_{f,\mathbf{Gr}(n,m)} V(n,m)$ as schemes, so

$$f^*V(n,m)(T) = \{(x,(\mathcal{V},v)) \in X(T) \times V(n,m)(T) \,:\, x^*V = \mathcal{V}\}.$$

On the other hand, for every $\mathbb{Z}$-scheme T, we have $V(T) = \{(x,v) \mid x \in X(T),\ v \in \Gamma(T,x^*V)\}$ (cf. the proof of Lemma 3.1). Thus, the map $(x,(\mathcal{V},v)) \mapsto (x,v) : f^*V(n,m)(T) \to V(T)$ determines a scheme isomorphism $a_V : f^*V(n,m) \to V$. It makes the diagram in the lemma commute because $f^*(p_0^\vee)$ is given section-wise by $(x,(\mathcal{V},v)) \mapsto (x,v)$. $\square$

We now show that $(V(n,m)^\vee, v_1^{(n,m)},\dots,v_m^{(n,m)})$ is universal among the rank-n vector bundles equipped with an m-tuple of generators.

THEOREM 6.4. *Let $m \geq n \geq 0$, let X be a scheme and let V be a rank-n vector bundle over X. There is a one-to-one correspondence between the following:*

(a) *tuples $(v_1,\dots,v_m) \in \Gamma(X,V)^m$ which generate V;*
(b) *pairs (f,φ) consisting of a morphism $f : X \to \mathbf{Gr}(n,m)$ and an $\mathcal{O}_X$-isomorphism $\varphi : f^*V(n,m)^\vee \to V$.*

The correspondence maps (f,φ) *to* $(\varphi(f^*v_1^{(n,m)}),\dots,\varphi(f^*v_m^{(n,m)}))$.

PROOF. The tuples in (a) are in one-to-one correspondence with

(a′) $\mathcal{O}_X$-epimorphisms $p:\mathcal{O}_X^m\to V$

via sending p to $(p(e_1),\dots,p(e_m))$, where $e_1,\dots,e_m$ is the standard basis of $\mathcal{O}_X^m$. One readily checks that, given (f,φ) as in (b), the corresponding map $p:\mathcal{O}_X^m\to V$ in (a′) is $\varphi\circ f^*p_0$. It is therefore enough to construct an inverse to the assignment $(f,\varphi)\mapsto p(f,\varphi):=\varphi\circ f^*p_0$

Let $p:\mathcal{O}_X^m\to V$ be an $\mathcal{O}_X$-epimorphism. Then $p^\vee:V^\vee\to\mathcal{O}_X^m$ is an $\mathcal{O}_X$-monomorphism. Let W be the image of $p^\vee$; it is a rank-n subbundle of $\mathcal{O}_X^m$, and so it determines a morphism $f=f(p):X\to\mathbf{Gr}(n,m)$. The morphism $p^\vee$ factors as

$$V^\vee\xrightarrow{j}W\hookrightarrow\mathcal{O}_X^m,$$

and by Lemma 6.3, we can complete this into a commutative diagram:

(1)

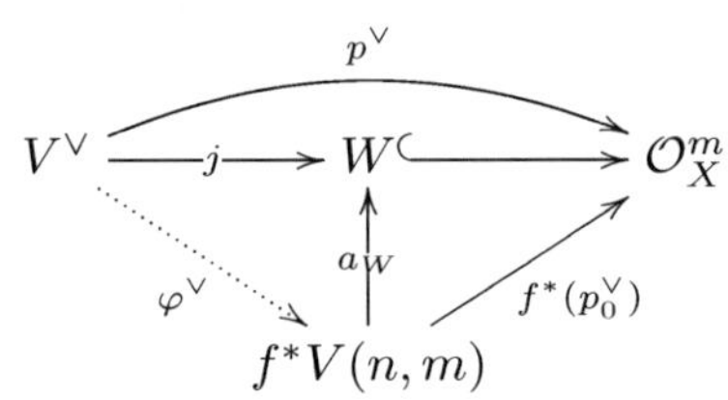

We define $\varphi^\vee:V^\vee\to f^*V(n,m)$ to be $a_W^{-1}\circ j$, and its dual is the desired isomorphism $\varphi=\varphi(p):f^*V(n,m)^\vee\to V$. We have therefore constructed a pair (f,φ) as in (b).

It remains to check that the assignments $p\mapsto(f(p),\varphi(p))$ and $(f,\varphi)\mapsto\varphi\circ f^*p_0$ are mutually inverse. Given an $\mathcal{O}_X$-epimorphism $p:\mathcal{O}_X^m\to V$, it follows readily from the diagram (1) that $p^\vee=f(p)^*(p_0^\vee)\circ\varphi(p)^\vee$, and therefore $p=\varphi(p)\circ f(p)^*p_0$. Conversely, suppose that $p=\varphi\circ f^*p_0$ for (f,φ) as in (b). Then $p^\vee=f^*(p_0^\vee)\circ\varphi^\vee$. Since $\varphi^\vee$ is an isomorphism, $f^*(p_0^\vee):f^*V(n,m)^\vee\to\mathcal{O}_X^m$ has the same image as $p^\vee:V^\vee\to\mathcal{O}_X^m$, which in turn has the same image as $f(p)^*(p_0^\vee)$ by the definition of $f(p)$; call this joint image W. By Lemma 6.3, both f and $f(p)$ correspond to $W\in\mathbf{Gr}(n,m)(X)$, so $f=f(p)$. Moreover, we have a commutative diagram:

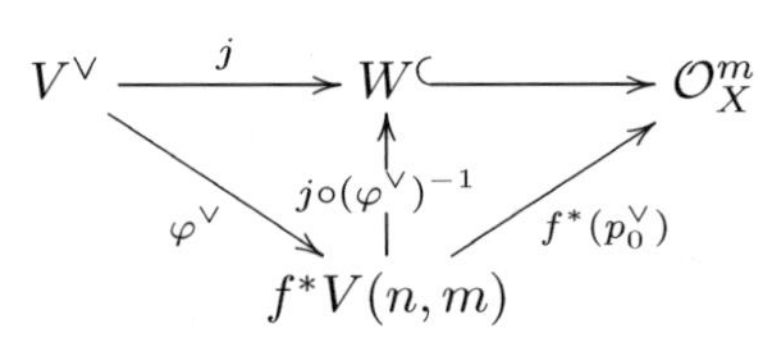

Lemma 6.3 tells us that $j\circ(\varphi^\vee)^{-1}$ coincides with a_W, so $\varphi^\vee=\varphi(p)^\vee$, meaning that $\varphi=\varphi(p)$. This completes the proof. □

Let $E(n,m)\to\mathbf{Gr}(n,m)$ denote the $\mathbf{GL}_n(\mathbb{Z})$-torsor corresponding to $V(n,m)^\vee$. We give two descriptions of $E(n,m)$ that will be needed in the sequel.

PROPOSITION 6.5. *Let* $m\geq n\geq 0$ *be integers. The functor of points of* $E(n,m)$ *(as a* $\mathbb{Z}$*-scheme) is isomorphic to the functor sending a scheme* X *to*

$$\{a\in\mathrm{M}_{n\times m}(\Gamma(X,\mathcal{O}_X))\ :\ \textit{the map } a:\mathcal{O}_X^m\to\mathcal{O}_X^n \textit{ is surjective}\}.$$

Under this isomorphism, the right action of $\mathbf{GL}_n(\mathbb{Z})$ *on* $E(n,m)$ *is given section-wise by* $(a,g)\mapsto g^{-1}a$*, and the map* $E(n,m)\to\mathbf{Gr}(n,m)$ *is given by sending* $a\in \mathrm{M}_{n\times m}(\Gamma(X,\mathcal{O}_X))$ *to the* $\mathcal{O}_X$*-submodule of* $\mathcal{O}_X^m$ *spanned by the rows of* a*.*

PROOF. Let $E'(n,m)$ denote the $\mathbf{GL}_n(\mathbb{Z})$-torsor corresponding to $V(n,m)$. By Lemma 3.1, for a $\mathbb{Z}$-scheme X, the set $E'(n,m)(X)$ is in natural bijection with the set of tuples $(f,v_1,\dots,v_n)$ consisting of a morphism $f:X\to\mathbf{Gr}(n,m)$ and a basis $v_1,\dots,v_n$ for $f^*V(n,m)$. The morphism f specifies a rank-n subbundle $\mathcal{V}$ of $\mathcal{O}_X^m$ and there is a canonical isomorphism $f^*V(n,m)\cong\mathcal{V}$ by Lemma 6.3. Thus, $E'(n,m)(X)$ can be naturally identified with the set of tuples $(\mathcal{V},v_1,\dots,v_n)$ consisting of a rank-n subbundle $\mathcal{V}$ of $\mathcal{O}_X^m$ and a basis $v_1,\dots,v_n\in\Gamma(X,\mathcal{O}_X^m)$ of $\mathcal{V}$. Let $a\in\mathrm{M}_{n\times m}(\Gamma(X,\mathcal{O}_X))$ be the matrix whose rows are $v_1,\dots,v_n$. Then a determines and is determined by $(\mathcal{V},v_1,\dots,v_n)$. Moreover, by checking at the stalks, one sees that the condition that $v_1,\dots,v_n$ is a basis of a subbundle of $\mathcal{O}_X^m$ is equivalent to the condition that $a:\mathcal{O}_X^m\to\mathcal{O}_X^n$ is an epimorphism. Thus, we have defined a natural isomorphism between $E'(n,m)(X)$ and the set of $\mathcal{O}_X$-epimorphisms $a:\mathcal{O}_X^m\to\mathcal{O}_X^n$. Lemma 3.1 also tells us that, under this isomorphism, the right $\mathbf{GL}_n(\mathbb{Z})$-action on $E'(n,m)$ is given by $(a,g)\mapsto g^{\mathrm{t}}a$ on sections. Lemma 3.2 then completes the proof. □

PROPOSITION 6.6. *Let* $m\geq n\geq 0$ *be integers and put* $d=m-n$*. The* $\mathbf{GL}_n(\mathbb{Z})$*-torsor* $E(n,m)\to\mathbf{Gr}(n,m)$ *corresponding to* $V(n,m)^\vee$ *is isomorphic to the quotient morphism*

$$\mathbf{GL}_m(\mathbb{Z})/\begin{bmatrix}1_{n\times n} & \mathbb{A}_{\mathbb{Z}}^{n\times d}\\ 0_{d\times n} & \mathbf{GL}_d(\mathbb{Z})\end{bmatrix}\to \mathbf{GL}_m(\mathbb{Z})/\begin{bmatrix}\mathbf{GL}_n(\mathbb{Z}) & \mathbb{A}_{\mathbb{Z}}^{n\times d}\\ 0_{d\times n} & \mathbf{GL}_d(\mathbb{Z})\end{bmatrix},$$

where here, $\mathbf{GL}_n(\mathbb{Z})$ *acts on* $\mathbf{GL}_m(\mathbb{Z})/\left[\begin{smallmatrix}1_{n\times n} & \mathbb{A}^{n\times d}\\ 0_{d\times n} & \mathbf{GL}_d(\mathbb{Z})\end{smallmatrix}\right]$ *via* $x\cdot g=x\left[\begin{smallmatrix}g & 0\\ 0 & 1_{d\times d}\end{smallmatrix}\right]$*.*

PROOF. We use the description of $E(n,m)$ in Proposition 6.5 to define left actions of $\mathbf{GL}_m(\mathbb{Z})$ on $E(n,m)$ and $\mathbf{Gr}(n,m)$ given by

$$h\cdot a=ah^{\mathrm{t}}\qquad\text{and}\qquad h\cdot\mathcal{V}=h\mathcal{V}$$

on sections. One readily checks that $E(n,m)\to\mathbf{Gr}(n,m)$ is $\mathbf{GL}_m(\mathbb{Z})$-equivariant. Put $t_0=\begin{bmatrix}1_{n\times n} & 0_{n\times d}\end{bmatrix}\in E(n,m)(\mathbb{Z})$ and let x_0 be the image of t_0 in $\mathbf{Gr}(n,m)(\mathbb{Z})$. It is routine to check that the stabilizers of t_0 and x_0 in $\mathbf{GL}_m(\mathbb{Z})$ are the subgroups $\left[\begin{smallmatrix}1_{n\times n} & \mathbb{A}^{n\times d}\\ 0_{d\times n} & \mathbf{GL}_d(\mathbb{Z})\end{smallmatrix}\right]$ and $\left[\begin{smallmatrix}\mathbf{GL}_n(\mathbb{Z}) & \mathbb{A}^{n\times d}\\ 0_{d\times n} & \mathbf{GL}_d(\mathbb{Z})\end{smallmatrix}\right]$, respectively. Provided that the action of $\mathbf{GL}_m(\mathbb{Z})$ on $E(n,m)$ is fppf transitive, we have isomorphisms

$$[h]\mapsto h\cdot t_0:\mathbf{GL}_m(\mathbb{Z})/\left[\begin{smallmatrix}1_{n\times n} & \mathbb{A}^{n\times d}\\ 0_{d\times n} & \mathbf{GL}_d(\mathbb{Z})\end{smallmatrix}\right]\to E(n,m)$$

and

$$[h]\mapsto h\cdot x_0:\mathbf{GL}_m(\mathbb{Z})/\left[\begin{smallmatrix}\mathbf{GL}_n(\mathbb{Z}) & \mathbb{A}^{n\times d}\\ 0_{d\times n} & \mathbf{GL}_d(\mathbb{Z})\end{smallmatrix}\right]\to \mathbf{Gr}(n,m),$$

and the proposition follows.

It remains to check that $\mathbf{GL}_m(\mathbb{Z})$ acts fppf transitively on $E(n,m)$. Let R be a ring and let $a\in E(n,m)(R)$. Then $a:R^m\to R^n$ is surjective, and thus $\ker a$ is a summand of R^m. By passing to a Zariski covering of $\operatorname{Spec}R$, we may assume that $\ker a$ is a free R-module. Fix an isomorphism $c:R^d\to\ker a\subseteq R^m$ and an R-homomorphism $b:R^n\to R^m$ such that $ab=1_{n\times n}$. Then $h=\begin{bmatrix}b & c\end{bmatrix}:R^m\to R^m$ lives in $\mathrm{GL}_m(R)$ and satisfies $ah=t_0$, so $h^{\mathrm{t}}\cdot a=t_0$. □

We finish with noting the following consequence of Theorem 6.4.

COROLLARY 6.7. *With notation as above, let V be a rank-n vector bundle over a scheme X and let E be the $\mathbf{GL}_n(\mathbb{Z})$-torsor corresponding to V. Then there is a one-to-one correspondence between:*

(a) *tuples $(v_1,\dots,v_m)\in\Gamma(X,V)^m$ which generate V;*
(c) *morphisms of $\mathbf{GL}_n(\mathbb{Z})$-torsors from $E\to X$ to $E(n,m)\to\mathbf{Gr}(n,m)$.*

The correspondence is compatible with pullbacks along morphisms $f:X'\to X$.

PROOF. This is a consequence of Theorem 6.4, the equivalence between $\mathbf{GL}_n(\mathbb{Z})$-torsors and rank-$n$ vector bundles, and Remark 2.1. □

REMARK 6.8. Using Proposition 6.5, one can see that Corollary 6.7 is in fact a special case of [**10**, Proposition 4.1]. (In [op. cit.], the base scheme is a field and not $\operatorname{Spec}\mathbb{Z}$, and X is assumed to be affine, but the proof works in our more general setting.) In fact, we could use Proposition 6.5 together with [**10**, Proposition 4.1] to prove Theorem 6.4, but this would have obscured the elementary nature of the correspondence in the theorem.

7. The Number of Generators of a Globally Generated Vector Bundle

In the previous section, we saw that for every $m\geq n\geq 0$, there are vector bundles which specialize to every rank-n vector bundle that is generated by m global sections. On the other hand, a theorem of Förster [**14**] says that every rank-n vector bundle over an *affine* noetherian scheme of dimension $\leq d$ can be generated by $n+d$ elements. By coupling these two observations, we see that the vector bundles $V_{n,n+d}$ and $V(n,n+d)^\vee$ considered in Section 6 specialize to every rank-n vector bundle over an affine noetherian scheme of dimension $\leq d$, that is, they are weakly versal for the class nAff_d of affine noetherian schemes of dimension $\leq d$. The same holds for the corresponding $\mathbf{GL}_n(\mathbb{Z})$-torsors. Thanks to Propositions 5.5 and 5.7, it follows that every flat locally of finite presentation linear group scheme G over a scheme S admits a torsor $E\to X$ (with X an algebraic space over S) that is weakly versal for nAff_d/S.

In this section, we shall strengthen Förster's Theorem in three ways that will allow us to construct torsors with stronger versality properties in the next section. Our arguments stem from the generalizations of Förster's Theorem appearing in [**32**], [**9**] and [**10**].

Throughout this section, if R is a ring, M is an R-module, and $\mathfrak{p}\in\operatorname{Spec}R$, then we let $M(\mathfrak{p})=M\otimes_R k(\mathfrak{p})$ and denote the image of $m\in M$ in $M(\mathfrak{p})$ by $m(\mathfrak{p})$. If we set $X=\operatorname{Spec}R$ and let $\mathcal{M}$ be the quasi-coherent $\mathcal{O}_X$-module corresponding to M, then $M(\mathfrak{p})$ is just $\mathcal{M}(\mathfrak{p})$. For a general scheme X and a coherent $\mathcal{O}_X$-module $\mathcal{M}$, Nakayama's lemma tells us that $m_1,\dots,m_t\in\Gamma(X,\mathcal{M})$ generate $\mathcal{M}$ if and only if $m_1(x),\dots,m_t(x)$ generate $\mathcal{M}(x)$ for every $x\in X_{\mathrm{cl}}$. We denote the restriction of $\mathcal{M}$ to an open subset $U\subseteq X$ by $\mathcal{M}_U$.

LEMMA 7.1. *Let R be a ring and let M be an R-module. Let S be a finite subset of* $\operatorname{Spec}R$ *and let $m_1,\dots,m_t\in M$. Suppose that for every $\mathfrak{p}\in S$, the elements $m_1(\mathfrak{p}),\dots,m_t(\mathfrak{p})$ do not generate $M(\mathfrak{p})$ as a $k(\mathfrak{p})$-vector space. Then there exists $m_{t+1}\in M$ such that for every $\mathfrak{p}\in S$, we have*

$$\dim\operatorname{span}_{k(\mathfrak{p})}\{m_1(\mathfrak{p}),\dots,m_{t+1}(\mathfrak{p})\}=\dim\operatorname{span}_{k(\mathfrak{p})}\{m_1(\mathfrak{p}),\dots,m_t(\mathfrak{p})\}+1.$$

PROOF. The proof is inspired by the proof of [**32**, Lemma 1]. We prove the lemma by induction on $|S|$. The case $S=\emptyset$ is clear. Suppose that $S\neq\emptyset$. Then there is some minimal element $\mathfrak{q}\in S$. Put $T=S-\{\mathfrak{q}\}$. By the induction hypothesis, there is $x\in M$ such that

$$(2)\quad \dim\operatorname{span}_{k(\mathfrak{p})}\{m_1(\mathfrak{p}),\dots,m_t(\mathfrak{p}),x(\mathfrak{p})\}=\dim\operatorname{span}_{k(\mathfrak{p})}\{m_1(\mathfrak{p}),\dots,m_t(\mathfrak{p})\}+1$$

for all $\mathfrak{p}\in T$. If $x(\mathfrak{q})\notin\operatorname{span}_{k(\mathfrak{q})}\{m_1(\mathfrak{q}),\dots,m_t(\mathfrak{q})\}$, then we can take $m_{t+1}=x$. Assume henceforth that $x(\mathfrak{q})\in\operatorname{span}_{k(\mathfrak{q})}\{m_1(\mathfrak{q}),\dots,m_t(\mathfrak{q})\}$. Since $\mathfrak{q}$ is minimal in S and prime, we have $\prod_{\mathfrak{p}\in T}\mathfrak{p}\not\subseteq\mathfrak{q}$. Choose some $r\in\prod_{\mathfrak{p}\in T}\mathfrak{p}-\mathfrak{q}$. Then replacing x with $x+ry$ for some $y\in M$ will not affect the validity of (2) for all $\mathfrak{p}\in T$. Since the image of M in $M(\mathfrak{q})$ generates $M(\mathfrak{q})$ as a $k(\mathfrak{q})$-module, and since $m_1(\mathfrak{q}),\dots,m_t(\mathfrak{q})$ do not generate $M(\mathfrak{q})$ as a $k(\mathfrak{p})$-vector space, there is some $y\in M$ such that $y(\mathfrak{q})\notin\operatorname{span}_{k(\mathfrak{q})}\{m_1(\mathfrak{q}),\dots,m_t(\mathfrak{q})\}$. Put $m_{t+1}:=x+ry$. Our assumptions on x and r imply that $m_{t+1}(\mathfrak{q})\notin\operatorname{span}_{k(\mathfrak{q})}\{m_1(\mathfrak{q}),\dots,m_t(\mathfrak{q})\}$. We therefore conclude that $\dim\operatorname{span}_{k(\mathfrak{p})}\{m_1(\mathfrak{p}),\dots,m_{t+1}(\mathfrak{p})\}=\dim\operatorname{span}_{k(\mathfrak{p})}\{m_1(\mathfrak{p}),\dots,m_t(\mathfrak{p})\}+1$ for all $\mathfrak{p}\in T\cup\{\mathfrak{q}\}=S$. □

The following is our first strengthening of Förster's Theorem.

THEOREM 7.2. *Let X be the spectrum of a noetherian ring, and let $\mathcal{M}$ be a coherent $\mathcal{O}_X$-module of rank $\leq n$ ($n\in\mathbb{N}\cup\{0\}$). Then, for every $d\in\mathbb{N}\cup\{0\}$, there exists an open subscheme $U\subseteq X$ such that* $\operatorname{codim}(X-U,X)>d$ *and $\mathcal{M}_U$ is generated by $d+n$ global sections that can be extended to all of X.*

PROOF. This is a modification of the proof of [**9**, Theorem 1.2]. We give a complete proof in order to introduce a variation later on, and also since it is simpler and clearer than listing the necessary changes.

We shall prove that for every $j\in\mathbb{N}\cup\{0\}$, there exist $a_1,\dots,a_j\in\Gamma(X,\mathcal{M})$ and a sequence of closed subsets $\emptyset=Y_{-1}^{(j)}\subseteq Y_0^{(j)}\subseteq Y_1^{(j)}\subseteq\cdots\subseteq Y_n^{(j)}=X$ such that the following conditions hold:[10]

(1) For every $i\in\{0,\dots,n\}$ and $x\in Y_i^{(j)}-Y_{i-1}^{(j)}$, there are $b_{i+1},\dots,b_n\in\mathcal{M}(x)$ such that $a_1(x),\dots,a_j(x),b_{i+1},\dots,b_n$ span $\mathcal{M}(x)$ as a $k(x)$-vector space.
(2) $\operatorname{codim}(Y_i^{(j)},X)\geq j-i$ for every $i\in\{0,\dots,n-1\}$.

Provided these data have been constructed, we can take $U=X-Y_{n-1}^{(n+d)}$. Indeed, U is open and condition (2) says that $\operatorname{codim}(X-U,X)>d$. Moreover, by (1), for every $x\in U$, the elements $a_1(x),\dots,a_{n+d}(x)$ generate $\mathcal{M}(x)=\mathcal{M}_U(x)$, so $\mathcal{M}_U$ is generated by the restriction of $a_1,\dots,a_{n+d}$ to U.

We prove the existence of the a_j and the $Y_i^{(j)}$ by induction on j. For the case $j=0$, take $Y_0^{(0)}=\cdots=Y_{n-1}^{(0)}=X$.

Suppose that $j\geq 0$ and $a_1,\dots,a_j$ and $Y_0^{(j)},\dots,Y_n^{(j)}$ were constructed. Fix $i\in\{0,\dots,n-1\}$. For every irreducible component X' of X and every irreducible component Y' of $Y_i^{(j)}\cap X'$ that is not contained in $Y_{i-1}^{(j)}$, choose a point $x\in Y'-Y_{i-1}^{(j)}$, and let S_i denote the finite set of points chosen in this manner. Note that $S_0,\dots,S_{n-1}$ are pairwise disjoint.

[10] The closed set $Y_i^{(j)}$ corresponds to the union of the locally closed sets $\mathcal{F}_0^{(j)},\dots,\mathcal{F}_i^{(j)}$ in [**9**, Theorem 1.2]. Note, however, that here we consider all the points of X, whereas in [op. cit.] we consider only the closed points. The reason for this is that, for a closed subset $Z\subseteq X$, one cannot bound from below $\operatorname{codim}(Z,X)$ by means of $\operatorname{codim}(Z_{\mathrm{cl}},X_{\mathrm{cl}})$.

By condition (1) and Lemma 7.1, there exists $a_{j+1} \in \Gamma(X, \mathcal{M})$ with the property that, for every $i \in \{0, \dots, n-1\}$ and every $x \in S_i$, there exist $b_{i+2}, \dots, b_n \in \mathcal{M}(x)$ such that $a_1(x), \dots, a_{j+1}(x), b_{i+2}, \dots, b_n$ span $\mathcal{M}(x)$ as a $k(x)$-vector space. Moreover, by [**9**, Lemma 2.3], x admits an open neighborhood U_x such that this condition holds for every point in U_x.

Let $i \in \{0, \dots, n-1\}$ and let $U_i = (\bigcup_{x \in S_i} U_x) - Y_{i-1}^{(j)}$. Then U_i is an open neighborhood of S_i not meeting $Y_{i-1}^{(j)}$. Define

$$Y_i^{(j+1)} = Y_i^{(j)} - U_i.$$

Since $U_i \cap Y_{i-1}^{(j)} = \emptyset$, we have $Y_i^{(j+1)} = Y_i^{(j)} - U_i \supseteq Y_{i-1}^{(j)} \supseteq Y_{i-1}^{(j+1)}$, so $Y_0^{(j+1)} \subseteq \dots \subseteq Y_{n-1}^{(j+1)}$. We set $Y_{-1}^{(j+1)} = \emptyset$ and $Y_n^{(j+1)} = X$.

We finish by checking that (1) and (2) hold for the filtration $Y_{-1}^{(j+1)} \subseteq Y_0^{(j+1)} \subseteq \dots \subseteq Y_n^{(j+1)}$ and $a_1, \dots, a_{j+1}$. For every $i \in \{0, \dots, n\}$, we have $Y_i^{(j+1)} - Y_{i-1}^{(j+1)} \subseteq (Y_i^{(j)} - Y_{i-1}^{(j)}) \cup U_{i-1}$ (with the convention $U_{-1} = \emptyset$), so (1) holds by the induction hypothesis and the construction of U_{i-1}. To prove (2), let X' be an irreducible component of X and let Y' be an irreducible component of $Y_i^{(j+1)} \cap X'$. If $Y' \subseteq Y_{i-1}^{(j)}$, then Y' is contained in some irreducible component Y'' of $Y_{i-1}^{(j)} \cap X'$ and we get $\operatorname{codim}(Y', X') \geq \operatorname{codim}(Y'', X') \geq j - (i-1) = (j+1) - i$. Otherwise, there is an irreducible component Y'' of $Y_i^{(j)} \cap X'$ containing Y'. Since U_i meets $Y'' - Y_{i-1}^{(j)}$ and does not meet $Y' - Y_{i-1}^{(j)}$, we must have $Y' \subsetneq Y''$, so $\operatorname{codim}(Y', X') \geq \operatorname{codim}(Y'', X') + 1 \geq j - i + 1 = (j+1) - i$. This completes the proof. □

We will also need the following variant of Theorem 7.2. The case $Z = \emptyset$ is a theorem of Swan [**32**, Theorem 1].

THEOREM 7.3. *Let X be an affine scheme such that the subspace X_{cl} is noetherian and has finite dimension $\leq d$, and let $\mathcal{M}$ be a coherent $\mathcal{O}_X$-module of rank $\leq n$ ($n \in \mathbb{N} \cup \{0\}$). Let $u : Z \to X$ be a closed immersion, let $\mathcal{M}_Z = u^*\mathcal{M}$ and let $b_1, \dots, b_{n+d} \in \Gamma(Z, \mathcal{M}_Z)$ be sections which generate $\mathcal{M}_Z$. Then there exist $a_1, \dots, a_{n+d} \in \Gamma(X, \mathcal{M})$ which generate X and satisfy $u^* a_i = b_i$ for all $i \in \{1, \dots, d+n\}$.*

PROOF. We may assume that Z is a closed subscheme of X. Endow $X_1 := X_{\mathrm{cl}} - Z$ with the topology induced from X. We claim that for every $j \in \{0, \dots, n+d\}$, there is a filtration of X_1 by closed subsets $\emptyset = Y_{-1}^{(j)} \subseteq Y_0^{(j)} \subseteq \dots \subseteq Y_n^{(j)} = X_1$ and $a_1, \dots, a_j \in \Gamma(X, \mathcal{M})$ such that:

(1) For every $i \in \{0, \dots, n\}$ and $x \in Y_i^{(j)} - Y_{i-1}^{(j)}$, there are $a'_{i+1}, \dots, a'_n \in \mathcal{M}(x)$ such that $a_1(x), \dots, a_j(x), a'_{i+1}, \dots, a'_n$ generate $\mathcal{M}(x)$ as a $k(x)$-vector space.
(2) $\operatorname{codim}(Y_i^{(j)}, X) \geq j - i$ for every $i \in \{0, \dots, n-1\}$.
(3) $u^* a_\ell = b_\ell$ for all $\ell \in \{1, \dots, j\}$.

Once the above data have been constructed, $a_1, \dots, a_{n+d}$ satisfy all the requirements. Indeed, condition (2) implies that $Y_{n-1}^{(d+n)} = \emptyset$, so by condition (1), the elements $a_1(x), \dots, a_{n+d}(x)$ generate $\mathcal{M}(x)$ for all $x \in X_1 = X_{\mathrm{cl}} - Z$. On the other hand, for all $x \in X_{\mathrm{cl}} \cap Z$, the elements $a_1(x), \dots, a_{n+d}(x)$ are $b_1(x), \dots, b_{n+d}(x)$ by (3), so they generate $\mathcal{M}(x)$ by assumption.

That $a_1, \dots, a_j$ and $Y_{-1}^{(j)}, \dots, Y_n^{(j)}$ exist is shown as in the proof of Theorem 7.2, with the following difference: Instead of using Lemma 7.1 to choose a_{j+1}, we use the Chinese Remainder Theorem to choose $a_{j+1} \in \Gamma(X, \mathcal{O}_X)$ such that $u^* a_{j+1} = b_{j+1}$ and such that $a_1(x), \dots, a_{j+1}(x)$ generate an $(i+1)$-dimensional subspace of $\mathcal{M}(x)$ for every $x \in S_i$ and $i \in \{0, \dots, n-1\}$. □

REMARK 7.4. Similarly to [**9**, Theorem 1.2], Theorems 7.2 and 7.3 remain true if we replace the $\mathcal{O}_X$-module $\mathcal{M}$ with a coherent $\mathcal{O}_X$-algebra (or even an *$\mathcal{O}_X$-multialgebra*), and interpret "generation" as generation of $\mathcal{O}_X$-algebras. The proof is the same, except one needs to replace Lemma 7.1 with a slightly more involved generalization to algebras. We omit the details as they will not be needed here.

Our third variant of Förster's Theorem applies to vector bundles over *general* schemes *over an infinite field.* The idea originates from [**10**, Proposition 5.2]. We begin with a consequence of Kleiman's Transversality Theorem [**19**, Theorem 2].

LEMMA 7.5. *Let k be a field, let X, X_1, X_2 be k-schemes of finite type, and let $f_i : X_i \to X$ $(i = 1, 2)$ be a k-morphism. Let H be a connected algebraic group over k acting fppf transitively on X. View X_1 as an X-scheme via f_1 and, given $h \in H(k)$, write hX_2 for X_2 viewed as an X-scheme via $h \circ f_2$ (here, h also denotes the automorphism of X that it induces). Then there exists a nonemtpy open subscheme U of H such that for all $h \in U(k)$, we have*

$$\dim X_1 \times_X hX_2 \leq \dim X_1 + \dim X_2 - \dim X.$$

PROOF. If X, X_1 and X_2 are integral and k is algebraically closed, then this is a consequence of Kleiman's Transversality Theorem. We reduce the lemma to the theorem by an argument similar to the proof of [**10**, Claim on p. 7289].

Consider the closed subscheme I of $H \times X_1 \times X_2$ cut by the equation $f_1(x_1) = h \cdot f_2(x_2)$, and let $p : I \to H$ be the restriction of the first projection. For $h \in H(k)$, the scheme-theoretic fiber of p over h, denoted I_h, is precisely $X_1 \times_X hX_2$. Let $Z = \{h \in H : \dim I_h > \dim X_1 + \dim X_2 - \dim X\}$ and put $U = U_{X_1,X_2} = H - \overline{Z}$. Then U satisfies all the requirements if it is nonempty.

Let $\overline{k}$ be an algebraic closure of k, and let q denote the morphism $H_{\overline{k}} := H \times_k \overline{k} \to H$. We claim that q is closed and open. Indeed, q is closed because it is integral [**31**, Tag 01WM]. To see that q is open, observe that every open $U' \subseteq H_{\overline{k}}$ arises as $U'' \times_\ell \overline{k}$ for some finite k-field $\ell \subseteq \overline{k}$ and open $U'' \subseteq H_\ell$, and the image of U'' in H is open because $H_\ell \to H$ is flat and locally of finite presentation.

Next, observe that X and $X_{\overline{k}} := X \times_k \overline{k}$ have the same dimension, and likewise for X_1 and X_2. Put $Z' = \{h \in H_{\overline{k}} : \dim(I_{\overline{k}})_h > \dim X_1 + \dim X_2 - \dim X\}$. By [**31**, Tag 02FY], we have $Z' = q^{-1}(Z)$. Since q is continuous and closed, this means that $q(\overline{Z'}) = \overline{q(Z')} = \overline{Z}$. Suppose now that there is $h' \in H_{\overline{k}} - \overline{Z'}$. If it were the case that $q(h') \in \overline{Z}$, then we would have $q(h') \in q(H_{\overline{k}} - \overline{Z'}) \cap \overline{Z}$. Since q is open, this would mean that $q(H_{\overline{k}} - \overline{Z'}) \cap Z \neq \emptyset$, which contradicts with $q^{-1}(Z) = Z'$. Thus, $q(h') \notin \overline{Z}$. This means that in order to prove that $H - \overline{Z} \neq \emptyset$, it is enough to prove that $H_{\overline{k}} - \overline{Z'} \neq \emptyset$. To that end, we may replace X, X_1, X_2 and H with their base change along $\operatorname{Spec} \overline{k} \to \operatorname{Spec} k$ and assume that $k = \overline{k}$. (Note that H remains connected.)

Now that $k = \overline{k}$, the scheme H_{red} is a subgroup of H [**22**, Corollary 1.39], and the action of H on X restricts to an action of H_{red} on X_{red} [**5**, Proposition 2.5.1(1)].

We may therefore replace X, X_1, X_2 and H with their reductions. Since H is connected, it is irreducible; and since and $H(k)$ acts transitively on $X(k)$, it follows that X is also irreducible.

Let $\{X_{1i}\}_{i=1}^{s}$ and $\{X_{2j}\}_{j=1}^{t}$ be the irreducible components of X_1 and X_2, respectively. Then $U_{X_1,X_2} \supseteq \bigcap_{i,j} U_{X_{1i},X_{2j}}$. Since X is irreducible, it is enough to show that $U_{X_1,X_2} \neq \emptyset$ when X_1 and X_2 are irreducible. We are now at the setting of Kleiman's Transversality Theorem (i.e., H, X, X_1, X_2 are integral and $k = \overline{k}$), which gives the desired conclusion $U_{X_1,X_2} \neq 0$. □

Recall from Section 6 that $E(n,m)$ $(m \geq n \geq 0)$ denotes the $\mathbf{GL}_n(\mathbb{Z})$-torsor associated to the dual of the tautological rank-n bundle $V(n,m)$ over $\mathbf{Gr}(n,m)$. Given a field k, we write $\mathbf{Gr}(n,m)_k$ for the base-change of $\mathbf{Gr}(n,m)$ along $\operatorname{Spec} k \to \operatorname{Spec} \mathbb{Z}$, and likewise for $V(n,m)$ and $E(n,m)$. By Proposition 6.5, for a ($\mathbb{Z}$-)scheme X, we may view $E(n,m)(X)$ as the collection of matrices $a = (a_{ij})_{i,j} \in \mathrm{M}_{n\times m}(\Gamma(X, \mathcal{O}_X))$ for which $a : \mathcal{O}_X^m \to \mathcal{O}_X^n$ is surjective.

LEMMA 7.6. *Let $r \geq m \geq n \geq 0$ be integers. With notation as above, let $Z = Z(n,m,r)$ denote the closed subscheme of $E(n,r)$ cut by the equations*

$$\det((a_{ij})_{i\in\{1,\dots,n\},j\in I}) = 0$$

as I ranges over the subsets of $\{1,\dots,m\}$ with $|I| = n$. Then:

(i) $\dim Z_k \leq nr - (m - n + 1)$.[11]
(ii) *$Z/\mathbf{GL}_n(\mathbb{Z})$ is a scheme and $Z \to Z/\mathbf{GL}_n(\mathbb{Z})$ is a $\mathbf{GL}_n(\mathbb{Z})$-torsor.*
(iii) *There is a $\mathbf{GL}_n(\mathbb{Z})$-equivariant morphism $(E(n,r) - Z) \to E(n,m)$ given by $(a_{ij})_{i\in\{1,\dots,n\},j\in\{1,\dots,r\}} \mapsto (a_{ij})_{i\in\{1,\dots,n\},j\in\{1,\dots,m\}}$ on sections.*

PROOF. (i) Let $W = W(n,m,r)$ be the closed subscheme of $\mathbb{A}_{\mathbb{Z}}^{n\times r}$ cut by the same equations as those defining Z. Then Z is an open subscheme of W. It is therefore enough to prove that $\dim W_k \leq nr-(m-n+1)$. Since $W \cong W(n,m,m)\times \mathbb{A}_{\mathbb{Z}}^{n\times(r-m)}$, we further reduce into proving that $\dim W(n,m,m)_k \leq nm-(m-n+1)$, and this well-known, e.g., see [**10**, Lemma 6.2(b), Example 6.4].

(ii) This can be checked directly using the standard affine covering of $\mathbf{Gr}(n,r)$, see [**31**, Tag 089T], but we can give a quicker (albeit less elementary) proof using results from Section 2. Since Z is a closed subscheme of $E(n,r)$, the group $\mathbf{GL}_n(\mathbb{Z})$ acts freely on Z. Thus, by Lemma 2.2, $Z/\mathbf{GL}_n(\mathbb{Z})$ is a $\mathbb{Z}$-algebraic space and $Z \to Z/\mathbf{GL}_n(\mathbb{Z})$ is a $\mathbf{GL}_n(\mathbb{Z})$-torsor. The induced morphism $Z/\mathbf{GL}_n(\mathbb{Z}) \to E(n,r)/\mathbf{GL}_n(\mathbb{Z}) = \mathbf{Gr}(n,r)$ is a closed immersion because its pullback along the fppf covering $E(n,r) \to \mathbf{Gr}(n,r)$ is the closed immersion $Z \to E(n,r)$ [**31**, Tag 0420] (see also [**31**, Tag 02YU]). In particular, the morphism of algebraic spaces $Z/\mathbf{GL}_n(\mathbb{Z}) \to \mathbf{Gr}(n,r)$ is representable, so $Z/\mathbf{GL}_n(\mathbb{Z})$ is a scheme.

(iii) Let X be a scheme and let $a = (a_{ij})$ be an X-section of $E(n,r) - Z$. We need to show that $\tilde{a} := (a_{ij})_{i\in\{1,\dots n\},j\in\{1,\dots,m\}} : \mathcal{O}_X^m \to \mathcal{O}_X^n$ is surjective. It is enough to check this after specializing to all the residue fields of X. We may therefore assume that $X = \operatorname{Spec} k$ for a field k and view a as an $n \times r$ matrix over k. Now, since the unique point of X is not mapped to Z under the morphism $X \to E(n,r)$, there is some $I \subseteq \{1,\dots,m\}$ with $\det((a_{ij})_{i,j\in I}) \neq 0$, so $\tilde{a} : k^m \to k^n$ is surjective. □

[11] In fact, equality holds, but we do not need that.

THEOREM 7.7. *Let k be an infinite field, let $d \in \mathbb{N} \cup \{0\}$, let X be a finite type k-scheme, and let V be a rank-n vector bundle over X. If V is generated by global sections, then there is an open subscheme U of X such that* $\operatorname{codim}(X - U, X) > d$ *and $V|_U$ is generated by $n + d$ global sections. In particular, if* $\dim X = d$, *then V is generated by $n + d$ global sections.*

PROOF. Since X is quasi-compact and V is coherent, V is generated by finitely many global sections. Say it is generated by r global sections with $r \geq n$. If $r \leq n + d$, we can take $U = X$ and finish, so assume $r > n + d$.

Let $p : E \to X$ denote the $\mathbf{GL}_n(k)$-torsor corresponding to V. By Corollary 6.7, the r global sections generating V give rise to a $\mathbf{GL}_n(k)$-torsor morphism $(\hat{f}, f)$ from $E \to X$ to $E(n,r)_k \to \mathbf{Gr}(n,r)_k$. Let $Z = Z(n, n + d, r)$ be as in Lemma 7.6, let $Y = Z/\mathbf{GL}_n(\mathbb{Z})$, and let $(\hat{j}, j)$ denote the evident $\mathbf{GL}_n(k)$-torsor morphism from $Z_k \to Y_k$ to $E(n,r)_k \to \mathbf{Gr}(n,r)_k$. Then $\hat{j}$ is a closed embedding, and therefore so is j [**31**, Tag 02L6]. Proposition 6.6 implies that $\mathbf{GL}_r(k)$ acts on $E(n,r)_k \to \mathbf{Gr}(n,r)_k$ and the action is fppf transitive on both source and target.

Let X' be an irreducible component of X and let f' be the restriction of $f : X \to \mathbf{Gr}(n,r)_k$ to X'. By Lemma 7.5 (applied with $H = \mathbf{GL}_r(k)$, $X_1 = Y_k$ and $X_2 = X'$) there is a nonempty open subscheme W' of $\mathbf{GL}_r(k)$ such that for all $h \in W'(k)$ we have

$$\begin{aligned}\dim Y_k \times_{\mathbf{Gr}(n,r)_k} hX' &\leq \dim Y_k + \dim X' - \dim \mathbf{Gr}(n,r)_k \\ &\leq [nr - (d+1) - n^2] + \dim X' - n(r-n) = \dim X' - d - 1,\end{aligned}$$

where in the second inequality we used Lemma 7.6(i) and the fact that $\dim Z_k = \dim Y_k + \dim \mathbf{GL}_n(k)$. Let W denote the (finite) intersection of all the W' as X' ranges over the irreducible components of X. Then W is a nonempty open subscheme of $\mathbf{GL}_r(k)$. As k is infinite, there exists $h \in W(k)$.

Put $X_1 = Y_k \times_{\mathbf{Gr}(n,r)_k} hX$. Since $j : Y \to \mathbf{Gr}(n,r)$ is a closed embedding, we may and shall view X_1 as a closed subscheme of X. By the previous paragraph, we have $\dim(X_1 \cap X') \leq \dim X' - d - 1$ for every irreducible component X' of X, so $\operatorname{codim}(X_1, X) > d$. Put $U = X - X_1$ and let $\hat{g} = h \circ \hat{f} : E \to E(n,r)_k$. Then $\hat{g}$ maps $E|_U$ to $E(n,r)_k - Z_k$. By Lemma 7.6(iii), we now have a well-defined $\mathbf{GL}_n(k)$-equivariant morphism $E|_U \to E(n,r)_k - Z_k \to E(n, n+d)_k$. Applying Corollary 6.7 again, we conclude that $V|_U$ is generated by $n+d$ global sections. □

EXAMPLE 7.8. Let k be an infinite field and let $n \in \mathbb{N}$. As usual, we let $\mathcal{O}(m) = \mathcal{O}_{\mathbb{P}^n_k}(m)$ ($m \in \mathbb{Z}$) denote the line bundle on $\mathbb{P}^n_k$ whose sections on an open $U \subseteq \mathbb{P}^n_k$ are the degree-m homogeneous regular functions on the cone of U in $\mathbb{A}^{n+1}_k$. Let $x_0, \dots, x_n$ denote the coordinate functions of $\mathbb{A}^{n+1}_k$. It is well-known that $\mathcal{O}(1)$ is generated by the global sections $x_0, \dots, x_n$, so for every $m \geq 0$, the line bundle $\mathcal{O}(m) \cong \mathcal{O}(1)^{\otimes m}$ is generated by the $\binom{m+n}{n}$ global sections $\{x_{i_1} \cdots x_{i_m} \mid 0 \leq i_1 \leq \cdots \leq i_m \leq n\}$. However, Theorem 7.7 tells us that $\mathcal{O}(m)$ can be generated by $\dim \mathbb{P}^n_k + \operatorname{rank} \mathcal{O}(m) = n + 1$ global sections, and indeed, $x_0^m, \dots, x_n^m$ generate $\mathcal{O}(m)$ as an $\mathcal{O}_{\mathbb{P}^n_k}$-module. This can be shown, for instance, by checking it at the stalks, or by looking at the graded module associated to $\mathcal{O}(m)$. (Note, however, that $x_0^m, \dots, x_n^m$ do *not* generate $\Gamma(\mathbb{P}^n_k, \mathcal{O}(m))$ as a $\Gamma(\mathbb{P}^n_k, \mathcal{O}_{\mathbb{P}^n_k})$-module if $m > 1$.)

8. d-Versal Torsors: Existence

We now put together the results of the last three sections to show that strongly d-versal torsors exist for all linear groups. We give three constructions with varying additional properties. In particular, we show that:

(1) Every linear reductive group scheme over an affine noetherian scheme S admits a torsor over a smooth affine base that is strongly versal for the class of affine S-schemes of dimension $\leq d$.
(2) Every linear reductive group scheme over a noetherian scheme S admits a torsor over a scheme that is strongly d-versal for all affine S-schemes.
(3) Every affine algebraic group over an infinite field k admits a torsor over a smooth quasi-projective k-scheme that is strongly d-versal for all quasi-projective k-schemes.

Consult Theorem 2.4 and [**15**] regarding the question of which reductive group schemes are linear. The torsors in (2) also have an additional property which allows one to extend specializations of these torsors from a closed subscheme to the ambient scheme, see Theorem 8.3.

Recall that if $\mathscr{C}$ is a class of S-schemes and S_1 is an S-scheme, then $\mathscr{C}/S_1$ denotes the class of S_1-schemes with underlying S-scheme in $\mathscr{C}$.

THEOREM 8.1. *Let $n, d \in \mathbb{N} \cup \{0\}$, let S be a scheme, and let G be a subgroup of $\mathbf{GL}_n(S)$ that is flat and locally of finite presentation over S. Let $\mathscr{C}_d$ denote the class of affine schemes Y such that Y_{cl} is noetherian and* $\dim Y_{\mathrm{cl}} \leq d$. *If $d > 0$, then the G-torsor given by the quotient map*

$$E := \mathbf{GL}_{n+d}(S)/\begin{bmatrix} 1_{n\times n} & 0_{n\times d} \\ 0_{d\times n} & \mathbf{GL}_d(S)\end{bmatrix} \to \mathbf{GL}_{n+d}(S)/\begin{bmatrix} G & 0_{d\times n} \\ 0_{n\times d} & \mathbf{GL}_d(S)\end{bmatrix} =: X$$

is strongly versal for $\mathscr{C}_d/S$. Here, G acts on the space $\mathbf{GL}_{n+d}(S)/[\begin{smallmatrix} 1 & 0 \\ 0 & \mathbf{GL}_d(S)\end{smallmatrix}]$ via $(x, g) \mapsto x[\begin{smallmatrix} g & 0 \\ 0 & 1_{d\times d}\end{smallmatrix}]$. If $d = 0$ and S is a scheme over an infinite field, then $E \to X$ is versal for $\mathscr{C}_0/S$.

Furthermore, the S-algebraic space X is smooth. If P is one of the properties: being an S-scheme, being an S-scheme locally of finite type, being an S-scheme of finite type, then X has P if $\mathbf{GL}_n(S)/G$ has P (cf. Theorems 2.3, 2.5(i)). If S is locally noetherian and G is an extension of a reductive S-group by a finite locally free S-group, then $X \to S$ is affine of finite type.

PROOF. We use notation from Section 6.

Suppose that $d > 0$. By Corollary 5.9, in order to prove the first part of the theorem, it is enough to show that the vector bundle $V_{n,n+d}$ is weakly versal for $\mathscr{C}_d$, and that its corresponding $\mathbf{GL}_n(\mathbb{Z})$-torsor (see Proposition 6.2),

$$E_{n,n+d} := \mathbf{GL}_{n+d}(\mathbb{Z})/[\begin{smallmatrix} 1 & 0 \\ 0 & \mathbf{GL}_d(\mathbb{Z})\end{smallmatrix}] \to \mathbf{GL}_{n+d}(\mathbb{Z})/[\begin{smallmatrix} \mathbf{GL}_n(\mathbb{Z}) & 0 \\ 0 & \mathbf{GL}_d(\mathbb{Z})\end{smallmatrix}] = X_{n,n+d},$$

admits an action by $\mathbf{SL}_{n+d}(\mathbb{Z})$ that is fppf transitive on $E_{n,n+d}$.

That $V_{n,d}$ is weakly versal for $\mathscr{C}_d$ follows from Proposition 6.1 and Theorem 7.3 (with $Z = \emptyset$). We get the required $\mathbf{SL}_{n+d}(\mathbb{Z})$-action on $E_{n,n+d} \to X_{n,n+d}$ by restricting the evident action of $\mathbf{GL}_{n+d}(\mathbb{Z})$ on $E_{n,n+d} \to X_{n,n+d}$. The action of $\mathbf{SL}_{n+d}(\mathbb{Z})$ on $E_{n,n+d}$ is fppf transitive because the action of $\mathbf{GL}_{n+d}(\mathbb{Z})$ is; and for every ring R, we have $\mathrm{SL}_{n+d}(R)[\begin{smallmatrix} 1_{n\times n} & 0 \\ 0 & \mathrm{GL}_d(R)\end{smallmatrix}] = \mathrm{GL}_{n+d}(R)$ (because $d > 0$). This proves the first part of the theorem in the case $d > 0$.

When $d = 0$ and S is a scheme over an infinite field, we prove the first part by using Remark 5.10 instead of Corollary 5.9, noting that $\mathbf{GL}_{n+d}(\mathbb{Z})$ acts fppf transitively on $E_{n,n+d}$ even when $d = 0$.

We proceed with the second part of the theorem, namely, the assertions about the structure of X. By Lemma 2.2, the morphism $\mathbf{GL}_{n+d}(S) \to X$ is a $G \times \mathbf{GL}_d(S)$-torsor, hence flat, locally of finite presentation and surjective (in the sense of [**31**, Tag 03MC]). Since the composition $\mathbf{GL}_{n+d}(S) \to X \to S$ is smooth, it follows that $X \to S$ is smooth [**31**, Tag 0AHE].

Since $E_{n,n+d}$ and $X_{n,n+d}$ are $\mathbb{Z}$-schemes of finite type (see the proof of Proposition 6.2), the spaces $E = E_{n,n+d} \times_{\mathbb{Z}} S$ and $X_{n,n+d} \times_{\mathbb{Z}} S$ are also S-schemes of finite type. Applying Proposition 3.3 to $E \to X_{n,n+d} \times_{\mathbb{Z}} S$ now tells us that $X = E/G$ has property P whenever $\mathbf{GL}_n(S)/G$ has it. Finally, if S is locally noetherian and G is an extension of a reductive S-group by a finite locally free S-group, then the same holds for $G \times \mathbf{GL}_d(S)$, and $X \to S$ is affine of finite type by Theorem 2.3. □

THEOREM 8.2. *Let n, d, S, G be as in Theorem 8.1. Let* nAff *denote the class of noetherian affine schemes. If $d > 0$, then the G-torsor given by the quotient map*

$$E := \mathbf{GL}_{n+d}(S)/\begin{bmatrix} 1_{n\times n} & \mathbb{A}_S^{n\times d} \\ 0_{d\times n} & \mathbf{GL}_d(S)\end{bmatrix} \to \mathbf{GL}_{n+d}(S)/\begin{bmatrix} G & \mathbb{A}_S^{n\times d} \\ 0_{d\times n} & \mathbf{GL}_d(S)\end{bmatrix} =: X$$

is strongly d-versal for nAff$/S$ *and strongly versal for the class $\mathscr{C}_d$ from Theorem 8.1. Here, G acts on $\mathbf{GL}_{n+d}(S)/[\begin{smallmatrix} 1_{n\times n} & \mathbb{A}_S^{n\times d} \\ 0_{d\times n} & \mathbf{GL}_d(S)\end{smallmatrix}]$ via $(x, g) \mapsto x[\begin{smallmatrix} g & 0 \\ 0 & 1_{d\times d}\end{smallmatrix}]$. If $d = 0$ and S is a scheme over an infinite field, then $E \to X$ is 0-versal for* nAff *and versal for $\mathscr{C}_0$.*

Furthermore, the S-algebraic space X is smooth. If P is one of the properties: being an S-scheme, being an S-scheme locally of finite type, being an S-scheme of finite type, then X has P if $\mathbf{GL}_n(S)/G$ has P. In particular, if S is locally noetherian and G is an extension of a reductive S-group by a finite locally free S-group, then X is an S-scheme of finite type.

PROOF. This is similar to the proof of Theorem 8.1 with the following differences: We use the vector bundle $V(n, n+d)^\vee$ over $\mathbf{Gr}(n, n+d)$ considered in Section 6 instead of $V_{n,n+d}$, Proposition 6.6 instead of Proposition 6.2 (to describe the $\mathbf{GL}_n(\mathbb{Z})$-torsor corresponding to $V(n, n+d)^\vee$), and Theorems 6.4 and 7.2 instead of Proposition 6.1 and Theorem 7.3 (to show that $V(n, n+d)^\vee$ is weakly d-versal for nAff). □

The torsor $E \to X$ of Theorem 8.2 enjoys an additional remarkable property which we call the *extension property.*

THEOREM 8.3 (Extension Property). *Let n, d, S, G and $E \to X$ be as in Theorem 8.2. Let $X' \in \mathscr{C}_d/S$, with $\mathscr{C}_d$ as in Theorem 8.1, and let $u : X'' \to X'$ be a closed embedding. Let $E' \to X'$ be a G-torsor and let $E'' = u^*E'$. Suppose there is a morphism $f : X'' \to X$ such that $f^*E \cong_{X''} E''$. Then f factors as $g \circ u$ for a morphism $g : X' \to X$ such that $g^*E \cong_{X'} E'$.*

PROOF. Put $F' = E' \times^G \mathbf{GL}_n(S)$, $F'' = E'' \times^G \mathbf{GL}_n(S)$ and recall that E is the S-scheme $E(n, n+d)_S$ (see the proof of Theorem 8.2). Let V' and V'' be the vector bundles corresponding to F' and F'', respectively. By Remark 2.1, the isomorphism $f^*E \cong_{X''} E''$ gives rise to a G-torsor morphism $(\hat{f}, f)$ from E'' to $E = E(n, n+d)_S$. Since $E(n, n+d)_S$ is also a $\mathbf{GL}_n(S)$-torsor, we have a $\mathbf{GL}_n(S)$-equivariant morphism $\hat{h}'' : F'' \to E$ given section-wise by $\hat{h}''([x'', a]) = \hat{f}(x'') \cdot a$. By Corollary 6.7,

$\hat{h}''$ determines a tuple $(v''_1, \dots, v''_{n+d}) \in \Gamma(X'', V'')^{n+d}$ which generates V''. By Theorem 7.3, there exist $v'_1, \dots, v'_{n+d} \in \Gamma(X', V')$ which generate V', and satisfy $u^* v'_i = v''_i$ for all $i \in \{1, \dots, n+d\}$. Applying Corollary 6.7 again, we see that $(v'_1, \dots, v'_{n+d})$ determines a $\mathbf{GL}_n(S)$-equivariant morphism $\hat{h}' : F' \to E$ and that $\hat{h}' \circ \hat{u} = \hat{h}''$, where $\hat{u} : F'' \to F'$ is the pullback of u along $F' \to X'$. Let $\hat{g}$ denote the composition of $\hat{h}' : F' \to E$ and the G-equivariant map $e' \mapsto [e', 1] : E' \to F'$. Then $\hat{g} : E' \to E$ descends to a morphism $g : X' \to X$ such that $g^* E \cong E'$, and $g \circ u = f$ because $\hat{h}' \circ \hat{u} = \hat{h}''$. □

Our last construction of strongly versal d-torsors applies only for algebraic groups over infinite fields, but the strong d-versality applies to non-affine schemes as well. Recall from Section 6 that $V(n, m)$ $(m \geq n \geq 0)$ denotes the tautological vector bundle over $\mathbf{Gr}(n, m)$. In the special case $n = 1$, we have $\mathbf{Gr}(1, m) = \mathbb{P}^{m-1}_{\mathbb{Z}}$ and $V(1, m) = \mathcal{O}_{\mathbb{P}^{m-1}_{\mathbb{Z}}}(-1)$.

THEOREM 8.4. *Let k be an infinite field, let $d, n \in \mathbb{N}$, and let G be a subgroup of $\mathbf{GL}_n(k)$. Let $Y = \mathbf{Gr}(n, n+d)_k \times_k \mathbb{P}^d_k$ and let V denote the vector bundle $p_1^* V(n, n+d)^\vee_k \otimes_{\mathcal{O}_Y} p_2^* \mathcal{O}(-1)$ on Y, where $p_1 : Y \to \mathbf{Gr}(n, n+d)_k$ and $p_2 : Y \to \mathbb{P}^d_k$ are the first and second projections, respectively. Let $E \to Y$ be the $\mathbf{GL}_n(k)$-torsor corresponding to V. Then the G-torsor*

$$E \to E/G$$

is strongly d-versal for the class of quasi-projective k-schemes. The k-algebraic space E/G is smooth, and if G is of finite type over k, it is also a quasi-projective k-scheme.

We remark that the quasi-projective k-schemes are precisely the finite type k-schemes admitting an ample line bundle [**31**, Tag 0B42].

PROOF. Let qProj/k denote the class of quasi-projective k-schemes. By Corollary 5.9, in order to show that $E \to E/G$ is strongly d-versal for qProj/k, it is enough to show that the rank-n vector bundle V is weakly versal for qProj/k and that $\mathbf{SL}_{n+d}(k) \times \mathbf{SL}_{1+d}(k)$ acts on $E \to Y$ in such a way that the action is fppf transitive on E.

To prove the first claim, let V' be a rank-n vector bundle on $X' \in \mathsf{qProj}/k$. Then X' admits an ample line bundle L. There exists some $\ell \in \mathbb{N}$ such that both $W := V' \otimes L^{\otimes \ell}$ and $L^{\otimes \ell}$ are generated by global sections [**31**, Tag 01Q3]. By Theorem 7.7, there is an open subscheme $U_1 \subseteq X'$ with $\operatorname{codim}(X' - U_1, X') > d$ such that $W|_{U_1}$ is generated by $n+d$ global sections, and by Theorem 6.4, this determines a k-morphism $f_1 : U_1 \to \mathbf{Gr}(n, n+d)_k$ such that $W|_{U_1} \cong f_1^* V(n, n+d)^\vee_k$. Similarly, there is an open subscheme U_2 of X' and a k-morphism $f_2 : U_2 \to \mathbf{Gr}(1, 1+d) = \mathbb{P}^d_k$ such that $\operatorname{codim}(X' - U_2, X') > d$ and $f_2^* \mathcal{O}_{\mathbb{P}^d_k}(1) = f_2^* V(1, 1+d)^\vee_k \cong L^{\otimes \ell}|_{U_2}$.

Put $U = U_1 \cap U_2$ and define $f : U \to \mathbf{Gr}(n, n+d)_k \times_k \mathbb{P}^d_k$ by $f(x) = (f_1(x), f_2(x))$ on sections. Then $(p_1 \circ f)^* V(n, n+d)^\vee_k \cong W|_U$ while $(p_2 \circ f)^* \mathcal{O}_{\mathbb{P}^d_k}(-1) \cong (L^{\otimes \ell})^\vee|_U$. This means that

$$f^* V \cong f^* p_1^* V(n, n+d)^\vee_k \otimes_{\mathcal{O}_U} f^* p_2^* \mathcal{O}(-1) \cong W|_U \otimes_{\mathcal{O}_U} (L^{\otimes \ell})^\vee|_U \cong V'|_U.$$

Since $\operatorname{codim}(X' - U, X') > d$, this proves that V is weakly d-versal for qProj/k.

To show that $E \to Y$ has the desired $\mathbf{SL}_{n+d}(k) \times \mathbf{SL}_{1+d}(k)$-action, we unfold the definition of E. Let $t_1 : E_1 \to X_1 := \mathbf{Gr}(n, n+d)_k$ be the $\mathbf{GL}_{n+d}(k)$-torsor corresponding to $V(n,n+d)_k^\vee$, and let $t_2 : E_2 \to X_2 := \mathbf{Gr}(1,1+d)_k = \mathbb{P}_k^d$ be the $\mathbf{GL}_1(k)$-torsor corresponding to $V(1,1+d)_k = \mathcal{O}(-1)$. Proposition 6.6 and Lemma 3.2 imply that $\mathbf{GL}_{n+d}(k)$ acts on $t_1 : E_1 \to X_1$ and $\mathbf{GL}_{1+d}(k)$ acts on $t_2 : E_2 \to X_2$. Since $d > 0$, these actions restrict to fppf transitive actions of $\mathbf{SL}_{n+d}(k)$ and $\mathbf{SL}_{1+d}(k)$ on E_1 and E_2, respectively. The $\mathbf{GL}_n(k)$-torsor corresponding to $p_1^*V(n,n+d)_k^\vee$ is $p_1^*E_1 = E_1 \times_k X_2 \to X_1 \times_k X_2 = Y$ and the $\mathbf{GL}_1(k)$-torsor corresponding to $p_2^*\mathcal{O}(-1)$ is $X_1 \times_k E_2 \to X_1 \times_k X_2 = Y$. Thus, $E_1 \times_k E_2 = (p_1^*E_1) \times_Y (p_2^*E_2)$ is a $\mathbf{GL}_n(k) \times_k \mathbf{GL}_1(k)$-torsor over Y corresponding to the pair of vector bundles $(p_1^*V(n,n+d)_k^\vee, p_2^*\mathcal{O}(-1))$. The induced morphism $E_1 \times_k E_2 \to X_1 \times_k X_2 = Y$ is just $t_1 \times t_2$ so $\mathbf{SL}_{n+d}(k) \times_k \mathbf{SL}_{1+d}(k)$ acts on this torsor, and the action on $E_1 \times_k E_2$ is fppf transitive.

Consider the k-group morphism $\varphi : G := \mathbf{GL}_n(k) \times_k \mathbf{GL}_1(k) \to \mathbf{GL}_n(k)$ given by $\varphi(g_1, g_2) = g_1 \otimes g_2$ on sections. If X is a k-scheme, (V_1, V_2) is a pair of vector bundles on X such that $\operatorname{rank} V_1 = n$ and $\operatorname{rank} V_2 = 1$, and $P \to X$ is the $\mathbf{GL}_n(k) \times_k \mathbf{GL}_1(k)$-torsor corresponding to the pair (V_1, V_2), then the $\mathbf{GL}_n(k)$-torsor corresponding to $V_1 \otimes_{\mathcal{O}_X} V_2$ is $P \times^{\varphi, G} \mathbf{GL}_n(k)$ up to X-isomorphism. This can be seen, for instance, by choosing a Zariski covering $\{U_i \to X\}_{i\in I}$ trivializing V_1 and V_2, and looking at the Čech 1-cocycles defining the torsors corresponding to V_1, V_2 and $V_1 \otimes_{\mathcal{O}_X} V_2$. Applying this to the $\mathbf{GL}_n(k) \times_k \mathbf{GL}_1(k)$-torsor $E_1 \times_k E_2 \to X_1 \times_k X_2 = Y$, we see that E from the theorem is $(E_1 \times_k E_2) \times^{\varphi, G} \mathbf{GL}_n(k)$. The $\mathbf{GL}_n(k)$-torsor $E \to Y$ inherits from $E_1 \times_k E_2 \to Y$ an $\mathbf{SL}_{n+d}(k) \times_k \mathbf{SL}_{1+d}(k)$-action such that the canonical map $E_1 \times_k E_2 \to E$ is $\mathbf{SL}_{n+d}(k) \times_k \mathbf{SL}_{1+d}(k)$-equivariant. Since $\varphi : G \to \mathbf{GL}_n(k)$ is k-fppf surjective, so is $E_1 \times_k E_2 \to E$, and it follows that $\mathbf{SL}_{n+d}(k) \times_k \mathbf{SL}_{1+d}(k)$ acts fppf transitively on E.

The smoothness of E/G over k is shown as in Theorem 8.1. To finish, suppose that G is of finite type. By Theorem 2.5(i), $\mathbf{GL}_n(k)/G$ is a k-scheme, and the same applies to E/G by Proposition 3.3. Note that E_1 and E_2 admit a k-point (see Proposition 6.5), and thus E/G has a k-point x. Moreover, x is closed because E/G is a k-scheme. Since $H := \mathbf{SL}_{n+d}(k) \times \mathbf{SL}_{1+d}(k)$ acts fppf transitively on E/G, the k-group H has *closed* subgroup H_1 — the stabilizer of x — such that $E/G \cong H/H_1$, and the latter is a quasi-projective k-scheme by Theorem 2.5(i). □

In light of the previous theorems, we ask:

QUESTION 8.5. Let S be a scheme, let G be a linear S-group and let $d \in \mathbb{N}$. Is there a G-torsor over an S-algebraic space that is strongly d-versal for the class of S-schemes admitting an ample line bundle?

QUESTION 8.6. Let G be an affine algebraic group over an infinite field k. Is there a G-torsor over a k-algebraic space that is strongly d-versal for the class of k-schemes of finite type?

9. Examples

In this section, we apply the results of Section 8 to particular group schemes in order to deduce the existence of strongly versal Galois extensions, Azumaya algebras and tori. We refer the reader to [**27**] or [**13**] for definitions concerning Galois extensions and Azumaya algebras, and to [**20**, III.§8] for information about Azumaya algebras with involution.

As in Theorem 8.1, we write $\mathscr{C}_d$ for the class of affine schemes X for which the subspace X_{cl} is noetherian and has dimension $\leq d$. It includes all noetherian affine schemes of dimension $\leq d$. We call a ring homomorphism $A \to B$ *universally injective* if $a' \mapsto 1_B \otimes a' : A' \to B \otimes_A A'$ is injective for every A-ring A'. By virtue of Proposition 4.3(ii), this is equivalent to $\operatorname{Spec} B \to \operatorname{Spec} A$ being universally schematically dominant. It is also equivalent to A being a *pure* A-submodule of B.

EXAMPLE 9.1. Let Γ be a finite group, and let G be the constant $\mathbb{Z}$-group corresponding to Γ. We may embed G as a closed subgroup of $\mathbf{GL}_n(\mathbb{Z})$ for some $n \in \mathbb{N}$, e.g., via the regular representation. Then Theorem 8.1 says that for every $d \in \mathbb{N}$, the G-torsor $E := \mathbf{GL}_{n+d}(\mathbb{Z})/[\begin{smallmatrix}1 & 0\\ 0 & \mathbf{GL}_d(\mathbb{Z})\end{smallmatrix}] \to \mathbf{GL}_{n+d}(\mathbb{Z})/[\begin{smallmatrix}G & 0\\ 0 & \mathbf{GL}_d(\mathbb{Z})\end{smallmatrix}] := X$ is strongly versal for the class $\mathscr{C}_d$, and X is affine and smooth over $\mathbb{Z}$. Since the category of Γ-Galois extensions of rings (morphisms are specializations) is (anti-) equivalent to the category of G-torsors over affine schemes, we can rephrase this result in terms of Γ-Galois extension as follows: Write $X = \operatorname{Spec} R$ and let S/R be the Γ-Galois extension corresponding to $E \to X$. Then R is a smooth $\mathbb{Z}$-ring and S/R is a Γ-Galois extension that is strongly versal for $\mathscr{C}_d$. That is, for every Γ-Galois extension S'/R' such that the space $\operatorname{Max} R'$ is notherian of dimension $\leq d$ (e.g., if R' is noetherian of Krull dimension $\leq d$), there are[12] $m \in \mathbb{N}$ and a ring homomorphism $\varphi : R \to R'[t_1, \dots, t_m]$ such that the specialization of S/R along φ is isomorphic to $S'[t_1, \dots, t_m]/R'[t_1, \dots, t_m]$ (as $R'[t_1, \dots, t_m]$-rings and Γ-modules), and the induced map $r \otimes r' \mapsto \varphi(r)r' : R \otimes_{\mathbb{Z}} R' \to R'[t_1, \dots, t_m]$ is universally injective. In particular, by specializing all the t_i to 0, we see that S/R specializes to any Γ-Galois extension S'/R' such that $\operatorname{Max} R$ is noetherian of dimension $\leq d$.

EXAMPLE 9.2. Let $G = \mathbf{PGL}_n(\mathbb{Z})$, the automorphism $\mathbb{Z}$-scheme of the $\mathbb{Z}$-algebra $\mathrm{M}_n(\mathbb{Z})$. Then G is linear and reductive. It is well-known that there is an equivalence of categories over $\mathbb{Z}$-algebraic spaces between $\mathsf{Tors}(G)$ (see Section 2) and the opposite category of degree-n Azumaya algebras over $\mathbb{Z}$-algebraic spaces. Thus, applying Theorem 8.1 to G implies that, for every $d \in \mathbb{N}$, there is a smooth $\mathbb{Z}$-algebra R and a degree-n Azumaya algebra A over R that is strongly d-versal for $\mathscr{C}_d$. That is, for every ring R' such that $\operatorname{Max} R'$ is noetherian of dimension $\leq d$ and every degree-n Azumaya R'-algebra A', there are $m \in \mathbb{N}$ and a ring homomorphism $\varphi : R \to R'[t_1, \dots, t_m]$ such that $A \otimes_R R'[t_1, \dots, t_m] \cong A'[t_1, \dots, t_m]$ as $R'[t_1, \dots, t_m]$-algebras; and moreover, $r \otimes r' \mapsto \varphi(r)r' : R \otimes_{\mathbb{Z}} R' \to R'[t_1, \dots, t_m]$ is universally injective. In particular, A/R specializes to any degree-n Azumaya algebra over any R' as above.

For a ring R, we write $\boldsymbol{\mu}_n(R)$ for the R-group of n-th roots of unity.

EXAMPLE 9.3. The conclusions of Example 9.2 continue to hold if we work over $\mathbb{Z}[\frac{1}{2}]$ instead of $\mathbb{Z}$ and replace "degree-n Azumaya algebra" with "degree-n Azumaya algebra with orthogonal (resp. symplectic) involution" (n is even in the symplectic case). To see this, apply Theorem 8.1 with $G = \mathbf{PO}_n(\mathbb{Z}[\frac{1}{2}]) = \mathbf{O}_n(\mathbb{Z}[\frac{1}{2}])/\boldsymbol{\mu}_2(\mathbb{Z}[\frac{1}{2}])$ (resp. $\mathbf{PSp}_n(\mathbb{Z}[\frac{1}{2}]) = \mathbf{Sp}_n(\mathbb{Z}[\frac{1}{2}])/\boldsymbol{\mu}_2(\mathbb{Z}[\frac{1}{2}])$ for n even) and note that $\mathsf{Tors}(G)$ is anti-equivalent to the category of degree-n Azumaya algebras with orthogonal (resp. symplectic) involution over $\mathbb{Z}[\frac{1}{2}]$-algebraic spaces; see [**20**, III.§8.5].

Recall that the period (also called exponent) of an Azumaya algebra over a ring R is the order of its Brauer class in the Brauer group of R. We say that an Azumaya algebra is *n-periodic* if its period divides n.

---

[12]By unfolding the proof of Theorem 8.1, one sees that m depends only on Γ and not on $R' \to S'$.

PROPOSITION 9.4. *Let $m, n, d \in \mathbb{N}$. Then there is a smooth $\mathbb{Z}$-ring R and an n-periodic degree-m Azumaya R-algebra that is strongly versal (among the n-periodic degree-m Azumaya algebras) for $\mathscr{C}_d$. That is, for any ring R' such that $\operatorname{Max} R'$ is noetherian of dimension $\leq d$ and every n-periodic degree-m Azumaya R'-algebra A', there are $\ell \in \mathbb{N}$ and a ring homomorphism $\varphi : R \to R'[t_1, \dots, t_\ell]$ such that $A \otimes_R R'[t_1, \dots, t_\ell] \cong A'[t_1, \dots, t_\ell]$ as $R'[t_1, \dots, t_\ell]$-algebras and $r \otimes r' \mapsto \varphi(r)r' : R \otimes_{\mathbb{Z}} R' \to R'[t_1, \dots, t_\ell]$ is universally injective.*

PROOF. Let $G = \mathbf{GL}_m(\mathbb{Z})/\boldsymbol{\mu}_n(\mathbb{Z})$, and let φ denote the evident $\mathbb{Z}$-group morphism $G \to \mathbf{GL}_m(\mathbb{Z})/\mathbf{G}_{\mathbf{m},\mathbb{Z}} = \mathbf{PGL}_m(\mathbb{Z})$. If $E' \to X'$ is a G-torsor, then $E' \times^{\varphi, G} \mathbf{PGL}_m(\mathbb{Z})$ is a $\mathbf{PGL}_m(\mathbb{Z})$-torsor which corresponds to a degree-m Azumaya algebra A' over X'. It is well-known that the period of A' divides n, and conversely, every n-periodic degree-m Azumaya algebra over X' is obtained in this manner (i.e., from a G-torsor over X').

The $\mathbb{Z}$-group G is reductive and linear (Theorem 2.4), and so Theorem 8.1 implies that there exists a G-torsor E over a smooth affine $\mathbb{Z}$-scheme X that is d-versal for $\mathscr{C}_d$. Write $X = \operatorname{Spec} R$, let $E_1 := E \times^{\varphi, G} \mathbf{PGL}_n(\mathbb{Z})$, and let A be the Azumaya R-algebra corresponding to E_1. Let R' and A' be as in the proposition, and let $E_1' \to X' := \operatorname{Spec} R'$ denote the $\mathbf{PGL}_n(\mathbb{Z})$-torsor corresponding to A'. Then there is a G-torsor $E' \to X'$ such that $E_1' \cong_{X'} E' \times^{\varphi, G} \mathbf{PGL}_n(\mathbb{Z})$. Since $E \to X$ is strongly versal for $\mathscr{C}_d$, there is $\ell \in \mathbb{N}$ and a morphism $f : \mathbb{A}^\ell_{\mathbb{Z}} \times X' \to X$ such that $f^*E \cong \mathbb{A}^\ell_{\mathbb{Z}} \times E'$, and $(f, p_2) : \mathbb{A}^\ell_{\mathbb{Z}} \times X' \to X \times X'$ is universally schematically dominant. The former means that $f^*E_1 \cong \mathbb{A}^\ell_{\mathbb{Z}} \times E_1'$. Writing $\varphi : R \to R'[t_1, \dots, t_\ell]$ for the adjoint of $f : \mathbb{A}^\ell_{\mathbb{Z}} \times X' \to X$, the proposition follows. $\square$

EXAMPLE 9.5. Let k be a field of characteristic 0. Given an abstract group Γ, let $\underline{\Gamma}$ denote the constant k-group corresponding to Γ. The k-group $\underline{\Gamma}$ is flat and locally of finite presentation, but it is not of finite type if Γ is infinite.

Let $n \in \mathbb{N}$ and let $G = \underline{\mathrm{GL}_n(\mathbb{Z})}$. There is a k-group monomorphism from $G = \bigsqcup_{\gamma \in \mathrm{GL}_n(\mathbb{Z})} \operatorname{Spec} k$ to $\mathbf{GL}_n(k)$ which restricts to the γ-section $\operatorname{Spec} k \to \mathbf{GL}_n(k)$ on the γ-copy of $\operatorname{Spec} k$ for every $\gamma \in \mathrm{GL}_n(\mathbb{Z})$. Thus, G is linear, and so Theorem 8.2 tells us that, for every $d \in \mathbb{N}$, there is a G-torsor $E_d \to X_d$ that is strongly d-versal for affine noetherian k-schemes. However, if $n > 1$, then the k-algebraic space $X_d = \mathbf{GL}_{n+d}(k)/\left[\begin{smallmatrix} G & \mathbb{A}^{n \times d}_k \\ 0_{n \times d} & \mathbf{GL}_{n+d}(k) \end{smallmatrix}\right]$, while smooth, is not a k-scheme, and also not separated (cf. [**31**, Tag 02Z7]).

It is well-known that the automorphism sheaf of the k-group $\mathbf{G}^n_{\mathbf{m},k}$ is isomorphic to $G = \underline{\mathrm{GL}_n(\mathbb{Z})}$. Thus, a standard argument shows that $\mathsf{Tors}(G)$ is equivalent as a category over k-algebraic spaces to the category of rank-n tori over k-algebraic spaces (morphisms are specializations). Let T_d be the rank-n torus over the k-algebraic space X_d corresponding to $E_d \to X_d$. Then T_d is strongly d-versal for affine noetherian k-schemes. In particular, T_d specializes to any rank-n torus T over an affine noetherian k-scheme X away from some codimesion-$(d+1)$ closed subscheme of X.

10. ∞-Versal Torsors

Throughout, Aff denotes the class of affine schemes. Let S be a scheme and let G be an S-group. This section is concerned with G-torsors that are versal for Aff/S. In particular, they specialize to every G-torsor over an affine scheme over

S. We will show that there are nontrivial S-groups admitting such torsors (over a finite type S-scheme), and characterize all these groups when S is the spectrum of a field of characteristic 0.

We write $\mathbf{U}_n(S)$ for the S-group of unipotent upper triangular $n \times n$ matrices over $\mathcal{O}_S$.

THEOREM 10.1. *Let S be a scheme and let G be a subgroup of $\mathbf{U}_n(S)$ that is flat and locally of finite presentation over S. Then the G-torsor $\mathbf{U}_n(S) \to \mathbf{U}_n(S)/G$ is strongly versal for Aff/S.*

Furthermore, the S-algebraic space $\mathbf{U}_n(S)/G$ is smooth. If S is the spectrum of field and G is of finite type over S, then $\mathbf{U}_n(S)/G \to S$ is quasi-projective. If G is finite and locally free, then $\mathbf{U}_n(S)/G \to S$ is affine and of finite type.

PROOF. Observe first that for every $X \in \mathsf{Aff}/S$, we have $\mathrm{H}^1_{\mathrm{fppf}}(X, \mathbf{U}_n(S)) = \{*\}$. Indeed, $\mathbf{U}_n(S)$ admits a filtration $\mathbf{U}_n(S) = U_0 \rhd U_1 \rhd \cdots \rhd U_r = 0$ such that $U_{i-1}/U_i \cong \mathbf{G}_{\mathbf{a},S}$ for all $i \in \{1, \dots, r\}$. Since X is affine, $\mathrm{H}^1_{\mathrm{fppf}}(X, \mathbf{G}_{\mathbf{a},S}) = \mathrm{H}^1_{\mathrm{fppf}}(X, \mathcal{O}_X) = 0$ [**23**, III, Lem. 2.15, Prop. 3.7]. Now, an inductive argument using the long cohomology exact sequence associated to $1 \to U_i \to U_{i-1} \to U_{i-1}/U_i \to 1$ shows that $\mathrm{H}^1_{\mathrm{fppf}}(X, U_i) = \{*\}$ for all i, and $\mathrm{H}^1_{\mathrm{fppf}}(X, \mathbf{U}_n(S)) = \{*\}$ in particular.

Put $E_0 := \mathbf{U}_n(S) \to \mathbf{U}_n(S)/G =: X_0$ and recall that this is indeed a G-torsor by Lemma 2.2. We first show that $E_0 \to X_0$ is weakly versal for Aff/S. Let $X \in \mathsf{Aff}/S$ and let $E \to X$ be a G-torsor. Then $E \to X$ corresponds to a class $\alpha \in \mathrm{H}^1_{\mathrm{fppf}}(X, G)$. It is well-known that the embedding $G \to \mathbf{U}_n(S)$ gives rise to an exact sequence of pointed sets:

$$(\mathbf{U}_n(S)/G)(X) \xrightarrow{\gamma} \mathrm{H}^1_{\mathrm{fppf}}(X, G) \to \mathrm{H}^1_{\mathrm{fppf}}(X, \mathbf{U}_n(S)),$$

in which γ maps a morphism $f : X \to \mathbf{U}_n(S)/G$ to the cohomology class of $f^* E_0$. By the previous paragraph, the image of α vanishes in $\mathrm{H}^1_{\mathrm{fppf}}(X, \mathbf{U}_n(S))$, so it lifts to some $f : X \to \mathbf{U}_n(S)/G$, and $f^* E_0 \cong E$.

To see that $E_0 \to X_0$ is strongly versal for Aff/S, observe that the left action of $\mathbf{U}_n(S)$ on E_0 and X_0 gives rise to an action of $\mathbf{U}_n(S)$ on the G-torsor $E_0 \to X_0$. This action is clearly fppf transitive on E_0. Since $\mathbf{U}_n(S) \cong \mathbb{A}_S^{n(n-1)/2}$ as S-schemes, Proposition 5.8(ii) tells us that $E_0 \to X_0$ is strongly versal for Aff/S.

The smoothness of $\mathbf{U}_n(S)/G \to S$ is shown as in Theorem 8.1. If S is the spectrum of a field and $G \to S$ is of finite type, then $\mathbf{U}_n(S)/G$ is a quasi-projective S-scheme by Theorem 2.5(i). If G is finite, then $\mathbf{U}_n(S)/G \to S$ is affine of finite type by Theorem 2.3. □

REMARK 10.2. Another sufficient condition for the existence of a G-torsor that is strongly versal for Aff/S is:

($*$) Every G-torsor $E \to X$ with $X \in \mathsf{Aff}/S$ is trivial.

Indeed, in this case, every G-torsor as in ($*$) is a specialization of the trivial G-torsor $G \to S$, making $G \to S$ strongly versal for Aff/S. Such S-groups may be deemed as "uninteresting" for the problem of finding groups with torsors that are weakly versal for Aff/S.

The proof of Theorem 10.1 can now be broken into the observations that $\mathbf{U}_n(S)$ satisfies ($*$), and that every flat locally of finite presentation subgroup of an S-group satisfying ($*$) admits a torsor that is weakly versal for Aff/S.

When $S = \operatorname{Spec} k$ for a field k of characteristic 0, it is well-known [**22**, Proposition 14.22, Remark 14.24] that every finite type subgroup G of $\mathbf{U}_n(k)$ admits a filtration $G = G_0 \rhd G_1 \rhd \cdots \rhd G_r = 0$ with $G_i/G_{i-1} \cong \mathbf{G}_{\mathbf{a},k}$, and thus already satisfies $(*)$. Consequently, in this case, Theorem 10.1 does not give "interesting" examples of k-groups admitting torsors that are strongly versal for Aff/S.

However, when $S = \operatorname{Spec} k$ for a field k of characteristic $p > 0$, the k-group $\mathbf{U}_n(k)$ admits subgroups which do not satisfy $(*)$, and so we get "interesting" examples of torsors that are strongly versal for Aff/S.

EXAMPLE 10.3. Let k be a field of characteristic $p > 0$ and let C_p denote the cyclic group of order p, viewed as a constant k-group. We identify C_p with the subgroup of $\mathbf{U}_2(k) \cong \mathbf{G}_{\mathbf{a},k}$ cut by the equation $t^p = t$ (with t being the coordiate of $\mathbf{G}_{\mathbf{a},k} \cong \mathbb{A}^1_k$). Then, by Theorem 10.1, $X = \mathbf{G}_{\mathbf{a},k}/C_p$ is an affine k-scheme of finite type, and $\mathbf{G}_{\mathbf{a},k} \to \mathbf{G}_{\mathbf{a},k}/C_p$ is strongly versal for Aff/k. In fact, $\mathbf{G}_{\mathbf{a},k} \to \mathbf{G}_{\mathbf{a},k}/C_p$ is isomorphic to the k-group homomorphism $\mathbf{G}_{\mathbf{a},k} \to \mathbf{G}_{\mathbf{a},k}$ given by $x \mapsto x^p - x$ on sections. As in Example 9.1(i), we can rephrase this in the language of C_p-Galois extensions: The C_p-Galois extension corresponding to $\mathbf{G}_{\mathbf{a},k} \to \mathbf{G}_{\mathbf{a},k}/C_p$ is the generic Artin–Schrier extension $k[t, x \,|\, x^p - x = t]/k[t]$ (here, $C_p = \langle \sigma \,|\, \sigma^p = 1 \rangle$ acts on the $k[t]$-algebra $k[t, x \,|\, x^p - x = t]$ by $\sigma(x) = x + 1$). Thus, for every C_p-Galois extension *of k-rings* S/R, there is $n \in \mathbb{N}$ (in fact, one can take $n = 1$) and a morphism $\varphi : k[t] \to R[t_1, \dots, t_n]$ such that $S[t_1, \dots, t_n]/R[t_1, \dots, t_n]$ is isomorphic to the specialization of $k[t, x \,|\, x^p - x = t]/k[t]$ along φ. Moreover, the induced map $R[t] = k[t] \otimes_k R \to R[t_1, \dots, t_n]$ is universally injective. Taking $k = \mathbb{F}_p$ and specializing $t_1, \dots, t_n$ to 0, we recover the well-known fact that every C_p-Galois extension S/R of characteristic-p rings is an Artin-Schrier extension, namely, there is $a \in R$ such that $S \cong R[x \,|\, x^p - x = a]$ as R-algebras and C_p-modules.

The last example generalizes to other Galois extensions as follows.

COROLLARY 10.4. *Let k be a field of characteristic $p > 0$ and let Γ be a finite p-group. Then there exists a Γ-Galois extension of k-rings S/R that is strongly versal for (spectra of) k-rings. In particular, it specializes to every other Γ-Galois extension of k-rings. The ring R can be taken to be a smooth k-algebra.*

PROOF. Let G denote the constant k-group corresponding to Γ. Since Γ is a p-group, it can be embedded in the abstract group $U_n(\mathbb{F}_p)$ for some $n \in \mathbb{N}$. Thus, G is isomorphic to a subgroup of the k-group $\mathbf{U}_n(k)$. By Theorem 10.1 and the equivalence between Γ-Galois extensions and G-torsors, the Γ-Galois extension S/R corresponding to the G-torsor $\mathbf{U}_n(k) \to \mathbf{U}_n(k)/G$ satisfies the requirements. □

REMARK 10.5. With notation as in Corollary 10.4, the existence of Γ-Galois extensions S/R that are weakly versal for Aff/k goes back to Saltman [**26**, Theorem 2.3], who moreover moreover gave explicit examples in which R is a polynomial ring over k (i.e., they are *generic*). Woodcock and Fleishmann [**11**], [**12**] gave other explicit constructions of weakly versal Γ-Galois extensions S/R such that both R and S are polynomial rings over k and which also enjoy additional cogeneration properties. They further investigate the minimal possible Krull dimension of such R. Our construction is very different from these sources, and it gives rise to strongly versal Γ-Galois extensions.

Let k be a field, and let ftAff/k denote the class of finite-type affine k-schemes. We finish this section with a partial converse to Theorem 10.1, showing that every affine algebraic group G over a characteristic-0 field k which admits a torsor over a finite type k-scheme that is weakly versal for ftAff/k is isomorphic to a subgroup of $\mathbf{U}_n(k)$, or equivalently, unipotent. Consequently, all such algebraic groups G are "uninteresting" in the sense that every G-torsor over an affine base is trivial. This contrasts with the case $\operatorname{char} k > 0$, where such "interesting" groups always exist.

THEOREM 10.6. *Let k be a field of characteristic 0 and let G be an affine algebraic group over k. Suppose that there exists a G-torsor $E \to X$ that is weakly versal for ftAff/k and such that X is a scheme of finite type over k. Then G is unipotent.*

We first prove the following lemma.

LEMMA 10.7. *Let X be a $\mathbb{C}$-scheme of finite type. Endow $X(\mathbb{C})$ with the analytic topology. Then there is $n \in \mathbb{N}$ such that the singular cohomology $\mathrm{H}^i(X(\mathbb{C}), \mathbb{Z})$ vanishes for all $i \geq n$.*

PROOF. When X is quasi-affine, this follows from a theorem of Lojasiewicz [**21**, Theorem 1], who showed that $X(\mathbb{C})$ has the homotopy type of a CW-complex of dimension $\leq 2 \dim X$. The general case follows from the quasi-affine case by taking a finite affine covering $\mathcal{U} = \{U_i\}_{i=1}^r$ of X and using the Čech spectral sequence $E_2^{p,q} = \mathrm{H}^p(\mathcal{U}, \underline{\mathrm{H}}^q(X(\mathbb{C}), \mathbb{Z}))) \Longrightarrow \mathrm{H}^i(X(\mathbb{C}), \mathbb{Z})$ [**31**, Tag 03OW]. □

PROOF OF THEOREM 10.6. We will use the fact that $\operatorname{char} k = 0$ to apply analytic tools such as the G-equivariant cohomology theory considered in [**10**, §11]. This requires k to be a subfield of $\mathbb{C}$, and reducing to this situation will necessitate some maneuvering.

For the sake of contradiction, suppose that G is not unipotent.

STEP 1. We may assume that $\mathbb{Q} \subseteq k$. Since G, E and X are of finite presentation over $\operatorname{Spec} k$, there is a finitely generated $\mathbb{Q}$-subfield k_0 of k, a k_0-group G_0, and a G_0-torsor $E_0 \to X_0$ such that $G_0 \times_{k_0} k = G$ and $E \to X$ is the base-change of $E_0 \to X_0$ along $\operatorname{Spec} k \to \operatorname{Spec} k_0$. (In what follows the subscript 0 indicates that the object at hand is a k_0-scheme or a k_0-morphism.)

Next, let $e \in \mathbb{N}$; the value of e will be determined in Step 3. It is well-known, e.g., see [**34**, Remark 1.4], that there exists a finite-dimensional k_0-vector space V_0 (which we also view as a k_0-scheme $\mathbb{A}_{k_0}^{\dim V_0}$), a linear representation $G_0 \to \mathbf{GL}(V_0)$ and an open subcheme $U_0 \subseteq V_0$ stable under G_0 such that U_0/G_0 is a quasi-projective k_0-scheme, $U_0 \to U_0/G_0$ is a G_0-torsor, and $\operatorname{codim}(V_0 - U_0, V_0) > e$. Jouanolou [**18**] showed that there also exists a vector bundle P_0 over U_0/G_0 and a P_0-torsor $q_0 : W_0 \to U_0/G_0$ such that W_0 is an affine k_0-scheme. Let $F_0 \to W_0$ be the pullback of the G_0-torsor $U_0 \to U_0/G_0$ along q_0, and put $F = F_0 \times_{k_0} k$ and $W = W_0 \times_{k_0} k$. Then $F \to W$ is a G-torsor over an *affine* k-scheme of finite type. Since $E \to X$ is weakly versal for ftAff/k, there exists a G-equivariant morphism $\hat{g} : F \to E$. As in the last paragraph, k admits a finitely generated k_0-subfield k_1 such that $\hat{g}$ is extended from a $G_0 \times_{k_0} k_1$-equivariant morphism $\hat{g}_1 : F_1 := F_0 \times_{k_0} k_1 \to E_0 \times_{k_0} k_1 := E_1$. We put $W_1 = W_0 \times_{k_0} k_1$, $U_1 = U_0 \times_{k_0} k_1$, $G_1 = G_0 \times_{k_0} k_1$ and so on, and let q_1 denote the k_1-morphism obtained from q_0 by base-change along $k_0 \to k_1$. The morphism $\hat{g}_1 : F_1 \to E_1$ gives rise to a k_1-morphism $g_1 : W_1 = F_1/G_1 \to E_1/G_1 = X_1$.

To conclude this step, we constructed finitely generated $\mathbb{Q}$-subfields $k_0 \subseteq k_1$ of k and a commutative diagram of finite type k_1-schemes

$$\begin{array}{ccccccc}
V_1 & \xleftarrow{\ \alpha\ } & U_1 & \xleftarrow{\ \hat{q}_1\ } & F_1 & \xrightarrow{\ \hat{g}_1\ } & E_1 \\
 & & \downarrow & & \downarrow & & \downarrow \\
 & & U_1/G_1 & \xleftarrow{\ q_1\ } & W_1 & \xrightarrow{\ g_1\ } & X_1
\end{array}$$

in which the columns are G_1-torsors and the squares represent morphisms of G_1-torsors. Moreover, q_1 carries the structure of a P_1-torsor for some vector bundle P_1 over U_1/G_1. The map denoted α is the base change of the G_0-equivariant open immersion $U_0 \to V_0$ to k_1. Thus, V_1 is a representation of G_1, and U_1 is a G_1-stable open subscheme of V_1 satisfying $\operatorname{codim}(V_1 - U_1, V_1) > e$.

Abusing the notation, we henceforth denote $G_1 = G_0 \times_{k_0} k_1$ as G.

STEP 2. Choose an embedding of k_0 in $\mathbb{C}$; this embedding will be used to choose the number e from Step 1 later on. Once e has been chosen, the field k_1 from Step 1 can be determined and we extend the embedding $k_0 \to \mathbb{C}$ into an embedding $k_1 \to \mathbb{C}$.

We are now in position to apply the G-equivariant cohomology theory considered in [**10**, §11]. We recall that this theory assigns to any finite-type k_1-scheme Y carrying a G-action cohomology groups $\{H^i_G(Y)\}_{i\in\mathbb{Z}}$, contravariantly in Y. It has the following properties:

(i) If $T \to Z$ is a G-torsor, then $\mathrm{H}^i_G(T) \cong \mathrm{H}^i(Z(\mathbb{C}), \mathbb{Z})$ canonically, with the right hand side denoting the singular cohomology of the analytic space $Z(\mathbb{C})$.

(ii) If V is a representation of G and U is a G-invariant open subvariety of V such that $\operatorname{codim}(V - U, V) > d$, U/G is a k-variety and $U \to U/G$ is a G-torsor, then the map $\mathrm{H}^i_G(Y) \to \mathrm{H}^i_G(Y \times_{k_1} U) \cong \mathrm{H}^i(((Y \times U)/G)(\mathbb{C}), \mathbb{Z})$ is an isomorphism for $i \le 2d$.

Letting G act trivially on $\operatorname{Spec} k_1$, we arrive at the cohomology groups $\mathrm{H}^*_G(\operatorname{Spec} k_1)$, which are denoted $\mathrm{H}^*(BG)$.

Since every k_1-variety Y carrying a G-action admits a unique G-equivariant morphism to $\operatorname{Spec} k_1$, we have a canonical morphism $\mathrm{H}^*(BG) \to \mathrm{H}^*_G(Y)$. The morphisms $\hat{q}_1$ and $\hat{g}_1$ from Step 1 now give rise to a commutative diagram

$$\begin{array}{ccccc}
\mathrm{H}^i((U_1/G)(\mathbb{C}), \mathbb{Z}) & \xrightarrow{\ \mathrm{H}^i(q_1)\ } & \mathrm{H}^i(W_1(\mathbb{C})) & \xleftarrow{\ \mathrm{H}^i(g_1)\ } & \mathrm{H}^i(X_1(\mathbb{C})) \\
\| & & \| & & \| \\
\mathrm{H}^i_G(U_1) & \xrightarrow{\ \mathrm{H}^i_G(\hat{q}_1)\ } & \mathrm{H}^i_G(F_1) & \xleftarrow{\ \mathrm{H}^i_G(\hat{g}_1)\ } & \mathrm{H}^i_G(E_1) \\
 & \nwarrow^{\alpha_i} & \uparrow{\scriptstyle \beta_i} & \nearrow_{\gamma_i} & \\
 & & \mathrm{H}^i(BG) & &
\end{array}$$

Applying (ii) to with $Y = \operatorname{Spec} k_1$ and $U = U_1$ from Step 1, we find that the map $\alpha_i : \mathrm{H}^i(BG) \to \mathrm{H}^i_G(U_1)$ is an isomorphism for all $i \le 2e$. Since the morphism $q_1 : W_1 \to U_1/G$ is a P_1-torsor, the map $W_1(\mathbb{C}) \to (U_1/G)(\mathbb{C})$ is a homotopy equivalence, and thus $\mathrm{H}^i(\hat{q}_1) : \mathrm{H}^i_G(U_1) \to \mathrm{H}^i_G(F_1)$ is an isomorphism for all $i \in \mathbb{Z}$.

This means that

$$\beta_i : \mathrm{H}^i(BG) \to \mathrm{H}^i_G(F_1)$$

is an isomorphism for all $i \leq 2e$.

STEP 3. We finally explain how to choose e and achieve the desired contraction. By applying Lemma 10.7 to $X_0 \times_{k_0} \mathbb{C}$, we see that there exists $n \in \mathbb{N}$ (depending only on X_0) such that

$$\mathrm{H}^i_G(E_1) = \mathrm{H}^i(X_1(\mathbb{C})) = \mathrm{H}^i(X_0(\mathbb{C})) = 0 \qquad \forall i > n.$$

Since G is not unipotent, by [**10**, Lemma 12.1], there exists $i > n$ such that $\mathrm{H}^i(BG) \neq 0$. We note that while the condition $\mathrm{H}^i(BG) \neq 0$ assumes that k_1 was chosen, one can choose i that works for every field k_1 lying between k_0 and $\mathbb{C}$, see [**10**, Remark 12.2]. Take $e = \lceil i/2 \rceil$.

Consider the diagram from Step 2. The map $\beta_i : \mathrm{H}^i(BG) \to \mathrm{H}^i_G(F_1)$ is an isomorphism, because $i \leq 2e$, and it factors via $\mathrm{H}^i_G(E_1)$, which is 0 because $i > n$. This forces $\mathrm{H}^i(BG) = 0$, which is absurd since $\mathrm{H}^i(BG) \neq 0$ by our choice of i. Having reached a contradiction, we conclude that the original group G must be unipotent. □

COROLLARY 10.8. *Let k be a field of characteristic 0 and let G be a nontrivial finite group. There is no G-Galois extension S/R, with R a k-ring of finite type, such that S/R specializes to any G-Galois extension of finite type k-rings.*

PROOF. View G as a constant algebraic group over $\operatorname{Spec} k$. The corollary follows from Theorem 10.6 and the equivalence between G-torsors and G-Galois extensions, because G is not unipotent. □

QUESTION 10.9. Let k be any field and let G be an affine algebraic group over k admitting a torsor $E \to X$ that is weakly versal for ftAff/k, and such that X is of finite type over k. Is G unipotent?

11. An Application to Symbol Length

We finish this paper with an application of strongly versal torsors to symbol length of Azumaya algebras over semilocal rings.

Hoobler [**17**] extended the Merkurjev–Suslin Theorem from fields to semilocal rings. As a corollary [**17**, Corollary 3], it follows that if $n \in \mathbb{N}$ and R is a semilocal $\mathbb{Z}[\frac{1}{n}, e^{2\pi i/n}]$-ring, then every Azumaya R-algebra A of period dividing n is Brauer equivalent to the tensor product of *symbol algebras*, i.e.,

$$A \sim_{\mathrm{Br}} (a_1, b_1)_{n,R} \otimes_R \cdots \otimes_R (a_\ell, b_\ell)_{n,R}.$$

Here, as usual, $(a,b)_{n,R} = R\langle x, y \,|\, x^n = a,\ y^n = b,\ xy = \rho_n yx\rangle$ where $a, b \in R^\times$ and ρ_n is the image of $e^{2\pi i/n}$ in R. The maximum possible ℓ required as A ranges over the n-periodic degree-m Azumaya R-algebras is denoted $\ell_R(n,m)$ and called the *(n,m)-symbol length* of R.

When R is a field k, it is well-known that $\ell_k(n,m)$ is finite. Classically, this is shown using the theory of versal (or generic) division algebras, or equivalently, versal torsors over appropriate algebraic groups. By using the strongly versal torsors constructed in Section 8, we can extend this finiteness result to semilocal rings in two ways.

THEOREM 11.1. *Let $n, m \in \mathbb{N}$ and let R be a semilocal $\mathbb{Z}[\frac{1}{n}, e^{2\pi i/n}]$-ring. Then there is $L \in \mathbb{N}$ such that for every semilocal R-ring R' with infinite residue fields, we have $\ell_{R'}(n, m) \leq L$*

THEOREM 11.2. *Let $n, m, s \in \mathbb{N}$. Then there is $L \in \mathbb{N}$ such that for every semilocal $\mathbb{Z}[\frac{1}{n}, e^{2\pi i/n}]$-ring R' having at most s maximal ideals, we have $\ell_{R'}(n, m) \leq L$.*

Using strongly versal torsors is key to the proof of Theorem 11.1. The proof of Theorem 11.2 only makes use of weakly versal torsors, but for group schemes over $\operatorname{Spec} \mathbb{Z}[\frac{1}{n}, e^{2\pi i/n}]$.

PROOF OF THEOREM 11.1. Let $\mathscr{C}_1$ denote the class of affine schemes X for which the subspace X_{cl} is noetherian of dimension ≤ 1. Note that $\mathscr{C}_1$ includes the spectra of all semilocal rings.

By Proposition 9.4, there is a smooth $\mathbb{Z}$-algebra R_0 and an n-periodic degree-m Azumaya algebra A_0 that is strongly versal for $\mathscr{C}_1$ (within the category of all n-periodic degree-m Azumaya algebras). Let $R_1 = R_0 \otimes_{\mathbb{Z}} R$ and $A_1 = A_0 \otimes_{\mathbb{Z}} R$. By Proposition 5.5 (see also Remark 5.3), the R_1-algebra A_1 is an n-periodic degree-m Azumaya R_1-algebra that is strongly versal for $\mathscr{C}_1 / \operatorname{Spec} R$.

Let $\mathfrak{m}$ be a maximal ideal of R. By Proposition 5.6(iv), $R_1/\mathfrak{m}R_1$ is an integral domain, so $\mathfrak{m}R_1 \in \operatorname{Spec} R_1$. Thus, $M = R_1 - \bigcup_{\mathfrak{m} \in \operatorname{Max} R} \mathfrak{m}R_1$ is a multiplicative subset of R_1. Let $\tilde{R}_1 = M^{-1}R$ and $\tilde{A}_1 = M^{-1}A_1$. Since R has only finitely many maximal ideals, the ring $\tilde{R}_1$ has finitely many maximal ideals, namely $\{\mathfrak{m}\tilde{R}_1 \,|\, \mathfrak{m} \in \operatorname{Max} R\}$. Now, by Hoobler's Theorem [**17**, Corollary 3], there is $L \in \mathbb{N}$ and $a_1, b_1, \dots, a_L, b_L \in \tilde{R}_1$ such that

$$\tilde{A}_1 \sim_{\mathrm{Br}} (a_1, b_1)_{n, \tilde{R}_1} \otimes_{\tilde{R}_1} \cdots \otimes_{\tilde{R}_1} (a_L, b_L)_{n, \tilde{R}_1}.$$

By spreading out, we see that there is some $u \in M$ such that $a_1, b_1, \dots, a_L, b_L$ live in $(R_1)_u := \{1, u, u^2, \dots\}^{-1} R_1$ and

$$(A_1)_u \sim_{\mathrm{Br}} (a_1, b_1)_{n, (R_1)_u} \otimes_{(R_1)_u} \cdots \otimes_{(R_1)_u} (a_L, b_L)_{n, (R_1)_u}.$$

Now let R' be a semilocal R-ring as in the theorem and let A' be an n-periodic degree-m Azumaya R'-algebra. By the strong versality of A_1, there is $s \in \mathbb{N}$ and an R-ring morphism $\varphi : R_1 \to R'[t_1, \dots, t_s]$ such that $A_1 \otimes_{R_1} R'[t_1, \dots, t_s] \cong A'[t_1, \dots, t_s]$ as $R'[t_1, \dots, t_s]$-algebras, and the induced map $R_1 \otimes_R R' \to R'[t_1, \dots, t_s]$ is universally injective. The former means that

$$(A_1)_u \otimes_{R_1} R'[t_1, \dots, t_s] \cong A' \otimes_{R'} R'[t_1, \dots, t_s]_{\varphi(u)},$$

and therefore $A' \otimes_{R'} R'[t_1, \dots, t_s]_{\varphi(u)}$ is Brauer-equivalent to the product of L symbol algebras over $R'[t_1, \dots, t_s]_{\varphi(u)}$. As a result, if there exists an R'-ring morphism $R'[t_1, \dots, t_s]_{\varphi(u)} \to R'$, then A' would be the product of L symbol algebras over R'. We will show that such a morphism exists.

We first claim that $(R_1)_u$ is faithfully flat over R. To see this, observe that R_1 is smooth over R, and therefore so is $(R_1)_u$. This means that $\operatorname{Spec}(R_1)_u \to \operatorname{Spec} R$ is flat and of finite presentation, hence open. Since $u \notin \mathfrak{m}R_1$ for every $\mathfrak{m} \in \operatorname{Max} R$, the image of $\operatorname{Spec}(R_1)_u \to \operatorname{Spec} R$ contains $\operatorname{Max} R$. As this map is also open, it follows that $\operatorname{Spec}(R_1)_u \to \operatorname{Spec} R$ is surjective, hence faithfully flat.

Next, we claim that $\varphi(u) \notin \mathfrak{m}'[t_1, \dots, t_s]$ for every $\mathfrak{m}' \in \operatorname{Spec} R'$. Indeed, since $R \to (R_1)_u$ is faithfully flat, so is $R' \to (R_1)_u \otimes_R R'$, and thus $((R_1)_u \otimes_R R')\mathfrak{m}' \neq (R_1)_u \otimes_R R'$. Since $R_1 \otimes_R R' \to R'[t_1, \dots, t_s]$ is universally injective, the same holds for $(R_1)_u \otimes_R R' \to R'[t_1, \dots, t_s]_{\varphi(u)}$. Therefore, the latter map remains injective after tensoring with $R'/\mathfrak{m}'$ over R'. Since $((R_1)_u \otimes_R R')\mathfrak{m}' \neq (R_1)_u \otimes_R R'$, this means that $R'[t_1,\dots,t_s]_{\varphi(u)} \cdot \mathfrak{m}' \neq R'[t_1,\dots,t_s]_{\varphi(u)}$, so $\varphi(u) \notin \mathfrak{m}'[t_1,\dots,t_s]$.

To finish, put $f = \varphi(u) \in R'[t_1,\dots,t_s]$. Then the image of f in $(R'/\mathfrak{m}')[t_1,\dots,t_s]$ is nonzero for every $\mathfrak{m}' \in \operatorname{Max} R'$. Our assumption that $R'/\mathfrak{m}'$ is infinite for all $\mathfrak{m}' \in \operatorname{Max} R'$ and the Chinese Remainder Theorem imply that there exist $r_1,\dots,r_s \in R'$ such that $f(r_1,\dots,r_s) \notin \mathfrak{m}'$ for every $\mathfrak{m}' \in \operatorname{Max} R'$, or rather, $f(r_1,\dots,r_s) \in R'^{\times}$. The R'-algebra morphism $R'[t_1,\dots,t_s] \to R'$ which specializes each t_i to r_i therefore gives rise to the desired R'-algebra morphism $R'[t_1,\dots,t_s]_{\varphi(u)} \to R'$. □

Proof of Theorem 11.2. Put $R = \mathbb{Z}[\frac{1}{n}, e^{2\pi i/n}]$ and then define R_1 and A_1 as in the proof of Theorem 11.1. Then A_1 is an n-periodic degree-m Azumaya R_1-algebra that is weakly versal for all semilocal R-rings.

Let P be a set of prime ideals in R_1 with $|P| \leq s$. As in the proof of Theorem 11.1, by localizing at the multiplicative set $R_1 - \bigcup_{\mathfrak{p}\in P} \mathfrak{p}$, we see that there is some $u = u(P) \in R_1 - \bigcup_{\mathfrak{p}\in P} \mathfrak{p}$ and $a_1, b_1, \dots, a_{\ell(P)}, b_{\ell(P)} \in (R_1)_u$, such that

$$(3) \quad (A_1)_{u(P)} \sim_{\mathrm{Br}} (a_1,b_1)_{n,(R_1)_{u(P)}} \otimes_{(R_1)_{u(P)}} \cdots \otimes_{(R_1)_{u(P)}} (a_{\ell(P)}, b_{\ell(P)})_{n,(R_1)_{u(P)}}.$$

Put $U(P) = \{\mathfrak{p} \in \operatorname{Spec} R_1 : u(P) \notin \mathfrak{p}\}$. By construction, the collection $\{U(P)^s \mid P \subseteq \operatorname{Spec} R_1,\ |P| \leq s\}$ is an open covering of $(\operatorname{Spec} R_1)^s$ endowed with the product topology. Since $\operatorname{Spec} R_1$ is quasi-compact, Tychonoff's theorem tells us that there are $P_1,\dots,P_t \subseteq \operatorname{Spec} R_1$ such that $(\operatorname{Spec} R_1)^s = \bigcup_{i=1}^t U(P_i)^s$. Otherwise said, for every $Q \subseteq \operatorname{Spec} R$ with $|Q| \leq s$, there is some $i \in \{1,\dots,t\}$ such that $Q \subseteq U(P_i)$, or equivalently, $u_i := u(P_i) \notin \bigcup_{\mathfrak{p}\in Q} \mathfrak{p}$.

Now let A' be an n-periodic degree-m Azumaya over a semilocal R-ring R' having at most s maximal ideals. By the construction of A_1, there is a ring homomorphism $\varphi : R_1 \to R'$ such that $A' \cong A_1 \otimes_{R_1} R'$. Let $Q = \{\varphi^{-1}(\mathfrak{m}') \mid \mathfrak{m}' \in \operatorname{Max} R'\}$. Then Q is a subset of $\operatorname{Spec} R_1$ of cardinality at most s. Thus, there is $i \in \{1,\dots,t\}$ such that $u_i \notin \bigcup_{\mathfrak{p}\in Q} \mathfrak{p}$. This means that $\varphi(u_i) \in R' - \bigcup_{\mathfrak{m}'\in \operatorname{Max} R'} \mathfrak{m}' = R'^{\times}$, so φ extends to a ring homomorphism $\psi : (R_1)_{u_i} \to R'$. Taking $P = P_i$ in (3) and tensoring both sides with R' now shows that the symbol length of A' is at most $\ell(P_i)$. As this holds for every choice of A', we conclude that $\ell_{R'}(n,m) \leq \max\{\ell(P_1),\dots,\ell(P_t)\}$. Take the right hand side to be the required number L. □

The following corollary of Theorem 11.2 is known to experts. It is also well-known if the characteristic of the field F is fixed in advance.

Corollary 11.3. *For every $n, m \in \mathbb{N}$, there is $L \in \mathbb{N}$ such that $\ell_F(n,m) \leq L$ for every field F containing a primitive n-th root of unity.*

Question 11.4. Does Theorem 11.1 continue to hold if we drop the requirement that the residue fields of R' are infinite? Does Theorem 11.2 continue to hold if we do not bound from above the number of maximal ideals?

Acknowledgments

We are grateful to A. Chapman, Z. Reichstein, D. Saltman and B. Williams for comments and suggestions regarding this work. We also thank A. Duncan for an inspiring conversation.

References

[1] S. A. Amitsur, *On central division algebras*, Israel J. Math. **12** (1972), 408–420, DOI 10.1007/BF02764632. MR318216

[2] S. Anantharaman, *Schémas en groupes, espaces homogènes et espaces algébriques sur une base de dimension 1* (French), Sur les groupes algébriques, Supplément au Bull. Soc. Math. France, Tome 101, Soc. Math. France, Paris, 1973, pp. 5–79, DOI 10.24033/msmf.109. MR335524

[3] A. Auel, U. A. First, and B. Williams, *Azumaya algebras without involution*, J. Eur. Math. Soc. (JEMS) **21** (2019), no. 3, 897–921, DOI 10.4171/JEMS/855. MR3908769

[4] H. Bass, *K-theory and stable algebra*, Inst. Hautes Études Sci. Publ. Math. **22** (1964), 5–60. MR174604

[5] M. Brion, *Some structure theorems for algebraic groups*, Algebraic groups: structure and actions, Proc. Sympos. Pure Math., vol. 94, Amer. Math. Soc., Providence, RI, 2017, pp. 53–126, DOI 10.1090/pspum/094/04. MR3645068

[6] M. Demazure and A. Grothendieck, *Schémas en groupes. III: Structure des schémas en groupes réductifs* (French), Lecture Notes in Mathematics, Vol. 153, Springer-Verlag, Berlin-New York, 1970. Séminaire de Géométrie Algébrique du Bois Marie 1962/64 (SGA 3). MR274460

[7] A. Duncan and Z. Reichstein, *Versality of algebraic group actions and rational points on twisted varieties*, J. Algebraic Geom. **24** (2015), no. 3, 499–530, DOI 10.1090/S1056-3911-2015-00644-0. MR3344763

[8] U. A. First, *The Brauer class of an Azumaya algebra with an orthogonal involution admitting a unitary element of reduced norm -1 is trivial*, `arXiv:2201.04921`, 2022.

[9] U. A. First and Z. Reichstein, *On the number of generators of an algebra* (English, with English and French summaries), C. R. Math. Acad. Sci. Paris **355** (2017), no. 1, 5–9, DOI 10.1016/j.crma.2016.11.015. MR3590278

[10] U. A. First, Z. Reichstein, and B. Williams, *On the number of generators of an algebra over a commutative ring*, Trans. Amer. Math. Soc. **375** (2022), no. 10, 7277–7321, DOI 10.1090/tran/8720. MR4491427

[11] P. Fleischmann and C. Woodcock, *Non-linear group actions with polynomial invariant rings and a structure theorem for modular Galois extensions*, Proc. Lond. Math. Soc. (3) **103** (2011), no. 5, 826–846, DOI 10.1112/plms/pdr016. MR2852290

[12] P. Fleischmann and C. Woodcock, *Free actions of p-groups on affine varieties in characteristic p*, Math. Proc. Cambridge Philos. Soc. **165** (2018), no. 1, 109–135, DOI 10.1017/S0305004117000317. MR3811548

[13] T. J. Ford, *Separable algebras*, Graduate Studies in Mathematics, vol. 183, American Mathematical Society, Providence, RI, 2017, DOI 10.1090/gsm/183. MR3618889

[14] O. Forster, *Über die Anzahl der Erzeugenden eines Ideals in einem Noetherschen Ring* (German), Math. Z. **84** (1964), 80–87, DOI 10.1007/BF01112211. MR163932

[15] P. Gille, *When is a reductive group scheme linear?*, Michigan Math. J. **72** (2022), 439–448, DOI 10.1307/mmj/20217208. MR4460258

[16] A. Grothendieck, *Éléments de géométrie algébrique. IV. Étude locale des schémas et des morphismes de schémas IV* (French), Inst. Hautes Études Sci. Publ. Math. **32** (1967), 361 pp. MR238860

[17] R. T. Hoobler, *The Merkuriev-Suslin theorem for any semi-local ring*, J. Pure Appl. Algebra **207** (2006), no. 3, 537–552, DOI 10.1016/j.jpaa.2005.10.020. MR2265538

[18] J. P. Jouanolou, *Une suite exacte de Mayer-Vietoris en K-théorie algébrique* (French), Algebraic K-theory, I: Higher K-theories (Proc. Conf., Battelle Memorial Inst., Seattle, Wash., 1972), Lecture Notes in Math., Vol. 341, Springer, Berlin-New York, 1973, pp. 293–316. MR409476

[19] S. L. Kleiman, *The transversality of a general translate*, Compositio Math. **28** (1974), 287–297. MR360616

[20] M.-A. Knus, *Quadratic and Hermitian forms over rings*, Grundlehren der mathematischen Wissenschaften [Fundamental Principles of Mathematical Sciences], vol. 294, Springer-Verlag, Berlin, 1991, DOI 10.1007/978-3-642-75401-2. MR1096299

[21] S. Lojasiewicz, *Triangulation of semi-analytic sets*, Ann. Scuola Norm. Sup. Pisa Cl. Sci. (3) **18** (1964), 449–474. MR173265

[22] J. S. Milne, *Algebraic groups: The theory of group schemes of finite type over a field*, Cambridge Studies in Advanced Mathematics, vol. 170, Cambridge University Press, Cambridge, 2017, DOI 10.1017/9781316711736. MR3729270

[23] J. S. Milne, *Étale cohomology*, Princeton Mathematical Series, No. 33, Princeton University Press, Princeton, NJ, 1980. MR559531

[24] M. Raynaud, *Solution d'un problème universal relatif aux modules projectifs de type fini*, C. R. Acad. Sci. Paris **260** (1965), 4391–4394. MR178027

[25] R. W. Richardson, *Affine coset spaces of reductive algebraic groups*, Bull. London Math. Soc. **9** (1977), no. 1, 38–41, DOI 10.1112/blms/9.1.38. MR437549

[26] D. J. Saltman, *Noncrossed product p-algebras and Galois p-extensions*, J. Algebra **52** (1978), no. 2, 302–314, DOI 10.1016/0021-8693(78)90240-5. MR480640

[27] D. J. Saltman, *Lectures on division algebras*, CBMS Regional Conference Series in Mathematics, vol. 94, Published by American Mathematical Society, Providence, RI; on behalf of Conference Board of the Mathematical Sciences, Washington, DC, 1999, DOI 10.1090/cbms/094. MR1692654

[28] David J. Saltman, *Mixed characteristic cyclic matters*, arXiv:2212.03713, 2022

[29] J.-P. Serre, *Cohomological invariants, Witt invariants, and trace forms*, Cohomological invariants in Galois cohomology, Univ. Lecture Ser., vol. 28, Amer. Math. Soc., Providence, RI, 2003, pp. 1–100. Notes by Skip Garibaldi. MR1999384

[30] C. S. Seshadri, *Geometric reductivity over arbitrary base*, Advances in Math. **26** (1977), no. 3, 225–274, DOI 10.1016/0001-8708(77)90041-X. MR466154

[31] *The Stacks Project Authors*, Stacks Project, https://stacks.math.columbia.edu, 2022.

[32] R. G. Swan, *The number of generators of a module*, Math. Z. **102** (1967), 318–322, DOI 10.1007/BF01110912. MR218347

[33] R. W. Thomason, *Equivariant resolution, linearization, and Hilbert's fourteenth problem over arbitrary base schemes*, Adv. in Math. **65** (1987), no. 1, 16–34, DOI 10.1016/0001-8708(87)90016-8. MR893468

[34] B. Totaro, *The Chow ring of a classifying space*, Algebraic K-theory (Seattle, WA, 1997), Proc. Sympos. Pure Math., vol. 67, Amer. Math. Soc., Providence, RI, 1999, pp. 249–281, DOI 10.1090/pspum/067/1743244. MR1743244

Department of Mathematics, University of Haifa, Haifa, 3498838, Israel
Email address: uriya.first@gmail.com

Contemporary Mathematics
Volume **800**, 2024
https://doi.org/10.1090/conm/800/16055

Optimal representability results of PI-algebras: nilpotency, growth and chain conditions

Be'eri Greenfeld, Louis Rowen, and Lance Small

ABSTRACT. We investigate representability results for PI-algebras satisfying various chain conditions and examine their optimality. We show that a Noetherian PI-algebra, whose quotient by the radical is finite over its center, is representable. We derive several applications and give a counterexample to this representability result when Noetherianity is relaxed to ACC on ideals, even within the class of affine algebras of linear growth. This example has radical cubed zero, and we show that if the radical squared is zero then such a counterexample cannot exist. We then modify our construction to yield a non-representable semiprimary PI-algebra with radical cubed zero, proving the tightness of a result of Amitsur, Rowen and Small, which was left open by Rowen and Small. We conclude with a discussion of the remaining open problems.

1. Introduction

One of the major problems in the theory of algebras satisfying polynomial identities (PI-algebras) is to determine whether a given PI-algebra is representable, namely, whether it embeds into a matrix ring over a field, or more generally a commutative ring[1]. For affine algebras, that is algebras which are finitely generated over a field, representability over a commutative ring and representability over a field coincide. For a comprehensive source for background, examples and historical references, see [**25**].

Every representable affine algebra satisfies a polynomial identity (the Amitsur-Levitzki standard identity), but not vice versa. The importance of representability is both *conceptual*, providing a reason for why a given algebra satisfies a polynomial identity, and *practical*, since representable algebras satisfy all the identities of matrices, which, though not fully understood, are more tangible than abstract polynomial identities. In particular, representable algebras satisfy the same identities as their Zariski closures, which are concrete objects studied and classified (see [**7**]).

2020 *Mathematics Subject Classification.* Primary 16R20, 16P40; Secondary 16N40, 16P90.
Key words and phrases. PI-algebra, Noetherian algebra, representable algebra, universal derivations.
This work was supported by the U.S.-Israel Binational Science Foundation (grant no. 2010149). The research of the second author was supported by the ISF grant 1994/20 and by the Anschel Pfeffer Chair.

[1]To distinguish between the two cases, the latter case is called 'weakly representable' (after [**25**]) or simply 'representable over a commutative ring'.

The representability question is strongly connected to other important problems in PI-theory such as Specht's problem (see [**6**]).

Bergman [**8**] gave an example of a non-representable finite ring; this ring does not contain a field, leading us to assume throughout that our rings are algebras over a field, when aiming to obtain any representability result. Throughout the paper, R is an algebra over a field F. By *radical* we refer to the nilradical, denoted as N, which is nilpotent if R is affine PI (by the Braun-Kemer-Razmyslov Theorem) or satisfies ACC on ideals. By convention, modules, annihilators, chain conditions such as Noetherianity etc. are on the left (unless otherwise mentioned).

A classical argument of Lewin [**17**] shows that there exist uncountably many affine PI-algebras which are not representable. Semiprime PI-algebras are representable [**5**, Corollary 1.6.10], naturally leading to the question of to what extent representability is affected by the nilpotency index of the radical.

There is a connection between representability and ascending chain conditions. Any algebra representable over a field satisfies ACC on annihilators of subsets. While the converse is not true in general (see [**14**]), Anan'in proved that every affine Noetherian PI-algebra is representable [**2**, Theorem (a)]; this was generalized by Rowen and Small [**25**, Theorem C] to the case of Noetherian PI-algebras which are finitely generated algebras over their centers.

How about non-affine Noetherian PI-algebras? This is an open question (even the Artinian case, see [**5**, Question 5, p. 388]). Our aim is to present both positive representability results and counterexamples, emphasizing the delicacy and sharpness of the existing state of the art results. We start with the following representability criterion for Noetherian algebras:

THEOREM 1.1. *Let R be a Noetherian algebra with radical N. If R contains a commutative subalgebra C such that R/N is finite (as a module) over the image of C modulo N, then R is representable.*

In fact, we prove a more general result, whose formulation is slightly technical (Theorem 2.3); virtually all of the left Noetherian PI-examples in the literature satisfy Theorem 2.3 (one notable exception is given in [**26**], which is representable a fortiori). Notice that Theorem 1.1 generalizes the Wehrfritz-Beidar Theorem (that a Noetherian algebra which is module finite over its center is representable).

As an application, a Noetherian PI-algebra whose quotient by the radical is affine of linear growth is representable (Corollary 2.2). This turns out to be sharp from the point of view of chain conditions, namely, relaxing Noetherianity to ACC(ideals) – there exist examples of non-representable affine PI-algebras of linear growth, satisfying ACC on ideals. In these examples the radicals cubed are zero (see Section 5). But such examples are again optimal, this time from the point of view of the nilpotency index of the radical (omitting any assumption on chain conditions):

PROPOSITION 1.2. *Let R be an affine PI-algebra whose radical squared is zero. If the quotient of R modulo its radical is finite over its center (in particular, if R has linear growth), then R is representable.*

Interestingly, our affine non-representable examples can be carried over to the following setting. We say that a ring R is *semiprimary* if its quotient modulo its radical is semisimple Artinian. Any Artinian ring is semiprimary, but not conversely. While the representability question of Artinian PI-algebras is open, the case of radical squared zero turns out to be feasible. Indeed, Amitsur, Rowen and Small proved that a semiprimary PI-algebra with radical squared zero is weakly representable; this appears in [**25**, Section 6].

In [**25**], the authors leave open whether arbitrary semiprimary PI-algebras are weakly representable, adding a warning on the (potential) difference with the square zero case. We resolve the general case, thereby proving that the Amitsur-Rowen-Small Theorem is best possible in the sharpest possible sense:

THEOREM 1.3. *Over any field, there exists a semiprimary PI-algebra R which is not weakly representable. Moreover, N^2 is a 1-dimensional vector space and $N^3 = (0)$.*

Along the way, we prove additional related representability results and review interesting related examples.

2. Representability of Noetherian algebras

2.1. Lifting representability.

LEMMA 2.1. *Let R be an algebra which is a finitely generated left module over a commutative Noetherian subalgebra C. Then R is representable.*

PROOF. The algebra R^{op} acts on itself by right multiplication, and this action is C-equivariant. Therefore:

$$R^{op} \hookrightarrow \mathrm{End}_C(R),$$

and by [**25**, Corollary 3.4] it is representable; considering the transpose of this representation, we obtain that R is representable as well. □

PROOF OF THEOREM 1.1. Let d be the nilpotency index of N. Consider R as a (left) C-module via the natural action of C, being a subalgebra of R.

Since R/N is finite over $\overline{C}$, the reduction of C modulo N, we can write:

$$R/N = \overline{C}\bar{r}_1 + \cdots + \overline{C}\bar{r}_q.$$

Pick arbitrary lifts $r_1, \ldots, r_q \in R$ of $\bar{r}_1, \ldots, \bar{r}_q$ modulo N. Therefore:

$$R = Cr_1 + \cdots + Cr_q + N.$$

We now prove by induction on e that N^{d-e} is a finite C-module. Fix $e = 1$. Since R is Noetherian, we can write N^{d-1} as a finite R-module:

$$\begin{aligned} N^{d-1} &= \sum_{i=1}^{k} Ru_i \\ &= \sum_{i=1}^{k} (Cr_1 + \cdots + Cr_q + N)u_i \\ &= \sum_{i=1}^{k} (Cr_1 + \cdots + Cr_q)u_i \end{aligned}$$

where the third equality follows since $Nu_1, \ldots, Nu_k \subseteq N^d = (0)$.

Now assume that the claim holds for $e' < e$. Again, since R is Noetherian we can write N^{d-e} as a finitely generated R-module:

$$\begin{aligned} N^{d-e} &= \sum_{i=1}^{l} Rv_i \\ &= \sum_{i=1}^{l} (Cr_1 + \cdots + Cr_q + N)v_i \\ &= \sum_{i=1}^{l} (Cr_1 + \cdots + Cr_q)v_i + N^{d-e+1}. \end{aligned}$$

But by the induction hypothesis, N^{d-e+1} is a finite C-module, so N^{d-e} is finite as well. In particular, we get that N itself is a finite C-module (for $e = d-1$), and therefore so is R. By the Formanek-Jategaonkar Theorem [**11**], it follows that C is a Noetherian ring and hence R is a Noetherian C-module. We are done by Lemma 2.1. □

The following is an immediate application of Theorem 1.1:

COROLLARY 2.2. *Suppose R is a Noetherian algebra over a field, and R/N is affine of linear growth. Then R is representable.*

PROOF. Since R/N is semiprime affine of linear growth, it is a finite module over a polynomial subring $F[x]$ by the Small-Stafford-Warfield Theorem [**27**]. Pick an arbitrary lift $r \in R$ of x. Then $C = F[r]$ satisfies the conditions of Theorem 1.1. □

We now give a more general (and more technical) form of Theorem 1.1:

THEOREM 2.3. *Let R be a Noetherian algebra over a field, which contains a subalgebra $W \subseteq R$ such that:*

- *R/N is a finitely generated module over $\overline{W}$, the reduction of W modulo N;*
- *W satisfies ACC on ideals;*
- *there exists an element $c \in Z(W)$ which is regular in R and $W[c^{-1}]$ is finite over its center.*

Then R is representable.

PROOF. Notice that the proof of Theorem 1.1 yields mutatis mutandis that R is a finite module over W.

Since c is regular, we can localize $M := R \otimes_W W[c^{-1}]$ (as a module), which is finite over $W[c^{-1}]$. Let C denote the center of W. Then, since W (and hence $W[c^{-1}]$) has ACC on ideals, it follows that $C[c^{-1}]$ is Noetherian by the Eakin-Formanek Theorem [**24**, Theorem 5.1.12]. In addition, $W[c^{-1}]$ is finite over $C[c^{-1}]$ and therefore M is finite over $C[c^{-1}]$ as well.

But then $\mathrm{End}_{C[c^{-1}]}M$ is representable by [**25**, Corollary 3.4]. Now R acts faithfully (since c is regular in R) on M, and this action is $C[c^{-1}]$-equivariant. Therefore R embeds into $\mathrm{End}_{C[c^{-1}]}M$, and is thus representable. □

3. Representability of algebras with radical squared zero

Since semiprime PI-algebras are representable and the radical of any affine or Noetherian PI-algebra is nilpotent, it is natural to consider representability in the presence of restrictions on the nilpotency index of the radical. In this section, we encounter the situation of algebras whose radicals squared are zero, and its special consequences. The squared zero case turns out to differ from the case of higher nilpotency indices, as demonstrated in the next sections.

QUESTION 3.1. Let R be an affine PI-algebra with $N^2 = (0)$. Is R representable?

It was conjectured by Bezushchak, Petravshuk and Zelmanov that the answer to this question is affirmative with potential interesting applications to automorphism groups of PI-algebras (see [**9**, p. 8]). We are not aware of a counterexample to Question 3.1 even if the the affineness assumption is removed, or replaced by Noetherianity.

Interestingly, Bergman's example [**8**] (see also [**10**, Example 3.16]) of a finite non-representable ring has radical squared zero (and, moreover, the quotient by the radical is an $\mathbb{F}_p$-algebra; the ring itself is not an algebra over a field, though).

The following is an extension of Lewin's result, that an algebra whose ideal generated by all commutators is squared zero is representable over a commutative ring [**18**, Theorem 14].

PROOF OF PROPOSITION 1.2. Let R be as in the statement, N its radical and C the center of R/N. As developed in [**18**] there exists a universal $R/N - R/N$-bimodule T, together with a universal derivation $\delta\colon R \to T$, fitting into the diagram:

$$\theta\colon R \to R' := \begin{pmatrix} R/N & T \\ 0 & R/N \end{pmatrix}$$

with the map:

$$\theta\colon r \mapsto \begin{pmatrix} \bar{r} & \delta(r) \\ 0 & \bar{r} \end{pmatrix}$$

The pair (T, δ) is universal among $R/N - R/N$-bimodules with derivations, with respect to morphisms which are compatible with the corresponding maps θ. By [**18**, Theorem 2], the kernel of the homomorphism θ attached to the universal bimodule is $N^2 = (0)$, namely, θ is injective. By [**18**, Lemma 3(ii)], T is a finitely generated $R/N - R/N$-bimodule, equivalently, as an $R/N \otimes_F (R/N)^{op}$-module. Notice that:

$$R' = \begin{pmatrix} R/N & T \\ 0 & R/N \end{pmatrix} \subseteq \begin{pmatrix} R/N \otimes_F C & T \\ 0 & R/N \end{pmatrix} =: \widehat{R}$$

where multiplication in $\widehat{R}$ is given by:

$$\begin{pmatrix} \bar{r} \otimes c & t \\ 0 & \bar{s} \end{pmatrix} \begin{pmatrix} \bar{r}' \otimes c' & t' \\ 0 & \bar{s}' \end{pmatrix} = \begin{pmatrix} \bar{r}\bar{r}' \otimes cc' & \bar{r}t'c + t\bar{s}' \\ 0 & \bar{s}\bar{s}' \end{pmatrix}$$

(where $\bar{r}, \bar{r}', \bar{s}, \bar{s}' \in R/N, c, c' \in C, t, t' \in T$) making it a well-defined associative algebra, since C is the center of R/N:

$$\begin{aligned}
&\left(\begin{pmatrix} \bar{r}\otimes c & t \\ 0 & \bar{s} \end{pmatrix}\begin{pmatrix} \bar{r}'\otimes c' & t' \\ 0 & \bar{s}' \end{pmatrix}\right)\begin{pmatrix} \bar{r}''\otimes c'' & t'' \\ 0 & \bar{s}'' \end{pmatrix} \\
&= \begin{pmatrix} \bar{r}\bar{r}'\otimes cc' & \bar{r}t'c + t\bar{s}' \\ 0 & \bar{s}\bar{s}' \end{pmatrix}\begin{pmatrix} \bar{r}''\otimes c'' & t'' \\ 0 & \bar{s}'' \end{pmatrix} \\
&= \begin{pmatrix} \bar{r}\bar{r}'\bar{r}''\otimes cc'c'' & \bar{r}\bar{r}'t''cc' + \bar{r}t'c\bar{s}'' + t\bar{s}'\bar{s}'' \\ 0 & \bar{s}\bar{s}'\bar{s}'' \end{pmatrix} \\
&= \begin{pmatrix} \bar{r}\otimes c & t \\ 0 & \bar{s} \end{pmatrix}\begin{pmatrix} \bar{r}'\bar{r}''\otimes c'c'' & \bar{r}'t''c' + t'\bar{s}'' \\ 0 & \bar{s}'\bar{s}'' \end{pmatrix} \\
&= \begin{pmatrix} \bar{r}\otimes c & t \\ 0 & \bar{s} \end{pmatrix}\left(\begin{pmatrix} \bar{r}'\otimes c' & t' \\ 0 & \bar{s}' \end{pmatrix}\begin{pmatrix} \bar{r}''\otimes c'' & t'' \\ 0 & \bar{s}'' \end{pmatrix}\right).
\end{aligned}$$

The benefit of $\widehat{R}$ is that it is a finitely generated left module over the following affine commutative subalgebra:

$$\begin{pmatrix} C\otimes_F C & 0 \\ 0 & C \end{pmatrix}.$$

This follows since R/N is a finitely generated module over C, and T is a finitely generated left module over $R/N \otimes_F (R/N)^{op}$, hence over $C \otimes_F C$. Moreover, C is affine (hence Noetherian) by the Montgomery-Small version of the Artin-Tate Lemma [**22**, Proposition 2]. It follows that $\widehat{R}$ is representable by Lemma 2.1; hence R, which embeds into $\widehat{R}$, is representable as well. □

COROLLARY 3.2. *Let R be an affine PI-algebra of linear growth whose radical squared is zero. Then R is representable.*

PROOF. This is evident from Proposition 1.2 together with the Small-Stafford-Warfield Theorem which ensures that an affine semiprime algebra of linear growth is finite over its center [**27**]. □

Notice that Corollary 3.2 also demonstrates the sharpness of Irving's examples [**14**], which are affine PI of linear growth and whose radicals cubed are zero.

4. Obstacles to representability

As a preparation for the next sections, in which we construct non-representable algebras showing the tightness of the representability results from the previous sections, let us consider some common 'obstacles' to non-representability.

If R is representable, then it satisfies the identities of $M_n(\mathbb{Z})$ for some n. Beyond that, which methods (which do not explicitly involve non-identities) can one use to show non-representability? Let us review three such tools.

4.1. Chains of annihilators. Any subring of a left and right Noetherian ring satisfies ACC on left annihilators and on right annihilators, and thus so does any affine algebra which is representable over a commutative ring. This fact plays a key role in establishing non-representability of many known examples. Two further examples deserve to be mentioned in this context:

- Irving [**14**] constructed affine PI-algebras which satisfy ACC on annihilators, yet are non-representable. Non-representability is established by the existence of *arbitrarily long* chains of annihilators, which is again impossible in any representable algebra.
- To complete the picture, Irving and Small [**15**] gave an example of an affine PI-algebra with a uniform bound on the lengths of chains of annihilators, which is still non-representable.

4.2. Growth. It is known (by means of Shirshov's height functions) that affine PI-algebras have polynomially bounded growth. Equivalently, they have finite Gel'fand-Kirillov dimension, which is defined as follows. Let V be a finite-dimensional generating subspace for R (i.e., $R = F\langle V\rangle$), then:

$$\mathrm{GKdim}(R) = \limsup_{n\to\infty} \frac{\log \dim_F V^n}{\log n}$$

which is independent of the choice of V. For more on growth of algebras and the GK-dimension, see [**16**].

By Markov's Theorem [**21**], the GK-dimension of any affine representable algebra is an *integer*. Thus if an algebra contains an affine subalgebra of non-integral GK-dimension, it cannot be representable.

Examples of affine graded (in fact, monomial) PI-algebras with non-integral GK-dimension were given by Warfield (see [**16**, Theorem 2.9]). Specifically,

$$F\langle x,y\rangle / I \quad \text{where} \quad I = \langle x\rangle^3 + \langle xy^i x \mid i \text{ is not a square}\rangle$$

satisfies the identity $[X_1,Y_1][X_2,Y_2][X_3,Y_3] = 0$ and has GK-dimension 2.5.

4.3. Residually finite and subdirectly irreducible rings. A theorem of Mal'cev [**20**] asserts that an affine algebra R over an infinite field F is representable if and only if for some $n \in \mathbb{N}$, the intersection of all ideals $I \triangleleft R$ such that $\dim_F R/I \leq n$ is zero. Indeed, if this intersection is zero then R embeds as a subdirect product of the corresponding at most n-dimensional quotients. Each such quotient embeds into $M_n(F)$, and therefore R embeds into $M_n(C)$ (for C a cartesian power of F). Conversely, an affine representable F-algebra R is representable over an affine commutative algebra C. But C has a set of ideals $\{I_i\}_{i\in\mathcal{S}}$ such that $\dim_F C/I_i \leq d$ for some uniform d and $\bigcap_{i\in\mathcal{S}} I_i = (0)$ (by a routine argument: one reduces to finitely many irreducible rings, uses Noether normalization and generic freeness to lift 1-codimensional ideals from a polynomial subalgebra). It follows that $M_n(I_i)\cap R \triangleleft R$ is at most dn^2-codimensional and $\bigcap_{i\in\mathcal{S}} (M_n(I_i)\cap R) = (0)$, and Mal'cev's theorem follows.

Notice that over finite fields, representability does not imply that the intersection of all at most n-codimensional ideals is zero for some n: in $R = \mathbb{F}_q[t]$, the intersection of all at most n-codimensional ideals contains the element $\prod_{0\leq i<j\leq q^n}(t^i - t^j)$. Nevertheless, if R is an affine representable algebra over a finite field then it is still residually finite-dimensional by the above argument combined with the fact that an affine commutative algebra over a field is always residually finite-dimensional.

In particular, if R is an affine F-algebra and:

$$\bigcap_{\substack{I\triangleleft R \\ \dim_F R/I<\infty}} I \neq (0)$$

then R is not representable. An extreme situation of this phenomenon is when the intersection of *all* non-zero ideals is non-zero; an algebra with this property is called 'subdirectly irreducible'.

- Irving [**14**, Theorem 2] showed that $F\langle x,y\rangle / \langle x^2, yxy-x\rangle$ is a PI-algebra of linear growth which is not representable. As observed by Irving, the image of xyx in the quotient ring is a non-zero element lying in the intersection of all non-zero ideals.
- Warfield's construction of a monomial algebra with non-integral GK-dimension, mentioned in the previous subsection, is not representable but, being non-negatively graded, it is residually finite-dimensional.

5. Non-representability: radical cubed zero

While affine Noetherian PI algebras are representable by Anan'in's Theorem, if one weakens the Noetherianity assumption to ACC(ideals), counter-examples exist, including some of linear growth. Let us review two such examples, which have already appeared in the past in different forms.

Let A be an affine F-algebra and $I \leq A$ a right ideal. Then the ring

$$R = \begin{pmatrix} F & A/I \\ 0 & A \end{pmatrix}$$

is an affine F-algebra. If A is PI then R, being a homomorphic image of a subring of $M_2(A)$, is also PI. Moreover, if A is a ring of the form

$$A = \begin{pmatrix} C_1 & C_2 \\ 0 & C_3 \end{pmatrix}$$

where C_1, C_2, C_3 are affine integral domains and $C_1, C_3 \subseteq C_2$, then the radical of R is cubed zero. Some variants built upon this construction have led to interesting counterexamples. Let us briefly introduce two of them:

- Let $C_1 = C_2 = F[t,t^{-1}]$, $C_3 = F[t]$, so $A = \begin{pmatrix} F[t,t^{-1}] & F[t,t^{-1}] \\ 0 & F[t] \end{pmatrix}$. Consider $I = \begin{pmatrix} 0 & F[t] \\ 0 & 0 \end{pmatrix}$, a right ideal of A. We claim that R is non-residually finite-dimensional (and hence non-representable). For $a \in A$, denote $\overline{a} = a + I$ the corresponding coset modulo I. Let

  $$e = \begin{pmatrix} 0 & 0 \\ 0 & \begin{pmatrix} 0 & 0 \\ 0 & 1 \end{pmatrix} \end{pmatrix} \in R.$$

  Let $J \triangleleft R$ be a finite-codimensional ideal. We have a surjective linear map:

  $$R/J \twoheadrightarrow e(R/J)e = \begin{pmatrix} 0 & 0 \\ 0 & \begin{pmatrix} 0 & 0 \\ 0 & F[t] \end{pmatrix} \end{pmatrix} \Big/ eJe$$

  and since $\dim_F R/J < \infty$, we deduce that $eJe \neq (0)$. Pick

  $$\begin{pmatrix} 0 & 0 \\ 0 & \begin{pmatrix} 0 & 0 \\ 0 & p(t) \end{pmatrix} \end{pmatrix} \in eJe \subseteq J$$

for some $p(t) \in F[t] \setminus \{0\}$. Write $p(t) = c_i t^i + \cdots + c_j t^j$ where $0 \leq i \leq j$ and $c_i \neq 0$. Then $c_i^{-1} t^{-i-1} p(t) = t^{-1} + f(t)$ for some $f(t) \in F[t]$. Now:

$$\begin{pmatrix} 0 & 0 \\ 0 & \begin{pmatrix} 0 & t^{-1}+f(t) \\ 0 & 0 \end{pmatrix} \end{pmatrix} = \begin{pmatrix} 0 & 0 \\ 0 & \begin{pmatrix} 0 & c_i^{-1}t^{-i-1} \\ 0 & 0 \end{pmatrix} \end{pmatrix} \cdot \begin{pmatrix} 0 & 0 \\ 0 & \begin{pmatrix} 0 & 0 \\ 0 & p(t) \end{pmatrix} \end{pmatrix} \in J,$$

so:

$$\begin{pmatrix} 0 & \overline{\begin{pmatrix} 0 & t^{-1} \\ 0 & 0 \end{pmatrix}} \\ 0 & 0 \end{pmatrix} = \begin{pmatrix} 0 & \overline{\begin{pmatrix} 1 & 0 \\ 0 & 1 \end{pmatrix}} \\ 0 & 0 \end{pmatrix} \cdot \begin{pmatrix} 0 & 0 \\ 0 & \begin{pmatrix} 0 & t^{-1}+f(t) \\ 0 & 0 \end{pmatrix} \end{pmatrix} \in J.$$

In particular, the intersection of all finite-codimensional ideals of R is non-zero, and hence R is non-representable.

- Let $C_1 = C_2 = F[t]$, $C_3 = F$, so $A = \begin{pmatrix} F[t] & F[t] \\ 0 & F \end{pmatrix}$. For any vector subspace $U \leq F[t]$, we have that $I = \begin{pmatrix} 0 & U \\ 0 & F \end{pmatrix}$ is a right ideal of A. Ideals of R take the form $\begin{pmatrix} V & I'/I \\ 0 & J \end{pmatrix}$ for $V \in \{(0), F\}$ (if $V = F$ then $I' = A$), some $J \lhd A$ and $I' \leq A$ a right ideal containing I and J. Moreover, any ideal of A takes the form $\begin{pmatrix} B & C \\ 0 & V' \end{pmatrix}$ for $B, C \lhd F[t], B \subseteq C$ and $V' \in \{(0), F\}$ (if $V' = F$ then $C = F[t]$).

  If U is taken to be a finite-codimensional subspace of $F[t]$, the resulting algebra R satisfies ACC on ideals. Let U be an arbitrary 1-codimensional subspace of $F[t]$ which does not contain any non-zero ideal. For concreteness, let $X = \{0, 1, 2, 4, 8, 16, \dots\}$ and let $U \leq F[t]$ be the (1-codimensional) subspace:

$$U = \mathrm{Span}_F\{t^i \mid i \notin X\} + \mathrm{Span}_F\{t-1, t^2-t, t^4-t^2, t^8-t^4, t^{16}-t^8, \dots\}.$$

That U contains no non-zero ideals follows since $2^{r+1} - 2^r \xrightarrow{r\to\infty} \infty$. Since U contains no non-zero ideals of $F[t]$ then I contains no non-zero ideals of A. Therefore, every non-zero ideal $K \lhd R$ contains an element x with zero diagonal and non-zero upper right corner $\bar{a}$, which, as a coset of A/I has a representative of the form $\begin{pmatrix} f(t) & g(t) \\ 0 & \alpha \end{pmatrix}$ where $f(t) \neq 0$ or $g(t) \notin U$. If $f(t) \neq 0$ then since U contains no non-zero ideals then for a suitable $h(t) \in F[t]$ we have that $f(t)h(t) \notin U$, so:

$$\begin{aligned} \begin{pmatrix} 0 & \overline{\begin{pmatrix} 0 & 1 \\ 0 & 0 \end{pmatrix}} \\ 0 & 0 \end{pmatrix} &= \begin{pmatrix} 0 & \overline{\begin{pmatrix} 0 & f(t)h(t) \\ 0 & 0 \end{pmatrix}} \\ 0 & 0 \end{pmatrix} \\ &= \begin{pmatrix} 0 & \overline{\begin{pmatrix} f(t) & g(t) \\ 0 & \alpha \end{pmatrix}} \\ 0 & 0 \end{pmatrix} \begin{pmatrix} 0 & 0 \\ 0 & \begin{pmatrix} 0 & h(t) \\ 0 & 0 \end{pmatrix} \end{pmatrix} \in K \end{aligned}$$

and similarly if $f(t) = 0$ and $g(t) \notin U$ then:

$$\begin{pmatrix} 0 & \overline{\begin{pmatrix} 0 & 1 \\ 0 & 0 \end{pmatrix}} \\ 0 & 0 \end{pmatrix} = \begin{pmatrix} 0 & \overline{\begin{pmatrix} 0 & g(t) \\ 0 & \alpha \end{pmatrix}} \\ 0 & 0 \end{pmatrix} \in K$$

Hence the intersection of all non-zero ideals of R is non-zero, so R is non-representable.

6. Non weakly representable semiprimary PI-algebras

6.1. Affine algebras associated with bilinear forms. We now give a different construction of a non-representable affine PI-algebra of linear growth satisfying ACC on ideals; this algebra has radical satisfying $N^3 = (0)$, which is best possible in view of Corollary 3.2. But more importantly, this construction can be modified to a *semiprimary* PI-algebra which is not representable over any commutative ring.

Let A be an F-algebra, and let M be an A-bimodule. Given an F-linear map $B\colon M \otimes_A M \to F$, we can define an F-algebra[2]:

$$R = \begin{pmatrix} F & M & F \\ 0 & A & M \\ 0 & 0 & F \end{pmatrix}$$

whose multiplication is given by:

$$\begin{pmatrix} \alpha_1 & v & \lambda \\ 0 & f & w \\ 0 & 0 & \alpha_2 \end{pmatrix} \begin{pmatrix} \alpha_1' & v' & \lambda' \\ 0 & f' & w' \\ 0 & 0 & \alpha_2' \end{pmatrix} = \begin{pmatrix} \alpha_1\alpha_1' & \alpha_1 v' + vf' & \alpha_1\lambda' + \alpha_2'\lambda + B(v, w') \\ 0 & ff' & fw' + \alpha_2' w \\ 0 & 0 & \alpha_2\alpha_2' \end{pmatrix}.$$

Since B is A-equivariant, namely B is defined over $M \otimes_A M$, this multiplication law endows R with a well-defined F-algebra structure. To check associativity, note that the only difficulty would be in checking the $(1,3)$ position products which involve the $(1,2)$ and $(2,3)$ positions, but:

$$\begin{aligned} &\alpha_1\alpha_1' \cdot \lambda'' + B(\alpha_1 v' + vf', w'') + (\alpha_1\lambda' + \alpha_2'\lambda + B(v, w')) \cdot \alpha_2'' \\ &\quad = \alpha_1 \cdot (\alpha_1'\lambda'' + \alpha_2''\lambda' + B(v', w'')) + B(v, f'w'' + \alpha_2'' w') + \lambda \cdot \alpha_2'\alpha_2'' \end{aligned}$$

We now specify A and M. Let $A = F[t]$ and let $M = Fu_1 \oplus Fu_2 \oplus \cdots$ be a countable dimensional F-vector space. We consider M as an $F[t]$-bimodule through

$$tu_i = ut_i = u_{i+1}.$$

We now specify B. Set:

$$B(u_i, u_j) = \begin{cases} 1, & \text{if } \exists\, m \geq 1 : \ i + j = 2^m \\ 0, & \text{otherwise.} \end{cases}$$

It is easy to verify that:

$$B(u_i t^j, u_k) = B(u_{i+j}, u_k) = B(u_i, u_{j+k}) = B(u_i, t^j u_k).$$

[2]We denote $B(v \otimes w)$ as $B(v, w)$.

So B is a well-defined F-linear map defined over $M \otimes_{F[t]} M$ and

$$R = \begin{pmatrix} F & M & F \\ 0 & F[t] & M \\ 0 & 0 & F \end{pmatrix}$$

is a well-defined F-algebra.

PROPOSITION 6.1. *The algebra R is an affine PI-algebra of linear growth whose radical cubed is zero, and $R/N \cong F \times F[t] \times F$.*

PROOF. The algebra R is generated over F by the following elements:

$$E_{1,1} = \begin{pmatrix} 1 & 0 & 0 \\ 0 & 0 & 0 \\ 0 & 0 & 0 \end{pmatrix}, \; E_{1,3} = \begin{pmatrix} 0 & 0 & 1 \\ 0 & 0 & 0 \\ 0 & 0 & 0 \end{pmatrix}, \; E_{2,2} = \begin{pmatrix} 0 & 0 & 0 \\ 0 & 1 & 0 \\ 0 & 0 & 0 \end{pmatrix}, \; E_{3,3} = \begin{pmatrix} 0 & 0 & 0 \\ 0 & 0 & 0 \\ 0 & 0 & 1 \end{pmatrix}$$

$$\begin{pmatrix} 0 & 0 & 0 \\ 0 & t & 0 \\ 0 & 0 & 0 \end{pmatrix}, \quad \begin{pmatrix} 0 & u_1 & 0 \\ 0 & 0 & 0 \\ 0 & 0 & 0 \end{pmatrix}, \quad \begin{pmatrix} 0 & 0 & 0 \\ 0 & 0 & u_1 \\ 0 & 0 & 0 \end{pmatrix}.$$

It is easy to see that if U is the F-linear span of the above elements then $\dim_F U^n$ grows linearly so $\operatorname{GKdim}(R) = 1$. Observe that the radical of R is

$$N = \begin{pmatrix} 0 & M & F \\ 0 & 0 & M \\ 0 & 0 & 0 \end{pmatrix};$$

so $N^3 = (0)$ and $R/N \cong F \times F[t] \times F$. Moreover, R satisfies a polynomial identity, e.g.,

$$[X_1, X_2][X_3, X_4][X_5, X_6] = 0,$$

as claimed. □

PROPOSITION 6.2. *The algebra R does not satisfy ACC on (left) annihilators. Consequently, R is not representable.*

PROOF. For a vector $w \in M$, denote $w^\perp = \{v \in M \mid B(v, w) = 0\}$. Notice that:

$$\operatorname{Ann}_R \begin{pmatrix} 0 & 0 & 0 \\ 0 & 0 & u_i \\ 0 & 0 & 0 \end{pmatrix} = \begin{pmatrix} F & u_i^\perp & F \\ 0 & 0 & M \\ 0 & 0 & F \end{pmatrix}.$$

Consider the sets:

$$S_n = \left\{ \begin{pmatrix} 0 & 0 & 0 \\ 0 & 0 & u_{2^n} \\ 0 & 0 & 0 \end{pmatrix}, \begin{pmatrix} 0 & 0 & 0 \\ 0 & 0 & u_{2^{n+1}} \\ 0 & 0 & 0 \end{pmatrix}, \dots \right\}.$$

It is easy to see that:

$$\begin{pmatrix} 0 & u_{2^n} & 0 \\ 0 & 0 & 0 \\ 0 & 0 & 0 \end{pmatrix} \in \operatorname{Ann}_R(S_{n+1}) \setminus \operatorname{Ann}_R(S_n)$$

since $2^n + 2^m$ is not a power of 2 for $m > n$; but $2^n + 2^n = 2^{n+1}$, so $u_{2^n} \notin u_{2^n}^\perp$. Therefore, we have an infinite strictly increasing ascending chain of annihilators:

$$\operatorname{Ann}_R(S_1) \subset \operatorname{Ann}_R(S_2) \subset \cdots;$$

so R does not satisfy ACC on left annihilators. □

We now consider two-sided ideals of R.

PROPOSITION 6.3. *The algebra R satisfies ACC on two-sided ideals.*

PROOF. Let $I \lhd R$, fix some $r \in I$, and write:

$$r = \begin{pmatrix} \alpha & v & \lambda \\ 0 & f(t) & w \\ 0 & 0 & \beta \end{pmatrix}.$$

Then:

$$E_{1,1}rE_{1,1} = \begin{pmatrix} \alpha & 0 & 0 \\ 0 & 0 & 0 \\ 0 & 0 & 0 \end{pmatrix}, \quad E_{2,2}rE_{2,2} = \begin{pmatrix} 0 & 0 & 0 \\ 0 & f(t) & 0 \\ 0 & 0 & 0 \end{pmatrix}, \quad E_{3,3}rE_{3,3} = \begin{pmatrix} 0 & 0 & 0 \\ 0 & 0 & 0 \\ 0 & 0 & \beta \end{pmatrix}$$

$$E_{1,1}rE_{3,3} = \begin{pmatrix} 0 & 0 & \lambda \\ 0 & 0 & 0 \\ 0 & 0 & 0 \end{pmatrix}, \quad (1 - E_{2,2})\,rE_{2,2} = \begin{pmatrix} 0 & v & 0 \\ 0 & 0 & 0 \\ 0 & 0 & 0 \end{pmatrix},$$

$$E_{2,2}r\,(1 - E_{2,2}) = \begin{pmatrix} 0 & 0 & 0 \\ 0 & 0 & w \\ 0 & 0 & 0 \end{pmatrix}$$

Moreover, for an arbitrary $p(t) \in F[t]$:

$$\begin{pmatrix} 0 & v & 0 \\ 0 & 0 & 0 \\ 0 & 0 & 0 \end{pmatrix} \begin{pmatrix} 0 & 0 & 0 \\ 0 & p(t) & 0 \\ 0 & 0 & 0 \end{pmatrix} = \begin{pmatrix} 0 & vp(t) & 0 \\ 0 & 0 & 0 \\ 0 & 0 & 0 \end{pmatrix},$$

$$\begin{pmatrix} 0 & 0 & 0 \\ 0 & p(t) & 0 \\ 0 & 0 & 0 \end{pmatrix} \begin{pmatrix} 0 & 0 & 0 \\ 0 & 0 & w \\ 0 & 0 & 0 \end{pmatrix} = \begin{pmatrix} 0 & 0 & 0 \\ 0 & 0 & p(t)w \\ 0 & 0 & 0 \end{pmatrix},$$

and

$$\begin{pmatrix} 0 & 0 & 0 \\ 0 & p(t) & 0 \\ 0 & 0 & 0 \end{pmatrix} \cdot r \cdot \begin{pmatrix} 0 & 0 & 0 \\ 0 & 1 & 0 \\ 0 & 0 & 0 \end{pmatrix} = \begin{pmatrix} 0 & 0 & 0 \\ 0 & f(t)p(t) & 0 \\ 0 & 0 & 0 \end{pmatrix}.$$

It follows that I contains a subspace of the form:

$$\begin{pmatrix} E & V & L \\ 0 & J & W \\ 0 & 0 & K \end{pmatrix}$$

where $E, K, L \in \{(0), F\}$, $J \lhd F[t]$, and V, W are $F[t]$-sub-bimodules of M. Moreover, V, W are non-zero if and only if v, w are non-zero, respectively, for some $r \in I$. Notice that any non-zero ideal of $F[t]$ is of finite codimension, and any non-zero sub-bimodule of M is of finite codimension. Therefore, we have that I is of finite codimension inside a subspace of the form:

$$\begin{pmatrix} E & V & L \\ 0 & J & W \\ 0 & 0 & K \end{pmatrix}$$

where $E, K, L \in \{(0), F\}$, $J \in \{(0), F[t]\}$ and $V, W \in \{(0), M\}$. It follows that any ascending chain of ideals of R stabilizes. $\square$

6.2. A semiprimary modification. We now extend the construction from Subsection 6.1. The resulting non weakly representable algebra is semiprimary (in fact, its quotient modulo the radical is a product of fields), and its radical cubed is zero. Moreover, N^2 is a 1-dimensional vector space; this shows, in the strongest possible sense, that the Amitsur-Rowen-Small result [**25**] is tight. Weak representability of arbitrary semiprimary PI-algebras was left open in [**25**], and is thereby settled here in the negative.

PROOF OF THEOREM 1.3. We take V to be a 1-dimensional $F(t)$-vector space, which we naturally identify with $F(t)$. We fix an F-linear basis for $F(t)$, say, $\mathfrak{B}$, containing $1, t, t^2, \ldots$, and define an F-linear map $\widetilde{B}\colon V \otimes_{F(t)} V \to F$ on basis elements $v \in \mathfrak{B}$ as follows:

$$\widetilde{B}(v,1) = \begin{cases} 1, & \text{if } \exists k \geq 1 : v = t^{2^k}, \\ 0, & \text{otherwise.} \end{cases}$$

We can therefore form, in the same manner as in Section 5, an F-algebra S. Notice that:

$$\begin{array}{ccc} V \otimes_{F(t)} V & \xrightarrow{\widetilde{B}} & F \\ \uparrow & & \| \\ M \otimes_{F[t]} M & \xrightarrow{B} & F \end{array}$$

where M and B are the same as in Subsection 6.1 and the left vertical embedding is the $F[t]$-equivariant map of vector spaces carrying $u_i \mapsto t^i$. Therefore, the F-algebra R constructed in Subsection 6.1 embeds into S:

$$R = \begin{pmatrix} F & M & F \\ 0 & F[t] & M \\ 0 & 0 & F \end{pmatrix} \hookrightarrow \begin{pmatrix} F & V & F \\ 0 & F(t) & V \\ 0 & 0 & F \end{pmatrix} = S.$$

Notice that, if $N(S) \lhd S$ is the radical of S, then $S/N(S) \cong F \times F(t) \times F$, and so S is a semiprimary PI-algebra. Since R is affine and non-representable (hence non weakly representable), S is not weakly representable either. Finally, $N(S)^2$ is a 1-dimensional subspace spanned by the upper-right corner of S. $\square$

7. Concluding remarks

7.1. Noetherian algebras. The principal remaining open problem is the following:

QUESTION 7.1. Let R be a Noetherian (or Artinian) PI-algebra. Must R be representable?

An important partial result can be found in [**10**, Theorem 3.23]: if a Noetherian PI-algebra is stably Noetherian (that is, remains Noetherian under scalar extensions), then it is representable. Combined with [**4**, Theorem 1.2], we immediately have:

COROLLARY 7.2. *A countably generated Noetherian PI-algebra over an uncountable algebraically closed field is representable.*

For more on stably Noetherian PI-algebras, see [**23**, Section 5]. The case of PI-algebras which are both left and right Noetherian in fact reduces to the case of Artinian PI-algebras (by [**13**]), but as already mentioned, even the Artinian case is open [**5**, Question 5, p. 388]. Notice that any fully bounded Noetherian algebra over an algebraically closed field whose cardinality is larger than its dimension is PI, and an arbitrary Artinian algebra over a field whose cardinality is larger than its dimension is algebraic (see [**1**]); if the base field is in addition algebraically closed, such an algebra must be finite-dimensional.

7.2. Module finiteness. Another open problem is whether an arbitrary algebra which is a finitely generated module over its center is weakly representable [**25**, Question 1.8]. While not able to resolve this problem, we would like to propose a finitary (equivalent) version of it.

QUESTION 7.3. Is there a function $f\colon \mathbb{N} \to \mathbb{N}$ satisfying the following: For any affine algebra A which is generated by d elements as a module over its center, A is embeddable into $f(d) \times f(d)$ matrices over a commutative ring.

This is indeed equivalent to the representability question for module-finite algebras presented above. Let R be an algebra which is generated as a module by d elements over its center, C, say $R = Cr_1 + \cdots + Cr_d$ (we may also assume that R is not affine, otherwise it is representable). Then any affine subalgebra $R_0 = F\langle a_1, \ldots, a_k\rangle \subseteq R$ is contained in an affine subalgebra which is generated by d elements as a module over its center. Indeed, for each $1 \le i \le k$ we can write a_i as $\sum_{j=1}^{d} c_{i,j} r_j$ for suitable $c_{i,j} \in C$. For each $1 \le p, q \le d$, we can write $r_p r_q = \sum_{l=1}^{d} \gamma_{p,q,l} r_l$ for suitable $\gamma_{p,q,l} \in C$. Now take $R_0 \subseteq A = F\langle r_1, \ldots, r_d, c_{1,1}, \ldots, c_{k,d}, \gamma_{1,1,1}, \ldots, \gamma_{d,d,d}\rangle$. Clearly A is generated by $r_1, \ldots, r_d$ as a module over its center. Assuming a positive answer to our finitary version, we could embed A and hence R_0 into $M_{f(d)\times f(d)}(K)$ for some commutative ring K (which depends on R_0). But R is embeddable into an ultraproduct of its affine subalgebras (Mal'cev's argument), hence:

$$R \hookrightarrow \prod_{\substack{R_0 \subseteq R \\ \text{affine}}} R_0 \Big/ \mathcal{F} \hookrightarrow \prod_{\substack{R_0 \subseteq R \\ \text{affine}}} M_{f(d)}(K_{R_0}) \Big/ \mathcal{F} \cong M_{f(d)}(\widehat{K})$$

for some non-principal ultrafilter $\mathcal{F}$ on the set of affine subalgebras of R_0 and a commutative ring $\widehat{K}$.

On the other hand, assume that there were affine algebras R_n which are all generated by d elements over their centers, R_n, and the minimum degree of any matrix embedding of R_n over a commutative ring is d_n, where $d_n \xrightarrow{n\to\infty} \infty$. Then $R = \prod_{n=1}^{\infty} R_n$ is generated by d elements over its center, $\prod_{n=1}^{\infty} C_n$. Hence by assumption, $R \hookrightarrow M_t(K)$ for some $t \in \mathbb{N}$ and some commutative algebra K, contradicting that $d_n \xrightarrow{n\to\infty} \infty$.

7.3. Ascending chain conditions. Any representable algebra has bounded length on chains of left (and right) annihilator ideals. Observe that weak representability does not imply any chain condition on annihilators. Indeed,

$$F[\lambda_1, \ldots, \mu_1, \ldots]/\langle \lambda_1 \cdots \lambda_i \mu_i \mid i = 1, 2, \ldots \rangle$$

is a commutative semiprime algebra which fails to satisfy ACC on annihilator ideals.

This discussion leads us to lower our sights and ask under what conditions does a PI-algebra satisfy ACC on right annihilators:

QUESTION 7.4. Does every left Noetherian PI-algebra satisfy ACC on right annihilators?

This question is a weakening of Question 7.1.

References

[1] S. A. Amitsur and L. W. Small, *Algebras over infinite fields, revisited.* part A, Israel J. Math. **96** (1996), no. part A, 23–25, DOI 10.1007/BF02785531. MR1432724

[2] A. Z. Anan′in, *Representability of Noetherian finitely generated algebras*, Arch. Math. (Basel) **59** (1992), no. 1, 1–5, DOI 10.1007/BF01199007. MR1166010

[3] K. I. Beĭdar, *Classical rings of quotients of PI-algebras* (Russian), Uspekhi Mat. Nauk **33** (1978), no. 6 (204), 197–198. MR526015

[4] J. P. Bell, *Noetherian algebras over algebraically closed fields*, J. Algebra **310** (2007), no. 1, 148–155, DOI 10.1016/j.jalgebra.2006.08.026. MR2307786

[5] A. Kanel-Belov, Y. Karasik, and L. H. Rowen, *Computational aspects of polynomial identities. Vol. 1: Kemer's theorems*, 2nd ed., Monographs and Research Notes in Mathematics, CRC Press, Boca Raton, FL, 2016. MR3445601

[6] A. Belov-Kanel, L. Rowen, and U. Vishne, *Full exposition of Specht's problem*, Serdica Math. J. **38** (2012), no. 1-3, 313–370. MR3014503

[7] A. Belov-Kanel, L. Rowen, and U. Vishne, *Structure of Zariski-closed algebras*, Trans. Amer. Math. Soc. **362** (2010), no. 9, 4695–4734, DOI 10.1090/S0002-9947-10-04993-7. MR2645047

[8] G. M. Bergman, *Some examples in PI ring theory*, Israel J. Math. **18** (1974), 257–277, DOI 10.1007/BF02757282. MR357488

[9] O. Bezushchak, A. Petravchuk, E. Zelmanov, *Automorphisms and derivations of affine commutative and PI-algebras*, preprint (https://doi.org/10.48550/arXiv.2211.10781).

[10] J.D.'Farina, *Stability properties in ring theory*, PhD thesis, University of California San Diego (2006), 75 pages. Available at: https://escholarship.org/uc/item/99f7p41m.

[11] E. Formanek and A. V. Jategaonkar, *Subrings of Noetherian rings*, Proc. Amer. Math. Soc. **46** (1974), 181–186, DOI 10.2307/2039890. MR414625

[12] R. Gordon, *Primary decomposition in right noetherian rings*, Comm. Algebra **2** (1974), 491–524, DOI 10.1080/00927877408822023. MR360691

[13] R. Gordon, *Artinian quotient rings of FBN rings*, J. Algebra **35** (1975), 304–307, DOI 10.1016/0021-8693(75)90052-6. MR384851

[14] R. S. Irving, *Affine PI-algebras not embeddable in matrix rings*, J. Algebra **82** (1983), no. 1, 94–101, DOI 10.1016/0021-8693(83)90175-8. MR701038

[15] R. S. Irving and L. W. Small, *The embeddability of affine PI-algebras in rings of matrices*, J. Algebra **103** (1986), no. 2, 708–716, DOI 10.1016/0021-8693(86)90162-6. MR864439

[16] G. R. Krause and T. H. Lenagan, *Growth of algebras and Gelfand-Kirillov dimension*, Revised edition, Graduate Studies in Mathematics, vol. 22, American Mathematical Society, Providence, RI, 2000, DOI 10.1090/gsm/022. MR1721834

[17] J. Lewin, *On some infinitely presented associative algebras*, J. Austral. Math. Soc. **16** (1973), 290–293. MR340329

[18] J. Lewin, *A matrix representation for associative algebras. I, II*, Trans. Amer. Math. Soc. **188** (1974), 293–308; ibid. 188 (1974), 309–317, DOI 10.2307/1996780. MR338081

[19] A. T. Ludgate, *A note on non-commutative Noetherian rings*, J. London Math. Soc. (2) **5** (1972), 406–408, DOI 10.1112/jlms/s2-5.3.406. MR313310

[20] A. Malcev, *On the representations of infinite algebras* (Russian, with English summary), Rec. Math. [Mat. Sbornik] N.S. **13(55)** (1943), 263–286. MR11084

[21] V.T. Markov, *Gelfand-Kirillov dimension: nilpotency of radical, representability, non-matrix varieties of algebras*, unpublished typescript (2 pages), Moscow, circa 1990.

[22] S. Montgomery and L. W. Small, *Fixed rings of Noetherian rings*, Bull. London Math. Soc. **13** (1981), no. 1, 33–38, DOI 10.1112/blms/13.1.33. MR599637

[23] D. Rogalski, *Stably Noetherian algebras of polynomial growth*, Algebr. Represent. Theory **24** (2021), no. 2, 519–540, DOI 10.1007/s10468-020-09958-w. MR4233506

[24] L. H. Rowen, *Polynomial identities in ring theory*, Pure and Applied Mathematics, vol. 84, Academic Press, Inc. [Harcourt Brace Jovanovich, Publishers], New York-London, 1980. MR576061

[25] L. Rowen and L. Small, *Representability of algebras finite over their centers*, J. Algebra **442** (2015), 506–524, DOI 10.1016/j.jalgebra.2014.10.028. MR3395071

[26] J. J. Sarraillé, *Noetherian PI rings not module-finite over any commutative subring*, Proc. Amer. Math. Soc. **84** (1982), no. 4, 457–463, DOI 10.2307/2044015. MR643729

[27] L. W. Small, J. T. Stafford, and R. B. Warfield Jr., *Affine algebras of Gel′fand-Kirillov dimension one are PI*, Math. Proc. Cambridge Philos. Soc. **97** (1985), no. 3, 407–414, DOI 10.1017/S0305004100062976. MR778674

Department of Mathematics, University of California, San Diego, La Jolla, California, 92093
Email address: bgreenfeld@ucsd.edu

Department of Mathematics, Bar-Ilan University, Ramat-Gan 52900, Israel
Email address: rowen@math.biu.ac.il

Department of Mathematics, University of California, San Diego, La Jolla, California, 92093
Email address: lwsmall@ucsd.edu

Contemporary Mathematics
Volume **800**, 2024
https://doi.org/10.1090/conm/800/16056

Transcendental splitting fields of division algebras

Daniel Krashen and Max Lieblich

ABSTRACT. We examine when division algebras can share common splitting fields of certain types. In particular, we show that one can find fields for which one has infinitely many Brauer classes of the same index and period at least 3, all nonisomorphic and having the same set of finite splitting fields as well as the same splitting fields of transcendence degree 1 and genus at most 1. On the other hand, we show that when one fixes any division algebra over a field, then any division algebras sharing the same splitting fields of transcendence degree at most 3 must generate the same cyclic subgroup of the Brauer group. In particular, there are only a finite number of such division algebras. We also show that a similar finiteness statement holds for splitting fields of transcendence degree at most 2.

1. Introduction

In this paper, we pursue the question of how much information about a Brauer class can be obtained from knowledge of its splitting fields of various types. In recent years, we have seen a number of examples in the literature both of division algebras sharing all of their maximal subfields, and even sharing all of their finite dimensional splitting fields without being isomorphic or without being powers of one another in the Brauer group (see [**3–7, 10, 15, 16, 19**]).

To describe these kinds of results, it is useful to consider partial orders on the collection of Brauer classes. For a field k and a collection of field extensions $\mathscr{K}$ of k, we write $\alpha_1 \prec_{\mathscr{K}} \alpha_2$ for Brauer classes $\alpha_1, \alpha_2 \in \mathrm{Br}(k)$ if for every $K \in \mathscr{K}$ splitting α_1, we have that K also splits α_2. We can then consider the "upper genus of α_1 with respect to $\mathscr{K}$", written $[\alpha_1, \infty)_{\mathscr{K}}$ consisting of all the classes α_2 such that $\alpha_1 \prec_{\mathscr{K}} \alpha_2$. Note that this is a subgroup of the Brauer group of k.

How does one expect this to behave in general? In the particular case of the number fields and finite field extensions, this is easy to describe – a field extension E/F of a number field F splits a Brauer class α whenever its composita EF_v with respect to completions F_v of F have local degree a multiple of the ramification degree (order of the Hasse invariant) of α at v. In particular, one then finds that $[\alpha, \infty)$ consists of exactly those Brauer classes whose Hasse invariants divide those of α. In particular, this is a finite group.

It is natural to ask in which general situations these upper genera are finite. We show that this holds if one considers the collection of finitely generated field

2020 *Mathematics Subject Classification.* Primary 12E30, 16K50.
Key words and phrases. Brauer groups, splitting fields.

extensions of transcendence degree at least 2 (Theorem 6.7) over a field of characteristic 0, and in the case of transcendence degree at least 3, it consists of precisely the cyclic subgroup generated by the Brauer class (Theorem 6.6).

From the other direction, it is natural to ask how large upper genera can be for field extensions of transcendence degree at most 1. For finite extensions, examples of infinite upper genera are given in [**15**, Theorem G]. In general, this question is still open (see Question 5.3); however, as these are represented by function fields of curves, one can subdivide this question according to the genus of the curves in question. For curves of genus 0, upper genera consist of singletons, as noted in Proposition 5.1. We show in Theorem 5.2 that one can construct examples of infinite collections of Brauer classes over a field which share all the same finite splitting fields and splitting fields which are function fields of curves of genus at most 1.

2. Preliminaries on obstructions and splittings

Splittings of Brauer classes by functions fields are governed by specializations of geometric Brauer classes. Therefore, in order to approach our results, it will be useful to first review this connection.

Suppose that X is a smooth, projective, geometrically connected variety over a field k. We may describe the relative Brauer group $\mathrm{Br}(X/k)$ in terms of obstruction classes for line bundles (see also [**8, 18**]). Note that this relative Brauer group can also be identified with $\mathrm{Br}(k(X)/k)$ since $\mathrm{Br}(X) \to \mathrm{Br}(k(X))$ is injective by [**12**, Cor. 1.8]. Throughout, for any scheme Y, we will write $\mathrm{Br}(Y)$ to denote the cohomological Brauer group $H^2(Y, \mathbb{G}_m)$.

We will have use to extend our discussion to the relative situation, so let us begin by assuming we have a projective morphism $\pi : \mathscr{X} \to S$ over some base scheme S such that π is cohomologically flat in dimension 0 (see [**2**, p. 206]), that is, so that $\pi_*\mathscr{O}_{\mathscr{X}} \simeq \mathscr{O}_S$ holds universally after base change. For any morphism $T \to S$, we may consider the Leray-Serre spectral sequence for étale cohomology (natural in T):

$$H^p(T, R^q\pi_*\mathbb{G}_m) \implies H^{p+q}(\mathscr{X}_T, \mathbb{G}_m).$$

This gives an exact sequence of low degree terms

$$\mathrm{Pic}_{\mathscr{X}_T/T}(T) = \mathrm{Pic}_{\mathscr{X}/S}(T) \to \mathrm{Br}(T) \to \mathrm{Br}(\mathscr{X}_T) \tag{1}$$

where $\mathrm{Pic}_{\mathscr{X}/S}$ denotes the sheafification of the Picard functor [**2**, p. 201] (note that one can use the étale topology instead of the flat topology since π is proper as discussed in [**2**, p. 203]). We may therefore describe the relative Brauer group $\mathrm{Br}(\mathscr{X}_T/T) = \ker(\mathrm{Br}(T) \to \mathrm{Br}(\mathscr{X}_T))$ as the image of the map from the sections of the Picard sheaf.

In particular, if we consider a smooth geometrically integral projective variety X over a field k, the Picard sheaf $\mathrm{Pic}_{X/k}$ is represented by a group scheme locally of finite type over k which we will denote Pic_X ([**9**, Ch. 9, Cor. 9.4.18.3]), and if we let $T = \mathrm{Pic}_X$, we obtain a natural Brauer class $\beta \in \mathrm{Br}(\mathrm{Pic}_X)$ such that the map $\mathrm{Pic}_X(k) \to \mathrm{Br}(k)$ in sequence (1) is given by specialization. This can be seen via the map of spectral sequences induced by pulling back with respect to a map

$x : \operatorname{Spec} k \to \operatorname{Pic}_X$ (i.e., $x \in \operatorname{Pic}_X(k)$):

$$\begin{array}{ccc} H^p(\operatorname{Pic}_X, R^q\pi_*\mathbb{G}_m) & \Longrightarrow & H^{p+q}(X_{\operatorname{Pic}_X}, \mathbb{G}_m) \\ \downarrow & & \downarrow \\ H^p(k, R^q\pi_*\mathbb{G}_m) & \Longrightarrow & H^{p+q}(X, \mathbb{G}_m), \end{array}$$

giving a commutative diagram with exact rows:

$$\begin{array}{ccccc} \operatorname{Pic}_X(\operatorname{Pic}_X) & \longrightarrow & \operatorname{Br}(\operatorname{Pic}_X) & \longrightarrow & \operatorname{Br}(X_{\operatorname{Pic}_X}) \\ \downarrow & & \downarrow & & \downarrow \\ \operatorname{Pic}_X(k) & \longrightarrow & \operatorname{Br}(k) & \longrightarrow & \operatorname{Br}(X). \end{array} \tag{2}$$

Since the pullback $x^* : \operatorname{Pic}_X(\operatorname{Pic}_X) \to \operatorname{Pic}_X(k)$ takes the identity morphism $\mathrm{id}_{\operatorname{Pic}_X}$ to x, it follows that the image of $\mathrm{id}_{\operatorname{Pic}_X}$ in $\operatorname{Br}(\operatorname{Pic}_X)$ describes a Brauer class on Pic_X whose specializations at rational points (i.e., pullbacks via $x^* : \operatorname{Br}(\operatorname{Pic}_X) \to \operatorname{Br}(k)$) exactly describe the Brauer classes split when the base is changed from k to X.

Definition 2.1. We call this class $\beta \in \operatorname{Br}(\operatorname{Pic}_X)$ the canonical obstruction class for X/k.

Remark 2.2. It follows that the map $\operatorname{Pic}_X(k) \to \operatorname{Br}(k)$ from the bottom of diagram (2) has the interpretation both as the homomorphism coming from the Artin-Leray spectral sequence as well as arising from taking at point $x \in \operatorname{Pic}_X(k)$ to $\beta|_x$, the specialization of the Brauer class β. Consequently, this class β has the property that specialization to rational points gives a homomorphism $\operatorname{Pic}_X(k) \to \operatorname{Br}(k)$.

Summarizing this discussion, we have:

Proposition 2.3. *Let X be a smooth projective variety, and let* Pic_X *be the Picard scheme for X. Then there is a Brauer class $\beta_X \in \operatorname{Br}(\operatorname{Pic}_X)$, called the canonical obstruction class, such that specialization*

$$\begin{aligned} \operatorname{Pic}_X(k) &\to \operatorname{Br}(k) \\ x &\mapsto \beta_X|_x \end{aligned}$$

induces a homomorphism from $\operatorname{Pic}_X(k)$ *to* $\operatorname{Br}(k)$ *whose image is exactly the relative Brauer group* $\operatorname{Br}(k(X)/k) = \operatorname{Br}(X/k)$.

3. Specializations of Brauer classes

As we have remarked in Proposition 2.3, splittings by function fields can be described by specializations of Brauer classes on schemes (i.e., the Picard scheme). For this reason, to understand how to construct situations where certain transcendental field extensions split a Brauer class α, we begin by investigating how to construct situations to make a class α a specialization of a particular geometric Brauer class.

Definition 3.1. Let k be a field. A **pointed Brauer class over** k is a triple (β, X, x) consisting of a geometrically integral variety X/k with a k-rational point x, and a Brauer class $\beta \in \operatorname{Br}(X)$ such that $\beta|_x = 0$.

REMARK 3.2. We note that if $b = (\beta, X, x)$ is a pointed Brauer class, then $\beta \in \operatorname{im}(\operatorname{Br}(k) \to \operatorname{Br}(X))$ if and only if $\beta = 0$, since evaluation at x gives a section, and by hypothesis $\beta|_x = 0$.

REMARK 3.3. If Pic^0_X denotes the identity component of the Picard variety of X, and if this is a smooth projective variety (for example, in the case that X is a curve) then since the identity element of $\operatorname{Pic}^0_X(k)$ is sent to the identity in $\operatorname{Br}(k)$ (by Remark 2.2), it follows that $(\beta_X, \operatorname{Pic}^0_X, 0)$ is a pointed Brauer class.

DEFINITION 3.4. We let $Б_k$ denote the set of all pointed Brauer classes over k. For $b = (\beta, X, x) \in Б_k$, we write $\operatorname{per}(b)$ for $\operatorname{per}(\beta)$ (the period of β), and if L/k is a field, we write b_L for (β_L, X_L, x_L).

DEFINITION 3.5. Let k be a field and $(\beta, X, x) \in Б_k$ a geometric Brauer class over k. We say that $\alpha \in \operatorname{Br}(k)$ is a specialization of (β, X, x) and write $\alpha \in (\beta, X, x)$ if $\alpha = \beta|_y$ for some $y \in X(k)$.

We note that this property is clearly preserved under arbitrary field extensions – if $\alpha \in b \in Б_k$, then $\alpha_L \in b_L \in Б_L$ for any L/k.

We will need to consider how a given Brauer class over k may be obtained by specialization of geometric Brauer classes. As nontrivial geometric Brauer classes are necessarily nonconstant by Remark 3.2, it is not clear, a priori, whether or not a general Brauer class over k can be obtained in this fashion.

PROPOSITION 3.6. *Let k be a field, and suppose that $\alpha, \beta \in \operatorname{Br}(k)$, $\operatorname{per}(\beta) | \operatorname{per}(\alpha)$ and $b \in Б_k$ with $\alpha \in b$. Then we may find an inert field extension E/k such that $\beta_E \in b_E$.*

This result follows immediately from the next lemma, observing that $\alpha \in b$ implies $\operatorname{per}\alpha | \operatorname{per} b$ and hence $\operatorname{per}\beta | \operatorname{per} b$ as well.

LEMMA 3.7. *Let k be a field, and suppose that $\beta \in \operatorname{Br}(k)$ is a Brauer class. Suppose that $b \in Б_k$ with $\operatorname{per}(\beta) \mid \operatorname{per}(b)$. Then we may find a geometrically integral variety $Y(\beta, b)$ with function field $k(\beta, b)$ such that $\operatorname{Br}(k) \to \operatorname{Br}(k(\beta, b))$ is injective and that $\beta_{k(\beta,b)} \in b_{k(\beta,b)}$.*

PROOF. Write $b = (\delta, X, x)$. Without loss of generality, we may assume that β and hence δ are nonzero. Since X has a smooth k-rational point, it follows that $\operatorname{Br}(k) \to \operatorname{Br}(X) \to \operatorname{Br}(k(X))$ is injective. Consider the Brauer class $\gamma = \delta - \beta_x \in \operatorname{Br}(X)$. Let $Y_{\beta,b} \to X$ be the Severi-Brauer scheme of γ and let $k(\beta, b)$ be its function field. Let $x' \in X(k(X))$ be the generic point of X, and $x \in X(k(\beta, b))$ the corresponding point defined by pullback. By Remark 3.2, as δ is nonzero, it and therefore also $\delta_{x'}$ are not in the image of $\operatorname{Br}(k)$. Since $\operatorname{per}(\beta_K) = \operatorname{per}(\beta) | \operatorname{per}(\delta_{x'})$, no nontrivial multiple of $\gamma = \delta_{x'} - \beta_{k(X)}$ is in the image of $\operatorname{Br}(k) \to \operatorname{Br}(k(X))$. As the kernel of the map $\operatorname{Br}(k(X)) \to \operatorname{Br}(k(\beta, b))$ is generated by the class γ by [**1**], it follows that the kernel of the composition $\operatorname{Br}(k) \to \operatorname{Br}(k(X)) \to \operatorname{Br}(k(\beta, b))$ is trivial. Finally, since $(\gamma)_{k(\beta,b)} = 0$ by construction, it follows that $\beta_{k(\beta,b)} = \delta_x$, as claimed. $\square$

4. Splitting Brauer classes with collections of field extensions

In this section we collect some useful language and observations for working with collections of field extensions and partial orders of Brauer classes by their splitting properties. We write $\operatorname{per}(\alpha)$ to denote the period of a Brauer class α.

DEFINITION 4.1. We say that a field extension L/k is **inert** if L/k is regular (that is, if $L \otimes_k \overline{k}$ is again a field) and if $\mathrm{Br}(L/k) = 0$.

NOTATION 4.2. Let F/k be regular and L/k any field extension. Then $F \otimes_k L$ is a domain, and we write $F \cdot_k L$ for the fraction field of $F \otimes_k L$ (or we will just write FL instead of $F \cdot_k L$ when k is clear from the context).

NOTATION 4.3. Let $\mathscr{K}$ be a collection of field extensions of k, and suppose F/k is regular. We write $\mathscr{K} \cdot_k L$ (or $\mathscr{K}L$) to denote the collection of fields of the form $K \cdot_k L$ for $K \in \mathscr{K}$.

DEFINITION 4.4. Let $\mathscr{A}_1, \mathscr{A}_2 \subset \mathrm{Br}(k)$ be nonempty collections of Brauer classes over k. We say that $\mathrm{per}(\mathscr{A}_1) | \mathrm{per}(\mathscr{A}_2)$ if $\mathrm{per}\, \alpha_1 | \mathrm{per}\, \alpha_2$ for all $\alpha_1 \in \mathscr{A}_1$ and $\alpha_2 \in \mathscr{A}_2$. We say that $\mathrm{per}\, \mathscr{A}_1 = \mathrm{per}\, \mathscr{A}_2$ if $\mathrm{per}\, \alpha_1 = \mathrm{per}\, \alpha_2$ for every $\alpha_i \in \mathscr{A}_i$ (and hence all periods of all Brauer classes in both sets coincide).

DEFINITION 4.5. For a collection of field extensions $\mathscr{K}$ and Brauer classes $\alpha_1, \alpha_2 \in \mathrm{Br}(k)$, we say $\alpha_1 \prec_{\mathscr{K}} \alpha_2$ if for every $L \in \mathscr{K}$, if $(\alpha_1)_L = 0$ then $(\alpha_2)_L = 0$. For $\mathscr{A}_1, \mathscr{A}_2 \subset \mathrm{Br}(k)$ we say $\mathscr{A}_1 \prec_{\mathscr{K}} \mathscr{A}_2$ if $\alpha_1 \prec_{\mathscr{K}} \alpha_2$ whenever $\alpha_1 \in \mathscr{A}_1$, $\alpha_2 \in \mathscr{A}_2$. Finally, we write $\mathscr{A}_1 \sim_{\mathscr{K}} \mathscr{A}_2$ if $\mathscr{A}_1 \prec_{\mathscr{K}} \mathscr{A}_2$ and $\mathscr{A}_2 \prec_{\mathscr{K}} \mathscr{A}_1$.

In other words, $\alpha_1 \prec_{\mathscr{K}} \alpha_2$ if every splitting field in $\mathscr{K}$ of α_1 also splits α_2. We will be interested in, for example, $\mathscr{K}$ being the finite extensions of k, or the finitely generated field extensions of a given transcendence degree, or the function fields of curves of genus 1. In each of these cases, the class $\mathscr{K}$ naturally can be thought of as "varying" with k in a natural way. This leads us to the following definition:

DEFINITION 4.6. Let k be a field. A class $\underline{\mathscr{K}}$ of field extensions of k is a rule which associates to every field extension F/k a set $\underline{\mathscr{K}}(F)$ whose elements are field extensions of F, such that if $E \in \underline{\mathscr{K}}(F)$ and L/F is any regular field extension, then every $E \cdot_F L$ is (isomorphic to) a field in $\underline{\mathscr{K}}(L)$.

The particular classes of fields we will consider, as have been described already above, satisfy an additional property. Essentially this property says that they are determined by finitely generated subfields.

DEFINITION 4.7. Let $\underline{\mathscr{K}}$ be a class of field extensions of k. We say that $\underline{\mathscr{K}}$ is **regularly compact** if for any sequence of field extensions $F_0 \subset F_1 \subset F_2 \subset \cdots$ of k, with F_{n+1}/F_n a regular extension for each n, and for every $K \in \underline{\mathscr{K}}(\bigcup F_n)$, there exists $K' \in \underline{\mathscr{K}}(F_m)$ for some m such that $K = K' \cdot_{F_m} (\bigcup F_n)$.

EXAMPLE 4.8. Let $\underline{\mathscr{K}}(F)$ be the collection of field extensions of the form $F(X)$, where X/F is a smooth geometrically integral curve of genus 1. Then $\underline{\mathscr{K}}(F)$ is regularly compact.

To see this, suppose we have $K \in \underline{\mathscr{K}}(\bigcup F_n)$. Then K is the function field of some curve X which may be written as the zeros to some system of polynomials $f_1, \ldots, f_N$ in some projective space. The coefficients of these polynomials lie in the field $\bigcup F_n$, but as there are only finitely many coefficients, they lie in a subfield F_m. We therefore see that the curve X is the base change of a curve X' defined over F_m, and if we set $K' = F_m(X')$, we have $K = K' \cdot_{F_m} (\bigcup F_n)$ as in the definition. Similar arguments yield the following examples as well:

EXAMPLE 4.9. Let $\underline{\mathscr{K}}(F)$ be the collection of field extensions of the form $F(X)$, where X/F is a smooth projective variety admitting a rational point over F. Then $\underline{\mathscr{K}}(F)$ is regularly compact.

EXAMPLE 4.10. Let $\underline{\mathscr{K}}(F)$ be the collection of finite field extensions of F. Then $\underline{\mathscr{K}}(F)$ is regularly compact.

LEMMA 4.11. *Let $\underline{\mathscr{K}}$ be any regularly compact class of fields and $\alpha \in \mathrm{Br}(k)$. Then the class $\underline{\mathscr{K}}_\alpha$ defined by $\underline{\mathscr{K}}_\alpha(F) = \{K \in \underline{\mathscr{K}}(F) \mid \alpha_K = 0\}$ is also regularly compact.*

PROOF. Indeed, if $K \in \underline{\mathscr{K}}(\bigcup F_n)$ and $\alpha_K = 0$, we find that by definition we may find m so that $K = K' \cdot_{F_m} (\bigcup F_n)$ for some m, and $K' \in \underline{\mathscr{K}}(F_m)$. But note that we can also write $K = \bigcup_{n \geq m} K' \cdot_{F_m} F_n$. Let X_α be a Severi-Brauer variety of some central simple algebra representing the class α. As $\alpha_K = 0$, we see that $X_\alpha(K) \neq \emptyset$. As the finitely many coordinates of such a rational point must lie on one of the fields $K' \cdot_{F_m} F_n$, we find that $K' \cdot_{F_m} F_n \in \underline{\mathscr{K}}_\alpha(F_n)$ and $K = K' \cdot_{F_m} (\bigcup F_n) = (K' \cdot_{F_m} F_n) \cdot_{F_n} (\bigcup F_n)$ as desired. □

LEMMA 4.12. *Suppose $\mathscr{A}, \mathscr{B} \subset \mathrm{Br}(k)$ are subsets of the Brauer group, $\underline{\mathscr{K}}$ and is a regularly compact class of field extensions of k. Suppose that for every F/k and $\alpha \in \mathscr{A}, \beta \in \mathscr{B}, K \in \underline{\mathscr{K}}(F)$ with $\alpha_K = 0$, we may construct an inert extension F' with $\beta_{K \cdot_F F'} = 0$. Then we may construct an inert extension $\widetilde{F}/F$ with $\mathscr{A}_{\widetilde{F}} \prec_{\underline{\mathscr{K}}(\widetilde{F})} \mathscr{B}_{\widetilde{F}}$.*

To prove this, we will make use of the following Lemma, which follows a well known pattern. We include a proof for convenience.

LEMMA 4.13. *Let k be a field, E/k a field extension, I a filtered index set (i.e., a directed partially ordered set), and suppose we have a collection of field extensions E_i for $i \in I$ such that $E_i \subset E_j$ for $i \leq j$ with $\bigcup_{i \in I} E_i = E$. If E_i/k is inert for each $i \in I$ then E/k is also inert.*

PROOF. Suppose the map $\mathrm{Br}(k) \to \mathrm{Br}(E)$ was not injective and let β be in the kernel. Choose a central simple k-algebra B representing the class of β, and an isomorphism $B \otimes_k E \simeq End(V)$. This isomorphism only will require finitely many field elements of E to express, and hence can only involve elements in finitely many of the subfields E_i. Letting j be the maximum this finite set of such i, we find that $\beta_{E_j} = 0$ contradicting the fact that E_j/k is inert. □

PROOF OF LEMMA 4.12. As in the statement, let us suppose we have constructed inert extensions $F' = F(K, \alpha, \beta)/F$ with $\beta_{K \cdot_F F'} = 0$ for every F/k and $\alpha \in \mathscr{A}, \beta \in \mathscr{B}, K \in \underline{\mathscr{K}}(F)$ with $\alpha_K = 0$. We first will show that we can construct an inert field extension $F(\mathscr{A}, \mathscr{B})/F$ such that for every $K \in \underline{\mathscr{K}}(F)$ with $\alpha_K = 0$ for some $\alpha \in \mathscr{A}$, we have $\beta_{K \cdot_F F(\mathscr{A}, \mathscr{B})} = 0$. We construct this field as follows. Let $\mathscr{T} \subset \underline{\mathscr{K}}(F) \times \mathscr{A} \times \mathscr{B}$ be those triples (K, α, β) such that $\alpha_K = 0$. Choose a well ordered set Ω with an initial element $\emptyset$, and a bijection

$$\begin{aligned} \phi : \Omega_+ &\to \mathscr{T}, \\ \omega &\mapsto (K_\omega, \alpha_\omega, \beta_\omega) \in \underline{\mathscr{K}}(F) \times \mathscr{A} \times \mathscr{B}, \end{aligned}$$

where $\Omega_+ = \Omega \setminus \{\emptyset\}$. We will define an increasing collection of inert field extensions F_ω (transfinite) inductively as follows. We begin by $F_\emptyset = F$. In general, we define $F'_\omega = \bigcup_{\lambda < \omega} F_\lambda$, and as these are inert and in particular regular over F, we find $K \cdot_F F'_\omega \in \underline{\mathscr{K}}(F'_\omega)$ and $(\alpha_\omega)_{K \cdot_F F'_\omega} = ((\alpha_\omega)_K)_{K \cdot_F F'_\omega} = 0$. Therefore, by hypothesis, there exists an inert extension $F_\omega = F'_\omega(K \cdot_F F'_\omega, \alpha_\omega, \beta_\omega)$ such that $(\beta_\omega)_{F_\omega} = 0$. We let $F(\mathscr{A}, \mathscr{B}) = \bigcup_{\omega \in \Omega} F_\omega$, which is inert by Lemma 4.13.

Let us check that whenever $K \in \underline{\mathscr{K}}(F)$ with $\alpha \in \mathscr{A}$ and $\alpha_K = 0$, we have $\beta_{K \cdot_F F(\mathscr{A},\mathscr{B})}$, as claimed. For such a choice $(K, \alpha, \beta) \in \mathscr{T}$, say $(K, \alpha, \beta) = (K_\omega, \alpha_\omega, \beta_\omega)$ for some $\omega \in \Omega$. By construction, we have $\beta_{K \cdot_F F'_\omega} = 0$ and $F'_\omega \subset F(\mathscr{A}, \mathscr{B})$. Therefore we have $K \cdot_F F'_\omega \subset K \cdot_F F(\mathscr{A}, \mathscr{B})$ showing $\beta_{K \cdot_F F(\mathscr{A},\mathscr{B})} = 0$.

We complete the proof with a second inductive procedure, this time constructing a sequence of fields F_n for $n \in \mathbb{Z}_{\geq 0}$, starting with $F_0 = F$ and with F_{n+1}/F_n inert by setting $F_{n+1} = F_n(\mathscr{A}_{F_n}, \mathscr{B}_{F_n})$. Let $\widetilde{F} = \bigcup F_n$, which is inert by Lemma 4.13. Let us now check that $\mathscr{A}_{\widetilde{F}} \prec_{\underline{\mathscr{K}}(\widetilde{F})} \mathscr{B}_{\widetilde{F}}$. For this, suppose $K \in \underline{\mathscr{K}}(\widetilde{F})$ with $\alpha_K = 0$ for some α and let $\beta \in \mathscr{B}$. We claim that $\beta_K = 0$ as well.

By the regular compactness of $\underline{\mathscr{K}}_\alpha$ from Lemma 4.11, we may find m and $K' \in \underline{\mathscr{K}}(F_m)$ with $K = K' \cdot_{F_m} \widetilde{F}$ and $\alpha_{K'} = 0$. But therefore $\beta_{K' \cdot_{F_m} F_m(\mathscr{A},\mathscr{B})} = \beta_{K' \cdot_{F_m} F_{m+1}} = 0$. But as $K' \cdot_{F_m} F_{m+1} \subset K' \cdot_{F_m} \widetilde{F} = K$, we have $\beta_K = 0$ as desired. □

5. Ordering of Brauer classes by splitting

In this section, we now complete our construction of collections of algebras with common finite splitting fields and common splitting fields arising as function fields of curves of genus at most 1 (see Theorem 5.2).

We consider the following classes of field extensions of a field k:

- $\underline{\mathscr{G}^0}(F) = \{F(X) \mid X/F \text{ is a smooth projective curve of genus } 0\}$,
- $\underline{\mathscr{G}^1}(F) = \{F(X) \mid X/F \text{ is a smooth projective curve of genus } 1\}$,
- $\underline{\mathscr{G}^{\leq 1}}(F) = \{F(X) \mid X/F \text{ is a smooth projective curve of genus at most } 1\}$,
- $\underline{\mathscr{F}in}(F) = \{K/F \text{ finite}\}$,
- $\underline{\mathscr{G}^1\mathscr{F}in}(F) = \underline{\mathscr{G}^1}(F) \cup \underline{\mathscr{F}in}(F)$,
- $\underline{\mathscr{G}^{\leq 1}\mathscr{F}in}(F) = \underline{\mathscr{G}^{\leq 1}}(F) \cup \underline{\mathscr{F}in}(F)$,

and following the ideas of Example 4.8 and Example 4.10, it is not hard to see that all of these classes are regularly compact.

Let us start by briefly describing the situation concerning splitting algebras by function fields of genus 0 curves.

PROPOSITION 5.1. *Let $\alpha, \beta \in \mathrm{Br}(k)$ with $\alpha \neq 0$, and suppose C is a genus 0 curve with $k(C)$ splitting α. Then α must be represented by a quaternion algebra and if $k(C)$ also splits β then either $\beta = 0$ or $\beta = \alpha$. In particular, we find that $\mathscr{A} \prec_{\underline{\mathscr{G}^0}(k)} \mathscr{B}$ always holds (trivially) when $\mathscr{A}$ consists entirely of classes of period greater than 2.*

PROOF. As curves of genus 0 are plane conics, they can be identified with Severi-Brauer varieties of quaternion algebras, say $C = X_Q$ for some quaternion algebra Q. As $\mathrm{Br}(k(X_Q)/k)$ consists of 0 and the class of Q, it follows that $[Q] = \alpha$ and $\beta \in \mathrm{Br}(k(X_Q)/k)$ is 0 or α, as claimed. □

THEOREM 5.2. *Let $\mathscr{A}_1, \mathscr{A}_2 \subset \mathrm{Br}(k)$ be collections of Brauer classes such that $\mathrm{per}(\alpha_2) | \mathrm{per}(\alpha_1)$ whenever $\alpha_1 \in \mathscr{A}_1, \alpha_2 \in \mathscr{A}_2$. Then we may find an inert field extension F/k such that $(\mathscr{A}_1)_F \prec_{\underline{\mathscr{G}^1\mathscr{F}in}(F)} (\mathscr{A}_2)_F$.*

Consequently, given any collection of Brauer classes $\mathscr{A}$, with all elements having the same period $n > 1$, we may find an inert extension F/k such that all the classes in $\mathscr{A}$ share the same set of splitting fields in $\underline{\mathscr{G}^1\mathscr{F}in}(F)$. Furthermore, if $n > 2$, we find that all the classes in $\mathscr{A}$ share the same set of splitting fields in $\underline{\mathscr{G}^{\leq 1}\mathscr{F}in}(F)$ as well.

Note that the latter statement follows from taking $\mathscr{A}_1 = \mathscr{A} = \mathscr{A}_2$, combined with Proposition 5.1.

QUESTION 5.3. Can Theorem 5.2 be extended to curves of higher genus as well?

To prove Theorem 5.2, we start with a few preliminary results.

LEMMA 5.4. *Suppose* $\alpha_1, \alpha_2 \in \mathrm{Br}(k)$ *with* $\operatorname{per}\alpha_2 | \operatorname{per}\alpha_1$*, and* C/k *is a genus 1 curve with* $(\alpha_1)_{k(C)} = 0$*. Then there exists an inert extension* E/k *with* $(\alpha_2)_{E(C)} = 0$*.*

PROOF. Write b for the pointed Brauer class $(\beta_C|_{J_C}, J_C, 0)$. By Proposition 2.3, since $\alpha_1 k(C) = 0$, it follows that $\alpha_1 \in b$ is a specialization, and hence $\operatorname{per}\alpha_1 | \operatorname{per} b$ and consequently $\operatorname{per}\alpha_2 | \operatorname{per} b$ as well. By Proposition 3.6, we may find an inert field extension E/k such that $(\alpha_2)_E \in b_E$. But $(\alpha_2)_E \in b_E$ means α_2 is a specialization of $(\beta_C)_E = \beta_{C_E}$ and hence $\alpha \in \mathrm{Br}(E(C)/E)$ again by Proposition 2.3. □

LEMMA 5.5. *Let* k *be a field,* $\alpha_1, \alpha_2 \in \mathrm{Br}(k)$ *with* $\operatorname{per}(\alpha_2) | \operatorname{per}(\alpha_1)$*. Let* K/k *be a finite field extension with* $(\alpha_1)_K = 0$*. Then we may find an inert field extension* E/k *such that* $(\alpha_2)_{K \cdot_k E} = (\alpha_2)_{K \otimes_k E} = 0$*.*

PROOF. Let K be a finite splitting field of α_1. By [**15**, Lem. 7.4, Cor. 7.6], we may find a regular field extension $k_K(\alpha_2)$ such that the compositum $Kk_K(\alpha_2)$ (equal in this case to the tensor product) splits α_2 and $\mathrm{Br}(k_K(\alpha_2)/k) = 0$, as claimed. □

PROOF OF THEOREM 5.2. By Lemma 4.12 we need only show that if we are given $\alpha_i \in \mathscr{A}_i$, $i = 1, 2$ and $K \in \underline{\mathscr{G}^1\mathscr{F}in}(F)$ for some field extension F/k, with $(\alpha_1)_K = 0$, we can find E/F inert with $(\alpha_2)_{K \cdot_F E} = 0$ as well. But in the case $K \in \underline{\mathscr{G}}^1(F)$, this follows from Lemma 5.4, and in the case $K \in \underline{\mathscr{F}in}(F)$ this follows from Lemma 5.5. □

6. Finiteness of higher upper genera

Let $\alpha_1, \alpha_2 \in \mathrm{Br}(k)$, and suppose $n = \operatorname{ind}(\alpha_1)$. Then $\alpha_1 \prec_{k(n-1)} \alpha_2$ if and only if α_2 is in the cyclic subgroup generated by α_1. This is because the function field of the Severi-Brauer variety of α_1 has transcendence degree $n-1$ and hence must also split α_2. But the only algebras split by this function field lie in the cyclic subgroup generated by α_1 by [**1**].

On the other hand, for a class $\alpha \in \mathrm{Br}(k)$, and a given integer i, we may still ask for which β we have $\alpha \prec_{k(i)} \beta$. This motivates the following definition:

DEFINITION 6.1. Let $\alpha \in \mathrm{Br}(k)$. We define the i'th upper genus of α, denoted $[\alpha, \infty)_i$ to be the set of $\beta \in \mathrm{Br}(k)$ such that $\beta \prec_{k(i)} \alpha$.

DEFINITION 6.2. Let $\alpha \in \mathrm{Br}(k)$. We say that α is i-minimal if $[\alpha, \infty)_i$ consists precisely of the cyclic subgroup of $\mathrm{Br}(k)$ generated by α.

PROPOSITION 6.3. *Given a class* $\alpha \in \mathrm{Br}(k)$*, there is a smooth projective variety* Y *of dimension at most 3 such that for every field extension* L/k*,* $\mathrm{Br}(L(Y)/L)$ *is exactly the cyclic subgroup generated by* α_L*.*

PROOF. If k is finite there is nothing to prove, as $\mathrm{Br}(k) = 0$. Thus, we assume that k is infinite. Consider the Severi-Brauer variety X for α. If $\dim X \leq 3$, we may set $Y = X$ and be done by [**1**]. Otherwise, let $Y \subset X$ be a smooth complete intersection of $\dim X - 3$ anti-canonical divisors in X, which is possible by Bertini's theorem (since k is infinite).

We will show that the map of group schemes

$$\mathrm{Pic}_X \to \mathrm{Pic}_Y \tag{3}$$

is an isomorphism. It follows that the map

$$\mathbb{Z} = \mathrm{Pic}_X(k) \to \mathrm{Pic}_Y(k)$$

is an isomorphism. Since the relative Brauer group is generated by obstructions associated to sections of the Picard scheme and $\mathrm{Br}(k(X)/k) = \langle \alpha \rangle$, we conclude that $\mathrm{Br}(k(Y)/k) = \langle \alpha \rangle$. Since our statements are independent of L, this gives the desired conclusion.

For technical reasons, we briefly recall that the deformation theory of invertible sheaves on a scheme Z has tangent space $H^1(Z, \mathscr{O}_Z)$ and obstruction space $H^2(Z, \mathscr{O}_Z)$. In particular, if $Z \to S$ is proper, of finite presentation and cohomologically flat in dimension 0, and if $H^i(Z_s, \mathscr{O}_{Z_s}) = 0$ for $i = 1, 2$ and all points $s \in S$, then $\mathrm{Pic}_{Z/S}$ is étale. Moreover, if, in addition, Z is smooth over S and S is integral, then $\mathrm{Pic}_{Z/S}$ is also universally closed over S.

This has two consequences. First, to show that (3) is an isomorphism of group schemes, it suffices to show that the map

$$\mathrm{Pic}(X \otimes \overline{k}) \to \mathrm{Pic}(Y \otimes \overline{k})$$

is an isomorphism of groups. This means that we can reduce the desired claim to the case that k is algebraically closed, and hence that $X = \mathbb{P}^n$.

Second, the deformation theory calculations allow us to reduce the question to characteristic 0. Assume that W is a complete local ring with fraction field of characteristic 0, and let $\mathscr{Y} \subset \mathbb{P}^n_W$ be a flat complete intersection of dimension at least 3. Consider the restriction map

$$\mathrm{Pic}_{\mathbb{P}^n/W} \to \mathrm{Pic}_{\mathscr{Y}/W}\,.$$

By the above considerations, this is an isomorphism if and only if it is an isomorphism over the algebraic closure of the fraction field of W. But this follows from the Grothendieck-Lefschetz Theorem [**14**, Cor. 3.3].

Finally, let W be the Witt vectors of k and assume that $X = \mathbb{P}^n$. Since Y is a complete intersection we can lift it to a smooth relative complete intersection $\mathscr{Y} \subset \mathbb{P}^n_W$. By the previous paragraph, we conclude that (3) is an isomorphism, as desired. □

PROPOSITION 6.4. *Let $\alpha \in \mathrm{Br}(k)$ be a Brauer class, and suppose that k is an uncountable field of characteristic 0. Then we may find a smooth projective variety Z of dimension at most 2 such that for every field extension L/k, $\mathrm{Br}(L(Z)/L)$ is exactly the cyclic subgroup generated by α_L.*

PROOF. Via Proposition 6.3, without loss of generality, we may assume that we have Y of dimension 3 such that $\mathrm{Pic}(Y_{\overline{k}}) = \mathbb{Z}$, whose generator corresponds to an α-twisted sheaf. If $Y_{\overline{k}}$ is a Severi-Brauer variety, then by Noether-Lefschetz, if $Z \subset Y$ is a very general hyperplane section with respect to a projective embedding of Y, pullback gives an isomorphism $\mathrm{Pic}\, Y \to \mathrm{Pic}\, Z$.

In the case Y is not a Severi-Brauer, choose an ample line bundle $\mathscr{L}$ on Y. By [**17**, Theorem 2], if we choose d sufficiently large, then Z cut out by a very general section of $\mathscr{L}^{\otimes d}$, we have $\operatorname{Pic} Y \to \operatorname{Pic} Z$ is again an isomorphism.

It follows again that $\operatorname{Pic} Z$ is generated by an α-twisted sheaf, and hence the relative Brauer group $\operatorname{Br}(k(Y)/k)$ is exactly the cyclic subgroup generated by α. □

PROPOSITION 6.5. *Suppose that α is a Brauer class over a field k of characteristic 0. Then we may find a smooth projective surface Y/k such that $\alpha \in \operatorname{Br}(Y/k)$ and $\operatorname{Br}(Y_L/L)$ is finite for every field extension L/k.*

PROOF. Without loss of generality, we may assume k is infinite. Consider the Severi-Brauer variety X for α. If $\dim X \leq 2$, we may set $Y = X$ and be done by [**1**]. Otherwise, fix a projective embedding $X \to \mathbb{P}^N$ for some N. Let $Y \subset X$ be a smooth linear section of X of dimension 2 (we can find such a subvariety since k is infinite). By the Lefschetz hyperplane Theorem [**11**, p. 156], the first Betti numbers of $X_{\mathbb{C}}$ and $Y_{\mathbb{C}}$ coincide and hence vanish (where $\mathbb{C}$ denotes an algebraic closure of k). It follows that the Albanese variety for Y and hence the Picard variety for Y must be trivial, which implies that the Picard group and Neron-Severi groups for Y agree. In particular, *Pic*(Y) is a finitely generated group [**13**, Theorem 5.1, Exp XIII, p. 650]. If β is the canonical obstruction class for Y, then for any field extension L/k, the class β_L is determined by its values on the generators of $\operatorname{Pic}_Y(L)$, which are finite. The relatives Brauer group $\operatorname{Br}(Y_L/L)$ is, by Proposition 2.3, generated by the values of β at these generators, and therefore is finite. □

THEOREM 6.6. *Let $\alpha \in \operatorname{Br}(k)$ be a nontrivial Brauer class and k a field of characteristic 0. Then $[\alpha, \infty)_3$ is always finite and equal to the cyclic subgroup generated by α. If k is also uncountable, then $[\alpha, \infty)_2 = \langle \alpha \rangle$ as well.*

PROOF. By Proposition 6.3, we may find Y of dimension at most 3 such that $\operatorname{Br}(k(Y)/k) = \langle \alpha \rangle$. But therefore $[\alpha, \infty)_3 \subset \operatorname{Br}(k(Y)/k)$. But since $[\alpha, \infty)_3$ is a subgroup of $\operatorname{Br}(k)$ containing α, the result follows. The case of an uncountable field follows similarly from Proposition 6.4. □

THEOREM 6.7. *Let $\alpha \in \operatorname{Br}(k)$ be a nontrivial Brauer class over a field k. Then $[\alpha, \infty)_2$ is always finite.*

PROOF. By Proposition 6.5, we may find Y of dimension at most 2 such that $\operatorname{Br}(k(Y)/k)$ is finite and contains α. But since $[\alpha, \infty)_2$ is contained in $\operatorname{Br}(k(Y)/k)$, the result follows. □

Acknowledgements

The authors are very grateful to the anonymous referee for many helpful comments and corrections, in particular in helping us find an error in the original version of this paper, which contained an incorrect proof of an affirmative answer to Question 5.3.

References

[1] S. A. Amitsur, *Generic splitting fields of central simple algebras*, Ann. of Math. (2) **62** (1955), 8–43, DOI 10.2307/2007098. MR70624

[2] S. Bosch, W. Lütkebohmert, and M. Raynaud, *Néron models*, Ergebnisse der Mathematik und ihrer Grenzgebiete (3) [Results in Mathematics and Related Areas (3)], vol. 21, Springer-Verlag, Berlin, 1990, DOI 10.1007/978-3-642-51438-8. MR1045822

[3] V. I. Chernousov, A. S. Rapinchuk, and I. A. Rapinchuk, *On the genus of a division algebra* (English, with English and French summaries), C. R. Math. Acad. Sci. Paris **350** (2012), no. 17-18, 807–812, DOI 10.1016/j.crma.2012.09.014. MR2989381

[4] V. I. Chernousov, A. S. Rapinchuk, and I. A. Rapinchuk, *The genus of a division algebra and the unramified Brauer group*, Bull. Math. Sci. **3** (2013), no. 2, 211–240, DOI 10.1007/s13373-013-0037-z. MR3084007

[5] V. I. Chernousov, A. S. Rapinchuk, and I. A. Rapinchuk, *Division algebras with the same maximal subfields* (Russian, with Russian summary), Uspekhi Mat. Nauk **70** (2015), no. 1(421), 89–122, DOI 10.4213/rm9649; English transl., Russian Math. Surveys **70** (2015), no. 1, 83–112. MR3353117

[6] V. I. Chernousov, A. S. Rapinchuk, and I. A. Rapinchuk, *On the size of the genus of a division algebra*, Tr. Mat. Inst. Steklova **292** (Algebra, Geometriya i Teoriya Chisel) (2016), 69–99, DOI 10.1134/S0371968516010052; English transl., Proc. Steklov Inst. Math. **292** (2016), no. 1, 63–93. MR3628454

[7] V. I. Chernousov, A. S. Rapinchuk, and I. A. Rapinchuk, *The finiteness of the genus of a finite-dimensional division algebra, and some generalizations*, Israel J. Math. **236** (2020), no. 2, 747–799, DOI 10.1007/s11856-020-1988-x. MR4093903

[8] M. Ciperiani and D. Krashen, *Relative Brauer groups of genus 1 curves*, Israel J. Math. **192** (2012), no. 2, 921–949, DOI 10.1007/s11856-012-0057-5. MR3009747

[9] B. Fantechi, L. Göttsche, L. Illusie, S. L. Kleiman, N. Nitsure, and A. Vistoli, *Fundamental algebraic geometry: Grothendieck's FGA explained*, Mathematical Surveys and Monographs, vol. 123, American Mathematical Society, Providence, RI, 2005, DOI 10.1090/surv/123. MR2222646

[10] S. Garibaldi and D. J. Saltman, *Quaternion algebras with the same subfields*, Quadratic forms, linear algebraic groups, and cohomology, Dev. Math., vol. 18, Springer, New York, 2010, pp. 225–238, DOI 10.1007/978-1-4419-6211-9_13. MR2648728

[11] P. Griffiths and J. Harris, *Principles of algebraic geometry*, Wiley Classics Library, John Wiley & Sons, Inc., New York, 1994. Reprint of the 1978 original, DOI 10.1002/9781118032527. MR1288523

[12] A. Grothendieck, *Le groupe de Brauer. II. Théorie cohomologique* (French), Dix exposés sur la cohomologie des schémas, Adv. Stud. Pure Math., vol. 3, North-Holland, Amsterdam, 1968, pp. 67–87. MR244270

[13] A. Grothendieck, *Théorie des intersections et théorème de Riemann-Roch* (French), Lecture Notes in Mathematics, Vol. 225, Springer-Verlag, Berlin-New York, 1971. Séminaire de Géométrie Algébrique du Bois-Marie 1966–1967 (SGA 6); Dirigé par P. Berthelot, A. Grothendieck et L. Illusie. Avec la collaboration de D. Ferrand, J. P. Jouanolou, O. Jussila, S. Kleiman, M. Raynaud et J. P. Serre. MR354655

[14] R. Hartshorne, *Ample subvarieties of algebraic varieties*, Lecture Notes in Mathematics, Vol. 156, Springer-Verlag, Berlin-New York, 1970. MR282977

[15] D. Krashen, E. Matzri, A. Rapinchuk, L. Rowen, and D. Saltman, *Division algebras with common subfields*, Manuscripta Math. **169** (2022), no. 1-2, 209–249, DOI 10.1007/s00229-021-01315-5. MR4462663

[16] D. Krashen and K. McKinnie, *Distinguishing division algebras by finite splitting fields*, Manuscripta Math. **134** (2011), no. 1-2, 171–182, DOI 10.1007/s00229-010-0390-0. MR2745259

[17] A. F. Lopez and C. Maclean, *Explicit Noether-Lefschetz for arbitrary threefolds*, Math. Proc. Cambridge Philos. Soc. **143** (2007), no. 2, 323–342, DOI 10.1017/S0305004107000126. MR2364653

[18] Q. Ma, *Specializing Brauer classes in Picard schemes*, J. Pure Appl. Algebra **226** (2022), no. 2, Paper No. 106832, 10, DOI 10.1016/j.jpaa.2021.106832. MR4287491

[19] A. S. Rapinchuk and I. A. Rapinchuk, *On division algebras having the same maximal subfields*, Manuscripta Math. **132** (2010), no. 3-4, 273–293, DOI 10.1007/s00229-010-0361-5. MR2652433

Department of Mathematics, University of Pennsylvania, Philadelphia, Pennsylvania
Email address: `dkrashen@upenn.edu`

Department of Mathematics, University of Washington, Seattle, Washington
Email address: `lieblich@uw.edu`

Contemporary Mathematics
Volume **800**, 2024
https://doi.org/10.1090/conm/800/16057

Groups versus rings

Michael Larsen and Aner Shalev

Dedicated to the memory and heritage of S.A. Amitsur on his centennial

Abstract. We compare groups with rings in a range of contexts in which ring theory problems tend to be more tractable than their group theory analogues. The main topics discussed in this paper are probabilistic identities in groups, rings which are almost PI, and random generation of groups and algebras. We show, for example, that almost PI algebras are PI, while the analogous problem for groups is very much open.

1. Introduction

In this paper we focus on interactions — old and new — between group theory and ring theory.

Challenging problems in group theory (including the Restricted Burnside Problem and understanding groups admitting automorphisms with few fixed points) have been solved by reducing them to problems in Lie rings; see Zelmanov's papers [**60–62**] and the survey [**57**]. In earlier work, Golod's construction of an infinite finitely generated residually finite p-group [**10**] is based on his negative solution to Levitzki's problem whether finitely generated associative nil algebras are nilpotent.

The classical ring-theoretic Waring problem solved by Hilbert (cf. [**48**]) sparked wide interest in the recent theory of word maps and word width in groups, see for instance [**22–25, 29–32, 46, 55, 58**] and the references therein. In the other direction, results as well as open problems in group theory have sparked interest in analogous problems for algebras (associative and non-associative), some of which have recently been fully solved.

For example, in [**27**] the notion of almost PI algebras, which is related to groups with a probabilistic identity [**26**], is studied. We show that an almost PI associative algebra (or Lie algebra or Jordan algebra) is actually PI. The analogous problem for groups is still open. Various related quantitative results for algebras are also given. This is done in Section 2.

Section 3 is devoted to random generation of finite and profinite groups and algebras. Random generation of finite simple groups was a major research topic in

2020 *Mathematics Subject Classification.* Primary 16R99.

Key words and phrases. Almost PI algebras, probabilistic identities, random generation.

The first author was partially supported by NSF grant DMS-2001349. The second author was partially supported by ISF grant 686/17 and the Vinik Chair of mathematics which he holds. Both authors were partially supported by BSF grants 2016072 and 2020037.

the 1990s, following Dixon's proof of Netto's 1882 conjecture on random generation of symmetric and alternating groups. We then discuss new results by Sercombe and the second named author [**56**] establishing random generation of various finite and profinite associative algebras.

2. Groups with probabilistic identities and almost PI rings

2.1. Probabilistic identities in groups. By a *word* we mean an element $w = w(x_1, \dots, x_n)$ of the free group F_n on $x_1, \dots, x_n$. Such a word w induces a word map $w_G : G^n \to G$ on any group G, induced by substituting group elements $g_i \in G$ for the free generators x_i.

If the word map w_G is identically 1, we say that w is an *identity* of G. If G is a finite or a profinite group, then we also view it as a probability space with respect to the uniform measure or the normalized Haar measure on G, respectively. In these cases, we let $P_{w,G}(g)$ where $g \in G$ denote the probability that $w_G(g_1, \dots, g_n) = g$. Let $\epsilon > 0$. We say that w is an *ϵ-probabilistic identity* of G if $P_{w,G}(1) \geq \epsilon$. For G infinite, we say that w is a probabilistic identity of G if $P_{w,G}(1) > 0$. More generally, we say that a word w is a non-homogeneous probabilistic identity of G if $P_{w,G}(g) > 0$ for some element $g \in G$.

We extend these notions to abstract infinite residually finite groups Γ by passing to their profinite completions $G = \hat{\Gamma}$.

See [**24, 25**] as well as [**26**] for a systematic study of probabilistic aspects of word maps.

The case where the word w is the commutator word $[x_1, x_2]$ was studied long ago by Peter Neumann [**49**]. He proved that if this word is an ϵ-probabilistic identity of a finite group G, then G has a normal subgroup H such that both $|G : H|$ and $|H'|$ are bounded above in terms of ϵ (so G is bounded-by-abelian-by-bounded). This implies that G satisfies an identity v which depends only on ϵ; indeed, we may take $v = [x_1^N, x_2^N]^N$ for a suitable $N = N(\epsilon)$.

See also Mann [**41**] for a similar result on finite groups in which x^2 is an ϵ-probabilistic identity.

It is a notoriously difficult open problem whether every probabilistic identity in a residually finite group Γ implies an actual identity; this problem is wide open even in the case where Γ is finitely generated. See [**25**], [**59**] and [**43**] for further discussion of this and related problems. A positive answer is given in [**59**] when the probabilistic identity w is a left normed commutator $[x_1, x_2, \dots, x_n]$ of arbitrary length.

There is one setting, however, where we have a satisfactory general result stating that any probabilistic identity implies an actual identity.

In [**25**], we gave a probabilistic analogue of the Tits alternative:

THEOREM 2.1. *Let Γ be a finitely generated linear group. Then Γ satisfies a probabilistic identity if and only if it is virtually solvable.*

Since every virtually solvable group satisfies an actual identity, we have the following corollary:

THEOREM 2.2. *A finitely generated linear group satisfies a probabilistic identity if and only if it satisfies an identity.*

The above result is extended in [**26**] as follows.

THEOREM 2.3. *Let G be a profinite group with infinitely many non-isomorphic non-abelian upper composition factors. Then G does not have a (homogeneous or non-homogeneous) probabilistic identity. Consequently, G is randomly free, namely, for any integer $n \geq 1$, the probability that randomly chosen elements $g_1, \ldots, g_n \in G$ freely generate a free subgroup (isomorphic to F_n) is 1.*

Recall that an upper composition factor of a profinite group G is a composition factor of some finite quotient of G. The second part of Theorem 2.3 follows from the first part using the σ-additivity of the Haar measure. This solves Problem 7 from [**6**], showing that a finitely generated linear group is either virtually solvable or randomly free.

2.2. Almost PI rings. In this subsection, we study the analogous problem for associative algebras and Lie algebras. Here, the possible identities are polynomials in a non-commuting set of variables (or Lie polynomials) rather than elements of a free group. Note that the notion of the degree of a polynomial in the free object of the respective category depends on this category (e.g., associative algebras, Lie algebras). We introduce the notion of an almost identity of an algebra as an analogue of a probabilistic identity. The key fact is that algebras with an almost identity satisfy an actual identity. We are interested in the case of infinite-dimensional algebras as well as in related quantitative results for finite dimensional simple algebras. The results in this section come from [**27**] and [**28**].

Throughout this section let k denote an algebraically closed field of arbitrary characteristic. We begin with the case that R is an associative k-algebra. Let n be a positive integer and $V = R^n$. For each non-commutative polynomial $Q \in k\langle x_1, \ldots, x_n\rangle$, evaluation of Q defines a map $e_Q\colon R^n \to R$. Let $R_Q := e_Q^{-1}(0)$ denote the set of $\mathbf{r} = (r_1, \ldots, r_n) \in V$ such that $e_Q(\mathbf{r}) = 0$. We say Q is a *c-almost identity* if R_Q contains a subset of the form $V_1 \times X_2$ where V_1 is a k-vector space, $V = V_1 \oplus V_2$, V_2 is finite dimensional as a vector space over k, and X_2 is an algebraic subset of V_2 of codimension $\leq c$. We say Q is an almost identity if it is a c-almost identity for some c or, equivalently, if R_Q contains an affine k-subspace of V of finite codimension.

The main result in this section is a strong form of the statement that every associative algebra satisfying an almost identity is PI. Namely, fixing the number of variables, the degree, and the codimension, there exists a universal identity for all associative algebras and all almost identities of the given type:

THEOREM 2.4. *Given c, d and n there exist an integer m and a non-zero non-commutative polynomial $P \in k\langle y_1, \ldots, y_m\rangle$ with the following property. If $Q \in k\langle x_1, \ldots, x_n\rangle$ is a non-commutative polynomial of degree d in n variables, and R is any associative k-algebra such that Q is a c-almost identity of R, then $P(r_1, \ldots, r_m) = 0$ for all $r_1, \ldots, r_m \in R$.*

For the matrix algebra $R = M_s(k)$, if P is an identity of R then $s \leq \deg P/2$. (This is, in fact, optimal by the Amitsur-Levitski theorem [**2**].) Thus, Theorem 2.4 implies that for any fixed Q, as $s \to \infty$, the codimension of R_Q grows without bound. Actually, much more is true:

THEOREM 2.5. *For each positive integer d there exist real numbers $a > 0$ and b such that if $Q \in k\langle x_1, \ldots, x_n\rangle$ is a non-commutative polynomial of degree $d \geq 0$, then the codimension of $M_{s,Q}$ in M_s^n is at least $as^2 + b$ for all positive integers s.*

An immediate corollary of Theorem 2.5 is that if R is an associative algebra generated by n elements, then the dimension of the representation variety $\mathrm{Hom}_{k-\mathrm{alg}}(R, M_s)$ is either ns^2 for all s (in which case R is freely generated by these elements) or at most $(n-\epsilon)s^2$ for s sufficiently large, where $\epsilon > 0$ depends only on the minimal degree of a relator. Combining Amitsur's theorem on embedding of PI rings [**1**] with Theorems 2.4 and 2.5, we can easily deduce the following:

COROLLARY 2.6. *For each $d > 0$ there exist real numbers $f = f(d), g = g(d)$ such that the following holds. Let R be an associative k-algebra having a c-almost identity of degree d. Suppose the Jacobson radical of R (i.e., the intersection of all primitive left ideals) is 0. Then R is a k-subalgebra of some algebra of the form $M_s(C)$, where C is a commutative k-algebra and $s \leq fc^{1/2} + g$.*

Combining Theorem 2.5 with the Lefschetz trace formula for varieties over finite fields and known upper bounds on Betti numbers for affine varieties defined by polynomials of bounded degree over finite fields (see [**18**]), we obtain the following result:

PROPOSITION 2.7. *For every $d \in \mathbb{N}$ there exist positive real numbers $\epsilon = \epsilon(d), N = N(d)$ depending only on d such that the following holds. If $\mathbb{F}$ is a finite field, $M = M_s(\mathbb{F})$ is a matrix algebra of dimension $\geq N$, and Q is a non-zero n-variable associative polynomial of degree d over $\mathbb{F}$, then*

$$|M_Q| \leq |M|^{n-\epsilon}.$$

In fact, all fibers of the evaluation map $e_Q : M^n \to M$ have size at most $|M|^{n-\epsilon}$.

This result may be viewed as a ring-theoretic analogue of [**24**, Theorem 1.2].

In particular, it follows that if M satisfies a probabilistic identity Q (namely, the probability that e_Q vanishes on a random n-tuple of elements of M is at least some fixed $\delta > 0$), then $\dim M$ is bounded from above (in terms of Q and δ).

Next we turn to almost nil algebras and algebras with an almost identity of the form x^d. The following lemma is an easy consequence of the structure theory of finite-dimensional associative algebras.

LEMMA 2.8. *Let R be a finite dimensional associative k-algebra, and let N be the subvariety of nilpotent elements of R. If $\operatorname{codim} N = c$ then $\dim R/J(R) \leq c^2$, so R has a nilpotent ideal of codimension at most c^2.*

The lemma can be extended as follows. For each monic polynomial $P(x) \in k[x]$ of degree n, the codimension of the variety of matrices in $M_s(k)$ with characteristic polynomial dividing P is at least s. It follows that if P is a c-almost identity of a finite dimensional associative k-algebra R, then $\dim R/J(R) \leq c^2$.

The well-known Nagata-Higman Theorem states that, if x^d is an identity of an associative (non-unital) k-algebra R, where the characteristic of k is zero or greater than d, then R is nilpotent. See [**7**, Chapter 6] for this and for explicit bounds on the degree of nilpotency of R in terms of d. For finitely generated algebras, we have an analogous result when x^d is an almost identity.

THEOREM 2.9. *Let R be an associative k-algebra, and suppose x^d is an almost identity of R. If R is finitely generated then R is virtually nilpotent, that is, it has a two-sided nilpotent ideal of finite codimension.*

The proof proceeds in two steps. First, we show that if x^d is a c-almost identity, then $\dim R/J(R) \leq c^2$. This part does not require finite generation. For finite-dimensional algebras, it is just Lemma 2.8, and it follows also for residually finite-dimensional algebras. On the other hand, it is not difficult to deduce residual finiteness of R from Corollary 2.6.

For the second step, we use Theorem 2.4 to prove R is a PI-algebra. If it is also finitely generated, the Kemer-Braun Theorem [**4**] implies that $J(R)$ is nilpotent.

For R residually finite-dimensional, we have a result [**28**] in the spirit of Theorem 2.9 for every homogeneous polynomial Q:

THEOREM 2.10. *If R is a residually finite dimensional associative k-algebra and Q is a homogeneous almost identity on R, then Q is an identity on some ideal I of finite codimension in R.*

A residually finite dimensional algebra R over a finite field is a dense subring of its profinite completion $\hat{R}$, which can be regarded as a probability space with respect to additive Haar measure. A polynomial Q in n variables is a *probabilistic identity* if $\hat{R}_Q$ has positive measure in $\hat{R}^n$. In [**28**] we prove:

THEOREM 2.11. *A homogeneous polynomial Q is a probabilistic identity for a residually finite dimensional algebra R over a finite field if and only if there exist a two-sided ideal I of finite codimension in R and elements $r_1, \ldots, r_n \in R$ such that the restriction of Q to $(r_1 + I) \times \cdots \times (r_n + I)$ is identically zero.*

A number of the results described in this section have Lie algebra analogues which can be proved with varying amounts of extra work. For Theorem 2.4, if we replace R by a Lie algebra L and non-commutative polynomials with Lie polynomials, the same result holds. In fact, this applies to Jordan algebras as well (cf. [**27**]). The additional ingredient needed is the fact that for fixed n, the dimension of the space of Lie polynomials in n variables of degree $\leq d$ grows faster than $(n-1)^d$.

The natural analogue of Theorem 2.5 seems to be a statement about general simple Lie algebras L over k. If we assume k is of characteristic 0, we can use the Killing-Cartan classification to show that every such L has a full matrix subalgebra M whose dimension can be bounded below by a constant multiple of $\dim L$. The codimension of L_Q in L^n is less than or equal to the codimension of $L_Q \cap M^n$ in M^n, which is less than or equal to the codimension of M_Q in M^n. This gives:

COROLLARY 2.12. *For each $d > 0$ there exist real numbers $a > 0$ and b such that if Q is a non-zero Lie polynomial of degree d in n variables and L is a finite dimensional simple Lie algebra over k of characteristic zero, then the codimension of the zero-set of Q in L^n is at least $a \dim L + b$.*

The Lie algebra version of Proposition 2.7 is straightforward.

When k has characteristic 0, Lemma 2.8 has the following Lie algebra version:

LEMMA 2.13. *If L is finite-dimensional a Lie algebra in characteristic zero, and the variety of ad-nilpotent elements of L has codimension c, then $\dim L/\mathrm{Rad}\,(L) \leq 4c^2 - c$.*

For L of type A_1 and E_8, this is sharp, and it is easy to check that it holds for all simple Lie algebras in characteristic zero.

Next, we turn to general (not necessarily associative) finite algebras. In [**28**] we show the following.

THEOREM 2.14. *Let R be any finite-dimensional algebra over a finite field, and $e_Q : R^n \to R$ a polynomial map associated with a polynomial Q of degree d in n variables. If $|e_Q^{-1}(0)| \geq 1 - 2^{-d}$ then Q is identically zero on R^n.*

Now, for $d \geq 2$ consider the degree d-Engel polynomial $E_{d-1} = [x, y, \cdots, y]$ a left-normed Lie product where y appears $d-1$ times, as an element of the free Lie algebra on x, y over the underlying field k. By Theorem 2.1 of Mann and Martinez [**44**], if L is a finite Lie algebra in which the Engel condition $E_{d-1} = 0$ holds with probability greater than $1 - 2^{-d}$ then E_{d-1} is an identity of L. Theorem 2.14 above extends this to any finite algebra and any polynomial.

Going back to arbitrary (possibly infinite-dimensional) Lie algebras, it would be interesting to know whether there is a Lie analogue of Theorem 2.9, specifically:

PROBLEM 2.15. Let L be a finitely generated Lie algebra. Suppose that for some $n \in \mathbb{N}$ the n-Engel polynomial $E_n = [x, y, \ldots, y]$ (where y occurs n times) is an almost identity of L. Does it follow that L is virtually nilpotent?

3. Random generation: from groups to algebras

3.1. Random generation of groups. We start with some background on random generation of finite groups with emphasis on finite simple groups. The following 140-year-old conjecture played a key role in many later developments.

CONJECTURE 3.1 (Netto, 1882). *Two randomly chosen permutations from S_n generate S_n or A_n with probability tending to 1 as $n \to \infty$.*

Key tools for proving this conjecture were provided by Erdős and Turán, in a series of seven papers published in 1965–1972, dealing with random permutations, see e.g., [**8**, **9**].

A useful lemma from [**9**] shows that the probability that a permutation in S_n has no cycles of length $a_1, \ldots, a_k$ is at most $(\sum_{i=1}^k a_i^{-1})^{-1}$.

In 1969 Dixon proved Netto's Conjecture, using the above lemma, some number theory, and Jordan's work on primitive permutation groups, see [**5**]. In the same paper Dixon posed the following

CONJECTURE 3.2 (Dixon, 1969). *Two randomly chosen elements of a finite simple group G generate G with probability tending to 1 as $|G| \to \infty$.*

This was a daring conjecture, since it was only proved for alternating groups, and it preceded the classification of finite simple groups (and even the discovery of some finite simple groups).

The proof of Dixon's conjecture was subsequently completed in [**33**], following [**17**] and [**5**].

More refined random generation results for finite simple groups were gradually obtained. The following conjecture was stated in [**17**].

CONJECTURE 3.3 (Kantor-Lubotzky, 1990). *A random involution and a random additional element generate a (nonabelian) finite simple group G with probability tending to 1 as $|G| \to \infty$.*

This conjecture was proved in [**34**]. What happens if we also restrict the order of the second generator? This question is partly motivated by a classical problem from the early 1900s: which finite simple groups are quotients of the modular group $\mathrm{PSL}_2(\mathbb{Z})$?

Since $\mathrm{PSL}_2(\mathbb{Z}) \cong C_2 \star C_3$, a group G is a quotient of the modular group if and only if it is $(2,3)$-generated, i.e., $G = \langle x, y \rangle$ such that $x^2 = y^3 = 1$. For example, the alternating groups $A_n, n \geq 9$ are $(2,3)$-generated as shown by G.A. Miller already in 1901.

Various $(2,3)$-generation results were subsequently obtained. At first the proofs were explicit in the sense that specific generating pairs of orders 2 and 3 were found. Then it was discovered that, using a probabilistic approach, more general results can be obtained. Let $P_{2,3}(G)$ denote the probability that random elements $x, y \in G$ satisfying $x^2 = y^3 = 1$ generate G.

The following theorem was proved by Liebeck and Shalev for alternating and classical groups [**35**], and subsequently by Lübeck and Malle for exceptional groups [**40**]. See also the recent paper [**11**] for related additional results by Guralnick, Liebeck, Lübeck and Shalev.

THEOREM 3.4. *For finite simple groups* $G \neq \mathrm{Sz}(2^k), \mathrm{PSp}_4(p^k)$, *we have* $P_{2,3}(G) \to 1$ *as* $|G| \to \infty$. *If* $G = \mathrm{PSp}_4(p^k)$ *where* $p \geq 5$, *we have* $P_{2,3}(G) \to 1/2$ *as* $|G| \to \infty$.

Consequently, excluding $\mathrm{Sz}(2^k), \mathrm{PSp}_4(2^k), \mathrm{PSp}_4(3^k)$ *and finitely many other groups, all finite simple groups are quotients of the modular group.*

The Suzuki groups are trivially not $(2,3)$-generated since they don't have elements of order 3. However, the groups $\mathrm{PSp}_4(2^k), \mathrm{PSp}_4(3^k)$ have elements of order 3, and it was surprising to find out that none of them is $(2,3)$-generated.

The modular group is an important example of a Fuchsian group. The theorem above studies the probability that a random homomorphism from it to a finite simple group G is surjective. See also [**36, 37**], where similar problems are solved in the wider context of Fuchsian groups.

We now turn to random generation in more general types of groups, which need not be simple or finite.

Let G be a profinite group, viewed as a probability space with the normalized Haar measure. For $d \geq 1$ let $P(G,d)$ denote the probability that d random elements of G generate G topologically. We say that G is *Positively Finitely Generated* (PFG) if $P(G,d) > 0$ for some $d \in \mathbb{N}$. A residually finite group is said to be PFG if its profinite completion is PFG.

PFG groups were first studied in 1992 by Jarden and Lubotzky [**16**] in the context of field arithmetic. In 1996, Mann proved that finitely generated prosolvable groups and some Chevalley groups over the integers are PFG [**42**]. This result was extended by Borovik, Pyber and Shalev [**3**] who proved that finitely generated profinite groups not having some alternating group A_d as an upper section (as in the celebrated Babai-Cameron-Pálfy Theorem) are PFG.

PFG groups were subsequently characterized by Mann and Shalev [**45**] in 1997, who showed that G is PFG if and only G has *Polynomial Maximal Subgroup Growth* (PMSG), i.e., the number $m_n(G)$ of maximal (open) subgroups of index n in G is bounded above by n^c for some fixed constant c. While the 'if' part is easy and was observed before, the proof of the 'only if' part relies on the classification of finite simple groups, the study of their maximal subgroups [**19**], and on tools from probability theory, including the Borel-Cantelli Lemma.

In 2002, Lubotzky [**38**] obtained an effective finitary version of this theorem and proved a related conjecture of Pak. We need some notation. Let G be a finitely generated profinite group. If we do not specify a base, log refers to base 2. Set

$$M(G) = \sup_{n>1} \log m_n(G)/\log n, \quad M^*(G) = \limsup_{n>1} \log m_n(G)/\log n,$$

which measure the degree of polynomial subgroup growth of G (and are infinite unless G has PMSG).

Define the *Pomerance invariant* of a profinite group G by

$$V(G) = \min\{d \geq 1 : P(G,d) \geq e^{-1}\},$$

introduced by Pomerance for finite abelian groups [**53**].

Let $x = (x_d)_{d\in\mathbb{N}}$ be a sequence of elements of G that are chosen randomly, uniformly and independently. Define a random variable τ_G by

$$\tau_G = \min\{d \geq 1 \,|\, \langle x_1, ..., x_d\rangle = G\} \in \mathbb{N} \cup \{+\infty\},$$

namely, the number of random elements of G required in order to generate G. Let $E(G)$ denote the expectation of τ_G. Observe that

$$E(G) = \sum_{d\geq 1} dP(\tau_G = d) = \sum_{d\geq 1}\Big(\sum_{c\geq d} P(\tau_G = c)\Big) = \sum_{d\geq 0}(1 - P(G,d)).$$

We can now state results from [**38**]. Let $d(G)$ denote the minimal number of generators of a group G.

THEOREM 3.5 (Lubotzky, 2002). *Let G be a finite group. Then*

(i) $M(G) \leq d(G) + 2\log\log|G| + 2$.
(ii) $V(G) \leq d(G) + 2\log\log|G| + 4.02$.
(iii) $E(G) \leq ed(G) + 2e\log\log|G| + 11$.

In 2011, Jaikin-Zapirain and Pyber provided a deep and more structural characterization of PFG groups with a wealth of applications to permutation groups, linear groups and related topics; see [**15**].

It is natural to ask whether $|M(G) - E(G)|$ can be arbitrarily large. A negative answer was recently provided by Lucchini and Moscatiello [**39**] who showed the following.

THEOREM 3.6 (Lucchini and Moscatiello, 2020). *For every finite group G we have $M(G) - 4 \leq E(G) \leq M(G) + 3$.*

3.2. Random generation of algebras. What can be said about random generation results for rings? We shall discuss recent results in this direction, mainly about finite (or sometimes profinite) algebras, recently obtained by Sercombe and the second named author [**56**].

We start with some background from ring theory, with emphasis on Wedderburn's classical results. Let q be a prime power, set $k = \mathbb{F}_q$, and let A be a finite algebra over k. The Wedderburn-Malcev Principal Theorem (Theorems 5.3.20 and 5.3.21 of [**54**]) states that there exists a semisimple subalgebra S of A such that $A = S \oplus J(A)$ as vector spaces, and that S is unique up to conjugation by elements of $1 + J(A)$. Wedderburn's Little Theorem (Theorem 7.1.11 of [**54**]) states that all finite division algebras are fields. Combining this with another theorem of Wedderburn (Theorem 2.1.8 of [**54**]), it follows that there is an algebra isomorphism $S \cong \prod_{i=1}^{r} M_{n_i}(q^{m_i})$ for positive integers r, n_i, m_i which are unique up to permutation of the simple factors.

We are interested in studying random generation of finite algebras A by two (or more) elements. Let $P(A)$ denote the probability that two elements of A chosen uniformly and independently will generate A as a (unital) k-algebra, i.e.,

$$P(A) = \frac{|\{(x,y) \in A \times A : \langle x, y\rangle = A\}|}{|A|^2}.$$

More generally, fix a natural number $d \in \mathbb{N}$, and let $P(A,d)$ be the probability that d random elements $a_1, \dots a_d \in A$ generate A as a unital k-algebra.

The maximal subalgebras of A provide essential information on the above probabilities. Indeed, the elements $a_1, \dots, a_d \in A$ don't generate A if and only if there is a maximal subalgebra B max A such that $a_i \in B$ $(1 \leq i \leq d)$. It follows that

$$1 - P(A,d) \leq \sum_{B \text{ max } A} |B|^d/|A|^d = \sum_{B \text{ max } A} |A:B|^{-d},$$

and

$$1 - P(A) = 1 - P(A,2) \leq \sum_{B \text{ max } A} |A:B|^{-2},$$

where the sums are over the maximal subalgebras of A.

We now discuss maximal subalgebras of finite simple algebras. Let $A = M_n(q^m)$, a simple k-algebra. For $1 \leq l \leq n-1$ let $P_{l,n-l}(q^m)$ be the subalgebra of A consisting of all block upper triangular matrices with two diagonal blocks of sizes l and $n-l$, respectively.

Let r be a positive integer. There is a natural embedding of $\mathbb{F}_{q^r}$ in $M_r(q)$ via the left regular representation. If r divides n, then this extends to an embedding of $M_{n/r}(q^{mr})$ in $M_n(q^m)$. If r divides m, then the subfield $\mathbb{F}_{q^{m/r}}$ of $\mathbb{F}_{q^m}$ extends naturally to a subalgebra $M_n(q^{m/r})$ of $M_n(q^m)$.

Set $\mathcal{P}(r) :=$ the set of prime divisors of r, and $\omega(r) := |\mathcal{P}(r)|$.

We define three sets of subalgebras of A;

$S1 := \{P_{l,n-l}(q^m) \mid l \in \mathbb{N}, l < n\}$,
$S2 := \{M_{n/a}(q^{ma}) \mid a \in \mathcal{P}(n)\}$, and
$S3 := \{M_n(q^{m/b}) \mid b \in \mathcal{P}(m)\}$.

A subalgebra of A that is conjugate to an element of $S1$ (resp. $S2$, $S3$) is said to be *of type* $(S1)$ (resp. $(S2)$, $(S3)$).

Iovanov and Sistko [**14**] showed that $S1 \cup S2 \cup S3$ is a set of representatives of the isomorphism classes of maximal subalgebras of A; we observed that the same holds for conjugacy classes (showing that isomorphic maximal subalgebras are conjugate).

Next, we discuss maximal subalgebras of general finite algebras.

By the Wedderburn-Malcev Principal Theorem, there exists a semisimple subalgebra S of A such that $A = S \oplus J(A)$. Decompose $S = \prod_{i=1}^{r} S_i$ where each S_i is simple. Let $i \in \{1, \dots, r\}$. Write $S_i = M_{n_i}(q^{m_i})$ for some integers m_i and n_i. Let $\mathcal{B}_i$ be the standard set of representatives of the conjugacy classes of maximal subalgebras of S_i. If $S_j \cong S_i$ for some $j \neq i$ then let $S_{ij} \subset S_i \times S_j$ denote the graph of an isomorphism.

We define three sets of subalgebras of A;

$$T1 := \{(B_j \times \prod_{i\neq j} S_i) \oplus J(A) \,|\, 1 \leq j \leq r; B_j \in \mathcal{B}_j\},$$
$$T2 := \{(S_{j_1 j_2} \times \prod_{i\neq j_1,j_2} S_i) \oplus J(A) \,|\, 1 \leq j_1 < j_2 \leq r, S_{j_1} \cong S_{j_2}\},$$
$$T3 := \{S \oplus H \,|\; H \text{ ideal of } A, \text{ maximal with respect to } H \subset J(A)\}.$$

It is shown in [**14**] that $T1 \cup T2 \cup T3$ is a set of representatives of the isomorphism classes of maximal subalgebras of A, and we observe that the same holds for conjugacy classes.

The study of random generation of finite simple associative algebras goes back to Peter Neumann and Cheryl Praeger [**50**]; among other things they show there that, if $A = M_n(q)$ considered as a k-algebra ($k = \mathbb{F}_q$), then $P(A) \to 1$ as $|A| \to \infty$.

In [**20**] R. Kravchenko, M. Mazur and B.V. Petrenko study generators of matrix algebras over finite fields. In particular, they show that, for $A = M_n(q^m)$ considered as an $\mathbb{F}_q$-algebra, we have

$$P(A) \to 1 \text{ as } |A| \to \infty.$$

It follows from the above results that finite simple algebras are randomly generated by two elements.

Since all finite simple algebras are 2-generated, it follows that there is an absolute constant $\delta > 0$ such that $P(A) \geq \delta$ for all finite simple algebras A.

We can now discuss recent results from [**56**].

THEOREM 3.7. *Let A be a finite simple algebra. Then $P(A) \geq 3/8$, with equality if and only if $A = M_2(2)$.*

This may be regarded as a ring-theoretic analogue of a theorem of N.E. Menezes, M. Quick and C.M. Roney-Dougal [**47**] showing that for finite simple groups G, $P(G) \geq 53/90$, with equality if and only if $G = A_6$.

Let $m(A)$ denote the minimal index of a proper subalgebra of a finite algebra A.

THEOREM 3.8. *Let A be a finite simple algebra that is not a field. Then $P(A) = 1 - \kappa(A)m(A)^{-1} + O(m(A)^{-4/3})$, where $1 < \kappa(A) < 4$.*

This result may be regarded as an analogue of a result from [**34**] for arbitrary finite simple groups.

Next, we discuss random generation of algebras by special elements.

Given a finite k-algebra A, let $P_N(A)$ denote the probability that two nilpotent elements of A chosen uniformly at random will generate A as a k-algebra.

THEOREM 3.9. *Let A be a finite simple algebra that is not a field. Then $P_N(A) \to 1$ as $|A| \to \infty$.*

Let $A = M_n(q^m)$ over $k = \mathbb{F}_q$. Let f be a monic polynomial of degree n over $\mathbb{F}_{q^m}$, and let $P_f(A)$ denote the probability that two random elements of A with characteristic polynomial f will generate A.

THEOREM 3.10. *Let A be a finite simple algebra that is not a field, say $A = M_n(q^m)$ for $n > 1$. Let f be a monic polynomial of degree n over $\mathbb{F}_{q^m}$. Then $P_f(A) \to 1$ as $|A| \to \infty$.*

The above result can be extended as follows. Let g be another monic polynomial of degree n over $\mathbb{F}_q$, and let $P_{f,g}(A)$ denote the probability that random elements x, y with characteristic polynomials f, g respectively generate A. Then we have $P_{f,g}(A) \to 1$ as $|A| \to \infty$.

Theorem 3.10 for $f(X) = X^n$ immediately implies Theorem 3.9. Note that Theorem 3.9 and therefore Theorem 3.10 do not hold when A is a field since then the only nilpotent element of A is 0.

Next we consider random generation of a finite simple algebra by two matrices of a given rank. Let $r \geq 0$, $A = M_n(q^m)$ where $n \geq r$. Let $P_r(A)$ denote the probability that two random matrices in A of rank r will generate A.

THEOREM 3.11. *Let $A = M_n(q^m)$, where $n \geq 2, m \geq 1$ and let p be the smallest prime divisor of n. Let $1 < r \leq n$ be an integer.*

(i) *If $r \leq n/p$ then $P_r(A) \leq 1 - \frac{81r^2}{256p^2q^{2m(p-1)r}}$.*

(ii) *If $r > n - \sqrt{n/8}$ then $P_r(A) \to 1$ as $|A| \to \infty$.*

For every fixed r the statement that $P_r(A) \to 1$ as $|A| \to \infty$ fails to be true. This follows from Theorem 3.11(i): fix q, m, p and let n tend to infinity.

Let $P^\times(A)$ be the probability that two random invertible elements of A will generate A as a k-algebra. Theorem 3.11(ii) for $r = n$ immediately yields:

COROLLARY 3.12. *Let A be a finite simple algebra. Then $P^\times(A) \to 1$ as $|A| \to \infty$.*

We now turn to the study of positively finitely generated profinite algebras A over $k = \mathbb{F}_q$; such an algebra is equipped with a normalized σ-additive Haar measure, turning it into a probability space.

Given such a profinite algebra A, define:

(1) $P(A, d) :=$ the probability that d randomly chosen elements of A generate A (topologically if A is infinite).
(2) A is *Positively Finitely Generated* (PFG) if $P(A, d) > 0$ for some d.
(3) A has *Polynomial Maximal Subalgebra Growth* (PMSG) if the number $m_n(A)$ of index n (open) maximal subalgebras of A is at most n^c for some fixed constant c.

In this case, since $1 - P(A, d) \leq \sum_{n>1} m_n(A)n^{-d} \leq \sum_{n>1} n^{c-d}$, for $d \geq c + 2$ we have $1 - P(A, d) \leq \zeta(2) - 1 = \pi^2/6 - 1 < 1$ where ζ is the Riemann zeta function. Hence $P(A, d) > 0$.

Therefore PMSG $\Longrightarrow$ PFG for algebras. Is it if and only if?

Set

$$M(A) = \sup_{n>1} \log m_n(A)/\log n, \quad M^*(A) = \limsup_{n>1} \log m_n(A)/\log n,$$

which measure the degree of polynomial subalgebra growth of A. These invariants are infinite unless A has PMSG.

Define the *Pomerance invariant* of a profinite algebra A by

$$V(A) = \min\{d \geq 1 : P(A, d) \geq e^{-1}\},$$

introduced in [**53**] for finite abelian groups.

Our main result on PFG algebras is the following.

THEOREM 3.13. *Let A be a profinite algebra. Then A is PFG if and only if A has PMSG. Moreover, if A is infinite we have $M^*(A) \leq V(A) + 1$.*

The bound above is better than related bounds obtained for profinite groups. Define the *Pak invariant* of a finite or profinite $\mathbb{F}_q$-algebra A by

$$E(A) = \sum_{d \geq 1} dP(A, d),$$

the expected number of random elements of A which generate A. This is a ring-theoretic analogue of a similar invariant defined by Pak for finite groups. We have

$$e^{-1}E(A) \leq V(A) \leq e(e-1)^{-1}E(A).$$

Indeed, this was proved by Pak for groups [**51**], and the same argument works for algebras.

Our next result establishes bounds on these invariants.

THEOREM 3.14. *Let A be a finite algebra. Then*
(i) $M(A) \leq 3 \log_q \log_q |A| + 5.69$.
(ii) $V(A) \leq \lceil 3 \log_q \log_q |A| + 7.71 \rceil$.
(iii) $E(A) \leq \lceil 3e \log_q \log_q |A| + 20.96 \rceil$.

Unlike in Theorem 3.5 for groups, in the above result for algebras there is no dependence on the minimal number of generators.

The proofs of Theorems 3.13 and 3.14 apply the Borel-Cantelli Lemma, as well as many of our previously stated results.

Define the *core* B_A of a subalgebra $B \leq A$ to be the (unique) maximal two-sided ideal of A which is contained in B. If $B_A = 0$ we say that B has trivial core.

A crucial result in the proofs of Theorems 3.13 and 3.14 is the following.

THEOREM 3.15. *Let A be a finite algebra. Then, for all $n > 1$, A has at most $6.93n$ maximal subalgebras of index n with trivial core.*

Lubotzky's proof of Theorem 3.5 relies on the following crucial result of Pyber (using CFSG): Let G be a finite group. Then, for all $n > 1$, G has at most cn^2 maximal subgroups of index n with trivial core, where c is some fixed non-explicit constant. Comparing the upper bounds in groups and in algebras, we encounter yet again the advantage of being an algebra.

References

[1] A. S. Amitsur, *An embedding of PI-rings*, Proc. Amer. Math. Soc. **3** (1952), 3–9, DOI 10.2307/2032446. MR45694

[2] A. S. Amitsur and J. Levitzki, *Minimal identities for algebras*, Proc. Amer. Math. Soc. **1** (1950), 449–463, DOI 10.2307/2032312. MR36751

[3] A. V. Borovik, L. Pyber, and A. Shalev, *Maximal subgroups in finite and profinite groups*, Trans. Amer. Math. Soc. **348** (1996), no. 9, 3745–3761, DOI 10.1090/S0002-9947-96-01665-0. MR1360222

[4] A. Braun, *The nilpotency of the radical in a finitely generated PI ring*, J. Algebra **89** (1984), no. 2, 375–396, DOI 10.1016/0021-8693(84)90224-2. MR751151

[5] J. D. Dixon, *The probability of generating the symmetric group*, Math. Z. **110** (1969), 199–205, DOI 10.1007/BF01110210. MR251758

[6] J. D. Dixon, L. Pyber, Á. Seress, and A. Shalev, *Residual properties of free groups and probabilistic methods*, J. Reine Angew. Math. **556** (2003), 159–172, DOI 10.1515/crll.2003.019. MR1971144

[7] V. Drensky and E. Formanek, *Polynomial identity rings*, Advanced Courses in Mathematics. CRM Barcelona, Birkhäuser Verlag, Basel, 2004, DOI 10.1007/978-3-0348-7934-7. MR2064082

[8] P. Erdős and P. Turán, *On some problems of a statistical group-theory. I*, Z. Wahrscheinlichkeitstheorie und Verw. Gebiete **4** (1965), 175–186 (1965), DOI 10.1007/BF00536750. MR184994

[9] P. Erdős and P. Turán, *On some problems of a statistical group-theory. II*, Acta Math. Acad. Sci. Hungar. **18** (1967), 151–163, DOI 10.1007/BF02020968. MR207810

[10] E. S. Golod, *On nil-algebras and finitely approximable p-groups* (Russian), Izv. Akad. Nauk SSSR Ser. Mat. **28** (1964), 273–276. MR161878

[11] R. M. Guralnick, M. W. Liebeck, F. Lübeck, and A. Shalev, *Zero-one generation laws for finite simple groups*, Proc. Amer. Math. Soc. **147** (2019), no. 6, 2331–2347, DOI 10.1090/proc/14404. MR3951415

[12] J. E. Humphreys, *Introduction to Lie algebras and representation theory*, Graduate Texts in Mathematics, Vol. 9, Springer-Verlag, New York-Berlin, 1972. MR323842

[13] J. E. Humphreys, *Conjugacy classes in semisimple algebraic groups*, Mathematical Surveys and Monographs, vol. 43, American Mathematical Society, Providence, RI, 1995, DOI 10.1090/surv/043. MR1343976

[14] M. C. Iovanov and A. H. Sistko, *Maximal subalgebras of finite-dimensional algebras*, Forum Math. **31** (2019), no. 5, 1283–1304, DOI 10.1515/forum-2019-0033. MR4000588

[15] A. Jaikin-Zapirain and L. Pyber, *Random generation of finite and profinite groups and group enumeration*, Ann. of Math. (2) **173** (2011), no. 2, 769–814, DOI 10.4007/annals.2011.173.2.4. MR2776362

[16] M. Jarden and A. Lubotzky, *Hilbertian fields and free profinite groups*, J. London Math. Soc. (2) **46** (1992), no. 2, 205–227, DOI 10.1112/jlms/s2-46.2.205. MR1182479

[17] W. M. Kantor and A. Lubotzky, *The probability of generating a finite classical group*, Geom. Dedicata **36** (1990), no. 1, 67–87, DOI 10.1007/BF00181465. MR1065213

[18] N. M. Katz, *Sums of Betti numbers in arbitrary characteristic*, Finite Fields Appl. **7** (2001), no. 1, 29–44, DOI 10.1006/ffta.2000.0303. Dedicated to Professor Chao Ko on the occasion of his 90th birthday. MR1803934

[19] P. Kleidman and M. Liebeck, *The subgroup structure of the finite classical groups*, London Mathematical Society Lecture Note Series, vol. 129, Cambridge University Press, Cambridge, 1990, DOI 10.1017/CBO9780511629235. MR1057341

[20] R. V. Kravchenko, M. Mazur, and B. V. Petrenko, *On the smallest number of generators and the probability of generating an algebra*, Algebra Number Theory **6** (2012), no. 2, 243–291, DOI 10.2140/ant.2012.6.243. MR2950154

[21] S. Lang and A. Weil, *Number of points of varieties in finite fields*, Amer. J. Math. **76** (1954), 819–827, DOI 10.2307/2372655. MR65218

[22] M. Larsen, *Word maps have large image*, Israel J. Math. **139** (2004), 149–156, DOI 10.1007/BF02787545. MR2041227

[23] M. Larsen and A. Shalev, *Word maps and Waring type problems*, J. Amer. Math. Soc. **22** (2009), no. 2, 437–466, DOI 10.1090/S0894-0347-08-00615-2. MR2476780

[24] M. Larsen and A. Shalev, *Fibers of word maps and some applications*, J. Algebra **354** (2012), 36–48, DOI 10.1016/j.jalgebra.2011.10.040. MR2879221

[25] M. Larsen and A. Shalev, *A probabilistic Tits alternative and probabilistic identities*, Algebra Number Theory **10** (2016), no. 6, 1359–1371, DOI 10.2140/ant.2016.10.1359. MR3544299

[26] M. Larsen and A. Shalev, *Words, Hausdorff dimension and randomly free groups*, Math. Ann. **371** (2018), no. 3-4, 1409–1427, DOI 10.1007/s00208-017-1635-y. MR3831276

[27] M. Larsen and A. Shalev, *Almost PI algebras are PI*, Proc. Amer. Math. Soc. **150** (2022), no. 4, 1369–1379, DOI 10.1090/proc/15818. MR4375729

[28] M. Larsen and A. Shalev, *Residually finite dimensional algebras and polynomial almost identities*, J. Algebra Appl. **21** (2022), no. 2, Paper No. 2250038, 9 pp., DOI 10.1142/S0219498822500384. MR4381296

[29] M. Larsen, A. Shalev, and P. H. Tiep, *The Waring problem for finite simple groups*, Ann. of Math. (2) **174** (2011), no. 3, 1885–1950, DOI 10.4007/annals.2011.174.3.10. MR2846493

[30] M. Larsen, A. Shalev, and P. H. Tiep, *Waring problem for finite quasisimple groups*, Int. Math. Res. Not. IMRN **10** (2013), 2323–2348, DOI 10.1093/imrn/rns109. MR3061941

[31] M. Larsen, A. Shalev, and P. H. Tiep, *Probabilistic Waring problems for finite simple groups*, Ann. of Math. (2) **190** (2019), no. 2, 561–608, DOI 10.4007/annals.2019.190.2.3. MR3997129

[32] M. W. Liebeck, E. A. O'Brien, A. Shalev, and P. H. Tiep, *The Ore conjecture*, J. Eur. Math. Soc. (JEMS) **12** (2010), no. 4, 939–1008, DOI 10.4171/JEMS/220. MR2654085

[33] M. W. Liebeck and A. Shalev, *The probability of generating a finite simple group*, Geom. Dedicata **56** (1995), no. 1, 103–113, DOI 10.1007/BF01263616. MR1338320

[34] M. W. Liebeck and A. Shalev, *Simple groups, probabilistic methods, and a conjecture of Kantor and Lubotzky*, J. Algebra **184** (1996), no. 1, 31–57, DOI 10.1006/jabr.1996.0248. MR1402569

[35] M. W. Liebeck and A. Shalev, *Classical groups, probabilistic methods, and the* $(2,3)$*-generation problem*, Ann. of Math. (2) **144** (1996), no. 1, 77–125, DOI 10.2307/2118584. MR1405944

[36] M. W. Liebeck and A. Shalev, *Fuchsian groups, coverings of Riemann surfaces, subgroup growth, random quotients and random walks*, J. Algebra **276** (2004), no. 2, 552–601, DOI 10.1016/S0021-8693(03)00515-5. MR2058457

[37] M. W. Liebeck and A. Shalev, *Fuchsian groups, finite simple groups and representation varieties*, Invent. Math. **159** (2005), no. 2, 317–367, DOI 10.1007/s00222-004-0390-3. MR2116277

[38] A. Lubotzky, *The expected number of random elements to generate a finite group*, J. Algebra **257** (2002), no. 2, 452–459, DOI 10.1016/S0021-8693(02)00528-8. MR1947971

[39] A. Lucchini and M. Moscatiello, *Generation of finite groups and maximal subgroup growth*, Adv. Group Theory Appl. **9** (2020), 39–49, DOI 10.32037/agta-2020-002. MR4123456

[40] F. Lübeck and G. Malle, $(2,3)$*-generation of exceptional groups*, J. London Math. Soc. (2) **59** (1999), no. 1, 109–122, DOI 10.1112/S002461079800670X. MR1688493

[41] A. Mann, *Finite groups containing many involutions*, Proc. Amer. Math. Soc. **122** (1994), no. 2, 383–385, DOI 10.2307/2161027. MR1242094

[42] A. Mann, *Positively finitely generated groups*, Forum Math. **8** (1996), no. 4, 429–459, DOI 10.1515/form.1996.8.429. MR1393323

[43] A. Mann, *Groups satisfying identities with high probability*, Internat. J. Algebra Comput. **28** (2018), no. 8, 1575–1584, DOI 10.1142/S021819671840009X. MR3899225

[44] A. Mann and C. Martinez, *Groups nearly of prime exponent and nearly Engel Lie algebras*, Arch. Math. (Basel) **71** (1998), no. 1, 5–11, DOI 10.1007/s000130050227. MR1627207

[45] A. Mann and A. Shalev, *Simple groups, maximal subgroups, and probabilistic aspects of profinite groups*. part B, Israel J. Math. **96** (1996), no. part B, 449–468, DOI 10.1007/BF02937317. MR1433701

[46] C. Martinez and E. Zelmanov, *Products of powers in finite simple groups*. part B, Israel J. Math. **96** (1996), no. part B, 469–479, DOI 10.1007/BF02937318. MR1433702

[47] N. E. Menezes, M. Quick, and C. M. Roney-Dougal, *The probability of generating a finite simple group*, Israel J. Math. **198** (2013), no. 1, 371–392, DOI 10.1007/s11856-013-0034-7. MR3096644

[48] M. B. Nathanson, *Additive number theory: The classical bases*, Graduate Texts in Mathematics, vol. 164, Springer-Verlag, New York, 1996, DOI 10.1007/978-1-4757-3845-2. MR1395371

[49] P. M. Neumann, *Two combinatorial problems in group theory*, Bull. London Math. Soc. **21** (1989), no. 5, 456–458, DOI 10.1112/blms/21.5.456. MR1005821

[50] P. M. Neumann and C. E. Praeger, *Cyclic matrices over finite fields*, J. London Math. Soc. (2) **52** (1995), no. 2, 263–284, DOI 10.1112/jlms/52.2.263. MR1356142

[51] I. Pak, *On the probability of generating a finite group*, preprint, 1999.

[52] R. S. Pierce, *Associative algebras*, Studies in the History of Modern Science, vol. 9, Springer-Verlag, New York-Berlin, 1982. Graduate Texts in Mathematics, 88. MR674652

[53] C. Pomerance, *The expected number of random elements to generate a finite abelian group*, Period. Math. Hungar. **43** (2001), no. 1-2, 191–198, DOI 10.1023/A:1015250102792. MR1830576

[54] L. H. Rowen, *Ring theory*, Student edition, Academic Press, Inc., Boston, MA, 1991. MR1095047

[55] J. Saxl and J. S. Wilson, *A note on powers in simple groups*, Math. Proc. Cambridge Philos. Soc. **122** (1997), no. 1, 91–94, DOI 10.1017/S030500419600165X. MR1443588

[56] D. Sercombe and A. Shalev, *Random generation of associative algebras*, submitted, arXiv:2009.01115.

[57] A. Shalev, *Lie methods in the theory of pro-p groups*, New horizons in pro-p groups, Progr. Math., vol. 184, Birkhäuser Boston, Boston, MA, 2000, pp. 1–54. MR1765116

[58] A. Shalev, *Word maps, conjugacy classes, and a noncommutative Waring-type theorem*, Ann. of Math. (2) **170** (2009), no. 3, 1383–1416, DOI 10.4007/annals.2009.170.1383. MR2600876

[59] A. Shalev, *Probabilistically nilpotent groups*, Proc. Amer. Math. Soc. **146** (2018), no. 4, 1529–1536, DOI 10.1090/proc/13867. MR3754339

[60] E. I. Zel′manov, *Solution of the restricted Burnside problem for groups of odd exponent* (Russian), Izv. Akad. Nauk SSSR Ser. Mat. **54** (1990), no. 1, 42–59, 221; English transl., Math. USSR-Izv. **36** (1991), no. 1, 41–60. MR1044047

[61] E. I. Zel′manov, *Solution of the restricted Burnside problem for 2-groups* (Russian), Mat. Sb. **182** (1991), no. 4, 568–592; English transl., Math. USSR-Sb. **72** (1992), no. 2, 543–565. MR1119009

[62] E. Zelmanov, *Lie algebras and torsion groups with identity*, J. Comb. Algebra **1** (2017), no. 3, 289–340, DOI 10.4171/JCA/1-3-2. MR3681577

Department of Mathematics, Indiana University, Bloomington, Indiana 47405
Email address: mjlarsen@indiana.edu

Einstein Institute of Mathematics, Hebrew University, Givat Ram, Jerusalem 9190, Israel
Email address: shalev@math.huji.ac.il

Contemporary Mathematics
Volume **800**, 2024
https://doi.org/10.1090/conm/800/16058

On the symbol length of symbols

Eliyahu Matzri

In memory of my father, Shlomo Matzri.

ABSTRACT. Fix a prime p and let F be a field with characteristic not p. Let G_F be the absolute Galois group of F, and let μ_{p^s} be the G_F-module of roots of unity of order dividing p^s in a fixed algebraic closure of F. Let $\alpha \in H^n(F, \mu_{p^s}^{\otimes n})$ be a symbol (i.e., $\alpha = a_1 \cup \cdots \cup a_n$ where $a_i \in H^1(F, \mu_{p^s})$) with effective exponent dividing p^{s-1} (that is $p^{s-1}\alpha = 0 \in \mathrm{H}^n(G_F, \mu_p^{\otimes n})$. In this work, we show how to write α as a sum of symbols coming from $H^n(F, \mu_{p^{s-1}}^{\otimes n})$, that is, symbols of the form $p\gamma$ for $\gamma \in H^n(F, \mu_{p^s}^{\otimes n})$. If $n > 3$ and $p \neq 2$, we assume F is prime to p closed and of characteristic zero. In the case $p = 2$, we also bound the symbol length of a sum of two symbols with effective exponent dividing p^{s-1}.

1. Introduction

Fix a prime p and let F be a field with characteristic not p. Let G_F be the absolute Galois group of F, and let μ_{p^s} be the G_F-module of roots of unity of order dividing p^s in a fixed algebraic closure of F. The Bloch-Kato conjecture states that for any n, m where $p \nmid m$, the norm residue map

$$h_m^n : K_n(F)/mK_n(F) \to \mathrm{H}^n(G_F, \mu_m^{\otimes n})$$

defined on generators by

$$h_m^n(\{a_1, \ldots, a_n\}) = a_1 \cup \cdots \cup a_n$$

is an isomorphism. The case of $n = 2$ was proved in [**10**] and is known as the Merkurjev-Suslin theorem and the general case was proved by Voevodsky in [**15**] and [**16**]. In particular, every element in $\mathrm{H}^n(G_F, \mu_m^{\otimes n})$ can be written as a sum of symbols of degree m, i.e., cup products $(a_1, \ldots, a_n)_{m,F} \doteqdot a_1 \cup \cdots \cup a_n$ where $a_i \in \mathrm{H}^1(G_F, \mu_m) \cong F^\times/mF^\times$ for all i. For $t < r$, there is a short exact sequence of G_F-modules,

$$1 \longrightarrow \mu_{p^t}^{\otimes n} \longrightarrow \mu_{p^r}^{\otimes n} \xrightarrow{p^t} \mu_{p^{r-t}}^{\otimes n} \to 1$$

2020 *Mathematics Subject Classification.* Primary 11R32, 19D45, 11E04.

Key words and phrases. Galois cohomology, higher symbols, Milnor K-theory, quadratic forms.

This research was supported by the Israel Science Foundation (grant No. 630/17).

where the left map takes the generator $\rho_{p^t} \otimes \cdots \otimes \rho_{p^t} \in \mu_{p^t}^{\otimes n}$ to $p^{r-t} \cdot \rho_{p^r} \otimes \rho_{p^r} \cdots \otimes \rho_{p^r} = \rho_{p^t} \otimes \rho_{p^r} \cdots \otimes \rho_{p^r} \in \mu_{p^r}^{\otimes n}$ where ρ_{p^r} is a root of unity of order p^r. And the right map is the natural one taking the generator $\rho_{p^r} \otimes \cdots \otimes \rho_{p^r} \in \mu_{p^r}^{\otimes n}$ to $\rho_{p^{r-t}} \otimes \cdots \otimes \rho_{p^{r-t}} \in \mu_{p^{r-t}}^{\otimes n}$. The above exact sequence induces the following part of a long exact sequence:

$$\mathrm{H}^n(G_F, \mu_{p^t}^{\otimes n}) \longrightarrow \mathrm{H}^n(G_F, \mu_{p^r}^{\otimes n}) \xrightarrow{p^t} \mathrm{H}^n(G_F, \mu_{p^{r-t}}^{\otimes n}). \tag{1}$$

Note: The left map takes a symbol $(a_1, \dots, a_n)_{p^t,F}$ to the symbol $p^{r-t}(a_1, \dots, a_n)_{p^r,F}$ and the right map takes the symbol $(a_1, \dots, a_n)_{p^r,F}$ to the symbol $(a_1, \dots, a_n)_{p^{r-t},F}$. Let $\alpha \in \mathrm{H}^n(G_F, \mu_{p^r}^{\otimes n})$. We say that α has effective exponent p^t where t is the minimal number such that α is in the kernel of the right map in the exact sequence (1) corresponding to t.

The Bloch-Kato theorem tells us that $\mathrm{H}^n(G_F, \mu_{p^t}^{\otimes n})$ is generated by symbols. Hence if α has effective exponent dividing p^t, then it can be written as the image of a sum of symbols in $\mathrm{H}^n(G_F, \mu_{p^t}^{\otimes n})$, that is, $\alpha = p^{r-t} \sum \alpha_i$ for $\alpha_i \in \mathrm{H}^n(G_F, \mu_{p^r}^{\otimes n})$. The minimal number of symbols needed is called the symbol length of the effective exponent-p^t degree-p^r class α.

Most of the known results on the symbol length problem are focused on the case $n = 2$ and F contains a root of unity of order $m = p^r$. In this case, we have $\mathrm{H}^2(G_F, \mu_{p^r}^{\otimes 2}) \cong \mathrm{H}^2(G_F, \mu_{p^r}) \cong {}_{p^r}\operatorname{Br}(F)$; hence one can think about the symbol length of central simple algebras of exponent dividing p^r. Probably the most famous result is due to Albert (and is prior to the Merkurjev-Suslin theorem, see [**1**]), stating that every central simple algebra of degree four and exponent two is isomorphic to the tensor product of two quaternion algebras. We refer the reader to [**9**] for more results in the case $n = 2$.

As a motivating example for this work, consider the case where $n = 2$ and F contains a root of unity of order $m = p^r$, ρ. Recall that for a pair of elements $a, b \in F^\times$, one can define a central simple algebra $D = F[x, y \mid x^m = a; y^m = b; yx = \rho xy]$ which is denoted $D = (a, b)_{m,F}$ and called a symbol algebra of degree m. Let $[D]$ denote the class of D in $\operatorname{Br}(F)$ and assume the exponent of $[D]$ is p^t for $t < r$, that is, $[D]^{\otimes p^t} = [F]$. Then the Merkurjev-Suslin theorem implies that $[D]$ can be written as a sum of symbol algebras of degree p^t. A theorem of Tignol (which applies to a more general situation than ours, see [**14**]) implies that the number of symbols needed is at most p^{r-t}. Inspired by this result, in this work we focus on the case where $\alpha \in \mathrm{H}^n(G_F, \mu_{p^r}^{\otimes n})$ is a single symbol of degree p^r and effective exponent p^t and bound its symbol length, i.e., $\alpha = a_1 \cup \cdots \cup a_n \in \mathrm{H}^n(G_F, \mu_{p^r}^{\otimes n})$ where $a_i \in \mathrm{H}^1(G_F, \mu_{p^r})$ for all i and $p^t \alpha = 0 \in \mathrm{H}^n(G_F, \mu_{p^{r-t}}^{\otimes n})$.

In Section 2, we give the necessary background for the work. In Section 3, we prove a technical result concerning the computation of norms from symbol algebras. In Section 4, we give an explicit computation of the corestriction of a symbol in $K_3^M(F)$ which will be used in the work. In Section 5, we deal with the case where $p = 2$. In Section 6, we deal with the case $n = 2$. In Section 7, we deal with the case $n = 3$. In Section 8, we deal with the case $n \geq 4$ and in section 9, we deal with the case of sums of two symbols where $p = 2$.

2. Background

Fix a number m and let F be a field with $\mathrm{char}(F) \nmid m$. Let G_F be the absolute Galois group of F, and let μ_m be the G_F-module of roots of unity of order dividing m in a fixed algebraic closure of F.

For $\alpha \in \mathrm{H}^n(G_F, \mu_{p^r}^{\otimes n})$, we will abuse notation and write $p^t\alpha$ to denote its image under the right map of sequence (1) and sometimes to denote the action of adding α p^t times.

DEFINITION 2.1. A symbol $\alpha \in \mathrm{H}^n(G_F, \mu_m^{\otimes n})$ is an element of the form $(a_1, \dots, a_n)_{m,F} \doteqdot a_1 \cup \cdots \cup a_n$ where $a_i \in \mathrm{H}^1(G_F, \mu_m)$.

DEFINITION 2.2. Let $\alpha \in \mathrm{H}^n(G_F, \mu_{p^r}^{\otimes n})$. The effective exponent of α is the minimal number p^t such that $p^t\alpha = 0 \in \mathrm{H}^n(G_F, \mu_{p^{r-t}}^{\otimes n})$, that is, when considering α as an element of $\mathrm{H}^n(G_F, \mu_{p^{r-t}}^{\otimes n})$ via the right map in the sequence (1). We denote the effective exponent of α by $e.exp(\alpha)$.

Having effective exponent p^t is stronger than having exponent p^t in the usual sense, because $\exp(\alpha) \leq e.exp(\alpha)$. However, they can be different; it all comes down to the image of a connecting map vanishing or not. However we have the following lemma due to Sivatski (see [**13**, Proof of Theorem 1.1]).

LEMMA 2.3. *If $\mu_{p^r} \subset F$ then the effective exponent is equal to the usual exponent.*

PROOF. Clearly, the claim is equivalent to showing the left map of sequence (1) is injective. If we continue the exact sequence (1) to the left using the assumption that $\mu_{p^r} \subset F$, we get the sequence

$$\mathrm{H}^{n-1}(G_F, \mu_{p^r}) \to \mathrm{H}^{n-1}(G_F, \mu_{p^{r-t}}) \to \mathrm{H}^n(G_F, \mu_{p^t}) \to \mathrm{H}^n(G_F, \mu_{p^r}). \quad (2)$$

But saying the left map in sequence (1) is injective is the same as saying the left map in sequence (2) is surjective, and this is clear in light of the Bloch-Kato Theorem. □

DEFINITION 2.4. Let $a, b \in F^\times$. We define the symbol algebra $(a, b)_{F,m}$ to be $F[x, y \mid x^m = a; y^m = b; yx = \rho xy]$, where ρ denotes a root of unity of order m. We call x, y standard generators for the symbol algebra $(a, b)_{F,m}$.

2.1. Twisted polynomials. We recall some known facts concerning twisted polynomials taken from [**5**, Chapter 1], [**6**, Chapter 3] and specialized to our needs. Let D be a division algebra with an automorphism of finite order and let $R = D[\lambda, \sigma] = D[\lambda \mid \lambda d = \sigma(d)\lambda]$ be the ring of (σ-) twisted polynomials over D. We denote the center of R by C. The twisted polynomial ring $R = D[\lambda, \sigma]$ is a left (and also right) Euclidean domain; that is, for any $f(\lambda), g(\lambda) \in R$, there are $q(\lambda), r(\lambda) \in R$ such that $f(\lambda) = q(\lambda)g(\lambda) + r(\lambda)$ and $\deg r(\lambda) < g(\lambda)$ or $r(\lambda) = 0$. As a consequence, we get that every left (and right) ideal is principal; so R is a (left and right) principle ideal domain. For left ideals of R, $Rf \subseteq Rg$ if and only if $f = hg$ for some $h \in R$; thus Rf is maximal if and only if f is irreducible, which happens if and only if R/Rf is simple as a (left) module over R.

If $I \trianglelefteq R$ is a two sided ideal, then we have that $I = (I \cap C)R$; in particular, I is of the form $Rf = fR$ where $f \in C$. For a left ideal Rf, the maximal two sided ideal contained in Rf (called the bound of Rf) is the annihilator $I = \mathrm{ann}\,(R/Rf)$.

Note that if Rf is a maximal left ideal (f is irreducible) and $I \neq 0$, then I is a maximal two sided ideal.

3. Computing norms from Kummer field extensions and symbol algebras

In this section, we consider Kummer field extensions and symbol algebras over a field F containing the necessary roots of unity and show how to conveniently (for our purposes) compute their norm forms.

LEMMA 3.1. *Let $p(x) = x^m + \sum_{i=1}^m a_i x^{m-i} \in F[x]$ be an irreducible polynomial. Then there is a field extension, L/F, of degree m such that $p(x) = \mathrm{N}_{L[x]/F[x]}(x-l)$ for $l \in L$.*

PROOF. This is clear as the Galois group of $p(x)$ acts transitively on its roots. □

LEMMA 3.2. *Let $K = F[x \mid x^n = a]$ be a Kummer field extension and let $x - f \in K$ where $f \in F$. Then $\mathrm{N}_{K/F}(x-f) = (-1)^n(f^n - a)$.*

PROOF. By direct computation, we have that $f^n - x^n = (f-x)(f-\rho_n x)\cdots(f-\rho_n^{n-1}x) = \mathrm{N}_{K/F}(f-x)$. Hence $\mathrm{N}_{K/F}(x-f) = (-1)^n(f^n-a)$. □

COROLLARY 3.3. *Let $D = (a,b)_{p,F}$ be a non trivial symbol algebra with standard pair of generators x, y. Then for $k \in K = F[x]$ we have that $\mathrm{Nrd}(y-k) = (-1)^p(\mathrm{N}_{K/F}(k) - b)$.*

PROOF. First notice that $(k^{-1}y)^p = \mathrm{N}_{K/F}(k)^{-1}b$ so that $L = F[k^{-1}y]$ is a Kummer field extension of degree p. Compute $\mathrm{Nrd}(y-k) = \mathrm{Nrd}(k)\,\mathrm{Nrd}(k^{-1}y-1) = \mathrm{N}_{K/F}(k)\,\mathrm{N}_{L/F}(k^{-1}y-1)$. Now by Lemma 3.2, we obtain $\mathrm{N}_{K/F}(k)\,\mathrm{N}_{L/F}(k^{-1}y-1) = \mathrm{N}_{K/F}(k)(-1)^p(1-\mathrm{N}_{K/F}(k)^{-1}b)$. Therefore, we conclude that $\mathrm{Nrd}(y-k) = (-1)^p(\mathrm{N}_{K/F}(k)-b)$, as claimed. □

THEOREM 3.4. *Let $D = (a,b)_{p,F}$ be a non trivial symbol algebra, $K = F[x \mid x^p = a]$, and $d = k_0 + k_1 y + \dots k_m y^m \in D$ where $0 < m < p$, $k_i \in K$, and $k_m \neq 0$. Then there exists a field extension L/F of dimension prime to p and bounded by $(p-1)!$ such that d can be written in the following way;*

$$d = k_m(y-r_1)\cdots(y-r_m)$$

where $r_i \in KL$.

PROOF. The main idea is to notice that D is a quotient of the twisted polynomial ring $A = K[\lambda,\sigma] = K[\lambda \mid \lambda k = \sigma(k)\lambda]$, where $\mathrm{Gal}(K/F) = \langle\sigma\rangle$. Indeed, sending $\lambda \to y$ and $x \to x$ gives an epimorphism $\phi : A \to D$. Let $d = k_0 + k_1 y + \dots + k_m y^m \in D$. Write $d = k_m(k_0' + k_1' y + \dots + y^m)$ and let $d' = k_0' + k_1' y + \dots + y^m$. Lift d' to a twisted polynomial $g = k_0' + k_1'\lambda + \dots + \lambda^m \in A$, let M be the left A module $M = A/Ag$, and notice that $\dim_F(M) = mp$. Now, since A is a principle ideal domain, g is irreducible if and only if M is a simple module. In particular, factoring g is the same as decomposing M, which is what we will do. First, we may assume g is irreducible; otherwise replace it by an irreducible component and notice that the sum of degrees of all the components is $m < p$.

Notice that A can be written as $A = (a, \lambda^p)_{C,p}$ where $C = F[\lambda^p]$ is the center of A. Now decomposing M as a module over A is the same as decomposing M as a module over $A/\operatorname{ann}(M)$, where $\operatorname{ann}(M)$ is the two sided ideal in A that annihilates M. As mentioned in subsection 2.1, we know that $\operatorname{ann}(M) = (\operatorname{ann}(M) \cap C)A$ so that $A/\operatorname{ann}(M) = (a, \lambda^p)_{R,p}$ where $R = C/(\operatorname{ann}(M) \cap C)$. Notice that by assumption $A/\operatorname{ann}(M)$ has a left module of dimension $mp < p^2$, and since p is a prime this tells us that $A/\operatorname{ann}(M)$ is split, that is, $A/\operatorname{ann}(M) \cong \mathrm{M}_p(R)$. Now the dimension over F of the minimal module for $A/\operatorname{ann}(M)$ is $p[R : F]$. On the other hand, by assumption M is that simple module and it has dimension pm. Thus we have that $[R : F] = m$.

Recall that there is a reduced norm map for $\mathrm{Nrd} : A \otimes F(\lambda^p) \to F(\lambda^p)$ which induces a reduced norm for A. Also as an element divides its reduced norm (see [**4**]), we see that $\mathrm{Nrd}(g) \in Ag$ and that $\mathrm{Nrd}(g) \in \operatorname{ann}(M) \cap C$. Recalling that $[R : F] = m$ and the fact that the degree of $\mathrm{Nrd}(g)$ as a polynomial in λ^p is m, we see that $\operatorname{ann}(M) \cap C = \langle \mathrm{Nrd}(g) \rangle$. Note that our assumption that M is simple implies that $\operatorname{ann}(M)$ is a maximal two sided ideal. Thus/ $\operatorname{ann}(M) \cap C = \langle \mathrm{Nrd}(g) \rangle$ is maximal in C, that is/ $\mathrm{Nrd}(g)$ is irreducible and R/F is a field extension of degree m. Hence in order to decompose M we need to split R, that is, extend scalars from F to a splitting field L of the polynomial $\mathrm{Nrd}(g)$ (again as a polynomial in λ^p). We see that $A \otimes_F L/\operatorname{ann}(M \otimes_F L) = \oplus_{i=1}^m \mathrm{M}_p(L)$ so that $M \otimes_F L \cong N^m$ where N is the unique simple left module for $M_p(L)$ which is of the form $A_L/(\lambda - r)A_L$ for $r \in LK$.

Gathering all of the above we get the following. Write $g = g_1 \cdots g_s$ where the g_i's are the (monic) irreducible components of g. For each $i = 1, \dots s$, we get a field extension L_i/F (which is the splitting field of $\mathrm{Nrd}(g_i)$ considered as a polynomial in λ^p) such that g_i splits into linear factors over KL_i, so that taking $L = L_1 \cdots L_s$ we get that g decomposes completely and clearly the dimension of L/F is at most $(p-1)!$ Now all that is left to do is to apply the map ϕ to get the decomposition of d over L as needed. □

COROLLARY 3.5. *Let $d \in D$ be as in Theorem 3.4. Then there are field extensions, $T_1, \dots, T_k/F$ such that $\sum_{i=1}^k [T_i : F] \le p - 1$ and elements $r_i \in KT_i$ such that*

$$\mathrm{Nrd}_D(d) = (-1)^{mp}\,\mathrm{N}_{K/F}(k_m) \cdot \prod_{i=1}^{k} \mathrm{N}_{T_i/F}(\mathrm{N}_{KT_i/T_i}(r_i) - b).$$

PROOF. The idea is to take a closer look at L mentioned above.

Let $d \in D$ let $g \in A$ be as above. Write $g = k_m \cdot g_1 \cdots g_s$ where each g_i is irreducible, this corresponds to $g = k_m \cdot d_1 \cdots d_s$. Let L_i be the splitting field of $\mathrm{Nrd}(g_i)$ considered as a polynomial in λ^p. For each i write $g_i = (\lambda - r_{i,1}) \cdots (\lambda - r_{i,m_i})$ where $m_i = \deg(g_i)$ and $r_{i,j} \in KL_i$.

Applying the reduced norm, we get

$$\mathrm{Nrd}(g_i) = \mathrm{Nrd}(\lambda - r_{i,1}) \cdots \mathrm{Nrd}(\lambda - r_{i,m_i}),$$

and by Lemma 3.2 we see that

$$\begin{aligned}\mathrm{Nrd}(g_i) &= (-1)^p(\mathrm{N}_{KT_i/T_i}(r_{i,1}) - \lambda^p) \cdots (-1)^p(\mathrm{N}_{KT_i/T_i}(r_{i,m_i}) - \lambda^p) \\ &= \mathrm{N}_{T_i/F}((-1)^p(\mathrm{N}_{KT_i/T_i}(r_{i,1}) - \lambda^p))\end{aligned}$$

where $T_i \cong F[\lambda^p]/\operatorname{Nrd}(g_i)$, which is a field extension of dimension $m_i = \deg(g_i)$. Specializing back to D, we get

$$\operatorname{Nrd}_D(d) = (-1)^{mp}\operatorname{N}_{K/F}(k_m)\cdot\prod_{i=1}^{k}\operatorname{N}_{T_i/F}(\operatorname{N}_{KT_i/T_i}(r_i)-b))$$

where $r_i = r_{i,1} \in KT_i$. Clearly, $\sum_{i=1}^{k}[T_i : F] = m \leq p-1$, and we are done. □

4. Computing corestriction for K_3^M

In this section, we compute the corestriction of a symbol (a,b,c) along a field extension K/F of degree n.

LEMMA 4.1. *We may assume $k = F(a) = F(b) = F(c)$.*

PROOF. Indeed, otherwise assume, for example, that $F(a) \subsetneq K$. Then using the projection formula we can compute: $\operatorname{Cor}_{K/F}(a,b,c) = \operatorname{Cor}_{F(a)/F}(a\cdot\operatorname{Cor}_{K/F(a)}(b,c)$ and we reduced to computing the corestriction along a field with lower dimension. □

PROPOSITION 4.2. *Let K/F be a field extension of degree n and $\alpha = (a,b,c)$ a symbol in $K_3^M(K)$. Let a_n denote the symbol length of $\operatorname{Cor}_{K/F}(\alpha)$. Then we have that $a_n \leq n + na_{n-1}$.*

PROOF. By Lemma 4.1, we may assume $K = F(a)$ so that we can think of $b \in K$ as polynomials in a. So let $b = fp(a)$ where $p(x)$ is a monic polynomial of degree at most $n-1$. Note that if $p(x)$ were reducible, we could split the computation and continue, but the worst case is when it is irreducible. Now let $L = F[x]/p(x)$, then $p(x) = \operatorname{N}_{L(x)/F(x)}\ (x-l)$ where $l \in L$ is a root of $p(x)$, specializing x to a gives $p(a) = \operatorname{N}_{LK/K}(a - l)$. Now we compute

$$\begin{aligned}
\operatorname{Cor}_{K/F}(a,b,c)_K &= \operatorname{Cor}_{K/F}(a,fp(a),c)_K = \operatorname{Cor}_{K/F}(a,f,c) + \operatorname{Cor}_{K/F}(a,p(a),c)_K\\
&= \operatorname{Cor}_{K/F}(f^{-1},a,c)_K + \operatorname{Cor}_{K/F}(a,\operatorname{N}_{LK/K}(a-l),c)_K\\
&\overset{p.f.+R.T.}{=} \sum_{i=1}^{n}(f^{-1},g_i,h_i)_F + \operatorname{Cor}_{KL/F}(a,a-l,c)_{KL}\\
&= \sum_{i=1}^{n}(f^{-1},g_i,h_i)_F + \operatorname{Cor}_{KL/F}(l,a-l/a,c)_{KL}\\
&\overset{p.f.+R.T.}{=} \sum_{i=1}^{n}(f^{-1},g_i,h_i)_F + \sum_{i=1}^{n}\operatorname{Cor}_{L/F}(l,u_i,v_i)_L.
\end{aligned}$$

Since $[L : F] = n-1$, we get that $a_n \leq n + na_{n-1}$ as claimed. □

COROLLARY 4.3. *$a_n \leq en! - 1 - O(1/n)$.*

REMARK 4.4. Notice that $a_n + a_m \leq a_{n+m}$, so that it is more efficient to split the corestriction along a field extension of degree $n+m$ to the sum of two corestrictions of degrees n and m.

5. The case $p = 2$

In this section, we consider symbols $\alpha \in \mathrm{H}^n(F, \mu_{2^s}^{\otimes n})$ where $e.exp(\alpha) \mid 2^{s-1}$. We know that α has the form $2\sum_{i=1}^{l} \alpha_i$ for α_i symbols in $\mathrm{H}^n(F, \mu_{2^s}^{\otimes n})$, and we want to find an upper bound for the symbol length of α, that is bound l. To achieve this goal, we use the fact (see [**15**]) that a symbol $(a_1, \cdots, a_n) \in \mathrm{H}^n(F, \mu_2)$ is trivial if and only if a_n is a value of the Pfister form $\langle\langle a_1, \cdots, a_{n-1}\rangle\rangle$.

THEOREM 5.1. *Let $\alpha = (a_1, \cdots, a_n) \in \mathrm{H}^n(F, \mu_{2^s}^{\otimes n})$ be a symbol with $e.exp(\alpha) = 2^{s-1}$. Then $\alpha = 2\sum_{i=1}^{2^{n-1}} \alpha_i$ where $\alpha_i \in \mathrm{H}^n(F, \mu_{2^s}^{\otimes n})$ are symbols, that is, the symbol length of α is at most 2^{n-1}.*

PROOF. We use induction on n. The base case $n = 1$ clearly holds, so we assume the theorem for $n - 1$ and prove for n. Write ϕ for the Pfister form $\langle\langle a_1, \cdots, a_{n-1}\rangle\rangle$ and ψ for the Pfister form $\langle\langle a_1, \cdots, a_{n-2}\rangle\rangle$. Recall that we have $\phi = \psi\perp(-a_{n-1})\psi$. By assumption, we have that $2^{s-1}\alpha = (a_1, \cdots, a_n)_2 = 0$; thus, a_n is a value of ϕ. That is, there exist vectors v_1, v_2 in the underlying vector space, V, of ψ such that $a_n = \psi(v_1) - a_{n-1}\psi(v_2) = \psi(v_2)(\psi(v_1)\psi(v_2^{-1}) - a_{n-1})$. But Pfister forms are round (see [**2**, p. 54, Cor. 9.9]) so that its set of values is a group. So there exist $v_3 \in V$ such that $a_n = \psi(v_2)(\psi(v_1)\psi(v_2^{-1}) - a_{n-1}) = \psi(v_2)(\psi(v_3) - a_{n-1})$. Now we compute:

$$\begin{aligned}
\alpha &= (a_1, \cdots, a_n) = (a_1, \cdots, \psi(v_2)(\psi(v_3) - a_{n-1})) \\
&= (a_1, \cdots, \psi(v_2)) + (a_1, \cdots, a_{n-1}, \psi(v_3) - a_{n-1}) \\
&= (a_1, \cdots, a_{n-2}, \psi(v_2), a_{n-1}^{-1}) + (a_1, \cdots, a_{n-2}, \psi(v_3), (a_{n-1} - \psi(v_3))/a_{n-1}) \\
&= \gamma_1 \cdot (a_{n-1}^{-1}) + \gamma_2 \cdot ((a_{n-1} - \psi(v_3))/a_{n-1}).
\end{aligned}$$

Notice that $\gamma_1, \gamma_2 \in \mathrm{H}^{n-1}(F, \mu_{2^s}^{\otimes n-1})$ have effective exponent dividing 2^{s-1}; hence, by induction, we get that for $j = 1, 2$ we have $\gamma_j = 2\sum_{i=1}^{2^{n-2}} \gamma_{j,i}$ where $\gamma_{j,i}$ are symbols in $\mathrm{H}^{n-1}(F, \mu_{2^s}^{\otimes n-1})$. All together, we see that

$$\alpha = 2\Bigg(\sum_{i=1}^{2^{n-2}} (\gamma_{1,i} \cdot (a_{n-1}^{-1}) + \gamma_{2,i} \cdot ((a_{n-1} - \psi(v_3))/a_{n-1}))\Bigg).$$

Writing $\alpha_i = \gamma_{1,i} \cdot (a_{n-1}^{-1})$ and $\alpha_{2^{n-2}+i} = \gamma_{2,i} \cdot ((a_{n-1} - \psi(v_3))/a_{n-1})$ for $i = 1, \cdots, 2^{n-2}$, we get that $\alpha = 2\sum_{i=1}^{2^{n-1}} \alpha_i$, completing the induction step and we are done. □

6. The case $n = 2$

In this section, we consider symbols $\alpha \in \mathrm{H}^2(F, \mu_{p^s}^{\otimes 2})$ where $e.exp(\alpha) | p^{s-t}$ for some $t \geq 1$.

THEOREM 6.1. *Let F be a field containing a root of unity of order p^t, ρ, and let $\alpha = (a, b) \in \mathrm{H}^2(G_F, \mu_{p^s}^{\otimes 2})$ be of effective exponent p^{s-t}. Then $\alpha = p^t \sum_{i=1}^{p^t} \beta_i$ for $\beta_i \in \mathrm{H}^2(F, \mu_{p^s}^{\otimes 2})$, that is, the symbol length of α is at most p^t.*

PROOF. This proof is really in the spirit of Tignol's proof (see [**14**]). By assumption we have that $p^{s-t}\alpha = 0$ in $\mathrm{H}^2(F, \mu_{p^t}^{\otimes 2})$. As we assume $\rho \in F$, we may think of $p^{s-t}\alpha$ as the symbol algebra $(a,b)_{p^t}$ in $\mathrm{Br}(F)$, so that Wedderburn's criterion tells us that $(a,b)_{p^t}=0$ if and only if $b=\mathrm{N}_{K/F}(u)$ for $u\in K=F[x|\ x^{p^t}=a]$. Now compute:

$$\begin{aligned}\alpha &= (a,b)_{F,p^s}=(a,\mathrm{N}_{K/F}(u))_{F,p^s}\\ &= \mathrm{Cor}_{K/F}(a,u)_{K,p^s}=\mathrm{Cor}_{K/F}(x^{p^t},u)_{K,p^s}\\ &= p^t\,\mathrm{Cor}_{K/F}(x,u)_{K,p^s}\overset{R.T.}{=}p^t\sum_{i=1}^{p^t}(a_i,b_i)_{F,p^s}.\end{aligned}$$ □

REMARK 6.2. If $\mu_{p^s}\subset F$ the above says that a symbol algebra of degree p^s and exponent p^{s-t} is a sum of at most p^t symbol algebras of degree p^{s-t}.

7. The case $n=3$

Let $\alpha=(a,b,c)$ be a symbol in $\mathrm{H}^3(G_F,\mu_p^{\otimes 3})$ and assume $\rho_p\in F$. The following variety is called the Merkurjev-Suslin variety for α, $X_\alpha=\{d\in(a,b)_{p,F}\mid \mathrm{Nrd}_{(a,b)_{p,F}}(d)=c\}$. In [**10**, Theorem 12.1], it is proved that X_α is a generic splitting variety α, that is, for every field extension L/F, $\alpha_L=0$ if and only if $X_\alpha(L)\neq\emptyset$. In this section, we will use this variety to bound the symbol length of a symbol.

THEOREM 7.1. *Let F be a field containing μ_p and let $\alpha=(a,b,c)\in\mathrm{H}^3(G_F,\mu_{p^s}^{\otimes 3})$ be a symbol with $e.exp(\alpha)|p^{s-1}$. Then the symbol length of α is a most $ep!-O(1)$.*

PROOF. By assumption we have that $p^{s-1}\alpha=0\in\mathrm{H}^3(G_F,\mu_p^{\otimes 3})$, so that the variety $X_{(a,b,c)}(F)\neq\emptyset$. In particular, there exist $d\in D=(a,b)_{p,F}$ such that $c=\mathrm{Nrd}_D(d)$. By Corollary 3.5, we have that there are field extensions, $K=F[x\mid x^p=a],T_1,\dots,T_t/F$ such that $m\div\sum_{i=1}^t[T_i:F]\le p-1$ and $\mathrm{Nrd}_D(d)=\mathrm{N}_{K/F}(k_m)\cdot\prod_{i=1}^t\mathrm{N}_{T_i/F}(b-\mathrm{N}_{KT_i/L_i}(k_i))$. Now we can compute

$$\begin{aligned}\alpha=(a,b,c)_F &\overset{Corollary\ 3.5}{=}(a,b,(-1)^{mp}\,\mathrm{N}_{K/F}(k_m)\cdot\prod_{i=1}^k\mathrm{N}_{T_i/F}(\mathrm{N}_{KT_i/T_i}(r_i)-b))_F\\ &=(a,b,\mathrm{N}_{K/F}(k_m))_F+(a,b,(-1)^{mp})_F+\sum_{i=1}^k(a,b,\mathrm{N}_{T_i/F}(\mathrm{N}_{KT_i/T_i}(r_i)-b))_F.\end{aligned}$$

Notice that $(a,b,(-1)^{mp})_F=0$, so we can continue

$$\begin{aligned}&=(a,b,\mathrm{N}_{K/F}(k_m))_F+\sum_{i=1}^k\mathrm{Cor}_{T_i/F}(a,b,\mathrm{N}_{KT_i/T_i}(r_i)-b)_{T_i}\\ &\overset{Remark\ 8.3}{=}(a,\mathrm{N}_{K/F}(k_m),b^{-1})_F+\sum_{i=1}^k\mathrm{Cor}_{T_i/F}(a,\mathrm{N}_{KT_i/T_i}(r_i),(b-\mathrm{N}_{KT_i/T_i}(r_i))/b)_{T_i}\end{aligned}$$

Now notice that $(a,\mathrm{N}_{K/F}(k_m))_F$ has effective exponent p^{s-1}; thus by Theorem 6.1, it is of the form $p\sum_{i=1}^p\beta_i$ for β_i symbols. Hence we have that $(a,\mathrm{N}_{K/F}(k_m),b^{-1})=p\sum_{i=1}^p\beta_i\cup b^{-1}$. The same argument for $(a,\mathrm{N}_{KT_i/T_i}(r_i))_{T_i}$ applies. Therefore,

$(a, \mathrm{N}_{KT_i/T_i}(r_i), (b - \mathrm{N}_{KT_i/T_i}(r_i))/b)_{T_i} = p\sum_{j=1}^{p} \gamma_j$ for γ_j symbols over T_i. Continuing the computation, we get that:

$$\alpha = p\sum_{i=1}^{p} \beta_i \cup b^{-1} + p\sum_{i=1}^{k}\sum_{j=1}^{p} \mathrm{Cor}_{T_i/F}\, \gamma_{i,j,T_i}.$$

Letting a_n be the symbol length of the corestriction of a symbol in K_3^M as in Proposition 4.2, we get that the symbol length of α is at most $p + \sum_{i=1}^{k}\sum_{j=1}^{p} a_{[T_i:F]} = p + p\sum_{i=1}^{k} a_{[T_i:F]}$. Now applying Remark 4.4 and the fact that $\sum_{i=1}^{k}[T_i : F] = p-1$, we get that the symbol length of α is at most $p + pa_{p-1} \leq ep! - O(1)$ as claimed. □

REMARK 7.2. It is possible to use Tignol's idea to prove Theorem 7.1 but the bound is worse than the one in Theorem 7.1.

8. The case $n \geq 4$

Let F be a prime to p closed field (in particular $\mu_p \subset F$), with $\mathrm{char}(F) = 0$ and $\alpha = (a_1, \ldots, a_n)_{F,p^{r+1}} \in \mathrm{H}^n(G_F, \mu_{p^{r+1}})$ be a symbol of effective exponent dividing p^r. Then by the Bloch-Kato Theorem, we know that α can be written as a sum of symbols of the form $p \cdot \mathrm{H}^n(G_F, \mu_{p^{r+1}})$.

Our objective is to bound the number of symbols needed. The main idea is to use the notion of a p-generic splitting variety for symbols in $\mathrm{H}^n(G_F, \mu_p^{\otimes n}) = \mathrm{H}^n(G_F, \mu_p)$.

DEFINITION 8.1. Let F be a field with $\mathrm{char}(F) = 0$ and $\alpha \in \mathrm{H}^n(G_F, \mu_p^{\otimes n})$ be a non trivial symbol. A p-generic splitting variety for α is an F variety V such that for every field extension L/F we have that $\alpha_L = 0$ if and only if there exists a prime to p extension M/L such that $V(M) \neq \emptyset$.

Recall the following construction of a p-generic splitting variety due to Rost (see [**12**],[**3**], [**7**]): Let F be a field and $a_1, \ldots, a_{n+1} \in F^\times$. Start with the form $\phi_0(t) = t^p$. Inductively define two forms where the X_i stand for a set of p^r variables:

- $\psi_{r+1}(X_1, \ldots, X_{p-1}) = \prod_{i=1}^{p-1}(1 - a_{r+1}\phi_r(X_i))$;
- $\phi_{r+1}(X_0, \ldots, X_{p-1}) = \phi_r(X_0)\psi_{r+1}(X_1, \cdots, X_{p-1})$.

Let $\alpha = (a_1, \ldots, a_{n+1})$ be a symbol and V be any smooth compactification of the affine hypersurface $V_\alpha \subset \mathbb{A}^{p^n}(F)$ defined by the equation:

$$\phi_n(X_0, \ldots, X_{p-1}) = a_{n+1}.$$

Then V is a p-generic splitting variety for α of dimension $p^n - 1$, also called a norm variety for α.

PROPOSITION 8.2. *We have the following equalities in $K_{r+1}^M(F(X))/p$ where X is a set of p^r variables (see* [**3**, Lemma 3.4]*) :*

(1) $a_1 \cdots a_r \cdot \psi_r = 0$;
(2) $a_1 \cdots a_r \cdot \phi_r = 0$.

REMARK 8.3. We have the following equalities in $K_2^M(F)$:

(1) $\{a, 1 - ta\} = \{t, (1 - at)^{-1}\}$;
(2) $\{a, b\} = \{-a/b, a + b\}$.

THEOREM 8.4. *Assume F is p special of characteristic zero and let $\alpha = (a_1, \dots, a_{n+1})_{F,p^{r+1}} \in \mathrm{H}^{n+1}(G_F, \mu_{p^{r+1}})$ be of effective exponent dividing p^r. Then, α can be written as a sum of p^n symbols coming from $\mathrm{H}^{n+1}(F, \mu_{p^r})$. That is, $\alpha = p\sum_{i=1}^{p^n} \alpha_i$ for $\alpha_i \in \mathrm{H}^{n+1}(G_F, \mu_{p^{r+1}})$.*

PROOF. We use induction on n. When $n = 0$, it is easy to see that the Theorem holds as any element in $\mathrm{H}^1(G_F, \mu_{p^{r+1}})$ is one symbol. For the induction step, let $\alpha = (a_1, \dots, a_{n+1})_{F,p^{r+1}} \in \mathrm{H}^{n+1}(G_F, \mu_{p^{r+1}})$ be of effective exponent dividing p^r. Then,

$$p^r \alpha = (a_1, \dots, a_{n+1})_{F,p} = 0 \in \mathrm{H}^{n+1}(G_F, \mu_p).$$

Hence, we know that there are $X_0, \dots, X_{p-1}$ where each of the X_i represent a set of p^r elements from F, such that

$$\begin{aligned} a_{n+1} &= \phi_n(X_0, \dots, X_{p-1}) \\ &= \phi_{n-1}(X_0)\psi_n(X_1, \cdots, X_{p-1}) = \phi_{n-1}(X_0) \prod_{i=1}^{p-1}(1 - a_n\phi_{n-1}(X_i)). \end{aligned}$$

Now compute,

$$\begin{aligned} \alpha &= (a_1, \dots, a_{n+1})_{F,p^{r+1}} \\ &= (a_1, \dots, a_n, \phi_{n-1}(X_0) \prod_{i=1}^{p-1}(1 - a_n\phi_{n-1}(X_i)))_{F,p^{r+1}} \\ &= (a_1, \dots, a_n, \phi_{n-1}(X_0))_{F,p^{r+1}} + (a_1, \dots, a_n, \prod_{i=1}^{p-1}(1 - a_n\phi_{n-1}(X_i)))_{F,p^{r+1}} \\ &= (a_1, \dots, a_n, \phi_{n-1}(X_0))_{F,p^{r+1}} + \sum_{i=1}^{p-1}(a_1, \dots, a_n, 1 - a_n\phi_{n-1}(X_i))_{F,p^{r+1}} \\ &\overset{\text{Remark 8.3}}{=} (a_1, \dots, a_{n-1}, \phi_{n-1}(X_0), a_n^{-1})_{F,p^{r+1}} \\ &\quad + \sum_{i=1}^{p-1}(a_1, \dots, a_{n-1}, \phi_{n-1}(X_i), (1 - a_n\phi_{n-1}(X_i))^{-1})_{F,p^{r+1}}. \end{aligned}$$

Notice that by Proposition 8.2. we have that:

- $p^r(a_1, \dots, a_{n-1}, \phi_{n-1}(X_0))_{p^{r+1}} = (a_1, \dots, a_{n-1}, \phi_{n-1}(X_0))_p = 0 \in \mathrm{H}^n(G_F, \mu_p)$.
- $p^r(a_1, \dots, a_{n-1}, \phi_{n-1}(X_i))_{p^{r+1}} = (a_1, \dots, a_{n-1}, \phi_{n-1}(X_i))_p = 0 \in \mathrm{H}^n(G_F, \mu_p)$.

Thus, by induction we see that:

- $(a_1, \dots, a_{n-1}, \phi_{n-1}(X_0))_{p^{r+1}} = p\sum_{i=1}^{p^{n-1}} \beta_i$ for symbols $\beta_i \in \mathrm{H}^n(G_F, \mu_{p^{r+1}})$.
- $(a_1, \dots, a_{n-1}, \phi_{n-1}(X_i))_{p^{r+1}} = p\sum_{j=1}^{p^{n-1}} \gamma_{i,j}$ for symbols $\gamma_{i,j} \in \mathrm{H}^n(G_F, \mu_{p^{r+1}})$.

Continuing the computation gives

$$\begin{aligned}\alpha &= (a_1,\dots,a_{n+1})_{F,p^{r+1}}\\ &= (a_1,\dots,a_{n-1},\phi_{n-1}(X_0),a_n^{-1})_{F,p^{r+1}}\\ &\quad+\sum_{i=1}^{p-1}(a_1,\dots,\phi_{n-1}(X_i),(1-a_n\phi_{n-1}(X_i))^{-1})_{F,p^{r+1}}\\ &= p\sum_{i=1}^{p^{n-1}}\beta_i\cup a_n^{-1}+p\sum_{i=1}^{p-1}\sum_{j=1}^{p^{n-1}}\gamma_{i,j}\cup((1-a_n\phi_{n-1}(X_i))^{-1})_{F,p^{r+1}}.\end{aligned}$$

All together, we get $p^{n-1}+(p-1)p^{n-1}=p^n$ symbols and we are done. □

9. Sums of two symbols in the case $p=2$

In this section, we use a known common slots/chain lemma for Pfister forms to get a bound for the symbol length of the sum of two symbols of degree 2^{t+1} and effective exponent 2^t. This method was first used in [**8**] in the context of the Brauer group.

DEFINITION 9.1. A chain isomorphism between two symbols is an isomorphism where only one slot is changed.

Example: $(a,b)\cong(a,c)$ is a chain isomorphism.

In [**11**], the following chain lemma is proved.

THEOREM 9.2. *Let $\langle\langle a_1,\cdots,a_n\rangle\rangle$ and $\langle\langle b_1,\cdots,b_n\rangle\rangle$ be two non trivial, isometric Pfister forms. Then there exist $x\in F^\times$ such that*

$$\langle\langle a_1,\cdots,a_{n-1},a_n\rangle\rangle\cong\langle\langle a_1,\cdots,a_{n-1},x\rangle\rangle\cong\langle\langle b_1,\cdots,b_{n-1},x\rangle\rangle\cong\langle\langle b_1,\cdots,b_n\rangle\rangle.$$

COROLLARY 9.3. *Let $\psi=\langle\langle a_1,\cdots,a_n\rangle\rangle$ and $\phi=\langle\langle b_1,\cdots,b_n\rangle\rangle$ be two non trivial, isometric Pfister forms. Then there exist a chain isomorphism of length $2n-1$ starting with $\langle\langle a_1,\cdots,a_n\rangle\rangle$ and ending with $\langle\langle b_1,\cdots,b_n\rangle\rangle$, i.e., in each step only one slot is changed.*

PROOF. We start with applying Theorem 9.2 to get $x_1\in F^\times$ such that

$$\begin{aligned}\langle\langle a_1,\cdots,a_{n-1},a_n\rangle\rangle&\cong\langle\langle a_1,\cdots,a_{n-1},x_1\rangle\rangle\cong\langle\langle b_1,\cdots,b_{n-1},x_1\rangle\rangle\\&\cong\langle\langle b_1,\cdots,b_n\rangle\rangle.\end{aligned}$$

Write $\psi_1=\langle\langle a_1,\cdots,a_{n-1},x_1\rangle\rangle\cong\langle\langle x_1^{(-1)^{n-1}},a_1,\cdots,a_{n-1}\rangle\rangle$ and $\phi_1=\langle\langle b_1,\cdots,b_{n-1},x_1\rangle\rangle\cong\langle\langle x_1^{(-1)^{n-1}},b_1,\cdots,b_{n-1}\rangle\rangle$; clearly, $\psi_1\cong\phi_1$. Now apply Theorem 9.2 again to get $x_2\in F^\times$ such that $\psi_2=\langle\langle x_1^{(-1)^{n-1}},x_2^{(-1)^{n-2}},\cdots,a_{n-2}\rangle\rangle$ and $\phi_2=\langle\langle x_1^{(-1)^{n-1}},x_2^{(-1)^{n-2}},\cdots,b_{n-2}\rangle\rangle$ and again $\psi_2\cong\phi_2$. After doing this n times, we obtain $\psi_n=\langle\langle x_1^{(-1)^{n-1}},x_2^{(-1)^{n-2}},\cdots,x_n\rangle\rangle$ and $\phi_n=\langle\langle x_1^{(-1)^{n-1}},x_2^{(-1)^{n-2}},\cdots,x_n\rangle\rangle$, and now we see that $\psi_n=\phi_n$. All together we see that we have a chain of isomorphisms

$$\psi\cong\psi_1\cong\cdots\cong\psi_n=\phi_n\cong\cdots\cong\phi_1\cong\phi,$$

and clearly in each isomorphism only one slot is changed and we are done. □

LEMMA 9.4. *Let $\gamma = \alpha - \beta \in \mathrm{H}^n(G_F, \mu_{2^s}^{\otimes n})$, where α, β are symbols, be such that $e.exp(\gamma) \mid 2^{s-1}$. Assume that there is a chain isomorphism of length m for the isomorphic Pfister forms $2^{s-1}\alpha$ and $2^{s-1}\beta$. Then $\alpha - \beta = \sum_{i=1}^{m} \gamma_i$ where for every i we have that $e.exp(\gamma_i) \mid 2^{s-1}$.*

PROOF. We prove by induction on m. For $m = 1$ we have that $2^{s-1}(\alpha - \beta) = 0 \in \mathrm{H}^n(G_F, \mu_2)$. Hence there is a chain:

$$2^{s-1}\alpha = 2^{s-1}(a_1, \dots, a_n) \cong 2^{s-1}(a_1, \dots, a_{n-1}, x) = 2^{s-1}\beta.$$

Write $\alpha - \beta = (a_1, \dots, a_n/x)$ and clearly $e.exp((a_1, \dots, a_n/x)) \mid 2^{s-1}$. Assume the Lemma is true for length m and consider the case where the length of the chain is $m + 1$. Then we have a chain of isomorphisms $2^{s-1}\alpha = (a_1, \dots, a_n)_2 \cong \alpha_1 = (a_1, \dots, x_1)_2 \cong \cdots \cong 2^{s-1}\beta$. Now write $\alpha - \beta = (a_1, \dots, a_n/x_1) + \alpha_1 - \beta$ and notice that $e.exp(a_1, \dots, a_n/x_1) \mid 2^{s-1}$ and $e.exp(\alpha_1 - \beta) \mid 2^{s-1}$. Now, there is a chain of length m for the isomorphism $2^{s-1}\alpha_1 \cong 2^{s-1}\beta$. By induction, we see that $\alpha_1 - \beta = \sum_{i=2}^{m+1} \gamma_i$. Writing $\gamma_1 = (a_1, \dots, a_n/x_1)$, we see that $\alpha - \beta = \sum_{i=1}^{m+1} \gamma_i$ where for each i we have that $e.exp(\gamma_i) \mid 2^{s-1}$ and we are done. □

THEOREM 9.5. *Let $\gamma = \alpha - \beta \in \mathrm{H}^n(G_F, \mu_{2^s}^{\otimes n})$ be such that $e.exp(\gamma) \mid 2^{s-1}$. Then the symbol length of γ is at most $(2n-1)2^{n-1}$.*

PROOF. Indeed, by Corollary 9.3 we see that there is a chain of length $2n-1$ for the isomorphism $2^{s-1}\alpha \cong 2^{s-1}\beta$. Hence by Lemma 9.4, we can write $\gamma = \sum_{i=1}^{2n-1} \gamma_i$ where for each i we have that $e.exp(\gamma_i) \mid 2^{s-1}$. Thus applying Theorem 5.1, we see that $\gamma = 2\sum_{i=1}^{(2n-1)2^{n-1}} \delta_i$ where each δ_i is a symbol in $\mathrm{H}^n(G_F, \mu_{2^s}^{\otimes n})$, that is, the symbol length of γ is at most $(2n-1)2^{n-1}$ and we are done. □

References

[1] A. A. Albert, *Structure of Algebras*, American Mathematical Society Colloquium Publications, Vol. 24, American Mathematical Society, New York, 1939. MR595

[2] R. Elman, N. Karpenko, and A. Merkurjev, *The algebraic and geometric theory of quadratic forms*, American Mathematical Society Colloquium Publications, vol. 56, American Mathematical Society, Providence, RI, 2008, DOI 10.1090/coll/056. MR2427530

[3] C. Haesemeyer and C. Weibel, *Norm varieties and the chain lemma (after Markus Rost)*, Algebraic topology, Abel Symp., vol. 4, Springer, Berlin, 2009, pp. 95–130, DOI 10.1007/978-3-642-01200-6_6. MR2597737

[4] D.E. Haile and L.H. Rowen, *Factorization of polynomials over division algebras*, Algebra Colloq. **2**, no. 1 (1995).

[5] N. Jacobson, *Finite-dimensional division algebras over fields*, Springer Science and Business Media, 2009.

[6] N. Jacobson, *The Theory of Rings*, American Mathematical Society Mathematical Surveys, Vol. II, American Mathematical Society, New York, 1943. MR8601

[7] A. Suslin and S. Joukhovitski, *Norm varieties*, J. Pure Appl. Algebra **206** (2006), no. 1-2, 245–276, DOI 10.1016/j.jpaa.2005.12.012. MR2220090

[8] E. Matzri, *$\mathbb{Z}_3 \times \mathbb{Z}_3$ crossed products*, J. Algebra **418** (2014), 1–7, DOI 10.1016/j.jalgebra.2014.06.035. MR3250437

[9] E. Matzri, *Symbol length in the Brauer group of a field*, Trans. Amer. Math. Soc. **368** (2016), no. 1, 413–427, DOI 10.1090/tran/6326. MR3413868

[10] A.S. Merkurjev and A. A. Suslin, *K-cohomology of Severi-Brauer varieties and the norm residue homomorphism*, Math. USSR Izv. **21** (1983), 307–340.

[11] E. Peyre, *Products of Severi-Brauer varieties and Galois cohomology*, K-theory and algebraic geometry: connections with quadratic forms and division algebras (Santa Barbara, CA, 1992), Proc. Sympos. Pure Math., vol. 58, Amer. Math. Soc., Providence, RI, 1995, pp. 369–401, DOI 10.1090/pspum/058.2/1327310. MR1327310

[12] M. Rost, *Norm varieties and algebraic cobordism*, Proceedings of the International Congress of Mathematicians, Vol. II (Beijing, 2002), Higher Ed. Press, Beijing, 2002, pp. 77–85. MR1957022

[13] A. S. Sivatski, *Applications of the Bloch-Kato conjecture to cohomological invariants and symbol length*, Math. Z. **299** (2021), no. 1-2, 459–472, DOI 10.1007/s00209-020-02678-3. MR4311609

[14] J.-P. Tignol, *Cyclic algebras of small exponent*, Proc. Amer. Math. Soc. **89** (1983), no. 4, 587–588, DOI 10.2307/2044587. MR718978

[15] V. Voevodsky, *Motivic cohomology with* $\mathbf{Z}/2$*-coefficients*, Publ. Math. Inst. Hautes Études Sci. **98** (2003), 59–104, DOI 10.1007/s10240-003-0010-6. MR2031199

[16] V. Voevodsky, *On motivic cohomology with* $\mathbf{Z}/l$*-coefficients*, Ann. of Math. (2) **174** (2011), no. 1, 401–438, DOI 10.4007/annals.2011.174.1.11. MR2811603

Department of Mathematics, Bar-Ilan University, Ramat-Gan 52900, Israel
Email address: elimatzri@gmail.com

Contemporary Mathematics
Volume **800**, 2024
https://doi.org/10.1090/conm/800/16059

Automorphisms and derivations of commutative and PI-algebras

Oksana Bezushchak, Anatoliy Petravchuk, and Efim Zelmanov

To the memory of S.A. Amitsur with admiration.

ABSTRACT. Let A be a finitely generated commutative (or PI) algebra over an associative commutative ring. We discuss extensions of classical theorems of Selberg and Engel to the groups $\mathrm{Aut}(A)$ and the Lie algebra $\mathrm{Der}(A)$.

Let Φ be an associative commutative ring with 1. Let V be an affine algebraic variety over Φ and let A be the algebra of regular (polynomial) functions on V.

The automorphism group $\mathrm{Aut}(A)$ and the Lie algebra of derivations $\mathrm{Der}(A)$ are referred to as the group of polynomial automorphisms of V and the Lie algebra of vector fields on V, respectively. Naturally, we talk about automorphisms and derivations that are Φ-linear.

Let the variety V be irreducible, i.e., the ring A is a domain. Let K be the field of fractions of A. Then the group $\mathrm{Aut}(K)$ is the group of birational automorphisms of V, and the Lie algebra $\mathrm{Der}(K)$ is the Lie algebra of rational vector fields on V.

Let $\mathbb{C}$ be the field of complex numbers. Let $A = \mathbb{C}[x_1, \ldots, x_n]$, $K = \mathbb{C}(x_1, \ldots, x_n)$ be the polynomial algebra and the field of rational functions, respectively. Then $\mathrm{Aut}(K)$, $\mathrm{Der}(K)$ (resp. $\mathrm{Aut}(A), \mathrm{Der}(A)$) are called the Cremona group and Cremona Lie algebra (resp. polynomial Cremona group and polynomial Cremona Lie algebra).

Recall that a group is said to be *linear* if it is embeddable into a group of invertible matrices over a commutative ring. Groups $\mathrm{Aut}(A)$ are, generally speaking, not linear. It has been an ongoing effort of many years to understand: which properties of linear groups can be carried over to groups $\mathrm{Aut}(A)$ and to Cremona groups?

J.P. Serre [**38**], [**39**] studied finite subgroups of Cremona groups. V.L. Popov [**30**] initiated the study of the question whether Jordan's celebrated theorem on finite subgroups of linear groups carries over to the groups $\mathrm{Aut}(A)$. For some important results in this direction see [**5**], [**8**], [**29**], [**30**], [**32**]. S. Cantat [**10**] proved the Tits Alternative for Cremona groups of rank 2.

In this paper, we prove analogs of A. Selberg's result [**37**] for finitely generated subgroups of $\mathrm{Aut}(A)$ and of the Engel theorem for subalgebras of $\mathrm{Der}(A)$. We say

2020 *Mathematics Subject Classification.* Primary 14R20, 14E07, 16W20, 16W25.
Key words and phrases. automorphism, derivation, affine algebraic variety, PI-algebra.

that a group is virtually torsion free if it has a subgroup of finite index that is torsion free.

THEOREM 1. *Let A be a finitely generated associative commutative Φ-algebra. Suppose that A does not have additive torsion. Then:*

(a) *An arbitrary finitely generated subgroup of the group* $\mathrm{Aut}(A)$ *is virtually torsion free;*

(b) *If A is a finitely generated ring (i.e., $\Phi = \mathbb{Z}$), then the group* $\mathrm{Aut}(A)$ *is virtually torsion free.*

COROLLARY 1 (an analog of the theorem of W. Burnside and I. Schur, see [**18**]). *Under the assumptions of Theorem 1(a), every torsion subgroup of* $\mathrm{Aut}(A)$ *is locally finite.*

COROLLARY 2. *Every torsion subgroup of a polynomial Cremona group* $\mathrm{Aut}\, F[x_1, \dots, x_n]$*, where F is a field of zero characteristic, has an abelian normal subgroup of finite index.*

This assertion immediately follows from Corollary 1 and from the Jordan property of the group $\mathrm{Aut}\, F[x_1, \dots, x_n]$ (see [**8**], [**32**]).

REMARK. If the group G in Corollary 1 is torsion of bounded degree, then we do not need any assumptions on additive torsion. Indeed, in [**6**] it was shown that the group $\mathrm{Aut}(A)$ is locally residually finite. Hence by the positive solution of the Restricted Burnside Problem (see [**42**], [**43**]), the group G is locally finite.

Recall that a derivation d of an algebra A is said to be *locally nilpotent* if for an arbitrary element $a \in A$ there exists an integer $n(a) \geqslant 1$ such that $d^{n(a)}(a) = 0$. Let $L \subseteq \mathrm{Der}(A)$ be a Lie algebra that consists of locally nilpotent derivations. The problem of whether it implies that the Lie algebra L is locally nilpotent was discussed in [**12**], [**27**], [**41**]. In particular, A. Skutin [**41**] proved local nilpotency of L for an algebra $A = F[x_1, \dots, x_n]$ of polynomials over a field F of zero characteristic.

THEOREM 2. *Let A be a finitely generated commutative algebra over a commutative ring. Let $L \subseteq \mathrm{Der}(A)$ be a subalgebra of* $\mathrm{Der}(A)$ *that consists of locally nilpotent derivations. Then the Lie algebra L is locally nilpotent.*

The assumption of finite generation of the algebra A is essential. If A is the algebra of polynomials in countably many variables over a field, then there exists a non locally nilpotent Lie algebra $L \subseteq \mathrm{Der}(A)$ that consists of locally nilpotent derivations. The following theorem, however, imposes a finiteness condition that is weaker than finite generation:

Let A be a commutative domain. Let K be the field of fractions of A. An arbitrary derivation of the domain A extends to a derivation of the field K, $\mathrm{Der}(A) \subseteq \mathrm{Der}(K)$. We have $K\,\mathrm{Der}(K) \subseteq \mathrm{Der}(K)$, hence $\mathrm{Der}(K)$ can be viewed as a vector space over the field K.

THEOREM 3. *Under the assumptions above, let $L \subseteq \mathrm{Der}(A)$ be a Lie ring that consists of locally nilpotent derivations. Suppose that* $\dim_K KL < \infty$*. Then the Lie ring L is locally nilpotent.*

A special case of Theorem 3 was proved by A.P. Petravchuk and K. Ya. Sysak [**27**].

The proof of Theorem 3 is based on a stronger version of Theorem 2, which is of independent interest.

Recall that a subalgebra B of a commutative algebra A is called an *order* in A if there exists a multiplicative subgroup $S \subset B$ such that

(1) every element from S is invertible in A;
(2) an arbitrary element $a \in A$ can be represented as $a = s^{-1}b$, where $s \in S$ and $b \in B$.

Let $L \subseteq \operatorname{Der}(A)$ be a subalgebra. The subset

$$A_L = \left\{ a \in A \;\middle|\; \begin{matrix} \text{for an arbitrary derivation } d \in L \text{ there exists} \\ \text{an integer } n(d) \geqslant 1 \text{ such that } d^{n(d)}(a) = 0 \end{matrix} \right\}$$

is a subalgebra of the algebra A.

PROPOSITION 1. *Let A be a finitely generated commutative domain. Let $L \subseteq \operatorname{Der}(A)$ be a Lie algebra of derivations. If the subalgebra A_L is an order in A, then the Lie algebra L is locally nilpotent.*

In order to achieve a natural generality and to expand to noncommutative cases, we extended Theorems 1 and 2 to rings with polynomial identities, i.e., PI-rings (see [**1**], [**7**]).

THEOREM 4. *Let A be a finitely generated PI-ring without additive torsion. Then the group* $\operatorname{Aut}(A)$ *is virtually torsion free.*

As in the case of Theorem 1, this theorem implies analogs of theorems of Burnside and Schur.

COROLLARY 3. *Let A be a finitely generated PI-algebra over an associative commutative ring. Suppose that A does not have additive torsion. Then every torsion subgroup of* $\operatorname{Aut}(A)$ *is locally finite.*

C. Procesi [**31**] proved local finiteness of torsion subgroups of multiplicative groups of PI-algebras.

REMARK. The analog of Selberg's theorem holds for automorphism groups of some algebras that are far from being PI. For example, it holds for the free associative algebra $F\langle x_1, \ldots, x_n\rangle$, where F is a field of zero characteristic.

THEOREM 5. *Let A be a finitely generated PI-algebra over an associative commutative ring. Let $L \subseteq \operatorname{Der}(A)$ be a subalgebra that consists of locally nilpotent derivations. Then the Lie algebra L is locally nilpotent.*

References

[1] E. Aljadeff, A. Giambruno, C. Procesi, and A. Regev, *Rings with polynomial identities and finite dimensional representations of algebras*, American Mathematical Society Colloquium Publications, vol. 66, American Mathematical Society, Providence, RI, 2020. MR4249615

[2] R. C. Alperin, *An elementary account of Selberg's lemma*, Enseign. Math. (2) **33** (1987), no. 3-4, 269–273. MR925989

[3] S. A. Amitsur, *On central division algebras*, Israel J. Math. **12** (1972), 408–420, DOI 10.1007/BF02764632. MR318216

[4] M. F. Atiyah and I. G. Macdonald, *Introduction to commutative algebra*, Addison-Wesley Publishing Co., Reading, Mass.-London-Don Mills, Ont., 1969. MR242802

[5] T. Bandman and Y. G. Zarhin, *Jordan properties of automorphism groups of certain open algebraic varieties*, Transform. Groups **24** (2019), no. 3, 721–739, DOI 10.1007/s00031-018-9489-2. MR3989688

[6] H. Bass and A. Lubotzky, *Automorphisms of groups and of schemes of finite type*, Israel J. Math. **44** (1983), no. 1, 1–22, DOI 10.1007/BF02763168. MR693651

[7] A. Kanel-Belov and L. H. Rowen, *Computational aspects of polynomial identities*, Research Notes in Mathematics, vol. 9, A K Peters, Ltd., Wellesley, MA, 2005. MR2124127

[8] C. Birkar, *Singularities of linear systems and boundedness of Fano varieties*, Ann. of Math. (2) **193** (2021), no. 2, 347–405, DOI 10.4007/annals.2021.193.2.1. MR4224714

[9] A. Braun, *The radical in a finitely generated P.I. algebra*, Bull. Amer. Math. Soc. (N.S.) **7** (1982), no. 2, 385–386, DOI 10.1090/S0273-0979-1982-15044-3. MR663790

[10] S. Cantat, *Sur les groupes de transformations birationnelles des surfaces* (French, with English and French summaries), Ann. of Math. (2) **174** (2011), no. 1, 299–340, DOI 10.4007/annals.2011.174.1.8. MR2811600

[11] J. Déserti, *The Cremona group and its subgroups*, Mathematical Surveys and Monographs, vol. 252, American Mathematical Society, Providence, RI, 2021, DOI 10.1090/surv/252. MR4256046

[12] G. Freudenburg, *Algebraic theory of locally nilpotent derivations*, Encyclopaedia of Mathematical Sciences, vol. 136, Springer-Verlag, Berlin, 2006. Invariant Theory and Algebraic Transformation Groups, VII. MR2259515

[13] H. Kraft, *Automorphism groups of affine varieties and a characterization of affine n-space*, Trans. Moscow Math. Soc. **78** (2017), 171–186, DOI 10.1090/mosc/262. MR3738084

[14] E. S. Golod, *On nil-algebras and finitely approximable p-groups* (Russian), Izv. Akad. Nauk SSSR Ser. Mat. **28** (1964), 273–276. MR161878

[15] R. I. Grigorchuk, *Degrees of growth of finitely generated groups and the theory of invariant means* (Russian), Izv. Akad. Nauk SSSR Ser. Mat. **48** (1984), no. 5, 939–985. MR764305

[16] I. Herstein, *Noncommutative Rings*, Carus Math. Monographs, Vol. 15, Math. Assoc. Amer., Washington, DC, 1968.

[17] N. Jacobson, *Structure of rings*, Revised edition, American Mathematical Society Colloquium Publications, Vol. 37, American Mathematical Society, Providence, RI, 1964. MR222106

[18] N. Jacobson, *Basic Algebra II*, W. H. Freeman and Co., San Franciso, CA, 1980.

[19] I. Kaplansky, *Rings with a polynomial identity*, Bull. Amer. Math. Soc. **54** (1948), 575–580, DOI 10.1090/S0002-9904-1948-09049-8. MR25451

[20] A. R. Kemer, *Identities of finitely generated algebras over an infinite field* (Russian), Izv. Akad. Nauk SSSR Ser. Mat. **54** (1990), no. 4, 726–753, DOI 10.1070/IM1991v037n01ABEH002053; English transl., Math. USSR-Izv. **37** (1991), no. 1, 69–96. MR1073084

[21] T. H. Lenagan and A. Smoktunowicz, *An infinite dimensional affine nil algebra with finite Gelfand-Kirillov dimension*, J. Amer. Math. Soc. **20** (2007), no. 4, 989–1001, DOI 10.1090/S0894-0347-07-00565-6. MR2328713

[22] J. Levitzki, *On the structure of algebraic algebras and related rings*, Trans. Amer. Math. Soc. **74** (1953), 384–409, DOI 10.2307/1990809. MR53089

[23] V. T. Markov, *The dimension of noncommutative affine algebras* (Russian), Izv. Akad. Nauk SSSR Ser. Mat. **37** (1973), 284–288. MR360683

[24] P. S. Novikov and S. I. Adjan, *Infinite periodic groups. I* (Russian), Izv. Akad. Nauk SSSR Ser. Mat. **32** (1968), 212–244. MR240178

[25] P. S. Novikov and S. I. Adjan, *Infinite periodic groups. II* (Russian), Izv. Akad. Nauk SSSR Ser. Mat. **32** (1968), 251–524. MR240179

[26] P. S. Novikov and S. I. Adjan, *Infinite periodic groups. III* (Russian), Izv. Akad. Nauk SSSR Ser. Mat. **32** (1968), 709–731. MR240180

[27] A. Petravchuk and K. Sysak, *On Lie algebras consisting of locally nilpotent derivations*, J. Lie Theory **27** (2017), no. 4, 1057–1068. MR3646031

[28] B. I. Plotkin, *Algebraic sets of elements in groups and Lie algebras* (Russian), Uspehi Mat. Nauk **13** (1958), no. 6(84), 133–138. MR103223

[29] V. L. Popov, *On the Makar-Limanov, Derksen invariants, and finite automorphism groups of algebraic varieties*, Affine algebraic geometry, CRM Proc. Lecture Notes, vol. 54, Amer. Math. Soc., Providence, RI, 2011, pp. 289–311, DOI 10.1090/crmp/054/17. MR2768646

[30] ———, *Jordan groups and automorphism groups of algebraic varieties, in: Automorphisms of Birational and Affine Geometry*, Springer Proceedings in Mathematics and Statistics, vol. 79, Springer, 2014.

[31] C. Procesi, *The Burnside problem*, J. Algebra **4** (1966), 421–425, DOI 10.1016/0021-8693(66)90031-7. MR212081

[32] Y. Prokhorov and C. Shramov, *Jordan property for Cremona groups*, Amer. J. Math. **138** (2016), no. 2, 403–418, DOI 10.1353/ajm.2016.0017. MR3483470

[33] Yu. P. Razmyslov, *Simple Lie algebras in varieties generated by Lie algebras of Cartan type* (Russian), Izv. Akad. Nauk SSSR Ser. Mat. **51** (1987), no. 6, 1228–1264, 1359, DOI 10.1070/IM1988v031n03ABEH001089; English transl., Math. USSR-Izv. **31** (1988), no. 3, 541–573. MR933963

[34] A. Regev, *Existence of polynomial identities in $A \otimes_F B$*, Bull. Amer. Math. Soc. **77** (1971), 1067–1069, DOI 10.1090/S0002-9904-1971-12869-0. MR284468

[35] L. Rowen, *Some results on the center of a ring with polynomial identity*, Bull. Amer. Math. Soc. **79** (1973), 219–223, DOI 10.1090/S0002-9904-1973-13162-3. MR309996

[36] L. H. Rowen, *Polynomial identities in ring theory*, Pure and Applied Mathematics, vol. 84, Academic Press, Inc. [Harcourt Brace Jovanovich, Publishers], New York-London, 1980. MR576061

[37] A. Selberg, *On discontinuous groups in higher-dimensional symmetric spaces*, Contributions to function theory (Internat. Colloq. Function Theory, Bombay, 1960), Tata Inst. Fund. Res., Bombay, 1960, pp. 147–164. MR130324

[38] J-P. Serre, *Le groupe de Cremona etses sous-groupes finis*, Séminaire Bourbaki (2008), no. 1000, 24 pp.

[39] ———, *A Minkowski-style bound for the orders of the finite subgroups of the Cremona group of rank 2 over an arbitrary field*, Moscow Math. J. **9** (2009), no. 1, 183–198.

[40] A. I. Širšov, *On rings with identity relations* (Russian), Mat. Sb. (N.S.) **43(85)** (1957), 277–283. MR95192

[41] A. A. Skutin, *Maximal Lie subalgebras among locally nilpotent derivations* (Russian, with Russian summary), Mat. Sb. **212** (2021), no. 2, 138–146, DOI 10.4213/sm9360; English transl., Sb. Math. **212** (2021), no. 2, 265–271. MR4223965

[42] E. I. Zel′manov, *Solution of the restricted Burnside problem for groups of odd exponent* (Russian), Izv. Akad. Nauk SSSR Ser. Mat. **54** (1990), no. 1, 42–59, 221; English transl., Math. USSR-Izv. **36** (1991), no. 1, 41–60. MR1044047

[43] ———, *Solution of the restricted Burnside problem for 2-groups*, Mat. Sb. **182** (1991), no. 4, 568–592.

[44] E. Zelmanov, *Lie algebras and torsion groups with identity*, J. Comb. Algebra **1** (2017), no. 3, 289–340, DOI 10.4171/JCA/1-3-2. MR3681577

FACULTY OF MECHANICS AND MATHEMATICS, TARAS SHEVCHENKO NATIONAL UNIVERSITY OF KYIV, UKRAINE

Email address: mechmatKNUbezushchak@gmail.com

FACULTY OF MECHANICS AND MATHEMATICS, TARAS SHEVCHENKO NATIONAL UNIVERSITY OF KYIV, UKRAINE

Email address: apetrav@gmail.com

DEPARTMENT OF MATHEMATICS, UNIVERSITY OF CALIFORNIA, SAN DIEGO, SAN DIEGO, CALIFORNIA 92093

Email address: efim.zelmanov@gmail.com

Contemporary Mathematics
Volume **800**, 2024
https://doi.org/10.1090/conm/800/16060

Tensor fundamental theorems of invariant theory

Claudio Procesi

To the memory of S.A. Amitsur who taught me PI theory

Abstract. The aim of this paper is to establish a first and second fundamental theorem for $GL(V)$ equivariant polynomial maps from k–tuples of matrix variables $End(V)^k$ to tensor spaces $End(V)^{\otimes n}$, in the spirit of H. Weyl's book *The Classical Groups* and of symbolic algebra.

Contents

1. Introduction

In this paper, V denotes a vector space over a field F of characteristic 0 of dimension d. In fact since all the formulas developed have rational coefficients, it is enough to assume $F = \mathbb{Q}$ and $End(V) = M_d(F) = M_d$, matrices in a basis. We introduce the main object of study.

Definition 1.1. Given a vector space V and $k, n \in \mathbb{N}$, we denote by the symbols $\mathcal{T}_k^n(V) = \mathcal{T}_X^n(V)$ the algebra of $GL(V)$ equivariant polynomial maps $f : End(V)^k \to End(V)^{\otimes n}$, from k–tuples $X = \{x_1, \ldots, x_k\}$ of matrix variables $End(V)^k$ to tensor spaces $End(V)^{\otimes n} = M_d^{\otimes n}$.

The aim of this paper is to establish a first and second fundamental theorem (FFT and SFT for short), in the spirit of H. Weyl's book *The Classical Groups* [**24**], for the algebras $\mathcal{T}_k^n(V) = \mathcal{T}_X^n(V)$.

From an algebraic point of view, in a given basis, so that $End(V) = M_d(\mathbb{Q})$ is the algebra of $d \times d$ matrices, the polynomial functions on $M_d(\mathbb{Q})^k$ are the polynomials $\mathbb{Q}[\xi_{a,b}^{(i)}]$ in the kd^2 variables $\xi_{a,b}^{(i)}$, $i = 1, \ldots, k;\ a, b = 1, \ldots, d$, which are thought of as the entries of the k *generic* $d \times d$ *matrices*

$$\Xi := \{\xi_1, \ldots, \xi_k\},\ \xi_i = (\xi_{a,b}^{(i)}) \in M_d(\mathbb{Q}[\xi_{a,b}^{(i)}]). \tag{1}$$

2020 *Mathematics Subject Classification.* Primary 16R30, 15A72.
Key words and phrases. Tensor polynomials, second fundamental theorem.

The group $GL(d,\mathbb{Q})$ acts on $M_d(\mathbb{Q})^k$ by simultaneous conjugation inducing an action on $\mathbb{Q}[\xi_{a,b}^{(i)}]$. The polynomial maps from $M_d(\mathbb{Q})^k$ to $M_d(\mathbb{Q})^{\otimes n}$ form the matrix algebra

$$M_d(\mathbb{Q})^{\otimes n} \otimes \mathbb{Q}[\xi_{a,b}^{(i)}] \simeq M_{d^n}[(\mathbb{Q}[\xi_{a,b}^{(i)}]). \tag{2}$$

On this algebra, we have the diagonal action of $GL(d,\mathbb{Q})$ and finally one has

$$\mathcal{T}_k^n(V) = \left[M_d(\mathbb{Q})^{\otimes n} \otimes \mathbb{Q}[\xi_{a,b}^{(i)}]\right]^{GL(d,\mathbb{Q})} \subset M_d(\mathbb{Q})^{\otimes n} \otimes \mathbb{Q}[\xi_{a,b}^{(i)}]. \tag{3}$$

The FFT, Theorem 2.22, describing generators of the algebra $\mathcal{T}_k^n(V)$ is fairly simple. It is based on the basic Formula (31), the *interpretation Formula,* from which follows that such maps are the evaluations in matrices of the symbolic twisted group algebra $T\langle X\rangle^{\otimes n} \ltimes \mathbb{Q}[S_n]$, Definition 2.11.

By $T\langle X\rangle$, we denote the free algebra with trace, Definition 2.3, in the variables $X = \{x_1, \dots, x_k\}$ and S_n commutes with $T\langle X\rangle^{\otimes n}$ by exchanging the tensor factors. In other words, we have a surjective map

$$\pi_n : T\langle X\rangle^{\otimes n} \ltimes \mathbb{Q}[S_n] \to \mathcal{T}_X^n(V) = \left[M_d(\mathbb{Q})^{\otimes n} \otimes \mathbb{Q}[\xi_{a,b}^{(i)}]\right]^{GL(d,\mathbb{Q})} \tag{4}$$

from the symbolic algebra to the algebra of equivariant maps.

The map π_n maps the variables x_i to the generic matrices ξ_i and the symmetric group S_n to its copy inside $M_d(\mathbb{Q})^{\otimes n}$, given by Formula (7).

The map π_n is compatible with the trace as defined, for the algebra $T\langle X\rangle^{\otimes n} \ltimes \mathbb{Q}[S_n]$, in Definition 2.11.

The SFT, Theorem 3.18, is the heart of the paper. In the spirit of T–ideals of universal algebra, cf. §3.4.1, it describes the kernels of the maps π_n.

By the classical method of polarization and restitution one reduces to study the restriction of π_n to the multilinear elements of $T\langle X\rangle^{\otimes n} \ltimes \mathbb{Q}[S_n]$ of degree k for all k, n. Here the main tool is the fact that these elements can be encoded by elements of $\mathbb{Q}[S_{n+k}]$ by a map called *interpretation* (see Definition 2.25).

In §2.17.1, we develop the basic formulas which this map satisfies and which are necessary for the proof of the main Theorem 3.8.

One defines in Formula (45), for each d, the $d+2$ *tensor Cayley Hamilton identities* $\mathfrak{C}_{k,d}(x)$, $k = 0, \dots, d+1$. These are elements of $T\langle X\rangle^{\otimes n} \ltimes \mathbb{Q}[S_n]$ homogeneous of degree k in the x and with $n = d+1-k$, which vanish when x is evaluated in a $d \times d$ matrix.

One first deduces all identities from these, see Theorem 3.8.

For $k = 0, d, d+1$, these have classical interpretation as, respectively, the antisymmetrizer on $d+1$ elements, the Cayley–Hamilton identity and the expression of $tr(x^{d+1})$ in terms of $tr(x^i)$, $i = 1, \dots, d$.

For the other k, they are new identities. For instance, $\mathfrak{C}_{1,3}(x)$, $\mathfrak{C}_{2,3}(x)$ are

$$\boxed{[(1,2,3)+(1,3,2)-(1,2)-(1,3)-(2,3)+1](x\otimes 1\otimes 1+1\otimes x\otimes 1+1\otimes 1\otimes x-tr(x))}$$

$$\boxed{(1-(1,2))\left([x^2\otimes 1+1\otimes x^2+x\otimes x]-tr(x)[x\otimes 1+1\otimes x]+\det(x)\right).}$$

Finally, using a further operation $\mathtt{t} : T\langle X\rangle^{\otimes n} \ltimes \mathbb{Q}[S_n] \to T\langle X\rangle^{\otimes n-1} \ltimes \mathbb{Q}[S_{n-1}]$, a *formal partial trace*, Proposition 3.12, corresponding to the natural trace contraction $M_d^{\otimes n} \to M_d^{\otimes n-1}$, one shows that

Formula (61) $\mathtt{t}(\mathfrak{C}_{k,d}(x)) = 0,\ \mathtt{t}(\mathfrak{C}_{k,d}(x) \cdot 1^{n-1} \otimes x) = -(k+1) \cdot \mathfrak{C}_{k+1,d}(x).$

From these facts follows the Main Theorem of this paper:

THEOREM (3.18). *All relations for equivariant maps, i.e., the elements of the kernels of the maps* $\pi_n : T\langle X\rangle^{\otimes n} \ltimes \mathbb{Q}[S_n] \to \mathcal{T}_X^n(V)$ *for all* n, *can be* formally deduced *from the antisymmetrizer* $\sum_{\sigma\in S_{d+1}} \epsilon_\sigma\sigma = \mathfrak{C}_{0,d}(x)$ *and* $tr(1) = d$.

Recall that $\sum_{\sigma\in S_{d+1}} \epsilon_\sigma\sigma \in \mathbb{Q}[S_{d+1}] \subset T\langle X\rangle^{\otimes d+1} \ltimes \mathbb{Q}[S_{d+1}]$.

The meaning of the expression *formally deduced* is that of Universal Algebra and T–ideals as explained in detail in Definition 3.6 and as modified in Definition 3.17. This method is at the heart of the present paper.

This sheds a new light even on the very classical Cayley–Hamilton identity which is recursively constructed from the antisymmetrizer by Formula (61).

Before approaching these new Theorems 1.9, 3.8, and 3.18, let us briefly recall how the first and second fundamental theorems of invariant theory appear classically.

1.2. The Classical Theory. The first and second fundamental theorems of invariant theory appear in H. Weyl's book *The Classical Groups* [**24**] as a basic tool to understand the Representation Theory of classical groups. One remarkable feature of these theorems is that they are ubiquitous; they appear classically in at least 3 different and apparently unrelated forms (see the next section).

The theorems were first developed in characteristic 0 and then extended, with considerable effort, to all characteristics, see [**9**], [**11**], [**25**], [**10**].

In this paper, we present yet another form of these theorems and will restrict to characteristic 0 (but see the Comments at the end of this paper). We will also develop the theory only for the general linear group $GL(V)$, that is, the group of all linear transformations of a vector space V finite dimensional over a field F (which one can assume to be $\mathbb{Q}$).

For the other classical groups, similar results hold (one may for instance approach them as in [**16**]) and will be treated elsewhere.

1.2.1. *The first fundamental theorem.* The group $GL(V)$ acts on V, its defining representation, and on its dual V^* by the dual action. By convention, we will write *vectors* in V with Roman letters while *covectors* in V^* with Greek letters.

We use the bracket notation; for $\varphi \in V^*$, $v \in V$, we often write $\langle\varphi \mid v\rangle := \varphi(v)$, so that the dual action is given by the formula

$$\langle g\varphi \mid v\rangle := \langle\varphi \mid g^{-1}v\rangle,\ \forall g \in GL(V),\ v \in V,\ \varphi \in V^*. \tag{5}$$

In other words, $\langle g\varphi \mid gv\rangle = \langle\varphi \mid v\rangle$, $\forall g \in GL(V)$ that is the function $\langle\varphi \mid v\rangle$ of v and φ is invariant. For this group, the *first fundamental theorem, FFT for short,* states that the previous functions generate all invariants of several copies of V and V^*. That is, denoting

$$(v_1, v_2, \dots, v_h; \varphi_1, \varphi_2, \dots, \varphi_k) \in V^{\oplus h} \oplus (V^*)^{\oplus k}.$$

THEOREM 1.3. *The polynomial functions on* $V^{\oplus h} \oplus (V^*)^{\oplus k}$ *which are* $GL(V)$ *invariant are generated by the* $h\cdot k$ *basic functions* $\langle\varphi_i \mid v_j\rangle$, $i = 1,\dots,k$; $j = 1,\dots,h$.

One of the remarkable features of the theory as presented by H. Weyl is the fact that this theorem is equivalent to a second theorem, with $h = k$.

THEOREM 1.4. *The algebra of linear operators on $V^{\otimes h}$ which commute with the diagonal action of $GL(V)$ is generated by the elements of the symmetric group S_h. The two actions are*

$$(6)\qquad g\cdot(v_1\otimes v_2\otimes\ldots\otimes v_h)=gv_1\otimes gv_2\otimes\ldots\otimes gv_h,\qquad g\in GL(V),$$

$$(7)\qquad \sigma\cdot(v_1\otimes v_2\otimes\ldots\otimes v_h)=v_{\sigma^{-1}(1)}\otimes v_{\sigma^{-1}(2)}\otimes\ldots\otimes v_{\sigma^{-1}(h)},\qquad \sigma\in S_h.$$

1.4.1. *The second fundamental theorem.* Together with a first fundamental theorem one has a *second fundamental theorem, SFT for short.* In this theorem, the relations among the invariants as generated by basic ones are described.

For the setting of Theorem 1.3, one should think of

$$V^{\oplus h}=\hom(F^h,V),\ (V^*)^{\oplus k}=\hom(V,F^k).$$

Then the invariants $\langle\varphi_i\mid v_j\rangle$ are the entries of the $h\times k$ matrix image of the map

$$\hom(F^h,V)\times\hom(V,F^k)\to\hom(F^h,F^k),\ (A,B)\mapsto B\circ A.$$

The image of this map is the variety of matrices of rank $\le d:=\dim V$.

The polynomial functions on $\hom(F^h,F^k)$ are in the variables $x_{i,j}$ which can be viewed as entries of a *generic matrix* X and the SFT is:

THEOREM 1.5. *The ideal of functions on the space of $h\times k$ matrices vanishing on the matrices of rank $\le d$ is nonzero if and only if $d<\min(h,k)$.*

In this case, it is generated by the determinants of all the $d+1\times d+1$ minors of the matrix X.

This is just the first of a long list of ideas and theorems of a geometric nature on special singularities.

Instead for the setting of Theorem 1.4, one has

THEOREM 1.6. *Consider the mapping $\pi: F[S_h]\to End(V^{\otimes h})$ from the group algebra to the linear operators, given by Formula (7).*

The kernel of π is nonzero only if $\dim V=d<h$ and then it is the two sided ideal of $F[S_h]$ generated by an antisymmetrizer $\sum_{\sigma\in S_{d+1}}\epsilon_\sigma\sigma$.

Here $S_{d+1}\subset S_h$ and $\epsilon_\sigma=\pm1$ is the sign of the permutation. This is just the first of a long list of ideas and theorems in Representation Theory.

1.7. Matrix invariants. There is still a third way in which the fundamental Theorems appear.

In this case we take as basic representation the direct sum of h copies of the matrix algebra $End(V)$ of linear maps of V, under simultaneous conjugation by $GL(V)$. We start with a remark.

REMARK 1.8. Given a vector space W, the symmetric group S_h acts on $W^{\otimes h}$ by the Formula (7);

thus, $S_h\subset End(W^{\otimes h})=End(W)^{\otimes h}$. Therefore, we have a priori two actions of S_h on $End(W)^{\otimes h}$, the one given by Formula (7) thinking of $End(W)^{\otimes h}$ as tensors and the conjugation action in $End(W)^{\otimes h}$ as algebra. If $\sigma\in S_h$ is a permutation of the tensor indices, we have for a tensor $a_1\otimes\ldots\otimes a_h\in M_d^{\otimes h}$.

$$(8)\qquad \sigma\circ a_1\otimes\ldots\otimes a_h\circ\sigma^{-1}=a_{\sigma^{-1}(1)}\otimes\ldots\otimes a_{\sigma^{-1}(h)}=\sigma\cdot a_1\otimes\ldots\otimes a_h.$$

Formula (8) states that these two actions coincide, cf. Formula (11).

We need to recall the multilinear invariants of m matrices, i.e., the invariant elements of the dual of $End(V)^{\otimes m}$. The theorem is, cf. [**15**]:

THEOREM 1.9. *The space of multilinear invariants of m endomorphisms $(x_1, x_2, \dots, x_m)$ of a d-dimensional vector space V is identified with the space $End_{GL(V)}(V^{\otimes m})$ and it will be denoted by $\mathcal{T}_d(m)$. It is linearly spanned by the functions:*

$$t_\sigma(x_1, x_2, \dots, x_m) := tr(\sigma^{-1} \circ x_1 \otimes x_2 \otimes \cdots \otimes x_m), \ \sigma \in S_m. \tag{9}$$

If $\sigma = (i_1 i_2 \dots i_h) \dots (j_1 j_2 \dots j_\ell)(s_1 s_2 \dots s_t)$ is the cycle decomposition of σ then we have that $t_\sigma(x_1, x_2, \dots, x_m)$ equals

$$= tr(x_{i_1} x_{i_2} \dots x_{i_h}) \dots tr(x_{j_1} x_{j_2} \dots x_{j_\ell}) tr(x_{s_1} x_{s_2} \dots x_{s_t}). \tag{10}$$

PROOF. We recall the standard proof since this is a basic ingredient of this paper.

First we note that the dual of $End(V)^{\otimes m}$ can be identified, in a $GL(V)$ equivariant way to $End(V)^{\otimes m}$ by the pairing formula:

$$\begin{aligned}\langle A_1 \otimes A_2 \cdots \otimes A_m | B_1 \otimes B_2 \cdots \otimes B_m \rangle &:= tr(A_1 \otimes A_2 \cdots \otimes A_m \circ B_1 \otimes B_2 \cdots \otimes B_m) \\ &= \prod tr(A_i B_i).\end{aligned}$$

Under this isomorphism, the multilinear invariants of matrices are identified with the $GL(V)$ invariants of $End(V)^{\otimes m}$ which in turn are spanned by the elements of the symmetric group, Theorem 1.4, hence by the elements of Formula (9).

As for Formula (10), since the identity is multilinear, in the variables x, it is enough to prove it on the decomposable tensors of $End(V) = V \otimes V^*$ which are the endomorphisms of rank 1, i.e., $u \otimes \varphi : v \mapsto \langle \varphi \,|\, v \rangle u$. We use the following basic formulas.

LEMMA 1.10.

$$\begin{aligned} \sigma^{-1} \circ u_1 \otimes \varphi_1 \otimes u_2 \otimes \varphi_2 \otimes \ldots \otimes u_m \otimes \varphi_m \\ = u_{\sigma(1)} \otimes \varphi_1 \otimes u_{\sigma(2)} \otimes \varphi_2 \otimes \ldots \otimes u_{\sigma(m)} \otimes \varphi_m \end{aligned} \tag{11}$$

$$u_1 \otimes \varphi_1 \otimes \ldots \otimes u_m \otimes \varphi_m \circ \sigma = u_1 \otimes \varphi_{\sigma(1)} \otimes u_2 \otimes \varphi_{\sigma(2)} \otimes \ldots \otimes u_m \otimes \varphi_{\sigma(m)}.$$

PROOF. Given $x_i := u_i \otimes \varphi_i$ and an element $\sigma \in S_m$ in the symmetric group, we have

$$\begin{aligned} &\sigma^{-1} \circ u_1 \otimes \varphi_1 \otimes u_2 \otimes \varphi_2 \otimes \ldots \otimes u_m \otimes \varphi_m (v_1 \otimes v_2 \otimes \ldots \otimes v_m) \\ &\quad = \prod_{i=1}^m \langle \varphi_i \,|\, v_i \rangle u_{\sigma(1)} \otimes u_{\sigma(2)} \otimes \ldots \otimes u_{\sigma(m)} \\ &\quad = u_{\sigma(1)} \otimes \varphi_1 \otimes u_{\sigma(2)} \otimes \varphi_2 \otimes \ldots \otimes u_{\sigma(m)} \otimes \varphi_m (v_1 \otimes v_2 \otimes \ldots \otimes v_m), \end{aligned}$$

and similarly for the other formula. □

So we need to understand in matrix formulas the invariants

$$tr(\sigma^{-1} \circ u_1 \otimes \varphi_1 \otimes u_2 \otimes \varphi_2 \otimes \ldots \otimes u_m \otimes \varphi_m) = \prod_{i=1}^m \langle \varphi_i \,|\, u_{\sigma(i)} \rangle. \tag{12}$$

We need to use the rules

$$u \otimes \varphi \circ v \otimes \psi = u \otimes \langle \varphi \,|\, v \rangle \psi, \quad tr(u \otimes \varphi) = \langle \varphi \,|\, u \rangle$$

from which Formula (10) easily follows by induction. □

REMARK 1.11.

$$(13)\quad t_{\tau\sigma\tau^{-1}}(x_1,\dots,x_m) \stackrel{(8)}{=} tr(\sigma^{-1}\circ x_{\tau(1)}\otimes\cdots\otimes x_{\tau(m)}) = t_\sigma(x_{\tau(1)},\dots,x_{\tau(m)}).$$

THEOREM 1.12 (FFT for matrices). *The ring $\mathcal{T}_d = \mathcal{T}_d(X)$ of invariants of $d\times d$ matrices under simultaneous conjugation is generated by the elements*

$$(14)\qquad tr(x_{i_1}x_{i_2}\dots x_{i_{k-1}}x_{i_k}).$$

This formula means that we take all possible noncommutative monomials in the x_i and form their traces.

PROOF. The ring of invariants of matrices contains the ring generated by the traces of monomials and both rings are stable under polarization and restitution; hence, by Aronhold method, see [**18**], Chapter 3, it is enough to prove that they coincide on multilinear elements and this is the content of the previous Lemma. □

Finally, for the second fundamental Theorem for matrices, one should first generalize that Theorem to a statement about the noncommutative algebra of equivariant maps $F : End(V)^h \to End(V)$. This noncommutative algebra is generated over the ring of invariants by the *coordinates* x_i, see [**2**].

In this case, the SFT is formulated in terms of universal algebra and the language of T–ideals §3.4.1; see [**17**] and [**22**] or Chapter 12 of [**2**].

THEOREM 1.13. *If $d = \dim_F V$, all relations among polynomial equivariant maps $F : End(V)^h \to End(V)$ are consequences of the Cayley–Hamilton identity and $tr(1) = d$.*

The reader may look at the proof of the second fundamental theorem 1.13 as in [**17**], [**18**], [**2**] or [**10**] since we have taken this as a model for the more general Theorem 3.8 of this paper.

This is just the first of a long list of ideas and theorems in noncommutative algebra, in particular, the Theory of Cayley–Hamilton algebras as presented in the two papers [**19**], [**20**].

REMARK 1.14. The first and second fundamental theorem for matrix invariants are not as precise as the other cases. In fact, from these Theorems for matrices, one can only infer estimates, see [**22**] or §12.2.4 of [**2**], and not a precise description of minimal generators and relations.

For 2×2 matrices the results are quite precise, due to the fact that in this case the action of $GL(2)$ on 2×2 matrices with 0 trace is equivalent to that of $SO(3)$ on its fundamental representation, and then one can apply the first and second fundamental theorems for this group, see Chapter 9 of [**2**].

For 3×3 matrices there are the computations in [**1**].

2. Equivariant tensor polynomial maps

2.1. Algebras with trace. Let us quickly recall this formalism which will be used throughout this paper.

DEFINITION 2.2. An algebra with trace is an associative algebra A over some commutative ring F together with an F linear map $tr : A \to A$ satisfying the following axioms, see [**2**] Chapter 2.3 or [**19**].

$$tr(a)\cdot b = b\cdot tr(a),\ tr(a\cdot b) = tr(b\cdot a),\ tr(tr(a)\cdot b) = tr(a)\cdot tr(b),\ \forall a,b\in A.$$

Then the image $tr(A)$ of tr is a central subalgebra, called the *trace algebra* and tr is $tr(A)$ linear.

We remark that in the definition, in the approach of *universal algebra*, we do not assume that trace takes values in F. The trace tr is to be thought of as a *unary operation.*

Algebras with trace form a category where maps are trace compatible homomorphisms.

If A is an algebra with trace, with trace algebra $T(A)$ then consider

$$A^{\otimes n} := A \otimes_{T(A)} A \otimes_{T(A)} \cdots \otimes_{T(A)} A, \tag{15}$$

that is, $A^{\otimes n}$ is the tensor product of n copies of A over its trace algebra $T(A)$. Furthermore, $A^{\otimes n}$ is an algebra with trace where

$$tr(a_1 \otimes \ldots \otimes a_n) := \prod_{i=1}^{n} tr(a_i)$$

and $T(A^{\otimes n}) = T(A)1^{\otimes n}$..

Algebras with trace admit free algebras.

DEFINITION 2.3. The free algebra with trace in some list of variables X is $T\langle X\rangle := F\langle X\rangle[tr(M)]$. That is, the polynomial algebra in the elements $tr(M)$ over the usual free algebra $F\langle X\rangle$. By $tr(M)$, we denote the class of a monomial M up to cyclic equivalence.

Its trace algebra is the polynomial algebra $F[tr(M)]$ in the elements $tr(M)$ over F (cf. [**19**] for a detailed definition).

In particular, apply Formula (15) to $A = T\langle X\rangle^{\otimes n}$; we then have:

REMARK 2.4. The algebra $T\langle X\rangle^{\otimes n}$ is a free $F[tr(M)]$ module with basis the *tensor monomials* $M_1 \otimes M_2 \otimes \ldots \otimes M_n$.

We finish this with a

DEFINITION 2.5. An *n-fold tensor trace identity of A* is an element of $T\langle X\rangle^{\otimes n}$ vanishing for all evaluations $\varphi : T\langle X\rangle^{\otimes n} \to A^{\otimes n}$ induced by evaluations $X \to A$.

2.6. Tensor polynomials. Let us go back to the algebras $\mathcal{T}_k^n(V) = \mathcal{T}_X^n(V)$, Definition 1.1, of $GL(V)$ equivariant polynomial maps $f : End(V)^k \to End(V)^{\otimes n}$.

Among these maps are the noncommutative polynomials in the tensor variables $x_j^{(i)} := 1^{\otimes i-1} \otimes x_j \otimes 1^{\otimes n-i}$, where x_j is a matrix variable.

DEFINITION 2.7. We call the $x_j^{(i)}$ *tensor variables* and the (noncommutative) polynomials they generate *tensor polynomials.*

The concept of *tensor polynomial or of tensor trace identities* as far as I know has not been considered by algebraists. I wish to thank Felix Huber for pointing out this notion which seems to play some role in Quantum Information Theory, see [**12**] and [**23**].

Given a free algebra $F\langle X\rangle$ and an associative algebra A over a field F, a map $f : X \to A$ induces a homomorphism $f : F\langle X\rangle \to A$ and conversely a homomorphism $f : F\langle X\rangle \to A$ is determined by its values on X.

The polynomial identities of A are the elements of $F\langle X\rangle$ vanishing under all evaluations.

Now, though, such a map f defines also, for all integers n, a homomorphism of the corresponding n–fold tensor products:

$$f^{\otimes n} : F\langle X\rangle^{\otimes n} \to A^{\otimes n}.$$

One can thus define as *n–fold tensor identity for A* an element $G \in F\langle X\rangle^{\otimes n}$ vanishing for all evaluations in A, i.e., $f^{\otimes n}(G) = 0$, $\forall f : X \to A$.

A similar notion holds for A an algebra with trace, Definition 2.5.

2.8. Equivariant maps and permutations. We now restrict to the case $A = M_d(F)$. Together with tensor polynomials, we also have the usual invariants of matrices, described in Theorem 1.12, which may be viewed as scalar valued maps to $M_d^{\otimes n}$, i.e., multiples of the identity of $M_d^{\otimes n}$.

Finally, one has the *constant* equivariant maps, that is, the $GL(V)$ invariant elements of $M_d^{\otimes n}$. They form the subalgebra $\Sigma_n(V) \subset End(V)^{\otimes n}$ spanned by the permutations $\sigma \in S_n$ (Formula (7)) described by Theorem 1.6.

Thus we have 3 types of objects to consider, each contained in the next:

DEFINITION 2.9.

(1) The *tensor polynomial maps*, i.e., the polynomial maps of $M_d(F)^X \to M_d(F)^{\otimes n}$ induced by $F\langle X\rangle^{\otimes n}$.
(2) The *trace tensor polynomial maps*, the maps of $M_d(F)^X \to M_d(F)^{\otimes n}$ induced by $T\langle X\rangle^{\otimes n}$.
(3) The *equivariant tensor polynomial maps*, i.e., the polynomial maps $f : M_d(F)^X \to M_d(F)^{\otimes n}$ equivariant under conjugation by $GL(d, F)$.

Under the multiplication of the algebra $M_d(F)^{\otimes n}$, each one of these spaces forms an algebra.

The way to understand the general form of such an equivariant map, item (3), is to associate an invariant to such a map.

Consider an equivariant polynomial map $H(x_1, \dots, x_k)$ of k, $d \times d$ matrix variables with values in $M_d^{\otimes n}$. To this associate the invariant scalar function of $k + n$, $d \times d$ matrix variables $x_1, \dots, x_k, y_1, y_2, \dots, y_n$:

$$T(H)(x_1, \dots, x_k, y_1, y_2, \dots, y_n) := tr(H(x_1, \dots, x_k)y_1 \otimes y_2 \otimes \dots \otimes y_n). \tag{16}$$

Since the trace form on $M_d^{\otimes n}$ is non degenerate, we can reconstruct H from Formula (16).

By Theorem 1.9, we have that Formula (16) is a linear combination of products of elements of type $tr(M)$ with M a monomial in the variables $x_1, \dots, x_k$ and $y_1, y_2, \dots, y_n$ and linear in these last variables.

So we say that H is *monomial* if it is of the following form. There exist monomials $M_i, i = 1, \dots, n$ and N_j in the variables $x_1, \dots, x_k$, possibly empty, that is, with value 1, such that, setting $z_i = M_i y_i$ we have:

$$T(H) = \prod_j tr(N_j) tr(z_{i_1} z_{i_2} \dots z_{i_{h_1}}) tr(z_{i_{h_1+1}} z_{i_{h_1+2}} \dots z_{i_{h_1+h_2}}) \\ \dots\dots tr(z_{i_{h_1+\dots h_{k-1}}} \dots z_{i_n}). \tag{17}$$

If $\sigma \in S_n$ is the permutation of cycles

$$\sigma = (i_1, i_2, \dots, i_{h_1})(i_{h_1+1}, i_{h_1+2}, \dots, i_{h_1+h_2}) \dots\dots (i_{h_1+\dots+h_{k-1}}, \dots, i_n),$$

then Formula (17) becomes

$$T(H) = \prod_j tr(N_j) tr(\sigma^{-1} \circ M_1 y_1 \otimes \cdots \otimes M_n y_n) \tag{18}$$

$$= tr\Big(\prod_j tr(N_j)(\sigma^{-1} \circ M_1 \otimes \cdots \otimes M_n) y_1 \otimes y_2 \otimes \dots \otimes y_n\Big)$$

$$\implies H = \prod_j tr(N_j) \sigma^{-1} \circ M_1 \otimes \cdots \otimes M_n. \tag{19}$$

THEOREM 2.10. *Equivariant tensor polynomial maps are linear combinations of maps given by Formula* (19).

In other words, using Definition 2.11 *below, the map* $\pi_n : T\langle X\rangle^{\otimes n} \ltimes \mathbb{Q}[S_n] \to \mathcal{T}^n_X(V)$, *from the symbolic algebra to the algebra of equivariant maps, is surjective.*

This Theorem may be viewed as a first fundamental theorem for tensor valued equivariant functions on matrices.

2.10.1. *The algebra* $T\langle X\rangle^{\otimes n} \ltimes \mathbb{Q}[S_n]$. We use the definitions and notations of §2.1 and Formula (15).

From the point of view of universal algebra, the FFT says that the n tensor valued equivariant maps are the evaluations in matrices of the elements of the *twisted* algebra $T\langle X\rangle^{\otimes n} \ltimes \mathbb{Q}[S_n]$.

DEFINITION 2.11. For every $n \in \mathbb{N}, n = 0, 1, \cdots$, we set:

(1) $T\langle X\rangle^{\otimes n} \ltimes \mathbb{Q}[S_n]$ is $T\langle X\rangle^{\otimes n} \otimes \mathbb{Q}[S_n]$, but with the commuting relations

$$\sigma \circ M_1 \otimes \dots \otimes M_n = M_{\sigma^{-1}(1)} \otimes \dots \otimes M_{\sigma^{-1}(n)} \circ \sigma, \quad \sigma \in S_n.$$

(2) The elements of $\mathbb{Q}[S_n]$ from the point of view of universal algebra are *constants* and are canonically evaluated by the map π of Formula (7) in $M_d^{\otimes n}$ (cf. Theorem 1.6).

(3) The algebra $T\langle X\rangle^{\otimes n} \ltimes \mathbb{Q}[S_n]$ is an *algebra with trace* (according to Definition 2.1) where trace is defined by Formulas (9) and (10), now thought of as definitions.

Its trace algebra coincides with the central trace algebra $\mathbb{Q}[tr(M)]$ of $T\langle X\rangle$ of Definition 2.5.

In particular, for $n = 1$, we have the free algebra with trace $T\langle X\rangle$ and, for $n = 0$ its trace algebra $\mathbb{Q}[tr(M)]$.

Given any map $\varphi : X \to M_d(\mathbb{Q})$, this extends to a unique trace compatible map

$$\varphi : T\langle X\rangle^{\otimes n} \ltimes \mathbb{Q}[S_n] \to M_d(\mathbb{Q})^{\otimes n}$$

which coincides with π on S_n. In this way the n tensor valued equivariant maps are the evaluations in matrices of the elements of the *twisted* algebra $T\langle X\rangle^{\otimes n} \ltimes \mathbb{Q}[S_n]$

REMARK 2.12. The algebra of tensor polynomial maps is identified with the algebra $F\langle X\rangle^{\otimes n}$ modulo the vanishing elements, that is, the tensor polynomial identities. We can make a similar statement for trace tensor polynomial maps, and $T\langle X\rangle^{\otimes n}$, and general equivariant maps and the twisted algebra $T\langle X\rangle^{\otimes n} \ltimes \mathbb{Q}[S_n]$. Observe that $F\langle X\rangle^{\otimes n} \subset T\langle X\rangle^{\otimes n} \subset T\langle X\rangle^{\otimes n} \ltimes \mathbb{Q}[S_n]$.

Splitting the cycles. In order to treat the SFT, it is useful to understand in a more precise way the multilinear case of the FFT.

If the map H, appearing in Formula (16), is linear also in the variables x_i then there is a permutation $\tau \in S_{n+k}$ such that Formula (17) equals

$$tr(H(x)y_1 \otimes y_2 \otimes \ldots \otimes y_n) := tr(\tau^{-1} \circ y_1 \otimes y_2 \otimes \ldots \otimes y_n \otimes x_1 \otimes x_2 \otimes \ldots \otimes x_k). \tag{20}$$

DEFINITION 2.13. We denote by $T_{k,\tau}^{(n)}(x_1, x_2, \ldots, x_k)$ the tensor valued map $H(x)$ associated to $\tau \in S_{n+k}$ by Formula (20).

We call the map $\tau \mapsto T_{k,\tau}^{(n)}(x_1, x_2, \ldots, x_k)$ the nth–interpretation of the permutation $\tau \in S_{n+k}$.

Our next task is to describe the elements $T_{k,\tau}^{(n)}(x_1, x_2, \ldots, x_k)$ in terms of the permutation τ. We will use a simple fact on permutations that we call *splitting the cycles*, used in [**17**] to prove the SFT for matrices, Theorem 1.13. By the *non trivial cycles* of a permutation, we mean the ones which move elements.

PROPOSITION 2.14. *[Splitting the cycles] Decompose* $\{1, 2, \ldots, m\} = A \cup B$ *as disjoint union of two subsets* A, B.

Every permutation $\tau \in S_m$ *can be uniquely decomposed as the product* $\tau = \tau_1\tau_2\tau_3$ *where:*

(1) *Each cycle of* τ_1 *contains exactly one element of* A.
(2) $\tau_2 \in S_B$ *is formed by the cycles of* τ *permuting only the indices in* B. *It commutes with* τ_1 *and* τ_3 *since the indices which it moves are disjoint from the ones moved by* τ_1, τ_3.
(3) $\tau_3 \in S_A$ *permutes only the indices in* A.

PROOF. First we may split $\tau = \bar{\tau}\tau_2$ where τ_2 collects all the cycles of τ permuting only the indices in B. Replacing τ with $\bar{\tau}$, we may assume that $\tau = \bar{\tau}_1\bar{\tau}_3$ where $\bar{\tau}_3$ collects all the cycles of τ permuting only the indices in A. Thus $\bar{\tau}_1$ collects all the cycles of τ involving both the indices in A and in B.

By construction, the 3 permutations $\bar{\tau}_1, \tau_2, \bar{\tau}_3$ commute with each other and the indices moved by τ_2 are disjoint from the ones moved by $\bar{\tau}_1$ and $\bar{\tau}_3$.

Then the construction is based, by induction on the number of cycles, on the following identity. For $a_1, a_2, a_3 \ldots, a_j$ numbers in A and $C_1, C_2, \ldots, C_k$ strings of numbers in B, each number appearing only once, consider the splitting of the permutation cycle:

$$\begin{aligned} \mathsf{c} &:= (\ C_1,\ a_1,\ C_2,\ a_2,\ C_3,\ a_3,\ \ldots,\ C_j,\ a_j) \\ &= (a_1,\ C_1)(\ a_2,\ C_2)(a_3,\ C_3)\ \ldots\ (a_j,\ C_j) \circ (a_1, a_2, a_3, \ldots, a_j). \end{aligned} \tag{21}$$

This we call *splitting a cycle* with respect to A, B.

e.g., $A = \{1, 2\},\ B = \{3, 4, 5, 6, 7, 8\};\ C_1 = \ 7, 8, 4; C_2 = 6, 3 :\ a_1 = 1,\ a_2 = 2$

$$\mathsf{c} = (1, 7, 8, 4, 2, 6, 3) = (\boxed{2}, 7, 8, 4)(\boxed{1}, 6, 3) \circ (1, 2).$$

We then split each cycle of $\bar{\tau}_1$ obtaining for each cycle c_i of $\bar{\tau}_1$ a product $\mathsf{c}_i = d_i \circ e_i$ with e_i permuting only the indices of A appearing in c_i and each cycle of d_i contains exactly one of the indices of A appearing in c_i.

Since the c_i involve moving disjoint indices, $\bar{\tau}_1 = \prod_i \mathsf{c}_i = \prod_i d_i \circ \prod_i e_i$. We finally set

$$\tau_1 := \prod_i d_i; \quad \tau_3 := \prod_i e_i \circ \bar{\tau}_3 \implies \bar{\tau}_1\bar{\tau}_3 = \tau_1\tau_3 \implies \tau = \tau_1\tau_2\tau_3. \tag{22}$$

The uniqueness is as follows. The factor τ_2 is well-defined from τ.

So assume we have $\sigma_1\sigma_3 = \tau_1\tau_3$ with σ_3, τ_3 permuting only the indices in A and each cycle of σ_1 and of τ_1 contains exactly one element of A. Multiplying both sides by τ_3^{-1}, we may assume $\tau_3 = 1$ so assume $\sigma_1\sigma_3 = \tau_1$.

If $\sigma_3 \neq 1$, the nontrivial cycles of σ_3 cannot be disjoint from those of σ_1; otherwise, they would be cycles of τ_1 a contradiction since the cycles of τ_1 contain some index in B.

But then there is a cycle c of σ_3 and some cycles (β_i, i) of σ_1 with β_i a non trivial string in B and i appears in c. The product $c\prod_i(\beta, i)$ gives rise (by gluing the cycles) to a cycle of τ_1 with more than one element in A, a contradiction. □

REMARK 2.15. Of course one could also split the cycle c from the *left* as:

$$= (a_1, a_2, a_3, \dots, a_j) \circ (a_j,\ C_1)(\ a_1,\ C_2)(a_2,\ C_3)\ \dots\ (a_{j-1},\ C_j). \tag{23}$$

DEFINITION 2.16. Denote by $U_A(B)$ the set of permutations in $S_m = S_{A\cup B}$ with the property that in each cycle appears at most one element of A.

We may say that Proposition 2.14, [Splitting the cycles], states that the product map $\pi : U_A(B) \times S_A \to S_{A\cup B}$, $(\tau, \sigma) \mapsto \tau \circ \sigma$ is a bijection with inverse $\gamma \mapsto (\gamma_1\gamma_2, \gamma_3)$.

REMARK 2.17. 1) By the uniqueness of the decomposition it follows that, if $\tau \in S_A$, we have $(\sigma \circ \tau)_3 = \sigma_3 \circ \tau$ for all $\sigma \in S_{A\cup B}$.

2) Assume $\tau = \gamma\rho$ with γ and ρ permutations on two disjoint sets of indices, $(A_1 \cup B_1), (A_2 \cup B_2)$ and $A = A_1 \cup A_2,\ B = B_1 \cup B_2,\ A \cap B = \emptyset$. Then

$$\tau_i = \gamma_i\rho_i,\ i = \{1, 2, 3\}\ \text{ for the respective decompositions.} \tag{24}$$

3) If $\sigma = \alpha \circ \beta,\ \alpha \in S_A,\ \beta \in S_B$ we have:

$$(\sigma\tau\sigma^{-1})_1 = \sigma\tau_1\sigma^{-1},\ (\sigma\tau\sigma^{-1})_2 = \beta\tau_2\beta^{-1},\ (\sigma\tau\sigma^{-1})_3 = \alpha\tau_3\alpha^{-1}. \tag{25}$$

4) $U_A(B)$ is stable under conjugation by elements of $S_A \times S_B$.

2.17.1. *Formulas.* We want to apply the previous Proposition 2.14 to Formula (20).

Decompose $\{1, 2, \dots, n+k\} = A \cup B$ with A the indices of type y and B the ones of type x:

$$A = \{1, 2, \dots, n\},\ B = \{n+1, \dots n+k\}.$$

We identify $S_B = S_k$. Denote, for simplicity of notations, with the symbols $Y_A := y_1 \otimes y_2 \otimes \dots \otimes y_n$; and $X_B := x_1 \otimes x_2 \otimes \dots \otimes x_k$. Thus, equation (20) is written as

$$tr(H(x)Y_A) := tr(\tau^{-1} \circ Y_A \otimes X_B). \tag{26}$$

If $\sigma \in S_A,\ \tau \in S_B$, we have

$$\sigma \circ \tau \circ Y_A \otimes X_B = (\sigma \circ Y_A) \otimes (\tau \circ X_B).$$

Notice next that Remark 1.11 gives:

LEMMA 2.18.

1) *If $\sigma \in S_B$ permutes the indices B, we have*

$$T^{(n)}_{k,\sigma\tau\sigma^{-1}}(x_1, x_2, \dots, x_k) = T^{(n)}_{k,\tau}(x_{\sigma(1)}, x_{\sigma(2)}, \dots, x_{\sigma(k)}). \tag{27}$$

2) *If $\gamma \in S_A$ permutes the indices A, we have*

(28) $$T^{(n)}_{k,\gamma\tau\gamma^{-1}}(x_1, x_2, \dots, x_k) = \gamma T^{(n)}_{k,\tau}(x_1, x_2, \dots, x_k)\gamma^{-1}$$

is a permutation of the tensor factors.

PROOF. 1)

$$tr((\sigma\tau\sigma^{-1})^{-1} \circ Y_A \otimes X_B) = tr(\tau^{-1} \circ Y_A \otimes (\sigma^{-1} \circ X_B \circ \sigma))$$

and, by Formula (8)

$$\sigma^{-1} \circ X_B \circ \sigma = x_{\sigma(1)} \otimes x_{\sigma(2)} \otimes \dots \otimes x_{\sigma(k)}.$$

2)

$$tr((\gamma\tau\gamma^{-1})^{-1} \circ Y_A \otimes X_B) = tr(\tau^{-1} \circ (\gamma^{-1} \circ Y_A \circ \gamma) \otimes X_B)$$
$$\gamma^{-1} \circ Y_A \circ \gamma = y_{\gamma(1)} \otimes y_{\gamma(2)} \otimes \dots \otimes y_{\gamma(n)}.$$

Thus

$$\begin{aligned} tr(T^{(n)}_{k,\gamma\tau\gamma^{-1}}(x_1, x_2, \dots, x_k)Y_A) &= tr(T^{(n)}_{k,\tau}(x_1, x_2, \dots, x_k)y_{\gamma(1)} \otimes y_{\gamma(2)} \otimes \dots \otimes y_{\gamma(n)}) \\ &= tr(\gamma T^{(n)}_{k,\tau}(x_1, x_2, \dots, x_k)\gamma^{-1}\gamma y_{\gamma(1)} \otimes y_{\gamma(2)} \otimes \dots \otimes y_{\gamma(n)}\gamma^{-1}) \\ &= tr(\gamma T^{(n)}_{k,\tau}(x_1, x_2, \dots, x_k)\gamma^{-1} y_1 \otimes y_2 \otimes \dots \otimes y_n). \end{aligned}$$

□

Assume that $\tau = \gamma_1\gamma_2$ with γ_1 and γ_2 permutations on two disjoint sets of indices, $(A_1 \cup B_1), (A_2 \cup B_2)$ and $A = A_1 \cup A_2$, $B = B_1 \cup B_2$. Then, up to conjugating with a permutation of $S_A \times S_B$, and using Lemma 2.18, we may assume $A_1 = \{1, 2, \dots, p\}$ the first p indices and similarly for $B_1 = \{n+1, n+2, \dots, n+h\}$ so that

(29) $$Y_A = Y_{A_1} \otimes Y_{A_2}, \quad X_B = X_{B_1} \otimes X_{B_2},$$
$$tr\left((\gamma_1\gamma_2)^{-1} \circ Y_A \otimes X_B\right) = tr\left(\gamma_1^{-1} \circ Y_{A_1} \otimes X_{B_1}\right) tr\left(\gamma_2^{-1} \circ Y_{A_2} \otimes X_{B_2}\right).$$

We finally deduce

(30) $$T^{(n)}_{k,\gamma_1\gamma_2}(x_1, x_2, \dots, x_k) = T^{(p)}_{h,\gamma_1}(x_1, x_2, \dots, x_h) \otimes T^{(n-p)}_{k-h,\gamma_2}(x_{h+1}, x_{h+2}, \dots, x_k).$$

Here if either A_1 or A_2 is empty, i.e., $p = 0$ or $p = n$, the corresponding element is a scalar and the tensor product is to be understood as multiplication.

We use the following notation.

DEFINITION 2.19. If $C = n + j_1, n + j_2, \dots, n + j_p$ is a string of indices in B, we say that the monomial $M := x_{j_1}x_{j_2} \dots, x_{j_p}$ is *associated to* C.

PROPOSITION 2.20. *If $\tau = \tau_1\tau_2\tau_3$ is the splitting relative to the decomposition $(1, 2, \dots, n) \cup (n+1, n+2, \dots, n+k)$ and h is the number of elements of B moved by τ_1, we have:*

(31) $$T^{(n)}_{k,\tau} = \tau_3^{-1} \circ t_{\tau_2}(x)T^{(n)}_{h,\tau_1}(x);$$
$$t_{\tau_2}(x) = \prod_\ell tr(N_\ell), \qquad T^{(n)}_{h,\tau_1}(x) = M_1 \otimes M_2 \otimes \dots \otimes M_n,$$

where $N_\ell = x_{j_1}x_{j_2} \dots x_{j_\ell}$ are the monomials associated to the cycles $(n + j_1, n + j_2, \dots, n+j_\ell)$ of τ_2. The $M_i = x_{f_1}x_{f_2} \dots x_{f_p}$ (as in Formula (17)) are the monomials associated to the cycle $(i, n + f_1, n + f_2, \dots, n + f_p)$ of τ_1.

PROOF. Here we use the notation $T^{(n)}_{h,\tau_1}(x)$ to mean a map in h of the variables x not necessarily the first.

Up to conjugating with a permutation of $S_A \times S_B$, we may apply Remark 2.17, and Formula (30). We thus reduce to τ a single cycle c.

If c consists only of indices in B, then it gives a contribution $tr(N)$; otherwise, for some $j \geq 1$, $A = \{1, 2, \dots, j\}$ and $B = \{j+1, j+2, \dots, j+h\}$ by conjugating with a permutation of S_A we may further assume

$$\begin{aligned}(32)\qquad \mathsf{c} &:= (\ C_1,\ 1,\ C_2,\ 2,\ C_3,\ 3,\ \dots,\ C_j,\ j)\\ &= (1,\ C_1)(\ 2,\ C_2)(3,\ C_3)\ \dots\ (j,\ C_j) \circ (1, 2, 3, \dots, j) = \mathsf{c}_1 \circ \mathsf{c}_3.\end{aligned}$$

If $C_i = j + a_1, j + a_2, \dots, j + a_{i_p}$, set $M_i := x_{a_1} x_{a_2} \dots, x_{a_{i_p}}$ its associated monomial, so

$$\begin{aligned}tr(\mathsf{c}^{-1} Y_j \otimes X_h) &= tr(M_1 y_1 M_2 y_2 M_3 y_3 \dots M_j y_j)\\ &\overset{(10)}{=} tr((1, 2, 3, \dots, j)^{-1} M_1 y_1 \otimes M_2 y_2 \otimes M_3 y_3 \otimes \dots \otimes M_j y_j)\\ &= tr((1, 2, 3 \dots, j)^{-1} M_1 \otimes M_2 \otimes M_3 \otimes \dots \otimes M_j \cdot y_1 \otimes y_2 \otimes y_3 \otimes \dots \otimes y_j)\end{aligned}$$

$$(33)\qquad \implies T^{(j)}_{h,\mathsf{c}}(X) = \mathsf{c}_3^{-1} M_1 \otimes M_2 \otimes M_3 \otimes \dots \otimes M_j.$$

Finally,

$$tr(\mathsf{c}_1^{-1} Y_j \otimes X_h) = \prod_{i=1}^{j} tr(M_i y_i) = tr(M_1 \otimes M_2 \otimes M_3 \otimes \dots \otimes M_j \cdot y_1 \otimes y_2 \otimes y_3 \otimes \dots \otimes y_j)$$

$$\implies T^{(j)}_{h,\mathsf{c}_1}(X) = M_1 \otimes M_2 \otimes M_3 \otimes \dots \otimes M_j.$$

□

THEOREM 2.21. *Assume* $A = \{1, 2, \dots, n\}$, $B = B_1 \cup B_2$ *with*

$$B_1 := \{n+1, n+2, \dots, n+h\},\ B_2 := \{n+h+1, n+h+2, \dots, n+k\}.$$

(1) *Take* $\rho \in S_{A \cup B_1}$ *and* $\gamma = (i, u)$ *a cycle, with* $i \in A$, *and* u *a string of all the indices of* B_2. *Set* $\ell := \rho_3^{-1}(i)$.

If M *is the monomial associated to* u *(Definition 2.19), we have*

$$(34)\qquad \begin{aligned}T^{(n)}_{k,\gamma\rho}(x_1, \dots, x_k) &= T^{(n)}_{h,\rho}(x_1, \dots, x_h) \cdot 1^{\otimes i-1} \otimes M \otimes 1^{\otimes n-i}\\ T^{(n)}_{k,\rho\gamma}(x_1, \dots, x_k) &= 1^{\otimes \ell-1} \otimes M \otimes 1^{\otimes n-\ell} \cdot T^{(n)}_{h,\rho}(x_1, \dots, x_h).\end{aligned}$$

(2) *Take* $\gamma = (i, u)$ *a cycle, with* $i \in B_1$, *and* u *a string of all the indices of* B_2. *If* M *is the monomial associated to* u *(Definition 2.19), we have*

$$(35)\qquad \begin{aligned}T^{(n)}_{k,\rho\gamma}(x_1, \dots, x_k) &= T^{(n)}_{h,\rho}(x_1, \dots, x_i M, \dots, x_h),\\ T^{(n)}_{k,\gamma\rho}(x_1, \dots, x_k) &= T^{(n)}_{h,\rho}(x_1, \dots, M x_i, \dots, x_h).\end{aligned}$$

(3) *If* $\gamma \in S_A$

$$(36)\qquad \begin{aligned}T^{(n)}_{k,\rho\gamma}(x_1, x_2, \dots, x_k) &= \gamma^{-1} T^{(n)}_{k,\rho}(x_1, x_2, \dots, x_k),\\ T^{(n)}_{k,\gamma\rho}(x_1, x_2, \dots, x_k) &= T^{(n)}_{k,\rho}(x_1, x_2, \dots, x_k)\gamma^{-1}.\end{aligned}$$

(4) *The inclusion* $i_n : S_{n-1+k} \subset S_{n+k}$ *as permutations fixing* n *gives, for* $\tau \in S_{n+k-1}$, *that* $T^{(n)}_{k,i_n(\tau)}(x_1, x_2, \dots, x_k) = T^{(n-1)}_{k,\tau}(x_1, x_2, \dots, x_k) \otimes 1$.

(5) *The inclusion* $i_k : S_{n+k-1} \subset S_{n+k}$ *as permutations fixing* $n+k$ *gives, for* $\tau \in S_{n-1+k}$*, that* $T^{(n)}_{k,i_k(\tau)}(x_1, x_2, \dots, x_k) = T^{(n)}_{k,\tau}(x_1, x_2, \dots, x_{k-1})tr(x_k)$.

PROOF. 1) Consider the splitting $\rho = \rho_1\rho_2\rho_3$ for $A \cup B_1$.

In order to understand $(i, u)\rho_1$, decompose $\rho_1 = \prod_{j=1}^n (v_j, j)$ into its cycles with the v_j's strings of indices of type B_1. Then

$$(i, u)\rho_1 = \prod_{j \neq i} (v_j, j)(i, u)(v_i, i) = \prod_{j \neq i} (v_j, j)(i, v_i, u).$$

Thus, the splitting of $\gamma\rho = (\gamma\rho)_1(\gamma\rho)_2(\gamma\rho)_3$ for $A \cup B$ is

$$(\gamma\rho)_1 = (i, u)\rho_1 = \prod_{j \neq i} (v_j, j)(i, v_i, u), \ (\gamma\rho)_2 = \rho_2, \ (\gamma\rho)_3 = \rho_3.$$

The cycle (i, v_i) has been replaced by the cycle (i, v_i, u).

Thus, if M_i denotes the monomial associated to the string v_i, we have that the monomial associated to the string v_i, u is M_iM. Formula (33) gives the first part of Formula (34). For the second part, we use the fact that $\rho_3(i, u) = (u, \rho_3^{-1}(i))\rho_3 = (u, h)\rho_3$ commutes with ρ_2 so $\rho(i, u) = \rho_1(u, h)\rho_2\rho_3 = \prod_{j \neq h}(v_j, j)(h, u, v_h)\rho_2\rho_3$, so M_h is replaced by MM_h.

$$\text{e.g.,} \ \ A = \{1, 2, 3\}, \quad B = \{4, 5, 6\} \cup \{7\}, \ \rho = (1, 2)(2, 5)(3, 4, 6),$$
$$T^{(3)}_{3,\rho} = (1, 2) \cdot x_2 \otimes 1 \otimes x_1x_3$$
$$\gamma = (7, 1), \gamma\rho = (1, 2)(2, 5, 7)(3, 4, 6), \quad T^{(3)}_{4,\gamma\rho} = (1, 2) \cdot x_2x_4 \otimes 1 \otimes x_1x_3.$$

2) This follows again from the fact that by multiplying ρ by γ, in $\gamma\rho$ we modify only the cycle (a, i) containing i by replacing i with the string u, i, obtaining (a, u, i) while in $\rho\gamma$ we replace i with the string i, u.

3) The first formula follows from Remark 2.17 1) and Formula (31).

As for the second, one can use the left splitting or conjugation by γ; and 4), 5) are clear. □

THEOREM 2.22. *The space* $\mathcal{T}_d^{\otimes n}(k)$ *of multilinear* $GL(V)$*-equivariant maps of* k *endomorphisms* $(x_1, x_2, \dots, x_k)$ *of a* d*–dimensional vector space* V *to* $End(V)^{\otimes n}$ *is identified with* $End_{GL(V)}(V^{\otimes n+k})$*. It is linearly spanned by the elements* $T^{(n)}_{k,\tau}$*,* $\tau \in S_{n+k}$ *of Formula* (31).

For instance, $n = 2, \ k = 1$:

$$tr(x_1y_2y_1) = tr((1, 2) \circ y_1 \otimes x_1y_2) \implies T^{(2)}_{1,(2,1,3)}(x_1) = (1, 2) \circ 1 \otimes x_1.$$
$$n = k = 3, \quad tr(x_3x_1y_2y_1x_2y_3) = tr((1, 2, 3) \circ y_1 \otimes x_3x_1y_2 \otimes x_2y_3)$$
$$\implies T^{(3)}_{3,(6,4,2,1,5,3)}(x_1, x_2, x_3) = (1, 2, 3) \circ 1 \otimes x_3x_1 \otimes x_2.$$

REMARK 2.23. An important remark is that all the previous Formulas developed using matrices still hold at the symbolic level.

In fact, for any given degrees k, n, we know that if $d \geq n + k$ the space $End_{GL(V)}(V^{\otimes n+k})$ is isomorphic to $\mathbb{Q}[S_{n+k}]$, i.e., there are no formal identities for $d \times d$ matrices of degree k in n^{th} tensors.

As a consequence, we have

REMARK 2.24. The elements $T_{k,\tau}^{(n)}(x_1, x_2, \ldots, x_k)$ should also be considered symbolically as elements of the algebra $T\langle X\rangle^{\otimes n} \ltimes \mathbb{Q}[S_n]$.

They form the space $\mathcal{T}_{mult}(k, n)$ of multilinear elements of degree k of the algebra $T\langle X\rangle^{\otimes n} \ltimes \mathbb{Q}[S_n]$.

DEFINITION 2.25. Given $\sum_{\tau\in S_{n+k}} a_\tau \tau$, the symbolic map

$$(37)\qquad T_k^{(n)}\Big(\sum_{\tau\in S_{n+k}} a_\tau \tau\Big) := \sum_{\tau\in S_{n+k}} a_\tau T_{k,\tau}^{(n)}(x_1, x_2, \ldots, x_k) \in T\langle X\rangle^{\otimes n} \ltimes \mathbb{Q}[S_n]$$

is called the *n–interpretation* of $\sum_{\tau\in S_{n+k}} a_\tau \tau$. It is a linear isomorphism between $\mathbb{Q}[S_{n+k}]$ and $\mathcal{T}_{mult}(k, n)$.

In particular, the explicit Formula for $T_{k,\tau}^{(n)}$ in Formula (31), is the *n interpretation of the permutation* τ.

By the classical method of polarization and restitution, one has that Formula (31) describes a general, not necessarily multilinear $GL(V)$-equivariant map.

For $n = 0$, this is the classical theorem of generation of invariants of matrices. For $n = 1$, this is the classical theorem of generation of equivariant maps from matrices to matrices. For $k = 0$, on the other hand, it is also the classical theorem that the span of the symmetric group is the centralizer of the linear group $G = GL(d, \mathbb{Q})$ on the nth tensor space, $V^{\otimes n}$ i.e.:

$$(M_d^{\otimes n})^G = End_G(V^{\otimes n}) = \Sigma_n(V) = \pi(\mathbb{Q}[S_n]).$$

3. The Second Fundamental Theorem

3.1. The $d + 2$ basic relations. Now, together with the first fundamental theorem, we have the second fundamental theorem giving the relations among the equivariant tensor valued maps.

In this case we want to describe, for given d and each n, the elements of the twisted algebras $T\langle X\rangle^{\otimes n} \ltimes \mathbb{Q}[S_n]$ vanishing under all evaluations $T\langle X\rangle^{\otimes n} \ltimes \mathbb{Q}[S_n] \to M_d^{\otimes n}$ induced by evaluations $X \to M_d$ of the variables X in $d \times d$ matrices, see Definition 2.11.

Again by the classical method of polarization and restitution, one reduces to study the multilinear relations, that is, the kernel of the interpretation map of Formula (37).

One starts from Theorem 1.6, that is, $\sum_{\sigma\in S_{d+1}} \epsilon_\sigma = 0$ as operator on $V^{\otimes d+1}$ or $\bigwedge^{d+1} V = \{0\}$. Hence the basic identity, for $d \times d$ matrices:

$$(38)\qquad tr\Big(\sum_{\sigma\in S_{d+1}} \epsilon_\sigma \sigma \circ z_1 \otimes z_2 \otimes z_3 \otimes \ldots \otimes z_{d+1}\Big) = 0.$$

If we use, for every $0 \le k \le d + 1$, for the variables $z_1, z_2, z_3, \ldots, z_{d+1}$, the $d \times d$ matrix variables $y_1, y_2, \ldots, y_{d+1-k}, x_1, \ldots, x_k$, Formula (16) produces from Formula (38), $d + 2$ relations

(39)

$$F_{k,d}(x_1, \ldots, x_k) := (-1)^k \sum_{\sigma\in S_{d+1}} \epsilon_\sigma T_{k,\sigma}^{(d+1-k)}(x_1, \ldots, x_k) = 0;\ k = 0, 1, \ldots, d + 1.$$

This is a multilinear relation of degree k in the variables $x_1, \dots, x_k$, for $d+1-k$ tensor valued equivariant maps:

$$tr(F_{k,d}(x_1, \dots, x_k) y_1 \otimes y_2 \otimes \dots \otimes y_{d+1-k}) = 0. \tag{40}$$

For $k = 0, d, d+1$ these relations have classical interpretations.

For $k = d$, this is the polarized form of the d-Cayley-Hamilton identity

$$x^d + \sum_{i=1}^{d} (-1)^i \sigma_i(x) x^{d-i}$$

and, for $k = d+1$, it is the polarized expression of Formula (41), which expresses $tr(x^{d+1})$ in terms of $tr(x^i)$, $i = 1, 2, \dots, d$. In other words, the expression of the $d+1$ Newton symmetric function $\psi_{d+1}(t_1, \dots, t_d) = \sum_{i=1}^{d} t_i^{d+1}$ in term of the Newton symmetric function $\psi_i(t_1, \dots, t_d)$, $i = 1, 2, \dots, d$.

$$tr(x^{d+1}) + \sum_{i=1}^{d} (-1)^i \sigma_i(x) tr(x^{d-i+1}). \tag{41}$$

In both cases, this is due to the symmetry of Formula (38) with respect to permuting the variables z_i.

For $k = 0$, it is the relation $\sum_{\sigma \in S_{d+1}} \epsilon_\sigma \sigma = 0$ as operator on $V^{\otimes d+1}$.

For intermediate $2 \le k \le d$, this is still symmetric in the variables $x_1, \dots, x_k$, so it can still be viewed as the polarized form of a tensor identity $\mathfrak{C}_{k,d}(x)$, in one variable x, for maps to $d+1-k$ tensors, obtained by restitution:

$$\mathfrak{C}_{k,d}(x) := \frac{1}{k!} F_{k,d}(x, \dots, x), \qquad \text{cf. Formula (45).}$$

For instance, $d = 2$, $k = 1$, we have:

$$\boxed{\mathfrak{C}_{1,2}(x) := (1 - (1,2)) \circ [x \otimes 1 + 1 \otimes x - tr(x) 1 \otimes 1] = 0.}$$

For $d = 3$, $k = 1$ and $d = 3$, $k = 2$

$$\mathfrak{C}_{1,3}(x) =$$

$$\boxed{[(1,2,3) + (1,3,2) - (1,2) - (1,3) - (2,3) + 1](x \otimes 1 \otimes 1 + 1 \otimes x \otimes 1 + 1 \otimes 1 \otimes x - tr(x) 1^{\otimes 3})}$$

$$\mathfrak{C}_{2,3}(x) = (1-(1,2)) \Big([x^2 \otimes 1 + 1 \otimes x^2 + x \otimes x] - tr(x)[x \otimes 1 + 1 \otimes x] + \tfrac{tr(x)^2 - tr(x^2)}{2} \Big)$$

$$= \boxed{(1 - (1,2)) \left([x^2 \otimes 1 + 1 \otimes x^2 + x \otimes x] - tr(x)[x \otimes 1 + 1 \otimes x] + \det(x) \right).}$$

We see a remarkable *factorization theorem* through two remarkable factors.

In order to see this in general, Theorem 3.2, let us first give a definition and recall some classical facts. Given two numbers $i, n \in \mathbb{N}$, $n > 0$, consider the set $\mathcal{P}(i, n)$ of all sequences

$$h_1, h_2, \dots, h_n, \qquad h_j \in \mathbb{N}, \quad \sum_j h_j = i.$$

We then define, for $i, n \in \mathbb{N}$, $n > 0$:

$$\boxed{\mathfrak{T}_{i,n}(x) := \sum_{\underline{h} \in \mathcal{P}(i,n)} x^{h_1} \otimes x^{h_2} \otimes \dots \otimes x^{h_n}.} \tag{42}$$

$$\text{e.g., } \mathfrak{T}_{0,2}(x) = 1 \otimes 1,\ \mathfrak{T}_{2,2}(x) = x^2 \otimes 1 + 1 \otimes x^2 + x \otimes x.$$

$$\mathfrak{T}_{i,1}(x) = x^i,\ \mathfrak{T}_{3,3}(x) = T_{3,0,0} + T_{2,1,0} + T_{1,1,1},\ \mathfrak{T}_{3,2}(x) = T_{3,0} + T_{2,1}.$$

Denote by

$$\det(t - x) = t^d + \sum_{i=1}^{d} (-1)^i \sigma_i(x) t^{d-i}, \quad \text{the characteristic polynomial of } x.$$

Recall the formulas (cf. [**10**])

$$\sigma_r(x) = \sum_{\substack{h_1+2h_2+\cdots+rh_r=r \\ h_1\geq 0,\ldots,h_r\geq 0}} (-1)^r \prod_{j=1}^{r} \frac{(-tr(x^j))^{h_j}}{h_j! j^{h_j}} \overset{(9)}{=} \frac{1}{r!} \sum_{\sigma\in S_r} \epsilon_\sigma t_\sigma(x). \tag{43}$$

Now given d, $0 \leq k \leq d+1$ set $n := d+1-k$, decompose

$$\{1, 2 \ldots, d+1\} = A \cup B, \quad A = \{1, 2, \ldots, n\}.$$

Recall that Definition 2.16, $U_A(B) \subset S_{A\cup B}$ denotes the set of permutations with the property that in each cycle appears at most one element of A.

Proposition 2.14, [Splitting the cycles] states that the product map $U_A(B) \times S_A \to S_{d+1}$, $(\tau, \sigma) \mapsto \tau \circ \sigma$ is a bijection.

By Theorem 2.21 3) we have, if $\tau \in S_A$ and $\sigma \in S_{d+1}$, the identity $T^{(n)}_{k,\sigma\tau} = \tau^{-1} T^{(n)}_{k,\sigma}$ so that Formula (39) becomes

$$\begin{aligned} F_{k,d}(x_1, \ldots, x_k) &= (-1)^k \sum_{\gamma\in S_{d+1}} \epsilon_\gamma T^{(n)}_{k,\gamma}(x_1, \ldots, x_k) \\ &= (-1)^k \Big(\sum_{\tau\in S_A} \epsilon_\tau \tau \Big) \circ \sum_{\sigma\in U_A(B)} \epsilon_\sigma T^{(n)}_{k,\sigma}(x_1, \ldots, x_k). \end{aligned} \tag{44}$$

THEOREM 3.2. *For all k with $0 \leq k \leq d$, the polynomial $F_{k,d}(x_1, \ldots, x_k)$ is the full polarization of the n–*tensor Cayley Hamilton polynomial, with *$(n := d+1-k,\ A = \{1, \ldots, n\},\ B = \{n+1, \ldots, n+k\})$:*

$$\mathfrak{C}_{k,d}(x) := \Big(\sum_{\tau\in S_A} \epsilon_\tau \tau \Big)\Big) \circ \Big[\mathfrak{T}_{k,n}(x) + \sum_{j=1}^{k} (-1)^j \sigma_j(x) \mathfrak{T}_{k-j,n}(x) \Big]. \tag{45}$$

PROOF. In order to prove Formula (45) consider, from Formula (44):

$$G_{k,d}(x_1, \ldots, x_k) := \sum_{\sigma\in U_A(B)} \epsilon_\sigma T^{(n)}_{k,\sigma}(x_1, \ldots, x_k). \tag{46}$$

The set $U_A(B)$ is stable under conjugation by the group S_B and such conjugation corresponds to a permutation of the variables x_i in $G_{k,d}(x_1, \ldots, x_k)$.

So, since $G_{k,d}(x_1, \ldots, x_k)$ is symmetric, it is the polarization of the element $\frac{1}{k!} G_{k,d}(x, \ldots, x)$ which we need to understand.

According to its definition, or Proposition 2.14, each permutation σ of $U_A(B)$ is a product $\sigma = \sigma_1\sigma_2 = \sigma_2\sigma_1$ with σ_2 the product of the cycles of σ involving only elements of B, and each cycle of σ_1 contains exactly one element of A.

These two permutations σ_1, σ_2 determine a partition of $B = B_1 \cup B_2$ in two subsets. With B_2 the indices moved by σ_1, and σ_2 is a product of cycles in B_1. Then σ_1 is a product of exactly n cycles c_j (possibly trivial) of some lengths $h_1 +$

$1, h_2+1, \dots, h_n+1,\ h_j \geq 0$ with c_j containing the index $j \in A$ and the remaining h_j indices in B_2.

(47) $\sigma_1 = \mathsf{c}_1\mathsf{c}_2\dots\mathsf{c}_n, \quad \mathsf{c}_j = (i_{j,1}, i_{j,2}, \dots, i_{j,h_j}, j),\ j \in A,\ i_{j,1}, i_{j,2}, \dots, i_{j,h_j} \in B_2.$

One has $|B_1| = k - \sum_i h_i$ and further, by Formula (31):

$$T^{(n)}_{k,\sigma}(X) = T^{(0)}_{\sigma_2}(X_1)T^{(n)}_{k,\sigma_1}(X_2)$$

where $T^{(0)}_{\sigma_2}(X_1)$ is an invariant product of traces of monomials, in the x variables X_1 indexed by B_1, while $T^{(n)}_{k,\sigma_1}(X_2)$ is of the form $M_1 \otimes \dots \otimes M_n$ with the M_j's monomials in the x variables X_2 indexed by B_2.

Given a partition of $B = B_1 \cup B_2$, denote by U_{A,B_2} the set of permutations of $A \cup B_2$ which decompose in exactly n cycles each containing one index $i = 1, \dots n$, or $i \in A$. So we have a decomposition

$$U_A(B) = \bigcup_{B=B_1\cup B_2} S_{B_1} \times U_{A,B_2}$$

and the following expansion of Formula (46) into the various decompositions $B = B_1 \cup B_2$:

$$G_{k,d}(x_1, \dots, x_k) = \sum_{B=B_1\cup B_2}\sum_{\sigma\in S_{B_1}} \epsilon_\sigma T^{(0)}_{j,\sigma}(X_1) \sum_{\tau\in U_{A,B_2}} \epsilon_\tau T^{(n)}_{k-j,\tau}(X_2), \tag{48}$$

where by (X_1) resp (X_2) we mean the variables among the $(x_1, \dots, x_k)$ relative to the indices of B_1, resp. B_2.

For a given j, consider the contribution to Formula (48) from all partitions $B = B_1 \cup B_2$ with $|B_1| = j,\ |B_2| = k - j$. When we evaluate all variables x_i in a single variable x, all the contributions relative to the subsets B_1 with the same cardinality j become equal so that

$$\text{If } |B_1| = j, \quad \sum_{\sigma\in S_{B_1}} \epsilon_\sigma T^{(0)}_\sigma(x) \overset{(43)}{=} j!\sigma_j(x).$$

Next compute $\sum_{\tau\in U_{A,B_2}} \epsilon_\tau T^{(n)}_{k,\tau}(x)$. An element $\tau \in U_{A,B_2}$ is uniquely of the form

$$\tau = \mathsf{c}_1\mathsf{c}_2\dots\mathsf{c}_n, \quad \mathsf{c}_a = (i_{a,1}, i_{a,2}, \dots, i_{a,h_j}, a),\ a = 1, \dots, n$$

with the elements $i \in B_2$ and B_2 has cardinality $k - j$.

Given n integers $h_1, \dots, h_n$ summing to $k - j$, we have exactly

$$\prod h_i! \binom{k-j}{h_1, \dots, h_n} = (k-j)!$$

such permutations which have sign $(-1)^{k-j}$. When we evaluate all variables x_i in a single variable x all the contributions become equal giving $(k-j)!$ times the summand $(-1)^{k-j}T_{k-j,\sigma}(x) = x^{h_1} \otimes \dots \otimes x^{h_n}$. The sequences $h_1, \dots, h_n$ vary in $\mathcal{P}(k-j, n)$, so

$$\text{If } |B_2| = k - j, \quad \sum_{\tau\in U_{A,B_2}} \epsilon_\tau T^{(n)}_{k-j,\tau}(x) \overset{(42)}{=} (-1)^{k-j}(k-j)!\mathfrak{T}_{k-j,n}(x).$$

Formula (48) for $G_{k,d}(x,\ldots,x)$ becomes

$$\sum_{j=0}^{k}\binom{k}{j}(-1)^{k-j}j!\sigma_j(x)(k-j)!\mathfrak{T}_{k-j},n(x) = k!(-1)^k\sum_{j=0}^{k}(-1)^j\sigma_j(x)\mathfrak{T}_{k-j,n}(x).$$

Substituting in Formula (44), we finally have

$$F_{k,d}(x) = k!(\sum_{\tau\in S_A}\epsilon_\tau\tau)\circ\left[\mathfrak{T}_{k,n}(x)+\sum_{j=1}^{k}(-1)^j\sigma_j(x)\mathfrak{T}_{k-j,n}(x)\right] \tag{49}$$

is the desired formula. □

REMARK 3.3. Formula (45) or (49) applies also to $k = d+1,\ n = 0$ provided we define $\mathfrak{T}_{i,0} := tr(x^i)$.

3.4. The second fundamental theorem.

3.4.1. *T–ideals.* Universal algebra is a concept first introduced by Garrett Birkhoff, see [**5**] and P. M. Cohn [**8**], or [**7**], for a more extensive treatment. If one has a class of algebras admitting free algebras $\mathcal{F}(X)$, in some set of variables X, a T–ideal is an ideal of $\mathcal{F}(X)$ closed under all algebra endomorphisms of $\mathcal{F}(X)$, which in turn are determined by *substitution maps* $X\to\mathcal{F}(X)$. T–ideals appear naturally as ideals of identities of algebras in the given class of algebras.

In this paper, we need a small generalization of this notion to take care of the tensor structure, see Definition 3.6 and 3.17.

We have already mentioned the relationship between the antisymmetrizer and the d-Cayley-Hamilton identity. A well-known result of Razmyslov and Procesi states that the T-ideal in the free algebra with trace, of relations for $d\times d$ matrices, is generated by the d–Cayley–Hamilton identity (and $tr(1) = d$) (see [**10**]).

We start from Remark 2.12 stating that the equivariant maps are the evaluations in matrices of the elements of the *twisted* algebra $T\langle X\rangle^{\otimes n}\ltimes\mathbb{Q}[S_n]$. Here by $X = \{x_1, x_2, \ldots, x_i, \ldots\}$, we indicate variables indexed by $\mathbb{N}$. Thus we have

DEFINITION 3.5. A *tensor identity* or *relation* for $d\times d$ matrices is an element of the algebra $T\langle X\rangle^{\otimes n}\ltimes\mathbb{Q}[S_n]$ vanishing under all evaluations of X in $d\times d$ matrices.

We denote this set of tensor identities by $I_d(n)\subset T\langle X\rangle^{\otimes n}\ltimes\mathbb{Q}[S_n]$.

Clearly, $I_d(n)$ is a two sided ideal of $T\langle X\rangle^{\otimes n}\ltimes\mathbb{Q}[S_n]$ and the algebra $\mathcal{T}^n_X(V)$, of $GL(V)$ equivariant polynomial maps, equals $T\langle X\rangle^{\otimes n}\ltimes\mathbb{Q}[S_n]/I_d(n)$.

Now there are certain operations under which tensor identities map to tensor identities.

First consider the endomorphisms as trace algebra, of $T\langle X\rangle$, which are given by *substitution maps* $g: X\to T\langle X\rangle$. Such a map g induces the map $g^{\otimes n}: T\langle X\rangle^{\otimes n}\to T\langle X\rangle^{\otimes n}$ which commutes with S_n and hence finally induces a map, identity on S_n

$$g^{\otimes n}\ltimes 1 : T\langle X\rangle^{\otimes n}\ltimes\mathbb{Q}[S_n]\to T\langle X\rangle^{\otimes n}\ltimes\mathbb{Q}[S_n].$$

The ideal $I_d(n)$ is clearly stable under these *substitution maps* $g^{\otimes n}\ltimes 1$.

Next the natural inclusion $S_m\times S_n\subset S_{m+n}$ induces a homomorphism of algebras, in fact an inclusion:

$$T\langle X\rangle^{\otimes m}\ltimes\mathbb{Q}[S_m]\otimes_{\mathbb{Q}[tr(M)]}T\langle X\rangle^{\otimes n}\ltimes\mathbb{Q}[S_n]\to T\langle X\rangle^{\otimes m+n}\ltimes\mathbb{Q}[S_{m+n}] \tag{50}$$

$$A\in T\langle X\rangle^{\otimes m},\ B\in T\langle X\rangle^{\otimes n},\ \sigma\in S_m,\ \tau\in S_n;\quad A\sigma\otimes B\tau\mapsto A\otimes B\sigma\tau,$$

and we have

(51)

$$I_d(m) \otimes_{\mathbb{Q}[tr(M)]} T\langle X\rangle^{\otimes n} \ltimes \mathbb{Q}[S_n] + T\langle X\rangle^{\otimes m} \ltimes \mathbb{Q}[S_m] \otimes_{\mathbb{Q}[tr(M)]} I_d(n) \subset I_d(m+n).$$

Denote for simplicity $\mathcal{T}(X,n) := T\langle X\rangle^{\otimes n} \ltimes \mathbb{Q}[S_n]$.

DEFINITION 3.6. A sequence $\{J(n)\}$ of ideals $J(n) \subset \mathcal{T}(X,n)$ will be called a *tensor–ideal* if

$$J(m) \otimes_{\mathbb{Q}[tr(M)]} \mathcal{T}(X,n) + \mathcal{T}(X,m) \otimes_{\mathbb{Q}[tr(M)]} J(n) \subset J(m+n),\ \forall m,n, \tag{52}$$

and a $T^{\otimes}$*–ideal* if furthermore

$$g^{\otimes n} \ltimes 1(J(n)) \subset J(n),\ \forall g : T\langle X\rangle \to T\langle X\rangle. \tag{53}$$

Clearly, the intersection of $T^{\otimes}$–ideals is still a $T^{\otimes}$–ideal, so we define:

A $T^{\otimes}$-ideal $\{J(n)\}$ is *generated* by a subset $S \subset \bigcup_n \mathcal{T}(X,n)$ if it is the minimal $T^{\otimes}$-ideal containing S.

We also say that the elements of each $\{J(n)\}$ are *deduced* from the elements S. We leave to the reader to understand how the previous Formulas translate into the *rules of deduction* of the elements in the $T^{\otimes}$-ideal $\{J(n)\}$ from the generating set S.

3.6.1. *The $T^{\otimes}$–ideal of tensor identities.* Clearly, the relations for $d\times d$ matrices $\{I_d(n)\}$ form a $T^{\otimes}$–ideal. By the classical method of polarization and restitution one can, studying relations or $T^{\otimes}$–ideals, restrict to multilinear elements $\mathcal{T}_{mult}(k,n)$. That is:

PROPOSITION 3.7. *If $\{J_1(n)\}$ and $\{J_2(n)\}$ are two $T^{\otimes}$–ideals having the same multilinear elements, they coincide.*

By definition, the space $\mathcal{T}_{mult}(k,n)$ of multilinear elements of degree k in $\mathcal{T}(X,n)$ is the span of the elements depending linearly only upon the first k variables $x_1,\dots,x_k$. We should remark that this subspace can be identified to $\mathbb{Q}[S_{k+n}]$ by the Formula (31), through the interpretation map $T_k^{(n)} : \tau \mapsto T_{k,\tau}^{(n)}$, Definition 2.25.

As for the $T^{\otimes}$-ideal $\{I_d(n)\}$ of tensor identities for $d\times d$ matrices, we start from the $d+2$ interpretations $F_{k,d}(x_1,\dots,x_k)$ of the antisymmetrizer as tensor identities for $d\times d$ matrices given by Formula (39). Equivalently, using polarization which is one of the rules of deduction, one could start with the 1–variable relations given by Formula (45). We claim

THEOREM 3.8. *$\{I_d(n)\}$ is generated, as $T^{\otimes}$–ideal, from the $d+2$ interpretations $\mathfrak{C}_{k,d}(x)$ of the antisymmetrizer and $tr(1) = d$.*

In other words, we may say that every relation for equivariant tensor valued polynomials maps from $d\times d$ matrices to tensor products of $d\times d$ matrices can be deduced *from the $d+2$ identities of Formula* (45) *and $tr(1) = d$.*

PROOF. From Proposition 3.7, it is enough to restrict to multilinear relations. The proof is then presented as an algorithm. From Remark 2.23, we know that all the formulas developed in §2.17.1 for multilinear equivariant maps hold also at the symbolic level.

By Theorem (3.8), one sees that the space $\mathcal{T}_{mult}(k,n) \cap I_d(n)$ of multilinear relations is 0 unless $m := k+n \ge d+1$.

In this case it is the image, under the mapping $T_k^{(n)} : \tau \mapsto T_{k,\tau}^{(n)}, \tau \in S_{k+n}$ of the two sided ideal of $\mathbb{Q}[S_{k+n}]$ generated by the antisymmetrizer A_{d+1}. Thus, it is formed by linear combinations of the k-interpretation (Definition 2.25) in terms of tensor valued maps of the linear generators, which we write as $\sigma \circ \tau \circ A_{d+1} \circ \tau^{-1}$, $\sigma, \tau \in S_{k+n}$, of this ideal.

For $m = d+1$ and each k with $0 \leq k \leq d+1$, setting $n = d+1-k$, we just have, up to scale, only the kth of the $d+2$ basic relations homogeneous of degree k. In other words, $I_d(n) \cap \mathcal{T}_{mult}(k,n) = \mathbb{Q} \cdot F_{k,d}(x_1, \dots, x_k)$.

For fixed $m > d+1$, we then decompose $\{1, 2, \dots, m\} = A \cup B$ with B the last k indices (the x indices) and A the first $n = m - k$ indices (the y indices). We may assume $n \geq 1$ otherwise we are in the case of invariants where the statement has already been proved.

Finally recall, by Lemma 2.18, that the conjugation action by elements of $S_A \times S_B$ commutes with the interpretation, where S_A permutes the tensor factors while S_B permutes the x variables.

We need thus to understand, for $m > d+1$ and $\sigma \circ \tau \circ A_{d+1} \circ \tau^{-1} \in \mathbb{Q}[S_m]$, the symbolic elements $T_k^{(n)}(\sigma \circ \tau \circ A_{d+1} \circ \tau^{-1})$, $k = 0, 1, \dots, m$, given by Formula (37), interpretations of $\sigma \circ \tau \circ A_{d+1} \circ \tau^{-1}$, and prove that they are deduced from the basic relations.

Given any set I of $d+1$ indices out of the set $\{1, 2, \dots, m\}$, we denote by $A_{d+1}(I) = \sum_{\sigma \in S_I} \epsilon_\sigma \sigma \in \mathbb{Q}[S_m]$ the antisymmetrizer in those indices.

The element $\tau \circ A_{d+1} \circ \tau^{-1}$ is, up to sign, the antisymmetrizer on the $d+1$ elements of $C := \tau(1, 2, \dots, d+1)$. Denote by $\mathfrak{A}_{d+1} := A_{d+1}(C)$.

We need to understand $T_k^{(n)}(\sigma \mathfrak{A}_{d+1})$.

Step 1 Decompose $\{1, 2, \dots, m\} = C \cup D$ and split $\sigma = \sigma_1 \sigma_2 \sigma_3$ by applying Proposition 2.14 to this decomposition. Since σ_3 is a permutation of the indices C, we have $\sigma_3 \mathfrak{A}_{d+1} = \pm \mathfrak{A}_{d+1}$ so we need only analyze $\sigma_1 \sigma_2 \mathfrak{A}_{d+1}$.

Now, since the indices of σ_2 are disjoint from those of $\sigma_1 \mathfrak{A}_{d+1}$, the interpretation of $\sigma_1 \sigma_2 \mathfrak{A}_{d+1}$ is, up to permuting the tensor factors, the tensor product of the two interpretations of σ_2 and of $\sigma_1 \mathfrak{A}_{d+1}$, Formula (30).

Therefore, the interpretation of $\sigma_1 \sigma_2 \mathfrak{A}_{d+1}$ is deduced from that of $\sigma_1 \mathfrak{A}_{d+1}$ and we may assume we are in this case from the start, denoting $\sigma_1 = \varphi$.

Step 2 We are left to understand the interpretation of $\varphi \mathfrak{A}_{d+1}$ where $\varphi = \prod_i \mathsf{c}^{(i)}$ is a product of its cycles $\mathsf{c}^{(i)}$ each containing exactly one element of C.

Use the same notations $\{1, 2, \dots, m\} = C \cup D = A \cup B$, $k = |B|$, $n = |A|$ for this relation $T_k^{(n)}(\varphi \mathfrak{A}_{d+1})$ (and $m = k + n$).

Assume first $\varphi = 1$ and let $h := |B \cap C|$, $p := |A \cap C|$, $h + p = d + 1$.
Then consider a permutation $\gamma = \gamma_1 \circ \gamma_2 \in S_A \times S_B$ such that

$$\gamma_1(A \cap C) = \{1, 2, \dots, p\}, \ \gamma_2(B \cap C) = \{p+1, p+2, \dots, p+h\},$$

and let $I := \{1, 2, \dots, p\} \cup \{p+1, p+2, \dots, p+h\}$.
Then $A_{d+1}(C) = \gamma^{-1} A_{d+1}(I) \gamma$, so, by Lemma 2.18, we have that

$$T_k^{(n)}(A_{d+1}(C))(x_1, \dots, x_k) = \gamma_1^{-1} T_k^{(n)}(A_{d+1}(I))(x_{\gamma_2(1)}, \dots, x_{\gamma_2(k)}) \gamma_1$$

is deduced from $T_k^{(n)}(A_{d+1}(I))(x_1,\dots,x_k)$. Again by Theorem 2.21 we have

$$T_k^{(n)}(A_{d+1}(I))(x_1,\dots,x_k) = (-1)^h \prod_{j=h+1}^{k} tr(x_j)F_{k,d}(x_1,\dots,x_h)\otimes 1^{\otimes n-p} \tag{54}$$

is deduced from the basic relation $F_{k,d}(x_1,\dots,x_h) = T_h^{(p)}(A_{d+1}(I))(x_1,\dots,x_h)$.

Step 3 Next split $\varphi = \varphi_1\varphi_2\varphi_3$, as in Proposition 2.14, with respect to the decomposition $A\cup B$. Recall that, by construction, $\varphi = \sigma_1 = \prod_j \mathsf{c}^{(j)}$ is a product of its cycles $\mathsf{c}^{(j)}$ each containing exactly one element of C.

Thus we split each cycle $\mathsf{c}^{(j)} = \mathsf{c}_1^{(j)}\mathsf{c}_2^{(j)}\mathsf{c}_3^{(j)}$ and for each $i = 1,2,3$ we have that $\varphi_i = \prod_j \mathsf{c}_i^{(j)}$ (cf. Formula (24)). We distinguish three cases:

(1) If $\mathsf{c}^{(j)}$ is formed only of elements of A we have $\mathsf{c}^{(j)} = \mathsf{c}_3^{(j)}$; let us call this set of indices j by S_3.
(2) If $\mathsf{c}^{(j)}$ is formed entirely of elements of B then $\mathsf{c}^{(j)} = \mathsf{c}_2^{(j)}$; let us call this set of indices j by S_2.
(3) Otherwise the splitting of the cycle, Formula (21), is $\mathsf{c}^{(j)} = \mathsf{c}_1^{(j)}\mathsf{c}_3^{(j)}$.
All elements of A appearing in $\mathsf{c}^{(j)}$ form the cycle $\mathsf{c}_3^{(j)}$ while each of these elements appears in one and only one of the cycles of $\mathsf{c}_1^{(j)}$; let us call this set of indices j by S_1.

Since $\varphi_3 = \prod_j \mathsf{c}_3^{(j)}$ is a permutation of indices of type y, setting

$$\overline{C} := \varphi_3(C),\ \overline{D} := \varphi_3(D),\ \overline{\mathfrak{A}}_{d+1} = A_{d+1}(\varphi_3(C)) = A_{d+1}(\overline{C})$$

by Theorem 2.21 (3), Formula (36) we have that

$$T_k^{(n)}(\varphi_1\varphi_2\varphi_3\mathfrak{A}_{d+1}) = T_k^{(n)}(\varphi_1\varphi_2\varphi_3\mathfrak{A}_{d+1}\varphi_3^{-1}\varphi_3) = \varphi_3^{-1}T_k^{(n)}(\varphi_1\varphi_2\overline{\mathfrak{A}}_{d+1}).$$

Step 4 We are thus reduced to study

$$T_k^{(n)}(\varphi_1\varphi_2\overline{\mathfrak{A}}_{d+1}) = T_k^{(n)}(\varphi_2\varphi_1\overline{\mathfrak{A}}_{d+1}) = T_k^{(n)}(\prod_{j\in S_2}\mathsf{c}^{(j)}\prod_{j\in S_1}\mathsf{c}_1^{(j)}\overline{\mathfrak{A}}_{d+1}).$$

Since $\overline{C} := \varphi_3(C)$ and $\varphi_3 \in S_A$ we have $\overline{C}\cap B = C\cap B$ so $B\cap D = B\cap\overline{D}$ is disjoint from $\overline{C}$.

By assumption, each cycle $\mathsf{c}^{(j)}$ contains a unique element h_j of C.

If $j\in S_2$ then $h_j\in B$, and $\mathsf{c}^{(j)} = (h_j, u_j)$ with u_j a string of elements of $\overline{D}\cap B$. The string u_j is also formed by elements of B which do not appear in $\prod_{j\in S_1}\mathsf{c}_1^{(j)}$.

Thus, by Formula (35), we have that $T_k^{(n)}(\prod_{j\in S_2}\mathsf{c}^{(j)}\prod_{j\in S_1}\mathsf{c}_1^{(j)}\overline{\mathfrak{A}}_{d+1})$ is obtained from $T_k^{(n)}(\prod_{j\in S_1}\mathsf{c}_1^{(j)}\overline{\mathfrak{A}}_{d+1})$ by replacing each variable x_{h_j}, $j\in S_2$ with the monomial $M_jx_{h_j}$ with M_j associated to the string u_j. This is one of the deduction rules.

Step 5 Up to permuting the variables, and renaming the values of n,k, we are finally reduced to analyze $T_k^{(n)}(\prod_{j\in S_1}\mathsf{c}_1^{(j)}\overline{\mathfrak{A}}_{d+1})$.

We have to distinguish two cases. The first for the indices S_1^B such that $h_j\in B\cap C = B\cap\overline{C}$ and the second for the indices S_1^A such that $h_j\in A\cap C$.

If $h_j \in A$ all the cycles decomposing $\mathsf{c}_1^{(j)}$ are of the form (a, v_a), $a \in A$ and v_a a string in $B \cap D$. Therefore to these elements we may apply either Formula (34) if $a \in \overline{C}$ or Formula (30) if $a \notin \overline{C}$.

Step 6 We are finally reduced to analyze $T_k^{(n)}(\prod_{j \in S_1^B} \mathsf{c}_1^{(j)} \overline{\mathfrak{A}}_{d+1})$.

Now for $j \in S_1^B$ we first remark that, since $h_j \in B$ is the only element in $\mathsf{c}^{(j)}$ belonging to C, we have that the elements $a \in A$ appearing in $\mathsf{c}^{(j)}$ are also in D.

Hence, for $j \in S_1^B$ we have $\mathsf{c}_3^{(j)}(C) = C$ and so $\varphi_3(C) = \prod_{j \notin S_1^B} \mathsf{c}_3^{(j)}(C) = \overline{C}$. Thus the elements $a \in A$ appearing in $\mathsf{c}^{(j)}, j \in S_1^B$ are also not in $\overline{C}$.

Thus for $j \in S_1^B$, we have that $\mathsf{c}_1^{(j)}$ is again a product of cycles (a, v_a), with $a \in A$, $a \notin \overline{C}$ and v_a a string in $B \cap D$ and finally of a single cycle $\mathsf{d}^{(j)} = (a, u_a, h_j, v_a)$, $a \in A$, $a \notin \overline{C}$ and u_a, v_a two strings in $B \cap D = B \cap \overline{D}$.

The first of these cycles are treated as before.

The cycles $\mathsf{d}^{(j)}$ correspond to some subset $\bar{A}$ of indices of A and we will write $h_a := h_j$ for $a \in \bar{A}$.

Step 7 Setting $\rho := \prod_{a \in \bar{A}}(a, u_a, h_a, v_a)$ we are reduced to analyze $T_k^{(n)}(\rho \overline{\mathfrak{A}}_{d+1})$, with $h_a \in B \cap \overline{C}$, $\bar{A} \cap \overline{C} = \emptyset$ and u_a, v_a two strings in $B \cap \overline{D}$. .

Split

$$(a, u_a, h_a, v_a) = (a, v_a)(h_a, u_a)(a, h_a) = (h_a, u_a)(a, v_a)(a, h_a).$$

Let $\gamma = \prod_{a \in \bar{A}}(a, h_a) = \gamma^{-1}$ so that $\rho\gamma = \prod_{a \in \bar{A}}(h_a, u_a)(a, v_a)$ and write

$$\rho \overline{\mathfrak{A}}_{d+1} = \rho\gamma(\gamma \overline{\mathfrak{A}}_{d+1} \gamma^{-1})\gamma.$$

Set $\widetilde{C} := \gamma(\overline{C})$, then $\gamma \overline{\mathfrak{A}}_{d+1} \gamma^{-1} = A_{d+1}(\gamma(\overline{C})) = A_{d+1}(\widetilde{C})$ is also an antisymmetrizer on $d+1$ indices, let us denote it by $\widetilde{\mathfrak{A}}_{d+1}$. Only now the x indices $h_a \in \overline{C} \cap B$ corresponding to the $a \in \bar{A}$ have been replaced by the y indices $a = \gamma(h_a) \in \bar{A}$ and $\rho \overline{\mathfrak{A}}_{d+1}$ has been replaced by

$$\rho\gamma \widetilde{\mathfrak{A}}_{d+1} \gamma = \prod_{a \in \bar{A}} (h_a, u_a) \cdot \prod_{a \in \bar{A}} (a, v_a) \widetilde{\mathfrak{A}}_{d+1} \prod_{a \in \bar{A}} (a, h_a).$$

The indices of $\prod_{a \in \bar{A}}(h_a, u_a)$ are all x indices, the indices u_a are disjoint from the indices in $\prod_{a \in \bar{A}}(a, v_j) \widetilde{\mathfrak{A}}_{d+1} \gamma$.

Therefore the interpretation of $\rho\gamma \widetilde{\mathfrak{A}}_{d+1} \gamma$ is obtained by Formula (35), from the interpretation of $\prod_{a \in \bar{A}}(a, v_a) \widetilde{\mathfrak{A}}_{d+1} \prod_{a \in \bar{A}}(a, h_a)$ by substituting each variable x_{h_a} with the monomial $M x_{h_a}$, with M associated to u_a. One of the rules of deduction.

We are thus left with $\prod_{a \in \bar{A}}(a, v_a) \widetilde{\mathfrak{A}}_{d+1} \prod_{a \in \bar{A}}(a, h_a)$, where $a \in \widetilde{C} \cap A$ and $v_a, h_a \notin \widetilde{C}$ and $v_a, h_a \in B$. Thus $\prod_{a \in \bar{A}}(a, v_a)$ and $\prod_{a \in \bar{A}}(a, h_a)$ are formed by a product of cycles for which we can apply the two Formulas (34).

We conclude that the interpretation of $\prod_a(a, v_a) \widetilde{\mathfrak{A}}_{d+1} \prod_a(a, h_a)$ is obtained from that of $\widetilde{\mathfrak{A}}_{d+1} = A_{d+1}(\widetilde{C})$, by multiplying from the right and from the left by tensor products of monomials.

Finally, the interpretation of $\widetilde{\mathfrak{A}}_{d+1} = A_{d+1}(\widetilde{C}) = A_{d+1}(\gamma\varphi_3(C))$ is treated by the discussion leading to Formula (54). □

Notice an interesting feature of this algorithm. The symbolic element $T_k^{(n)}(\sigma \circ \tau \circ A_{d+1} \circ \tau^{-1})$, interpretation of $\sigma \circ \tau \circ A_{d+1} \circ \tau^{-1}$, is deduced from just one of the basic relations $F_{\ell,d}(x_1,\dots,x_h)$,. On the other hand, we discover the value of ℓ only at the end of the algorithm.

If we start from a relation which is not multilinear, we may assume it is homogeneous in all its variables. Then we first fully polarize this relation in all its variables obtaining a multilinear relation, and then we apply the algorithm and finally we restitute all the variables obtaing the way in which this relation is deduced from the basic ones.

3.9. The final theorem.

3.9.1. *Symbolic operations on equivariant maps.* Some operations on equivariant maps from matrices to tensors can be interpreted as operations on the symbolic algebra and in particular on permutations.

Consider the following basic operations on elements of $M_d^{\otimes n}$:

$$m : X_1 \otimes X_2 \otimes \dots \otimes X_{n-1} \otimes X_n \mapsto X_1 \otimes X_2 \otimes \dots \otimes X_n X_{n-1} \in M_d^{\otimes n-1} \tag{55}$$

$$\mathtt{t} : X_1 \otimes X_2 \otimes \dots \otimes X_{n-1} \otimes X_n \mapsto tr(X_n) X_1 \otimes \dots \otimes X_{n-1} \in M_d^{\otimes n-1}. \tag{56}$$

One obtains many similar operations by combining these basic ones.

LEMMA 3.10. *For each $1 \le i < n$, we have:*

$$\mathtt{t}((n,i) \circ X_1 \otimes X_2 \otimes \dots \otimes X_n) = X_1 \otimes \dots \otimes X_n X_i \otimes \dots \otimes X_{n-1}. \tag{57}$$

PROOF. We may assume that $X_j := u_j \otimes \varphi_j,\ j = 1,\dots,n$ be n decomposable endomorphisms,

$$\begin{aligned}\mathtt{t}((n,i) \circ X_1 \otimes X_2 \otimes \dots \otimes X_n) &\overset{(11)}{=} \mathtt{t}(u_1 \otimes \varphi_1 \otimes u_2 \otimes \varphi_2 \otimes \dots \otimes u_n \otimes \varphi_i \otimes \dots \otimes u_i \otimes \varphi_n)\\ &= u_1 \otimes \varphi_1 \otimes u_2 \otimes \varphi_2 \otimes \dots \otimes \langle \varphi_n \mid u_i \rangle u_n \otimes \varphi_i \otimes \dots \otimes u_{n-1} \otimes \varphi_{n-1}\\ &= X_1 \otimes \dots \otimes X_n X_i \otimes \dots \otimes X_{n-1}.\end{aligned}$$

□

In particular, $m = \mathtt{t} \circ (n, n-1)$. Then we remark that, if $\sigma \in S_n$ fixes n, we have $\sigma \circ \mathtt{t} = \mathtt{t} \circ \sigma$. So consider $S_{n-1} \subset S_n$ the permutations fixing n.

We have the coset decomposition

$$S_n = S_{n-1} \cup \bigcup_{i=1}^{n-1} S_{n-1}(n,i).$$

From the previous Lemma we deduce, for $n \ge 2$:

PROPOSITION 3.11. *If $\sigma \in S_{n-1}$ then $\mathtt{t} \circ \sigma = \sigma \circ \mathtt{t}$ and*

$$\mathtt{t}(\sigma \circ X_1 \otimes X_2 \otimes \dots \otimes X_n) = \sigma \circ tr(X_n) X_1 \otimes \dots \otimes X_{n-1}; \tag{58}$$

in particular, $\mathtt{t}(\sigma) = \sigma \cdot tr(1)$.

If $\sigma = \tau(n,i),\ \tau \in S_{n-1}$, then

$$\mathtt{t}(\sigma \circ X_1 \otimes X_2 \otimes \dots \otimes X_n) = \tau \circ X_1 \otimes \dots \otimes X_n X_i \otimes \dots \otimes X_{n-1}. \tag{59}$$

In particular, $\mathtt{t}(\sigma) = \tau$.

PROPOSITION 3.12. *Using Formulas* (58) *and* (59), *and $\mathtt{t}(1) := tr(1)$, one can define* $\mathtt{t}$ *as a* formal operation $\mathtt{t} : T\langle X\rangle^{\otimes n} \ltimes \mathbb{Q}[S_n] \to T\langle X\rangle^{\otimes n-1} \ltimes \mathbb{Q}[S_{n-1}]$ *extending the formal trace $tr : T\langle X\rangle \to \mathbb{Q}[tr(M)]$.*

This is a partial trace *which is linear with respect to multiplication by the scalars $\mathbb{Q}[tr(M)]$ and preserves the degree in X.*

From Formulas (58) and (59) we have, setting $tr(1) = d$, for the identities $\mathtt{t}(I_d(n)) \subset I_d(n-1)$.

Recall that the element, with $A = \{1, 2, \ldots, n\}$,

$$\mathfrak{C}_{k,d}(x) := \Big(\sum_{\tau \in S_A} \epsilon_\tau \tau\Big) \circ \mathfrak{U}_{k,d}(x)$$

(60)
$$\mathfrak{U}_{k,d}(x) := \mathfrak{T}_{k,n}(x) + \sum_{j=1}^{k} (-1)^j \sigma_j(x) \mathfrak{T}_{k-j,n}(x).$$

of Formula (45) is an n–tensor identity of degree k when evaluated in $d \times d$ matrices, $d = n + k - 1$.

REMARK 3.13. From Theorem 3.8 follows in particular that there are no identities in degree k on $s < d+1-k$ tensors and furthermore, up to a scalar constant, $\mathfrak{C}_{k,d}(x)$ is the unique identity in degree k on $n = d+1-k$ tensors.

THEOREM 3.14. *Upon specialyzing $tr(1) = d$ we have, for $n \geq 1 \iff k \leq d$:*

(61)
$$\mathtt{t}(\mathfrak{C}_{k,d}(x)) = 0, \ \mathtt{t}(\mathfrak{C}_{k,d}(x) \cdot 1^{n-1} \otimes x) = -(k+1) \cdot \mathfrak{C}_{k+1,d}(x).$$

PROOF. For $n = 1$, $k = d$, we have $\mathfrak{C}_{d,d}(x) = x^d + \sum_{i=1}^{d} (-1)^i \sigma_i(x) x^{d-i}$ is the Cayley–Hamilton element and

$$\mathtt{t}(x^d + \sum_{i=1}^{d} (-1)^i \sigma_i(x) x^{d-i}) = tr(x^d) + \sum_{i=1}^{d} (-1)^i \sigma_i(x) tr(x^{d-i}) = 0,$$

is the recursive formula expressing Newton symmetric functions in term of elementary ones. Finally

$$\mathtt{t}(x^{d+1} + \sum_{i=1}^{d} (-1)^i \sigma_i(x) t^{d-i+1}) = tr(x^{d+1}) + \sum_{i=1}^{d} (-1)^i \sigma_i(x) tr(x^{d-i+1})$$

is the Formula $\mathfrak{C}_{d+1,d}(x)$ expressing the $d+1$ Newton symmetric function in d variables in term of the preceding ones.

So assume $n \geq 2$, both elements $\mathtt{t}(\mathfrak{C}_{k,d}(x))$ and $\mathtt{t}(\mathfrak{C}_{k,d}(x) \cdot 1^{n-1} \otimes x)$ are tensor identities on $n - 1 = d - k$ tensors, respectively of degree k and $k+1$ for $d \times d$ matrices.

Thus by the previous remark, on degree of identities, we have $\mathtt{t}(\mathfrak{C}_{k,d}(x)) = 0$ and $\mathtt{t}(\mathfrak{C}_{k,d}(x) \cdot 1^{n-1} \otimes x) = \alpha \cdot \mathfrak{C}_{k+1,d}(x)$ for some scalar α.

Observe that, by Proposition 3.11,

$$\sum_{\tau \in S_{1,2,\ldots,n}} \epsilon_\tau \tau = \Big(\sum_{\tau \in S_{1,2,\ldots,n-1}} \epsilon_\tau \tau\Big)\Big(1 - \sum_{i=1}^{n-1} (i, n)\Big)$$

$$\implies \mathtt{t}\Big(\Big(\sum_{\tau \in S_{1,2,\ldots,n}} \epsilon_\tau \tau\Big) \circ \mathfrak{U}_{k,d}(x) \cdot 1^{n-1} \otimes x\Big)$$

$$\overset{(58)}{=} \Big(\sum_{\tau \in S_{1,2,\ldots,n-1}} \epsilon_\tau \tau\Big) \mathtt{t}\Big(\Big(1 - \sum_{i=1}^{n-1} (i, n)\Big) \circ \mathfrak{U}_{k,d}(x) \cdot 1^{n-1} \otimes x\Big).$$

From Formula (57) it follows that

$$\mathtt{t}\Big(\Big(1-\sum_{i=1}^{n-1}(i,n)\Big)\circ\mathfrak{U}_{k,d}(x)\cdot 1^{n-1}\otimes x\Big)\in T\langle X\rangle^{\otimes n-1}.$$

Thus, since $T\langle X\rangle^{\otimes n-1}\ltimes\mathbb{Q}[S_{n-1}]$ is a free right module over $T\langle X\rangle^{\otimes n-1}$ with basis the elements $\tau\in S_{n-1}$, it follows that

$$\mathtt{t}\Big(\Big(1-\sum_{i=1}^{n-1}(i,n)\Big)\circ\mathfrak{U}_{k,d}(x)\cdot 1^{n-1}\otimes x\Big)=\alpha\cdot\mathfrak{U}_{k+1,d}(x).$$

We compute α as a coefficient of the leading term $1^{\otimes n-2}\otimes x^{k+1}$ in the previous Formula.

This term arises only in $\mathtt{t}((1-\sum_{i=1}^{n-1}(i,n)))\circ\mathfrak{T}_{k,n}(x)\cdot 1^{n-1}\otimes x)$. In fact, $\mathfrak{T}_{k,n}(x)=\sum_{\underline{h}\in\mathcal{P}(k,n)}x^{h_1}\otimes x^{h_2}\otimes\ldots\otimes x^{h_n}$ and we see, from Formula (57), that the only contributions to the term $1^{\otimes n-2}\otimes x^{k+1}$ can arise from $\mathtt{t}(-(n-1,n)A)$ with A the terms

$$\Big(\sum_{i+j=k,}1^{\otimes n-2}\otimes x^i\otimes x^j\Big)\cdot 1^{n-1}\otimes x=\sum_{i=0}^{k}1^{\otimes n-2}\otimes x^i\otimes x^{k-i+1}$$

$$\mathtt{t}\Big(-(n-1,n)\circ\sum_{i=0}^{k}1^{\otimes n-2}\otimes x^i\otimes x^{k-i+1}\Big)=-(k+1)\cdot 1^{\otimes n-2}\otimes x^{k+1}.$$

□

The specialization $tr(1)=d$ is necessary since for instance formally

$$\begin{aligned}\mathtt{t}(\mathfrak{C}_{1,2}(x))&=\mathtt{t}((1-(1,2))\circ[x\otimes 1+1\otimes x-tr(x))\\&=tr(1)x+tr(x)-tr(x)tr(1)-2x+tr(x)=(tr(1)-2)(x-tr(x)).\end{aligned}$$

$$\mathtt{t}\Big(\sum_{\tau\in S_{d+1}}\epsilon_\tau\tau\Big)\overset{3.11}{=}(tr(1)-d)\sum_{\tau\in S_d}\epsilon_\tau\tau.$$

EXERCISE 3.15. $\mathtt{t}(\mathfrak{C}_{k,d}(x))=(tr(1)-d)\mathfrak{C}_{k,d-1}(x),\ \forall k\le d.$

REMARK 3.16. For the multilinear identities of Formula (39), we have

$$F_{k+1,d}(x_1,\ldots,x_{k+1})=\mathtt{t}(F_{k,d}(x_1,\ldots,x_k)\cdot 1^{\otimes d-k}\otimes x_{k+1}).$$

At this point, one should introduce the operation $\mathtt{t}$ in the definition of the *algebras* to be used to deduce an identity from another. So we change Definition 3.6 of a T-ideal to:

DEFINITION 3.17. A $T^{\otimes}$-ideal $J(n)\subset\mathcal{T}(X,n)$ will be called a $T^{\otimes}_{\mathtt{t}}$*-ideal* if it is also stable under $\mathtt{t}$.

Under this new definition, we finally have the conclusive result.

THEOREM 3.18. *[SFT for equivariant maps] The ideal $\{I_d(n)\}$ is generated, as $T^{\otimes}_{\mathtt{t}}$–ideal, by the antisymmetrizer $\sum_{\sigma\in S_{d+1}}\epsilon_\sigma\sigma$ and $tr(1)=d$.*

PROOF. Formula (61) gives recursively the $d+2$ formulas $\mathfrak{C}_{k,d}(x)$ from the antisymmetrizer $\mathfrak{C}_{0,d}(x)=\sum_{\sigma\in S_{d+1}}\epsilon_\sigma\sigma$ and then we apply Theorem 3.8. □

There is a final remarkable fact.

Assume we take the algebras $T\langle X\rangle^{\otimes n} \ltimes \mathbb{Q}[S_n]/\bar{I}_d(n)$ modulo the $T_{\mathtt{t}}^{\otimes}$–ideal $\bar{I}_d$ generated by the antisymmetrizer A_{d+1} and no condition on $tr(1)$.

From Exercise 3.15, we have

$$\mathtt{t}(A_{d+1}) = (tr(1) - d)A_d \implies \mathtt{t}^d(A_{d+1}) = \prod_{i=1}^{d}(tr(1) - i) \in \bar{I}_d.$$

The algebra $\mathbb{Q}[\lambda]/\prod_{i=1}^{d}(\lambda - i) = \oplus_{i=1}^{d}\mathbb{Q}$ and so $T\langle X\rangle^{\otimes n} \ltimes \mathbb{Q}[S_n]/\bar{I}_d(n)$ decomposes as a direct sum of d summands, in the ith summand we have $tr(1) = i$. But now by the same formula $\mathtt{t}(A_{d+1}) = (tr(1) - d)A_d$, we deduce from A_{d+1} and $tr(1) = i$ that in the the ith summand we have also $A_{i+1} = 0$. Therefore we deduce the decomposition of the algebra $T\langle X\rangle^{\otimes n} \ltimes \mathbb{Q}[S_n]/\bar{I}_d(n)$ as a direct sum of the d algebras of equivariant maps for $i \times i$ matrices, $i = 1, \dots, d$.

THEOREM 3.19.

$$T\langle X\rangle^{\otimes n} \ltimes \mathbb{Q}[S_n]/\bar{I}_d(n) = \oplus_{i=1}^{d} T\langle X\rangle^{\otimes n} \ltimes \mathbb{Q}[S_n]/I_i(n) = \oplus_{i=1}^{d} \mathcal{T}_X^n(\mathbb{Q}^i).$$

4. The algebra of equivariant maps

4.1. The structure of $\mathcal{T}_X^n(V)$. What can we say about the algebra $\mathcal{T}_X^n(V)$? We assume that X has at least 2 elements, the case of just one variable being special and left to the reader. Let us first recall the theory for $n = 1$, for a detailed study we refer to the book [**2**].

The algebra $T_d\langle\Xi\rangle := \mathcal{T}_X^1(V)$ is the free algebra with trace $T\langle X\rangle$ modulo the d–Cayley Hamilton identity and $tr(1) = d$. This algebra is a domain generated by k generic matrices $\Xi = \{\xi_1, \dots, \xi_k\}$ of Formula (1), and the traces of their monomials.

Its center is the algebra of invariants of matrices $T_d^{(0)}\langle\Xi\rangle$. If $Q_d^{(0)}\langle\Xi\rangle$ is the field of fractions of $T_d^{(0)}\langle\Xi\rangle$, then $Q_d\langle\Xi\rangle := T_d\langle\Xi\rangle \otimes_{T_d^{(0)}\langle\Xi\rangle} Q_d^{(0)}\langle\Xi\rangle$ is a division algebra of dimension d^2 over its center *the equivariant rational functions.*

The subalgebra $\mathbb{Q}[\xi_1, \dots, \xi_k] \subset T_d\langle\Xi\rangle$ is called the *algebra of generic matrices* and it is the free algebra modulo the polynomial identities of $d \times d$ matrices. One of the remarkable theorems of the theory is that $\mathbb{Q}[\xi_1, \dots, \xi_k]$ has a non trivial center $\mathfrak{Z}_d(X) \subset T_d^{(0)}\langle\Xi\rangle$.

An element $c \in \mathfrak{Z}_d(X)$ with no constant coefficient is called a *central polynomial.* Moreover, the fields of fractions of $\mathfrak{Z}_d(X)$ and $T_d^{(0)}\langle\Xi\rangle$ coincide.

In fac,t from a strong theorem of M. Artin [**4**], [**3**] one has that (cf. [**2**], Theorem 10.3.2) if c is a central polynomial, then

$$\mathbb{Q}[\xi_1, \dots, \xi_k][c^{-1}] = T_d\langle\Xi\rangle[c^{-1}] \tag{62}$$

is an Azumaya algebra of rank d^2 over its center $T_d^{(0)}\langle\Xi\rangle[c^{-1}]$.

Take the tensor product

$$Q_d\langle\Xi\rangle^{\otimes n} := Q_d\langle\Xi\rangle \otimes_{Q_d^{(0)}\langle\Xi\rangle} Q_d\langle\Xi\rangle \dots \otimes Q_d\langle\Xi\rangle \otimes_{Q_d^{(0)}\langle\Xi\rangle} Q_d\langle\Xi\rangle$$

of n copies of $Q_d\langle\Xi\rangle$ over its center $Q_d^{(0)}\langle\Xi\rangle$. This is a central simple algebra contained in the matrix algebra

$$M_d(\mathbb{Q}(\xi^{(i)}_{h,k}))^{\otimes n} :=$$

$$M_d(\mathbb{Q}(\xi^{(i)}_{h,k})) \otimes_{\mathbb{Q}(\xi^{(i)}_{h,k})} \otimes M_d(\mathbb{Q}(\xi^{(i)}_{h,k})) \ldots \otimes M_d(\mathbb{Q}(\xi^{(i)}_{h,k})) \otimes_{\mathbb{Q}(\xi^{(i)}_{h,k})} M_d(\mathbb{Q}(\xi^{(i)}_{h,k}))$$

and

$$M_d(\mathbb{Q}(\xi^{(i)}_{h,k}))^{\otimes n} = M_d(\mathbb{Q})^{\otimes n} \otimes_{\mathbb{Q}} Q(\xi^{(i)}_{h,k}).$$

LEMMA 4.2.
(63)

$$Q_d\langle\Xi\rangle^{\otimes n} \otimes_{Q_d^{(0)}\langle\Xi\rangle} \mathbb{Q}(\xi^{(i)}_{h,k})) = M_d(\mathbb{Q}(\xi^{(i)}_{h,k}))^{\otimes n} \implies Q_d\langle\Xi\rangle^{\otimes n} = \left(M_d(\mathbb{Q}(\xi^{(i)}_{h,k}))^{\otimes n}\right)^{GL(d,\mathbb{Q})}.$$

PROOF. The natural map of $Q_d\langle\Xi\rangle^{\otimes n} \otimes_{Q_d^{(0)}\langle\Xi\rangle} \mathbb{Q}(\xi^{(i)}_{h,k})$ to $M_d(\mathbb{Q}(\xi^{(i)}_{h,k}))^{\otimes n}$ is an isomorphism since they are both central simple algebras of the same dimension d^{2n} over the field $\mathbb{Q}(\xi^{(i)}_{h,k}))$.

Since $Q_d^{(0)}\langle\Xi\rangle = \mathbb{Q}(\xi^{(i)}_{h,k}))^{GL(d,\mathbb{Q})}$, the second claim follows. □

In the same way, we have the universal faithfully flat splitting ([**2**], Corollary 10.4.3).

$$\mathbb{Q}[\xi_1,\ldots,\xi_k][c^{-1}]^{\otimes n} \otimes_{T_d^{(0)}\langle\Xi\rangle[c^{-1}]} \mathbb{Q}[\xi^{(i)}_{h,k}][c^{-1}] = M_d(\mathbb{Q}[\xi^{(i)}_{h,k}][c^{-1}])^{\otimes n} \tag{64}$$

and an isomorphism at the level of Azumaya algebra

$$\begin{aligned} \mathbb{Q}[\xi_1,\ldots,\xi_k][c^{-1}]^{\otimes n} &= \left(M_d(\mathbb{Q}[\xi^{(i)}_{h,k}][c^{-1}])^{\otimes n}\right)^{GL(d,\mathbb{Q})} \\ &= \left(M_d(\mathbb{Q})^{\otimes n} \otimes_{\mathbb{Q}} \mathbb{Q}[\xi^{(i)}_{h,k}][c^{-1}]\right)^{GL(d,\mathbb{Q})}. \end{aligned} \tag{65}$$

From Formula (3), we have that $\mathcal{T}^n_X(V) \subset Q_d\langle\Xi\rangle^{\otimes n}$ and we claim

THEOREM 4.3. *$\mathcal{T}^n_X(V)$ is a prime algebra, if $|X| > 1$ its center is $T_d^{(0)}\langle\Xi\rangle$ and, for all central polynomial c, we have*

$$\mathcal{T}^n_X(V)[c^{-1}] = \mathbb{Q}[\xi_1,\ldots,\xi_k][c^{-1}]^{\otimes n}. \tag{66}$$

The tensor power is with respect to the center $T_d^{(0)}\langle\Xi\rangle[c^{-1}]$ of $\mathbb{Q}[\xi_1,\ldots,\xi_k][c^{-1}]$.

If $|X| > 1$, we have that two generic matrices generate $d \times d$ matrices and their corresponding tensor variables generate the tensor power which is a matrix algebra with center $\mathbb{Q}$. Therefore, the center of $\mathcal{T}^n_X(V)$ is formed by the scalar valued equivariant maps, that is, the invariants $T_d^{(0)}\langle\Xi\rangle$.

In order to prove the remaining part of the Theorem let us recall a theorem which is attributed to Oscar Goldman in the book of M. A. Knus and M. Ojanguren, *Théorie de la descente et algèbres d'Azumaya* ([**14**], p. 112).

If R is a rank n^2 Azumaya algebra over its center A, the map

$$\pi : R \otimes_A R^{op} \to End_A(R),\ \pi\Big(\sum_i a_i \otimes b_i\Big)(x) = \sum_i a_i x b_i$$

is an isomorphism.

Then define $\mathtt{s} \in R \otimes_A R$ by $\pi(\mathtt{s})(x) = tr(x)$. The element $\mathtt{s}$ is called the *Goldman element.*

THEOREM 4.4. *We have*

$$\mathtt{s}^2 = 1, \quad \mathtt{s}(a \otimes b)\mathtt{s}^{-1} = b \otimes a. \tag{67}$$

Moreover, for every map $A \to B$ *so that* $B \otimes_A R \simeq M_n(B)$*, the element* $\mathtt{s}$ *maps to the permutation operator* $(1,2) \in M_n(B) \otimes_B M_n(B)$.

PROOF OF THEOREM 4.3. We have a homomorphism π_n of Formula (4)

$$\pi_n : T\langle X\rangle^{\otimes n} \ltimes \mathbb{Q}[S_n] \to M_d(\mathbb{Q})^{\otimes n} \otimes_{\mathbb{Q}} \mathbb{Q}[\xi_{h,k}^{(i)}],\ x_i \mapsto \xi_i,\ S_n \mapsto S_n \subset M_d(\mathbb{Q})^{\otimes n}$$

which factors through the algebra $\mathcal{T}_X^n(V)$ and then extends to the localized algebra of Formula (66):

$$\pi_n : \mathcal{T}_X^n(V)[c^{-1}] = \mathbb{Q}[\xi_1, \dots, \xi_k][c^{-1}]^{\otimes n} \xrightarrow{\mathtt{j}} M_d(\mathbb{Q})^{\otimes n} \otimes_{\mathbb{Q}} \mathbb{Q}[\xi_{h,k}^{(i)}][c^{-1}].$$

By Theorem 4.4 we have the Goldman element $\mathtt{s}_1 \in \mathbb{Q}[\xi_1, \dots, \xi_k][c^{-1}]^{\otimes 2}$ and in the same way elements $\mathtt{s}_i \in \mathbb{Q}[\xi_1, \dots, \xi_k][c^{-1}]^{\otimes n}$ which in the splitting become the generators $\mathtt{j}(\mathtt{s}_i) = (i, i+1)$ of the symmetric group S_n.

Therefore, since $\mathtt{j}\pi_n(i, i+1) = (i, i+1)$ we have $\pi_n(i, i+1) = \mathtt{s}_i$ (by uniqueness). The algebra $\mathbb{Q}[\xi_1, \dots, \xi_k][c^{-1}]$, of Formula (62), is Azumaya; hence, it is closed under trace. Since the algebra $\mathcal{T}_X^n(V)$ is generated by the generic matrices, the traces of their monomials and the symmetric group, we have

$$\pi_n(\mathcal{T}_X^n(V)[c^{-1}]) = \mathbb{Q}[\xi_1, \dots, \xi_k][c^{-1}]^{\otimes n}.$$

Since π_n on $\mathcal{T}_X^n(V)[c^{-1}]$ is injective we have the second claim.

The fact that $\mathcal{T}_X^n(V) \subset M_d(\mathbb{Q})^{\otimes n} \otimes_{\mathbb{Q}} \mathbb{Q}[\xi_{h,k}^{(i)}]$ is a prime algebra follows from the fact that it is torsion free over its center $T_d^{(0)}\langle \Xi \rangle \subset \mathbb{Q}[\xi_{h,k}^{(i)}]$ and localizes to the Azumaya algebra $\mathbb{Q}[\xi_1, \dots, \xi_k][c^{-1}]^{\otimes n}$ of rank d^{2n} over its center which is a domain. □

REMARK 4.5. It is an interesting problem, cf. [**23**], to understand formulas for the canonical element $\mathtt{s}$ as a fraction of a tensor polynomial by a central element. In the paper [**23**], the numerator of this expression is called a *swap polynomial.* This problem is treated to some extent in the paper [**21**] where we construct *balanced* swap polynomials.

4.6. The spectrum. Since the algebra $\mathcal{T}_X^n(V) \subset M_{d^n}[\mathbb{Q}[\xi_{a,b}^{(i)}]$, Formula (2), is closed under trace, then it is a d^n–Cayley–Hamilton algebra, according to the theory developed in [**19**]. Its trace algebra equals its center $T_d^{(0)}\langle \Xi \rangle$, the same for all n.

Therefore, when we extend the scalars from $\mathbb{Q}$ to $\mathbb{C}$, the variety with coordinate ring $T_d^{(0)}\langle \Xi \rangle \otimes \mathbb{C}$ parametrizes isomorphism classes of semisimple representations of dimension d^n of the algebra $\mathcal{T}_X^n(V) \otimes \mathbb{C}$.

Now by the same reason, when $n = 1$, the variety with coordinate ring $T_d^{(0)}\langle \Xi \rangle \otimes \mathbb{C}$ parametrizes the isomorphism classes of semisimple representations of dimension d of the free algebra $\mathbb{C}\langle X \rangle$ or of the free d–Cayley–Hamilton algebra $T_d\langle X \rangle := \mathcal{T}_X^1(V) \otimes \mathbb{C}$. That is, if X has k elements, the conjugacy classes of k–tuples of $d \times d$ matrices $(a_1, \dots, a_k) \in M_d(\mathbb{C})^k$ generating a semisimple algebra.

To this k-tuple is then associated a map $\mathcal{T}_X^n(V) \to M_d(\mathbb{C})^{\otimes n}$ and the subalgebra of $M_d(\mathbb{C})^k$ generated by the elements $1^{\otimes i} \otimes a_j \otimes 1^{\otimes n-i-1}$ and S_n is semisimple. This matrix algebra is the entire algebra $M_d(\mathbb{C})^{\otimes n}$ if and only if the representation is irreducible which means it is a point of the spectrum of one of the Azumaya algebras $\mathbb{Q}[\xi_1,\ldots,\xi_k][c^{-1}]$.

When the representation is semisimple and not irreducible, we have a decomposition $V = \oplus_{j=1}^a V_j$ into irreducibles which induces a decomposition $V^{\otimes n} = \oplus_{i_1,\ldots,i_n} V_{i_1} \otimes \ldots \otimes V_{i_n}$ whose terms are permuted by the group S_n. In each orbit, there is a term $W := V_1^{\otimes h_1} \otimes V_2^{\otimes h_2} \otimes \ldots \otimes V_k^{\otimes h_a}$, $\sum h_i = n$, which is stabilized by a Young subgroup $H = S_{h_1} \times S_{h_1} \times \ldots \times S_{h_a}$ giving the semisimple representation $Ind_H^{S_n} W$.

Comments

1) Most of the results of this paper hold in a characteristic free way. In particular, all identities with integer coefficients continue to hold. Theorem 2.10 still holds, from the theory of Donkin [**11**], provided in Formula (19) one replaces the factors $tr(N_i)$ by $\sigma_j(N_i)$. The theorem of M. Artin has been generalized by Procesi to all rings, [**2**].

The only result which should require particular care is Theorem 3.8.

In fact, in order to carry out the proof in positive characteristic, one would need to follow closely the rather difficult and non trivial calculations of Zubkov, see [**25**] or [**10**].

I have not tried to do this since it would have made the treatment extremely technical and very hard to follow but I believe that the argument can be generalized to this setting.

2) The algebra $T\langle X\rangle^{\otimes n} \ltimes \mathbb{Q}[S_n]$ contains the two subalgebras $T\langle X\rangle^{\otimes n}$ and $\mathbb{Q}\langle X\rangle^{\otimes n}$. The identities belonging to the first subalgebra are the *tensor trace identities*; the ones belonging to the second subalgebra are the *tensor polynomial identities.*

Although it is true that these can be deduced from the antisymmetrizer, their structure is far from being understood. A start in the study of tensor polynomial identities appears in the paper of the author with F. Huber [**13**].

As for tensor trace identities, we know that for $n = 1$ they are generated by the d-Cayley Hamilton identity.

For higher n the situation is more complex since the algebra of equivariant maps is not flat over its trace algebra except for the special case $d = |X| = 2$. This is actually interesting and partially treated in [**21**]. Let us explain what happens in general.

We have a map $j : T\langle X\rangle^{\otimes n} \to Q_d\langle \Xi\rangle^{\otimes n}$. A simple argument shows that its kernel is formed by the tensor trace identities since we may view this as a specialization to generic matrices. Now when we use the d–Cayley Hamilton identity this map factors through a map $j_d : T_d\langle X\rangle^{\otimes n} \to Q_d\langle \Xi\rangle^{\otimes n}$.

By simple localization arguments, we then see that if $a \in T_d\langle X\rangle^{\otimes n}$ is in the kernel of j_d then there is an invariant b so that $ba = 0$. In fact, if b is any central polynomial we have $b^k a = 0$ for some k.

We can also take any nonzero discriminant $\delta := \det(tr(b_i b_j))$ of d^2 elements $b_i \in T_d\langle X\rangle$. Then after localizing, $T_d\langle X\rangle[\delta^{-1}]$ is a free module with basis the b_i over the localized trace algebra $T_d^{(0)}\langle X\rangle[\delta^{-1}]$, so its nth tensor power is also a free module and embeds in $Q_d\langle \Xi\rangle^{\otimes n}$.

So we may say that up to multiplication by some power of this discriminant, a tensor trace identity can be deduced from the d–Cayley Hamilton identity.

On the other hand, the situation is similar to that of functional identities; therefore, as in the paper [**6**], one can have tensor trace identities not deduced from the d-Cayley Hamilton identity. For instance for $d = 2$, the tensor polynomial identity $St(2) = Alt_X x_1 x_2 \otimes x_3 x_4$, see [**13**], is not a consequence of the 2-Cayley Hamilton identity.

One can try to see which of the tensor polynomial identities discussed in [**13**] are not a consequence of the d-Cayley Hamilton identity.

References

[1] S. Abeasis and M. Pittaluga, *On a minimal set of generators for the invariants of* 3×3 *matrices*, Comm. Algebra **17** (1989), no. 2, 487–499, DOI 10.1080/00927878908823740. MR978487

[2] E. Aljadeff, A. Giambruno, C. Procesi, and A. Regev, *Rings with polynomial identities and finite dimensional representations of algebras*, American Mathematical Society Colloquium Publications, vol. 66, American Mathematical Society, [Providence], RI, 2020. MR4249615

[3] M. Artin, *On Azumaya algebras and finite dimensional representations of rings*, J. Algebra **11** (1969), 532–563, DOI 10.1016/0021-8693(69)90091-X. MR242890

[4] G. Azumaya, *On maximally central algebras*, Nagoya Math. J. **2** (1951), 119–150. MR40287

[5] G. Birkhoff, *Universal algebra*, Proc. First Canadian Math. Congress, Montreal, 1945, Univ. Toronto Press, Toronto, ON, 1946, pp. 310–326. MR19586

[6] M. Brešar, C. Procesi, and Š. Špenko, *Quasi-identities on matrices and the Cayley-Hamilton polynomial*, Adv. Math. **280** (2015), 439–471, DOI 10.1016/j.aim.2015.03.021. MR3350226

[7] S. Burris and H. P. Sankappanavar, *A course in universal algebra*, Graduate Texts in Mathematics, vol. 78, Springer-Verlag, New York-Berlin, 1981. MR648287

[8] P. M. Cohn, *Universal algebra*, 2nd ed., Mathematics and its Applications, vol. 6, D. Reidel Publishing Co., Dordrecht-Boston, Mass., 1981. MR620952

[9] C. de Concini and C. Procesi, *A characteristic free approach to invariant theory*, Advances in Math. **21** (1976), no. 3, 330–354, DOI 10.1016/S0001-8708(76)80003-5. MR422314

[10] C. De Concini and C. Procesi, *The invariant theory of matrices*, University Lecture Series, vol. 69, American Mathematical Society, Providence, RI, 2017, DOI 10.1090/ulect/069. MR3726879

[11] S. Donkin, *Invariants of several matrices*, Invent. Math. **110** (1992), no. 2, 389–401, DOI 10.1007/BF01231338. MR1185589

[12] F. Huber, *Positive maps and matrix contractions from the symmetric group*, `arXiv:2002.12887` 28–2–2020.

[13] F. Huber and C. Procesi, *Tensor polynomial identities*, Israel J. Math. **247** (2022), no. 1, 125–147, DOI 10.1007/s11856-021-2262-6. MR4425330

[14] M.-A. Knus and M. Ojanguren, *Théorie de la descente et algèbres d'Azumaya* (French), Lecture Notes in Mathematics, Vol. 389, Springer-Verlag, Berlin-New York, 1974. MR417149

[15] B. Kostant, *A theorem of Frobenius, a theorem of Amitsur-Levitski and cohomology theory*, J. Math. Mech. **7** (1958), 237–264, DOI 10.1512/iumj.1958.7.57019. MR92755

[16] G. Lehrer and R. Zhang, *The second fundamental theorem of invariant theory for the orthogonal group*, Ann. of Math. (2) **176** (2012), no. 3, 2031–2054, DOI 10.4007/annals.2012.176.3.12. MR2979865

[17] C. Procesi, *The invariant theory of* $n \times n$ *matrices*, Advances in Math. **19** (1976), no. 3, 306–381, DOI 10.1016/0001-8708(76)90027-X. MR419491

[18] C. Procesi, *Lie groups: An approach through invariants and representations*, Universitext, Springer, New York, 2007. MR2265844

[19] C. Procesi, *T-ideals of Cayley Hamilton algebras*, Indag. Math. (N.S.) **32** (2021), no. 6, 1190–1228, DOI 10.1016/j.indag.2021.06.002. MR4334165

[20] C. Procesi, *Norms and Cayley-Hamilton algebras*, Atti Accad. Naz. Lincei Rend. Lincei Mat. Appl. **32** (2021), no. 1, 23–61, DOI 10.4171/rlm/925. MR4251239

[21] C. Procesi, *A construction of swap or switch polynomials*, Adv. Math. **400** (2022), Paper No. 108308, 24, DOI 10.1016/j.aim.2022.108308. MR4395947

[22] Yu. P. Razmyslov, *Trace identities of matrix algebras via a field of characteristic zero*, Math. USSR Izvestia (translation) **8** (1974), 727–760.

[23] David Trillo, Benjamin Dive, Miguel Navascués, *Translating uncontrolled systems in time*, Quantum 4, 374 (2020), https://doi.org/10.22331/q-2020-12-15-374

[24] H. Weyl, *The classical groups: Their invariants and representations*, Princeton Landmarks in Mathematics, Princeton University Press, Princeton, NJ, 1997. MR1488158

[25] A. N. Zubkov, *On a generalization of the Razmyslov-Procesi theorem* (Russian, with Russian summary), Algebra i Logika **35** (1996), no. 4, 433–457, 498, DOI 10.1007/BF02367026; English transl., Algebra and Logic **35** (1996), no. 4, 241–254. MR1444429

Dipartimento di Matematica, Univrsità di Roma, La Sapienza, Piazzale A. Moro, 00185 Roma, Italia

Email address: procesi@mat.uniroma1.it

Contemporary Mathematics
Volume **800**, 2024
https://doi.org/10.1090/conm/800/16061

Algebraic groups with good reduction and the genus problem

Igor A. Rapinchuk

Abstract. We first provide an overview of several results dealing with the genus of a division algebra and highlight the role of ramification in its analysis. We then give a survey of recent developments on the genus problem for simple algebraic groups and its connections to the analysis of groups with good reduction.

1. Introduction

This paper is an expanded version of the author's talk at the Amitsur Centennial Symposium. Our goal is to give an overview of some recent work on the surprising and exciting connections between the study of algebraic groups with good reduction and the genus problem, which is concerned with characterizing simple algebraic groups in terms of their maximal tori over the field of definition. These developments in fact comprise just one aspect of the emerging arithmetic theory of algebraic groups over higher-dimensional fields in which good reduction plays a central role – we refer the reader to [**24**] for an in-depth account of good reduction for linear algebraic groups and its connections with various other directions, including local-global principles and weak-commensurability of Zariski-dense subgroups and applications to Riemann surfaces. The reader can also consult [**25**] for a more concise and less technical survey of these topics.

The initial focus of the genus problem was on the study of division algebras having the same maximal subfields. So, to establish the appropriate context, we will begin by discussing the genus of a division algebra and the role of ramification in its analysis. We will then give a precise statement of the genus problem for algebraic groups. In this setting, it appears that groups with good reduction are an adequate substitute for unramified division algebras. We will therefore present a brief discussion of good reduction and formulate our main Finiteness Conjecture for groups with good reduction over finitely generated fields. We will then indicate the connection between the Finiteness Conjecture and the genus problem. To conclude the paper, we will survey some of the available results.

2020 *Mathematics Subject Classification.* Primary 11E72, 16K20, 20G15.

Key words and phrases. Division algebras, Galois cohomology, linear algebraic Groups.

The author was partially supported by a Collaboration Grant for Mathematicians from the Simons Foundation and by NSF grant DMS-2154408.

2. The genus problem for division algebras

Let K be a field and D_1 and D_2 be two central division K-algebras of degree n. We say that D_1 and D_2 *have the same maximal subfields* if a degree n field extension P/K admits a K-embedding $P \hookrightarrow D_1$ if and only if it admits a K-embedding $P \hookrightarrow D_2$. Then one can ask the following natural question:

($*$) *Let D_1 and D_2 be central division algebras of the same degree. How are D_1 and D_2 related if they have the same maximal subfields?*

We note that this problem is similar in spirit to the following famous theorem of Amitsur [**1**]:

THEOREM 2.1 (Amitsur). *Let D_1 and D_2 be finite-dimensional central division algebras over a field K that have the same* splitting fields, *i.e., for a field extension F, the algebra $D_1 \otimes_K F$ is F-isomorphic to a matrix algebra $\mathrm{M}_{n_1}(F)$ if and only if the algebra $D_2 \otimes_K F$ is isomorphic to a matrix algebra $\mathrm{M}_{n_2}(F)$. Then $n_1 = n_2$ and the classes $[D_1]$ and $[D_2]$ in the Brauer group $\mathrm{Br}(K)$ generate the same subgroup, $\langle[D_1]\rangle = \langle[D_2]\rangle$.*

The crucial point is that Amitsur's proof relies in a very essential way on *infinite* (non-algebraic) extensions of K – namely, so-called *generic splitting fields* (concrete examples of which are function fields of Severi-Brauer varieties). So, it is natural to ask if Amitsur's Theorem can be proved using only *finite* extensions of K. In other words, does the theorem's conclusion still hold if one only assumes that that D_1 and D_2 have the same *finite-dimensional* splitting fields or just the same maximal subfields? It turns out that the answer is (strongly) negative already over global fields. Indeed, this follows from the observation that, using the Albert-Brauer-Hasse-Noether theorem (see [**20**, Ch. 18, §18.4]), one can construct arbitrarily large collections of pairwise non-isomorphic cubic division algebras having the same maximal subfields over number fields (the same construction actually works for division algebras of any degree $d > 2$, cf. [**4**, §1]). On the other hand, two quaternion division algebras over a number field that have the same quadratic subfields are necessarily isomorphic. Thus, even over number fields, question ($*$) appears to be interesting.

Until about 10 years ago, no information at all was available on ($*$) over any fields other than global. The following question along these lines was first asked in [**21**, Remark 5.4]:

Are quaternion division algebras over $\mathbb{Q}(x)$ determined uniquely up to isomorphism by their maximal subfields?

Shortly after it was formulated, this question was answered in the affirmative by D. Saltman. In subsequent work, he and S. Garibaldi [**13**] showed that the answer is still affirmative over the field of rational functions $k(x)$, where k is any number field, and also in some other situations. This marked the starting point of the investigation of question ($*$) over fields more general than global (we note that a similar question, formulated in terms of finite-dimensional splitting fields, was considered in [**17**]).

To quantify our discussion, it is convenient to introduce the notion of the *genus* of a division algebra (this terminology was suggested by L.H. Rowen).

DEFINITION 2.2. Let D be a finite-dimensional central division algebra over a field K. Then the genus $\mathbf{gen}(D)$ of D is defined to be the set of classes $[D'] \in \mathrm{Br}(K)$ represented by central division K-algebras D' having the same maximal subfields as D.

Broadly speaking, the general goal of the *genus problem* is to characterize the genus of a given division algebra – note that this is essentially a reformulation of our original question $(*)$. Most recent work in this direction has dealt with the following two more precise questions:

- *When does* $\mathbf{gen}(D)$ *reduce to a single element?* (Note that this is the case if and only if D is determined uniquely up to isomorphism by its maximal subfields.)
- *When is* $\mathbf{gen}(D)$ *finite?*

We should point out that over a number field K, the Albert-Brauer-Hasse-Noether Theorem enables one to resolve both questions: namely, the genus of every quaternion division algebra is trivial (i.e., reduces to a single element), while the genus of any division algebra of higher degree is nontrivial but always finite (see [**4**, Proposition 3.1] for the details).

Turning now to more general fields, we would first like to mention the following theorem for rational function fields that was established in [**23**].

THEOREM 2.3 (Stability Theorem). *Assume that* char $k \neq 2$. *If* $|\mathbf{gen}(\Delta)| = 1$ *for any central division quaternion algebra* Δ *over* k, *then* $|\mathbf{gen}(D)| = 1$ *for any quaternion algebra* D *over* $k(x)$.

(An analogous statement also holds for all division algebras having exponent two in the Brauer group, cf. [**3**].) Note that a consequence of Theorem 2.3 is that the genus of a quaternion algebra over the purely transcendental extension $k(x_1, \ldots, x_r)$ of a number field k of any (finite) transcendence degree reduces to a single element. On the other hand, at this point, it is not known whether there exists a central quaternion division algebra D over a finitely generated field K of characteristic $\neq 2$ having nontrivial genus.

Next, observe that $|\mathbf{gen}(D)| > 1$ whenever D does not have exponent two since in that case, the opposite algebra D^{op} is not isomorphic to D, but clearly has the same maximal subfields as D. We will therefore focus on finiteness properties of the genus. Let us first point out that over general fields, the genus $\mathbf{gen}(D)$ can be infinite. Indeed, adapting a construction that has been suggested by a number of people, including M. Schacher, A. Wadsworth, M. Rost, S. Garibaldi and D. Saltman, J. Meyer [**19**] produced examples of quaternion algebras over "large" fields with infinite genus. (By construction, these fields have infinite transcendence degree over the prime subfield.) Subsequently, S. Tikhonov [**30**] extended this approach to construct examples of division algebras of any prime degree having infinite genus. On the other hand, over finitely generated fields, we have the following.

THEOREM 2.4. *Let* K *be a finitely generated field. Then for any finite-dimensional central division* K*-algebra* D, *the genus* $\mathbf{gen}(D)$ *is finite.*

There are two proofs of Theorem 2.4, which can be found in [**5**] and [**8**]. Both arguments depend on an analysis of ramification, but require somewhat different amounts of information about the unramified Brauer group. Before indicating some of the main points, we first recall that if F is a field equipped with a discrete valuation v, then a central simple F-algebra A is said to be *unramified* at v if there exists an Azumaya algebra $\mathcal{A}$ over the valuation ring $\mathcal{O}_v$ of the completion K_v such that

$$\mathcal{A} \otimes_{\mathcal{O}_v} K_v \simeq A \otimes_K K_v.$$

Furthermore, if F is equipped with a set V of discrete valuations, the corresponding *unramified Brauer group* is defined as

$$\mathrm{Br}(F)_V = \{x \in \mathrm{Br}(F) \mid x \text{ is unramified at all } v \in V\}.$$

We also recall that any finitely generated field K possesses natural sets of discrete valuations called *divisorial.* More precisely, let $\mathfrak{X}$ be a *model* of K, i.e., a normal separated irreducible scheme of finite type over $\mathbb{Z}$ (if $\mathrm{char}\, K = 0$) or over a finite field (if $\mathrm{char}\, K > 0$) such that K is the function field of $\mathfrak{X}$. It is well-known that to every prime divisor $\mathfrak{Z}$ of $\mathfrak{X}$, there corresponds a discrete valuation $v_{\mathfrak{Z}}$ on K. Then

$$V(\mathfrak{X}) = \{v_{\mathfrak{Z}} \mid \mathfrak{Z} \text{ prime divisor of } \mathfrak{X}\}$$

is called the divisorial set of places of K corresponding to the model $\mathfrak{X}$. Any set of places V of K of this form (for some model $\mathfrak{X}$) will be simply called *divisorial.*

The proof of Theorem 2.4 given in [**5**] requires the additional assumption that the degree n of the division algebra D is relatively prime to $\mathrm{char}\, K$. For the argument, we fix a divisorial set of places V of K. Note that since $\mathrm{char}\, K$ is prime to n, we can assume without loss of generality that for each $v \in V$, the characteristic of the residue field $K^{(v)}$ is prime to n. Then one of the essential ingredients in the proof is the following observation:

LEMMA 2.5. *If D and D' are central division K-algebras of degree n having the same maximal subfields, then for any $v \in V$, either both algebras are unramified at v or both are ramified.*

(See [**3**, Lemma 2.5] for the details.) This ultimately enables one to reduce the proof of Theorem 2.4 to showing that the n-torsion subgroup ${}_n\mathrm{Br}(K)_V$ is finite and leads to the estimate

$$|\mathbf{gen}(D)| \leq |{}_n\mathrm{Br}(K)_V| \cdot \varphi(n)^r,$$

where r is the number of $v \in V$ where D ramifies (which is necessarily finite for a divisorial set).

Our second proof of Theorem 2.4, given in [**8**], also uses the analysis of ramification, but does not impose any restrictions on the characteristic of K. The reason is that the argument does not require the finiteness of the full n-torsion subgroup of $\mathrm{Br}(K)$, but only the finiteness of certain of its subgroups. On the other hand, since these subgroups depend on the division algebra at hand, we do not obtain a general estimate on the size of the genus as provided by our first proof.

Without going into the details, let us mention that an interesting generalization of the genus, termed the *upper genus*, was analyzed in [**18**], where the study of the genus problem for algebras with involution was also begun.

3. The genus problem for algebraic groups

We now turn our attention to the genus problem for algebraic groups. Roughly speaking, to define the genus in this context, we replace maximal subfields with maximal tori in the definition of the genus of a division algebra.

More precisely, given two reductive algebraic groups G_1 and G_2 over a field K, we say that G_1 and G_2 *have the same isomorphism classes of maximal K-tori* if every maximal K-torus T_1 of G_1 is K-isomorphic to some maximal K-torus T_2 of G_2, and vice versa.

DEFINITION 3.1. Let G be an absolutely simple simply connected algebraic group over a field K. The *genus* $\mathbf{gen}_K(G)$ of G is the set of K-isomorphism classes of (inner) K-forms G' of G that have the same isomorphism classes of maximal K-tori as G.

(We recall that if G is an algebraic group over K, then a K-group G' is a called a *K-form of G* if G and G' become isomorphic over a separable closure K^{sep}.)

In analogy to the case of division algebras, the following two questions have received the most attention so far:

- *When does $\mathbf{gen}_K(G)$ reduce to a single element?*
- *When is $\mathbf{gen}_K(G)$ finite?*

The basic case where K is a number field was considered in [**21**, Theorem 7.5], where the following result was established (although the term "genus," which appeared later, was not used).

THEOREM 3.2. *Let G be an absolutely almost simple simply connected algebraic group over a number field K. Then*

(1) *$\mathbf{gen}_K(G)$ is finite;*
(2) *if G is not of type A_n, D_{2n+1} $(n > 1)$, or E_6, we have $|\mathbf{gen}_K(G)| = 1$.*

Given that this result resolves both questions over number fields, the next natural problem is to investigate the behavior of the genus over more general (finitely generated) fields. On the basis of Theorem 3.2 (as well a number of other results that we will discuss in §5), in conjunction with the statements for division algebras mentioned in the previous section, we have been led to the following conjecture.

CONJECTURE 3.3.

(1) *Let $K = k(x)$ be the field of rational functions in one variable over a number field k. If G is an absolutely almost simple simply connected algebraic K-group with center $Z(G)$ of order ≤ 2, then the genus $\mathbf{gen}_K(G)$ reduces to a single element.*
(2) *Let G be an absolutely almost simple simply connected algebraic group over a finitely generated field of "good" characteristic. Then the genus $\mathbf{gen}_K(G)$ is finite.*
(Here, char $K = p$ *is said to be "good" if either $p = 0$ or $p > 0$ and does not divide the order of the Weyl group of G.)*

As we saw in the previous section, the study of the genus of a division algebra is based on a careful analysis of ramification. In the setting of algebraic groups, it appears that a suitable substitute for unramified division algebras are algebraic groups with good reduction. In the next section, we will describe the precise connection between the genus and groups with good reduction, which underlies most of the progress that has been achieved on Conjecture 3.3 so far.

4. Groups with good reduction

We begin this section by first recalling some of the basic facts concerning groups with good reduction and then formulating our main Finiteness Conjecture for forms with good reduction.

Let K be a field equipped with a discrete valuation v, and suppose G is a reductive affine algebraic group over K. We say that G has *good reduction at* v if there exists a reductive group scheme[1] $\mathcal{G}$ over the valuation ring $\mathcal{O}_v$ of the completion K_v whose generic fiber $\mathcal{G} \otimes_{\mathcal{O}_v} K_v$ is isomorphic to $G \otimes_K K_v$. The special fiber (or *reduction*) $\mathcal{G} \otimes_{\mathcal{O}_v} K^{(v)}$, where $K^{(v)}$ is the residue field of K_v, is then denoted by $\underline{G}^{(v)}$ – it is a connected algebraic group over $K^{(v)}$ of the same type as G. Furthermore, given a set V of discrete valuations of K, we will say that G has *good reduction with respect to* V if G has good reduction at all $v \in V$.

Informally, having good reduction means that the group $G \times_K K_v$ has a "nice" $\mathcal{O}_v$-structure whose reduction modulo $\mathfrak{p}_v$ yields a connected reductive group. As the following examples demonstrate, in various situations of interest, this condition can be characterized in very concrete terms.

EXAMPLE 4.1.

(a) (cf. [**9**, Example 2.2]) If $G = \mathrm{SL}_{1,A}$ is the algebraic group associated with the elements of reduced norm 1 in a central simple algebra A over K, then G has good reduction at v if and only if A is unramified at v.

(b) (cf. [**9**, Example 2.3]) Suppose q is a non-degenerate quadratic form in n variables over K and the residue field $K^{(v)}$ has characteristic $\neq 2$. Then the spinor group $G = \mathrm{Spin}_n(q)$ has good reduction at v if and only if q is equivalent over K_v to a quadratic form

$$\lambda(a_1x_1^2 + \cdots + a_nx_n^2), \quad \text{with } \lambda \in K_v^{\times}, a_i \in \mathcal{O}_v^{\times}.$$

Historically, the study of algebraic groups with good reduction can be traced back to the work of Harder [**15**], Colliot-Thélène and Sansuc [**10**], and Gross [**14**]. In [**15**], the focus was mainly on algebraic groups over a number field K having good reduction with respect to sets V consisting of almost all nonarchimedean places of K. Groups with good reduction have also been analyzed extensively over $K = k(x)$ (the field of rational functions in one variable over a field k). In [**22**], Raghunathan and Ramanathan considered the case where V consists of the discrete valuations $v_{p(x)}$ corresponding to *all* irreducible polynomials $p(x) \in k[x]$. Later, groups having good reduction at all $v \in V \setminus \{v_x\}$ were analyzed by Chernousov, Gille and Pianzola [**2**]; these results then played a crucial role in their proof of the conjugacy of analogues of Cartan subalgebras in certain infinite-dimensional Lie algebras.

A key point is that in all of these cases, K is the fraction field of a Dedekind ring R, and V consists of discrete valuations associated with the nonzero prime ideals of R, making the situation "1-dimensional." By contrast, our recent and ongoing work addresses the analysis of groups with good reduction in the higher-dimensional setting of arbitrary finitely generated fields. More precisely, the following Finiteness Conjecture is one of the central elements of our current investigations.

[1] Let R be a commutative ring and $S = \mathrm{Spec}\, R$. Recall that a *reductive R-group scheme* is a smooth affine group scheme $\mathcal{G} \to S$ such that the geometric fibers $\mathcal{G}_{\bar{s}}$ are connected reductive algebraic groups (see [**12**, Exp. XIX, Definition 2.7] or [**11**, Definition 3.1.1]).

CONJECTURE 4.2. *Let G be a reductive algebraic group over a finitely generated field K and V be a divisorial set of discrete valuations of K. Then the set of K-isomorphism classes of K-forms G' of G that have good reduction at all $v \in V$ is finite (at least when the characteristic of K is "good.").*

(When G is an absolutely almost simple algebraic group, "good" characteristic is used here in the same sense as in Conjecture 3.3. For algebraic tori, by good characteristic, we simply mean char $K = 0$.)

We refer the reader to [**24**] for an extensive discussion of groups with good reduction and, in particular, the key role of the Finiteness Conjecture in the current development of the arithmetic theory of algebraic groups over higher-dimensional fields. For our present purposes, we will only mention some connections to the genus problem.

THEOREM 4.3 ([**6**, Theorem 5], [**9**, Theorem 1.1]). *Let G be an absolutely almost simple linear algebraic group over a field K and let v be a discrete valuation of K. Assume that the residue field $K^{(v)}$ is finitely generated and that* char $K^{(v)} \neq 2$ *if G is of type B_ℓ ($\ell \geq 2$). If G has good reduction at v, then any $G' \in \mathbf{gen}_K(G)$ also has good reduction at v. Moreover, the reduction $\underline{G'}^{(v)}$ lies in the genus $\mathbf{gen}_{K^{(v)}}(\underline{G}^{(v)})$ of the reduction $\underline{G}^{(v)}$.*

As we already mentioned above, one should view groups with good reduction as an analogue of unramified division algebras (this point of view is justified, for instance, by Example 4.1). From this perspective, Theorem 4.3 can be thought of as a partial analogue of Lemma 2.5. Although we refer the reader to [**9**] for the details, we would like to point out that the proof of Theorem 4.3 is based on an entirely new approach to good reduction of simple algebraic groups that shows that the existence of good reduction can be characterized in terms of the presence of maximal tori with certain specific properties.

In the case of finitely generated fields, we have the following consequence.

COROLLARY 4.4. *Let G be an absolutely almost simple algebraic group over an infinite finitely generated field K, and let V be a divisorial set of places of K. Assume that* char $K \neq 2$ *if G is of type B_ℓ ($\ell \geq 2$). Then there exists a finite subset $S \subset V$ such that every $G' \in \mathbf{gen}_K(G)$ has good reduction at all $v \in V \setminus S$.*

In particular, it follows that the truth of Conjecture 4.2 for all divisorial sets V would yield the finiteness of $\mathbf{gen}_K(K)$ for any absolutely almost simple algebraic K-group G. In other words, the Finiteness Conjecture provides a uniform approach for resolving one important aspect of the genus problem for algebraic groups. Continuing the parallel with division algebras, we thus see that, in this context, the Finiteness Conjecture plays a role analogous to that of our finiteness results for the unramified Brauer group in the study of the genus of a division algebra.

5. Overview of results

In this section, we will give a brief overview of some available results on Conjecture 4.2 and the genus problem. The reader can consult [**24**] for a more detailed account.

To begin with, Conjecture 4.2 has been settled completely for algebraic tori.

THEOREM 5.1 (cf. [**26**, Theorem 1.1], [**27**, Corollary 5.2]). *Let K be a finitely generated field and V be a divisorial set of places of K. Then for any $d \geq 1$, the set of K-isomorphism classes of d-dimensional K-tori that have good reduction at all $v \in V$ and for which the degree $[K_T : K]$ of the minimal splitting field is prime to the characteristic exponent of K, is finite.*

Turning now to semisimple groups, we first note that in the classical setting when K is a number field and V is any set consisting of almost all nonarchmidean places of K, Conjecture 4.2 can be reduced to the consideration of absolutely almost simple groups, in which case the assertion follows from well-known results on the Galois cohomology of algebraic groups and the description of groups with good reduction over p-adic fields (see [**24**, Proposition 5.2]). The general case, however, presents a number of new challenges.

Here are some representative results. First, we have the following statement for inner forms of type A_n over (essentially arbitrary) finitely generated fields, which is derived from our finiteness results for the unramified Brauer group (discussed in §2).

THEOREM 5.2 ([**5**]). *Let K be a finitely generated field, V a divisorial set of discrete valuations of K, and $n \geq 2$ an integer prime to* char K. *Then the number of K-isomorphism classes of groups of the form $\mathrm{SL}_{1,A}$, where A is a central simple K-algebra of degree n, that have good reduction at all $v \in V$, is finite.*

Using this statement, together with Theorems 2.3 and 4.3 and some additional considerations involving generic tori, we obtain the next result concerning the genus.

THEOREM 5.3 (cf. [**3**, Theorem 5.3] and [**4**, Theorem 6.3]).

(1) *Let D be a central division algebra of exponent two over the field of rational functions $K = k(x_1, \ldots, x_r)$, where k is either a number field or a finite field of characteristic $\neq 2$. Then for $G = \mathrm{SL}_{m,D}$ $(m \geq 1)$, the genus $\mathbf{gen}_K(G)$ reduces to a single element.*
(2) *Let $G = \mathrm{SL}_{m,D}$, where D is a central division algebra over a finitely generated field K of degree prime to* char K. *Then $\mathbf{gen}_K(G)$ is finite.*

Next, following Kato [**16**], we say that a field K is a *2-dimensional global field* if it is either the function field $k(C)$ of a smooth geometrically integral curve C over a number field k or the funcion field $\mathbb{F}_q(S)$ of a smooth geometrically integral surface S over a finite field $\mathbb{F}_q$.

THEOREM 5.4 (cf. [**7**, Theorem 1.1]). *Let K be a two-dimensional global field of characteristic $\neq 2$ and let V be a divisorial set of discrete valuations of K. Fix an integer $n \geq 5$. Then the set of K-isomorphism classes of spinor groups $G = \mathrm{Spin}_n(q)$ of nondegenerate quadratic forms in n variables over K that have good reduction at all $v \in V$ is finite.*

Using Voevodksy's resolution of Milnor's conjecture on quadratic forms, we reduce the proof of this result to the analysis of finiteness properties of unramified cohomology with μ_2-coefficients, which we verify in the case of 2-dimensional global fields. Similar statements are also available for some special unitary groups of types A_n and C_n and for groups of type G_2. We note that if n is odd, then all K-forms of $G = \mathrm{Spin}_n(q)$ are of the form $G' = \mathrm{Spin}_n(q')$, with q' a non-degenerate quadratic form over K in n variables, Theorem 5.4 yields Conjecture 4.2 in this case.

Turning now to the genus, we have the following statements for spinor groups, groups of type G_2, and groups of type F_4 that split over a quadratic extension.

THEOREM 5.5 (cf. [**7**, Theorem 1.2], [**26**, Theorem 5.5]). *Suppose K is either a 2-dimensional global field of characteristic $\neq 2$ or the field of rational functions $k(x, y)$ in two variables over a number field k. Let $G = \mathrm{Spin}_n(q)$ be the spinor group of a nondegenerate quadratic form q over K of* odd *dimension $n \geq 5$. Then $\mathbf{gen}_K(G)$ is finite.*

THEOREM 5.6 ((cf. [**7**, Theorems 9.1 and 9.3] and [**26**, Proposition 5.3]). *Let G be a simple algebraic K-group of type G_2.*

(1) *If K is the field of rational functions $k(x)$, where k is a number field, then $|\mathbf{gen}_K(G)| = 1$.*
(2) *If k is a number field and K is one of the following:*
    - *$K = k(x_1, \ldots, x_r)$ is the field of rational functions in any (finite) number of variables;*
    - *$K = k(C)$ is the function field of a smooth geometrically integral curve C over k;*
    - *$K = k(X)$ is the function field of a Severi-Brauer variety X over k associated with a central division algebra D over k of degree ℓ, where ℓ is either odd or $\ell = 2$,*

then $\mathbf{gen}_K(G)$ is finite.

THEOREM 5.7 (cf. [**9**, Theorems 1.10 and 1.11]).

(1) *Let k be a number field, and set $K = k(x)$. Then for any simple algebraic k-group G of type F_4 that splits over a quadratic extension of K, the genus $\mathbf{gen}_K(G)$ is trivial.*
(2) *Let K be either a 2-dimensional global field of characteristic $\neq 2, 3$ or the field of rational functions $k(x, y)$ in two variables over a number field k. Then for any simple K-group G of type F_4 that splits over a quadratic extension of K, the genus $\mathbf{gen}_K(G)$ is finite.*

To conclude this section, we would like to mention a couple of results dealing with a newly-discovered phenomenon that we refer to as "killing the genus by a purely transcendental extension" – the reason for this choice of terminology is that, in the two cases considered below, the genus essentially becomes as small as possible after passing to a suitable purely transcendental extension.

THEOREM 5.8 (cf. [**9**, Theorem 1.5]). *Let A be a central simple algebra of degree n over a finitely generated field k, and let $G = \mathrm{SL}_{1,A}$. Assume that* $\mathrm{char}\, k$ *is prime to n, and let $K = k(x_1, \ldots, x_{n-1})$ be the field of rational functions in $(n-1)$ variables. Then $\mathbf{gen}_K(G \times_k K)$ consists of (the isomorphism classes of) groups of the form $H \times_k K$, where $H = \mathrm{SL}_{1,B}$ and B is a central simple algebra of degree n such that its class $[B]$ in the Brauer group $\mathrm{Br}(k)$ generates the same subgroup as the class $[A]$.*

The proof uses Amitsur's theorem on generic splitting fields [**1**], and a result of D. Saltman [**28**], [**29**] on function fields of Severi-Brauer varieties.

THEOREM 5.9 (cf. [**9**, Theorem 1.6]). *Let G be a group of type G_2 over a finitely generated field k of characteristic $\neq 2, 3$, and let $K = k(x_1, \ldots, x_6)$ be the field of rational functions in 6 variables. Then $\mathbf{gen}_K(G \times_k K)$ reduces to a single element.*

The proof relies on properties of Pfister forms.

Acknowledgments

I would like to thank the conference organizers for the opportunity to give a talk at the Amitsur Centennial Symposium. I would also like to thank the anonymous referee for a careful reading of the paper and suggestions that helped to improve the exposition.

References

[1] S. A. Amitsur, *Generic splitting fields of central simple algebras*, Ann. of Math. (2) **62** (1955), 8–43, DOI 10.2307/2007098. MR70624

[2] V. Chernousov, P. Gille, and A. Pianzola, *Torsors over the punctured affine line*, Amer. J. Math. **134** (2012), no. 6, 1541–1583, DOI 10.1353/ajm.2012.0051. MR2999288

[3] V. I. Chernousov, A. S. Rapinchuk, and I. A. Rapinchuk, *The genus of a division algebra and the unramified Brauer group*, Bull. Math. Sci. **3** (2013), no. 2, 211–240, DOI 10.1007/s13373-013-0037-z. MR3084007

[4] V.I. Chernousov, A.S. Rapinchuk, I.A. Rapinchuk, *Division algebras with the same maximal subfields*, Russian Math. Surveys **70** (2015), no. 1, 91-122.

[5] V. I. Chernousov, A. S. Rapinchuk, and I. A. Rapinchuk, *On the size of the genus of a division algebra*, Tr. Mat. Inst. Steklova **292** (2016), no. Algebra, Geometriya i Teoriya Chisel, 69–99, DOI 10.1134/S0371968516010052; English transl., Proc. Steklov Inst. Math. **292** (2016), no. 1, 63–93. MR3628454

[6] V. I. Chernousov, A. S. Rapinchuk, and I. A. Rapinchuk, *On some finiteness properties of algebraic groups over finitely generated fields* (English, with English and French summaries), C. R. Math. Acad. Sci. Paris **354** (2016), no. 9, 869–873, DOI 10.1016/j.crma.2016.07.012. MR3535336

[7] V. I. Chernousov, A. S. Rapinchuk, and I. A. Rapinchuk, *Spinor groups with good reduction*, Compos. Math. **155** (2019), no. 3, 484–527, DOI 10.1112/S0010437X1900705X. MR3920326

[8] V. I. Chernousov, A. S. Rapinchuk, and I. A. Rapinchuk, *The finiteness of the genus of a finite-dimensional division algebra, and some generalizations*, Israel J. Math. **236** (2020), no. 2, 747–799, DOI 10.1007/s11856-020-1988-x. MR4093903

[9] V. I. Chernousov, A. S. Rapinchuk, and I. A. Rapinchuk, *Simple algebraic groups with the same maximal tori, weakly commensurable Zariski-dense subgroups, and good reduction*, Adv. Math. **438** (2024), Paper No. 109437, 78 pp., DOI 10.1016/j.aim.2023.109437. MR4677076

[10] J.-L. Colliot-Thélène and J.-J. Sansuc, *Fibrés quadratiques et composantes connexes réelles* (French), Math. Ann. **244** (1979), no. 2, 105–134, DOI 10.1007/BF01420486. MR550842

[11] B. Conrad, *Reductive group schemes*, Autour des schémas en groupes, École d'été "Schémas en groupes," vol. I (Luminy 2011), Soc. Math. France, Paris 2014.

[12] M. Demazure and A. Grothendieck, *Schémas en groupes*, Séminaire de géométrie algébrique du Bois Marie 1962/64 (SGA 3). Lect. Notes Math. **151**, **152**, **153**, Springer, 1970.

[13] S. Garibaldi and D. J. Saltman, *Quaternion algebras with the same subfields*, Quadratic forms, linear algebraic groups, and cohomology, Dev. Math., vol. 18, Springer, New York, 2010, pp. 225–238, DOI 10.1007/978-1-4419-6211-9_13. MR2648728

[14] B. H. Gross, *Groups over* **Z**, Invent. Math. **124** (1996), no. 1-3, 263–279, DOI 10.1007/s002220050053. MR1369418

[15] G. Harder, *Halbeinfache Gruppenschemata über Dedekindringen* (German), Invent. Math. **4** (1967), 165–191, DOI 10.1007/BF01425754. MR225785

[16] K. Kato, *A Hasse principle for two-dimensional global fields*, J. Reine Angew. Math. **366** (1986), 142–183, DOI 10.1515/crll.1986.366.142. With an appendix by Jean-Louis Colliot-Thélène. MR833016

[17] D. Krashen and K. McKinnie, *Distinguishing division algebras by finite splitting fields*, Manuscripta Math. **134** (2011), no. 1-2, 171–182, DOI 10.1007/s00229-010-0390-0. MR2745259

[18] D. Krashen, E. Matzri, A. Rapinchuk, L. Rowen, and D. Saltman, *Division algebras with common subfields*, Manuscripta Math. **169** (2022), no. 1-2, 209–249, DOI 10.1007/s00229-021-01315-5. MR4462663

[19] J. S. Meyer, *Division algebras with infinite genus*, Bull. Lond. Math. Soc. **46** (2014), no. 3, 463–468, DOI 10.1112/blms/bdt104. MR3210701

[20] R. S. Pierce, *Associative algebras*, Studies in the History of Modern Science, vol. 9, Springer-Verlag, New York-Berlin, 1982. Graduate Texts in Mathematics, 88. MR674652

[21] G. Prasad and A. S. Rapinchuk, *Weakly commensurable arithmetic groups and isospectral locally symmetric spaces*, Publ. Math. Inst. Hautes Études Sci. **109** (2009), 113–184, DOI 10.1007/s10240-009-0019-6. MR2511587

[22] M. S. Raghunathan and A. Ramanathan, *Principal bundles on the affine line*, Proc. Indian Acad. Sci. Math. Sci. **93** (1984), no. 2-3, 137–145, DOI 10.1007/BF02840656. MR813075

[23] A. S. Rapinchuk and I. A. Rapinchuk, *On division algebras having the same maximal subfields*, Manuscripta Math. **132** (2010), no. 3-4, 273–293, DOI 10.1007/s00229-010-0361-5. MR2652433

[24] A. S. Rapinchuk and I. A. Rapinchuk, *Linear algebraic groups with good reduction*, Res. Math. Sci. **7** (2020), no. 3, Paper No. 28, 66, DOI 10.1007/s40687-020-00226-3. MR4149442

[25] A. S. Rapinchuk and I. A. Rapinchuk, *Recent developments in the theory of linear algebraic groups: good reduction and finiteness properties*, Notices Amer. Math. Soc. **68** (2021), no. 6, 899–910, DOI 10.1090/noti. MR4270402

[26] A. S. Rapinchuk and I. A. Rapinchuk, *Some finiteness results for algebraic groups and unramified cohomology over higher-dimensional fields*, J. Number Theory **233** (2022), 228–260, DOI 10.1016/j.jnt.2021.07.001. MR4356851

[27] A. S. Rapinchuk and I. A. Rapinchuk, *Properness of the global-to-local map for algebraic groups with toric connected component and other finiteness properties*, Math. Res. Lett. **30** (2023), no. 3, 913–943. MR4696434

[28] D. J. Saltman, *Norm polynomials and algebras*, J. Algebra **62** (1980), no. 2, 333–345, DOI 10.1016/0021-8693(80)90186-6. MR563232

[29] D. J. Saltman, *Lectures on division algebras*, CBMS Regional Conference Series in Mathematics, vol. 94, Published by American Mathematical Society, Providence, RI; on behalf of Conference Board of the Mathematical Sciences, Washington, DC, 1999, DOI 10.1090/cbms/094. MR1692654

[30] S. V. Tikhonov, *Division algebras of prime degree with infinite genus* (Russian, with Russian summary), Tr. Mat. Inst. Steklova **292** (2016), no. Algebra, Geometriya i Teoriya Chisel, 264–267, DOI 10.1134/S0371968516010167. English version published in Proc. Steklov Inst. Math. **292** (2016), no. 1, 256–259. MR3628465

Department of Mathematics, Michigan State University, East Lansing, Michigan
Email address: rapinchu@msu.edu

Contemporary Mathematics
Volume **800**, 2024
https://doi.org/10.1090/conm/800/16062

Genus 1 curves in Severi–Brauer surfaces

David J. Saltman

Shimshon Amitsur, my mentor and friend, who we all miss very much

ABSTRACT. In a talk at the Banff International Research Station in 2015, Asher Auel asked questions about genus one curves in Severi-Brauer varieties $SB(A)$. More specifically, he asked about the smooth cubic curves in Severi-Brauer surfaces, that is, in $SB(D)$ where D/F is a degree three division algebra. Even more specifically, he asked about the Jacobian, E, of these curves. In this paper we give a version of an answer to both these questions, describing the surprising connection between these curves and properties of the algebra A. Let F contain ρ, a primitive third root of one. Since D/F is cyclic, it is generated over F by x, y such that $xy = \rho yx$ and we call x, y a skew commuting pairs. The connection mentioned above is between the Galois structure of the three torsion points $E[3]$ and the Galois structure of skew commuting pairs in extensions $D \otimes_F K$. Given a description of which E arise, we then describe, via Galois cohomology, which C arise.

1. Introduction

In algebraic geometry, perhaps the simplest object is projective space. The next simplest object might be a Severi–Brauer variety, which is only interesting when the ground field F is not algebraically closed. Severi–Brauer varieties are defined by a central simple algebra A/F. In fact, the Severi–Brauer variety, $\mathrm{SB}(A)$, is defined as the variety of minimal right ideals of A. More precisely, if A/F has degree n (i.e., dimension n^2 over F), then $\mathrm{SB}(A)$ is a closed subvariety of the Grassmann variety $G_n(A)$ consisting of the n dimensional subspaces of A which are right ideals. It follows that $\mathrm{SB}(A)$ has a rational point if and only if $A \cong M_n(F)$, or in words, if A is split. If $A = \mathrm{End}_F(V)$ is split, then $\mathrm{SB}(A)$ is isomorphic to the projective space $\mathbb{P}^{n-1}$, since any such right ideal I can be identified with the line $L \subset V$ which is the range of all nonzero elements of I.

This explains why we view $\mathrm{SB}(A)$ as "almost" projective space. After a finite extension of F, $\mathrm{SB}(A)$ is projective space. That is, $\mathrm{SB}(A)$ is a form of projective space.

If A/F has degree two, that is, A/F is a quaternion algebra, then $\mathrm{SB}(A)$ is a curve and a form of the projective line. Quite a bit is known about these curves, since they are just the smooth conics in $\mathbb{P}^2$.

2020 *Mathematics Subject Classification.* Primary 16K20, 14H52.
Key words and phrases. Severi-Brauer, Genus one.

Thus, perhaps, it pays to consider the case of next highest dimension, the Severi–Brauer variety of a degree-three division algebra D/F. Note that the division algebra case is the interesting one. If A/F is degree three and not a division algebra, then A is split and $\mathrm{SB}(A)$ is $\mathbb{P}^2$.

Of course $\mathrm{SB}(D)$ is a form of $\mathbb{P}^2$, and to understand it one might want to understand the curves in $\mathrm{SB}(D)$. To start with, suppose $L \subset \mathrm{SB}(D)$ is a line, by which we mean it becomes a line when we split D. If we extend scalars to the separable closure $\bar{F}$, then $D \otimes_F \bar{F} = \mathrm{End}_{\bar{F}}(V)$ and the line L is identified with a subspace $W \subset V$ of dimension 2. To such a W we can associate the ideal, J, of all elements with range in W, which is of dimension 6 over $\bar{F}$. If L is defined over F, then J is Galois invariant, and so yields an ideal of D of dimension 6. This implies D is split, the case we are avoiding.

Suppose, then, that $C \subset SB(D)$ is an absolutely irreducible smooth degree two curve or a conic. (The other cases are easy.) C has no rational point, so C is of the form $\mathrm{SB}(D')$ for D'/F a quaternion division algebra. That is, there is a stalk $R = \mathcal{O}_C$ of the sheaf of regular functions on $\mathrm{SB}(D)$ such that R/M is the field of fractions, K', of $\mathrm{SB}(D')$.

Almost by definition, $\mathrm{SB}(D)$ splits D. Since R is a stalk, R splits D. By the functionality of the Brauer group, K' splits D. But the kernel of $\mathrm{Br}(F) \to \mathrm{Br}(K')$ is generated by D', implying that D is trivial or Brauer equivalent to D', and both are a contradiction. Thus $\mathrm{SB}(D)$ does not contain any absolutely irreducible conics.

There is a better way to view the above result. Let $\bar{F}$ be the separable closure of F, so $\mathrm{SB}(D) \times_F \bar{F} = \mathbb{P}^2_{\bar{F}}$. The Picard group of $\mathbb{P}^2_{\bar{F}}$ consists of the standard line bundles $\mathcal{O}(n)$ for $n \in \mathbb{Z}$. Artin ([**2**, p. 203]) observed that the Picard group of $\mathrm{SB}(D)$ is generated by $\mathcal{L}$, the unique line bundle that pulls back to $\mathcal{O}(3)$ on $\mathbb{P}^2_{\bar{F}}$. Viewing the Picard group as the divisor class group, we have that all curves on $\mathrm{SB}(D)$ have degree a multiple of three.

If C is an reducible curve of degree 3, then since no lines are Galois invariant it must be a union a three distinct lines which do not all intersect at one point. That is, C is a triangle determined by three distinct but Galois conjugate points. This means that $\mathrm{SB}(D)$ has plenty of triangles, namely, one for each maximal subfield. If C is an absolutely irreducible but not smooth curve, then C has one or two singular points which must be Galois conjugate and so D is split by a degree one or two extension, which is impossible.

Thus we can turn to the situation we are really interested in, where $C \subset \mathrm{SB}(D)$ is a cubic smooth absolutely irreducible curve, which is a curve of genus one. All this discussion motivates the question asked by Asher Auel, namely, to characterize the genus one curves in $\mathrm{SB}(D)$.

It is not hard to see from basic considerations that the j invariants that appear form a dense set in $\mathbb{P}^1_F$. Using the techniques of [**9**], Várilly-Alvarado and Viray have given an explicit subset of such j invariants which appear. Asher himself observed in unpublished work that cubic curves with $j = 0$ appear. The techniques developed here can distinguish between curves of the same j invariant, which are not isomorphic. Instead of j invariants, what is important for our main result is the group of three torsion points as a Galois module. Given that the elliptic curve E appears as the Jacobian of a curve C, we describe which C arise via Galois cohomology. Note the connection between these results and the result of [**5**, p. 331], which, for fixed E, gives the connection between cohomology and maps from genus one curves to Severi-Brauer varieties. The author would like to thank Asher Auel for the original question but most importantly for his detailed responses to earlier versions of this paper. We would also like to thank the referee for comments which motivated changes that, we hope, improve the exposition and clarity of the paper.

2. Preliminaries

Let A/F be a central simple algebra of degree three. Set $\bar{F}$ to be the separable closure of F, and $\bar{G}$ the Galois group of $\bar{F}/F$. Define $\mathrm{SB}(A)$ to be the Severi–Brauer variety of the central simple algebra A. Recall that $\mathrm{SB}(A)$ is the variety of right ideals of A of dimension 3 over F. For simplicity, we assume the characteristic of F is not two or three.

We know that $\mathrm{SB}(A)$ is a form of $\mathbb{P}^2$ and that the line bundle $\mathcal{O}(3)$ is defined on $\mathrm{SB}(A)$, which is also clear because this is the anticanonical bundle. We want to be more concrete, however. The tensor power $A^3 = A \otimes_F A \otimes_F A$ has an action of the symmetric group S_3 and we consider B, the algebra of S_3 fixed elements. By [**7**, p. 33], this action of S_3 is induced by conjugation by a group $S_3 \subset (A^3)^*$. That is, B is the centralizer of $F[S_3] \subset A^3$. But $F[S_3]$ is the direct sum $F \oplus F \oplus M_2(F)$ where the first F corresponds to the trivial representation. If $e \in F[S_3]$ is the associated idempotent to this first summand, we set $S^3(A) = eBe$. Note that if we extend scalars to $\bar{F}$, so $A \otimes_F \bar{F} = \mathrm{End}_{\bar{F}}(\bar{V})$, then $S^3(A \otimes_F \bar{F}) = \mathrm{End}_{\bar{F}}(S^3(\bar{V}))$, where $S^3(\bar{V})$ is the symmetric power of the vector space $\bar{V}$. For this reason we call $S^3(A)$ the symmetric power of A. Note that this implies that $S^3(A)$ has degree 10. Also note that as $A^3 = M_{27}(F)$ is split, $S^3(A)$ is also split and so $S^3(A) = M_{10}(F)$.

In particular, $\mathrm{SB}(S^3(A)) = \mathbb{P}^9$ is the nine dimensional projective space. There is an embedding $\mathrm{SB}(A) \to \mathrm{SB}(S^3(A))$ defined by $I \to e(I \otimes I \otimes I)^{S_3}e \subset S^3(A)$. Extending scalars to to $\bar{F}$ again, this is just the Segre embedding $\mathbb{P}^2(\bar{F}) \to \mathbb{P}^9(\bar{F})$. Finally, the line bundle $\mathcal{O}(1)_{\mathbb{P}^9}$ is defined on $\mathbb{P}^9 = \mathrm{SB}(S^3(A))$. If L is the restriction of this bundle to $\mathrm{SB}(A)$, then after extending scalars to $\bar{F}$, L becomes $\mathcal{O}(3)_{\mathbb{P}^2}$ which shows this bundle is defined over $\mathrm{SB}(A)$.

LEMMA 1. *There is a line bundle L on $\mathrm{SB}(A)$ which becomes $\mathcal{O}(3)_{\mathbb{P}^2}$ after scalar extension to $\bar{F}$ and is the restriction of $\mathcal{O}(1)_{\mathbb{P}^9}$ on $\mathrm{SB}(S^3(A))$.*

The bundle $\mathcal{O}_{\mathbb{P}^2}(3)$ has a ten dimensional space, $\bar{W}$, of global sections which are the cubic forms in three variables. That is, if $s \in \bar{W}$, then the zeroes of s are a cubic curve in $\mathbb{P}^2$, and for all s in a Zariski open set this is a smooth cubic curve of genus one. Now L has a space of global sections of dimension 10, W, over F and so a $s \in W$ defines a cubic curve C in $SB(A)$. Once again, for a Zariski open set of s this C is smooth and of genus one. Of course, except in the trivial case A is split, $\mathrm{SB}(A)$ has no rational points and so C has no rational points. However, we can define $E(C)$ as the Jacobian of C and $E(C)$ is an elliptic curve. The question we attack in this paper is the question of which $E(C)$ appear for a given A, and given $E(C)$ which C appear.

Of course, the group $\mathrm{PGL}_3(\bar{F})$ acts on $\mathbb{P}^2(\bar{F})$ and this action extends to an action on $\mathbb{P}^9(\bar{F})$. Viewing this as an action of $\mathrm{GL}_3(\bar{F})$, this induces an action on the global sections of $\mathcal{O}(3)$ which is just the change of variables action on the cubic forms.

We need to descend this group action to F. Let $\bar{G}$ be the Galois group of $\bar{F}/F$. Of course $A \otimes_F \bar{F} = M_3(\bar{F})$, so there is a Galois action of $\bar{G}$ on $\mathrm{GL}_3(\bar{F})$ such that the $\bar{G}$ fixed elements are A^*. Hilbert's Theorem 90 shows that G fixed elements of $\mathrm{PGL}_3(\bar{F})$ lift to $\bar{G}$ fixed elements of $GL_3(\bar{F})$, so A^*/F^* is the group of $\bar{G}$ fixed elements of $\mathrm{PGL}_3(\bar{F})$. Moreover, A^*/F^* is the group of F automorphisms of A and hence A^*/F^* acts on $\mathrm{SB}(A)$ and $\mathrm{SB}(S^3(A))$. Of course, after extending scalars to $\bar{F}$, this is just the action of $PGL_3(\bar{F})$ on $\mathbb{P}^2$ and $\mathbb{P}^9$.

3. Classical Invariant Theory

Let us review the classical invariant theory of cubic curves over $\bar{F}$. This material is very far from new, but it is useful to review so we can treat the non-algebraically closed case. As a general reference we suggest [**3**, Chapter 3].

Recall that $\bar{W}$ is the space of global sections of $\mathcal{O}_{\mathbb{P}^2}(3)$ over $\bar{F}$. Let $\bar{T}$ be the projective space of $\bar{W}$, that is, $\bar{T} = \mathbb{P}^9$. Then $\bar{T}$ can be viewed as the space of cubic curves. We begin by recalling some basic facts about a smooth cubic curve. Any line in $\mathbb{P}^2$ intersects $\bar{C}$ in three points, counting multiplicity. Recall that a point P on $\bar{C}$ is called an inflection point if the tangent line to $\bar{C}$ at P has an intersection with $\bar{C}$ that is at least multiplicity 3 (any tangent line has multiplicity 2). For the moment fix one inflection point $[0]$ on $\bar{C}$. Looking at all the intersections of lines with $\bar{C}$, we see that the embedding $\bar{C} \to \mathbb{P}^2$ is defined by the line bundle $\mathcal{L}(3[0])$. Because of this we have:

LEMMA 2. *Any embedding $\iota : \bar{C} \to \mathbb{P}^2(\bar{F})$ as a cubic curve is determined by the image of $[O]$, $[P]$, and $[Q]$ where $[P] - [0]$ and $[Q] - [0]$ generate the group of three torsion points of $\bar{C}$.*

PROOF. This follows because the global sections of $\mathcal{L}(3[0])$ are spanned by s_0, s_1, s_2 where s_1/s_0 has divisor $3[P] - 3[0]$ and s_2/s_0 has divisor $3[Q] - 3[0]$. $\square$

The Jacobian $E(\bar{C})$ of $\bar{C}$ is the group of degree 0 divisors modulo principal divisors. In particular, $E(\bar{C})$ is an abelian group. If we fix $[0]$ as above, then $\bar{C} \cong E(\bar{C})$ via $[P] \to [P] - [0]$. Moreover, there is a natural action $E(\bar{C}) \times \bar{C} \to \bar{C}$ in which, for example, it is true that $[P] - [0]$ acting on $[0]$ is $[P]$. Thus $E(\bar{C})$ is a subgroup of the automorphism group of $\bar{C}$ which we call the translations and $E(\bar{C})$ acts transitively on $\bar{C}$.

Since $\bar{C} \cong E(\bar{C})$, a choice of $[0]$ imparts an additive group structure on $\bar{C}$. It will be useful to recall the classical construction of this structure. If P, Q, R are three points on $\bar{C}$, then we say these points "sum to zero" if they are colinear and hence the intersection points of some line with $\bar{C}$. To make a group with zero $[0]$, for any point $[P]$, one defines $[-P]$ by the condition that $[P]$, $[-P]$, and $[0]$ are colinear. Then $[P] + [Q] = [R]$ if $[P]$, $[Q]$, and $[-R]$ are colinear. Thus "summing to zero" is independent of the choice of $[0]$.

The automorphisms of $\bar{C}$ have the form $T \rtimes \mathrm{Aut}$ where $T = E(\bar{C})$ is the group of translations and Aut is the group of automorphisms that fix $[0]$. Thus if $\bar{C}$ and $\bar{C}'$ are two isomorphic smooth cubic curves in $\mathbb{P}^2$, there is an isomorphism between them that preserves the embedding implying that there is an isomorphism between them that extends to $\mathbb{P}^2$. That is, $\bar{C}$ and $\bar{C}'$ are in the same $PGL_3(\bar{F})$ orbit in $\mathbb{P}^9$. The quotient $\mathbb{P}^9/\mathrm{PGL}_3(\bar{F})$ is birationally $\mathbb{P}^1$ and we will later remind the reader that there is an induced map to the j line which is generically 12 to 1. The action of $PGL_3(\bar{F})$ on cubic curves will be key to our argument.

Let $S_t \subset PGL_3(\bar{F})$ be the stabilizer of some $\bar{C}$. Then S_t consists of the automorphisms of $\bar{C}$ which preserve the embedding in $\mathbb{P}^2$. All the elements of

Aut preserve the embedding, but only the elements of T of order 3 preserve the embedding. It follows that $S_t = S \rtimes \mathrm{Aut}$ where S is isomorphic to $\mathbb{Z}/3\mathbb{Z} \oplus \mathbb{Z}/3\mathbb{Z}$ and is the group of translations by three torsion elements of $E(\bar{C})$. If the j invariant of $\bar{C}$ is not 0 or 1728, then $\mathrm{Aut} = S_t/S = \mathbb{Z}/2\mathbb{Z}$ and Aut is generated by the -1 map on C viewed as an elliptic curve with $[0]$ as the zero element. If $j = 0$ then $S_t/S = C_6$ the cyclic group of order 6, and if $j = 1728$ then $S_t/S = C_4$.

Given $\bar{C}$, we let I be the set of its nine inflection points. In the following discussion we have two goals. First, we observe that we can define a nine point "Jacobian", $E(I)$, for I that does not involve $\bar{C}$. We accomplish this by restricting the argument, reviewed above for all of $\bar{C}$, to only these nine points. Second, we want to relate this $E(I)$ to the stabilizer of $\bar{C}$ in $\mathrm{PGL}_3(\bar{F})$ and, in particular, to the group of translations. What we will show is that if we pick one point in I to be the identity, then the other 8 points correspond to points on the elliptic curve $E(\bar{C})$ of order 3. The next result describes some properties of I where we assume some $\bar{C}$ exists but not anything specific about it.

LEMMA 3. *Suppose P, $Q \in I$ are distinct. Then there is exactly one third point $R \in I$ such that P, Q, R are colinear in $\mathbb{P}^2$ and R is distinct from both P and Q.*

PROOF. We identify $\bar{C}$ with $E(\bar{C})$ via a choice of $[0]$. A line in $\mathbb{P}^2$ meets any $\bar{C}$ in three points. Thus, $R \in \bar{C}$ is well-defined. Moreover, $P + Q + R$ is zero in $\bar{C}$, and so R must also be a three torsion point and hence an inflection point and hence in I. If, say, $R = Q$ then $P + 2Q = 0$ in $\bar{C}$ which implies $P = Q = R$. □

Because of the above, we say three points of I are colinear if they are distinct and colinear in $\mathbb{P}^2$, or are all the same.

We can form a group from this relationship on I, by repeating for I the construction of $E(\bar{C})$ but restricted to these nine points. Let $E(I)$ be the set of equivalences classes of all pairs (P,Q), $P,Q \in I$, where we say $(P,Q) \sim (R,S)$ if there is a single $T \in I$ such that P,S,T and R,Q,T are colinear. Note that all pairs (P,P) are equivalent. We let $\{(P,Q)\}$ be the equivalence class containing (P,Q).

LEMMA 4.

(a) *If $(P,Q) \sim (R,S)$ then $(P,R) \sim (Q,S)$.*
(b) *If $(P,Q) \sim (R,Q)$ then $P = R$.*
(c) *For any $P,Q,R \in I$, there are unique $S,S' \in I$ such that $(P,Q) \sim (R,S) \sim (S',R)$.*
(d) *If $(P,Q) \sim (Q,P)$ then $P = Q$.*
(e) *If we fix Q then the map $P \to \{(P,Q)\}$ is a bijection between I and $E(I)$.*

PROOF. Part (a) is direct from the definition. For (b), we have P,Q,T and R,Q,T colinear and so $P = R$. Turning to (c), we have R,Q,T colinear for a unique T and define S by taking P,T,S colinear (defining S' is similar). For (d), only P is colinear with P and P (P is an inflection point). Part (e) follows from (a) and (b). □

We can now proceed to define the group structure on $E(I)$. We say $\{(P,Q)\} + \{(R,S)\} = \{(P,U)\}$ if $(R,S) \sim (Q,U)$. One can check this is well-defined with identity the class of (P,P). Also $\{(P,Q)\} + \{(Q,R)\} = \{(P,R)\}$, so $\{P,Q\} + \{Q,P\} = \{P,P\}$. Since colinearity is a symmetric relationship, this defines an abelian group. Finally, and importantly, there is a natural action of $E(I)$ on I defined by $P + \{(Q,P)\} = Q$.

LEMMA 5.

(a) *If* $\alpha \in E(I)$ *then* $3\alpha = 0$.
(b) *If* $\alpha \in E(I)$ *then* $(P,Q) \sim (P+\alpha, Q+\alpha)$.
(c) P,Q,R *are colinear if and only if for some (hence all)* S, $\{(P,S)\} + \{(Q,S)\} + \{(R,S)\} = 0$. *In particular,* P, Q, *and* R *are colinear if and only if* $(Q,P) \sim (P,R)$.
(d) *If* P,Q,R *are colinear, then so are* $P+\alpha, Q+\alpha, R+\alpha$ *for any* $\alpha \in E(I)$.
(e) *If* $P \in I$ *and* $\alpha \in E(I)$, *then* P, $P+\alpha$ *and* $P+2\alpha$ *are colinear.*

PROOF. To begin with (a), if P,Q,R are colinear then $(P,Q) \sim (Q,R) \sim (R,P)$ and $3\{P,Q\} = \{(P,Q)\} + \{(Q,R)\} + \{(R,P)\} = \{P,P\}$. As for (b), if α contains (S,P) and (T,Q) then $(S,P) \sim (T,Q)$ so $(P,Q) \sim (S,T)$. Turning to (c), the independence from the choice of S is clear from adding $3\{(S',S)\}$ to both sides. Thus we need only consider $\{(P,P)\} + \{(Q,P)\} + \{(R,P)\} = 0$ or $\{(Q,P)\} + \{(R,P)\} = 0$ or $(Q,P) \sim (P,R)$ which means the line through Q and R also goes through P. Part (d) is now clear. As for (e), let $\alpha = \{(Q,P)\} = \{(R,Q)\}$ so P,Q,R are colinear. Then $P+\alpha = Q$ and $2\alpha = \{(Q,P)\} + \{(R,Q)\} = \{(R,P)\}$ so $P + 2\alpha = R$. □

Let us, briefly, assume F is not necessarily equal to $\bar{F}$. Suppose $C \supset I$ is a cubic curve defined over F with Jacobian E, and I are the inflection points. Then $E \times C \cong C \times C$ via the morphism $(\alpha, P) \to (P+\alpha, P)$. Thus there is a projection $C \times C \to E$ that maps $(P,Q) \to [P]-[Q]$ where $[P]-[Q]$ is the degree-zero divisor on C. The diagram:

$$\begin{array}{ccc} I \times I & \subset & C \times C \\ \downarrow & & \downarrow \\ E(I) & \subset & E(C) \end{array}$$

commutes. All of this makes sense over $\bar{F}$ where $I \times I$ is the zero dimensional reduced variety. Of course, $I \times I$ is defined over any field where I is defined. The above diagram imparts to $E(I)$ a variety structure defined over any field F where C and hence I are defined, though of course when $F \neq \bar{F}$, then $K_1 = \operatorname{Spec}(E(I))$ and $K_2 = \operatorname{Spec}(I \times I)$ are direct sums of fields, perhaps strictly containing F. If I is defined over F but C is not, we can proceed as follows. The $\bar{F}$ points of $I \times I$ are morphisms $\phi_{P,Q} : K_2 \to \bar{F}$ and there is one for each P,Q. Then we can define K_1 to be $\{f \in K_2 | \phi_{P,Q}(f) = \phi_{R,S}(f)$ whenever $(P,Q) \sim (R,S)\}$. Since this is a linear condition on f, this is compatible with the definition of K_1 over F's where C is not defined. All of which is saying:

LEMMA 6. $I \times I \to E(I)$ *is a morphism of zero dimensional varieties whenever* I *is defined.*

COROLLARY 1. *Suppose* $\bar{C}'$ *also contains* I. *Then* I *forms the set of inflection points of* $\bar{C}'$ *also.* $E(I)$ *can be identified with the three torsion points of both the Jacobians of* $\bar{C}$ *and* $\bar{C}'$.

We note that the phrase "can be identified" does not mean this identification is unique. The nonuniqueness of the identification will be a point we will have to handle carefully.

We know classically that the action of $E(I)$ on I extends first to an action on $\bar{C}$, namely translation by three torsion points, and further therefore to an action on $\mathbb{P}^2$. Thus $E(I)$ corresponds to a subgroup $S \subset \mathrm{PGL}_3(\bar{F})$. We can give more detail about S, and this is well known (see [**4**, p. 131]). Let F_3 be the field of three elements. From the definition, S is generated by x, y each of order three which commute in $PGL_3(\bar{F})$ and represent a choice of basis, α, β over $F_3 = \mathbb{Z}/3\mathbb{Z}$ of $E(I)$. Let $\tilde{x}, \tilde{y} \in GL_3(\bar{F})$ be preimages of x, y where we can assume (over $\bar{F}$), that $\tilde{x}^3 = \tilde{y}^3 = 1$. Up to conjugation, there are two possibilities. Either $\tilde{x}$ and $\tilde{y}$ commute or $\tilde{x}\tilde{y} = \rho\tilde{y}\tilde{x}$ for a nontrivial 3 root of one ρ. We will argue that the latter holds, a classical fact related to the existence of the Weil pairing.

Fix $P \in I$. Then $P, P+\alpha, P+2\alpha$ are colinear; as are $P+\beta, P+\beta+\alpha$, and $P+\beta+2\alpha$ as well as $P+2\beta, P+2\beta+\alpha, P+2\beta+2\alpha$. All of these lines are left invariant by α. Note that these are three distinct lines because any line contains at most three points of I. Lifting to $GL_3(\bar{F})$ acting on $\mathbb{A}^3$, there are three distinct two-dimensional subspaces preserved by $\tilde{x}$. But $\tilde{x}$ has three distinct eigenvalues and so these are precisely all the two-dimensional subspaces preserved by $\tilde{x}$. We can make the same argument for $\tilde{y}$ and conclude that $\tilde{y}$ preserves the spaces associated to the lines $P, P+\beta, P+2\beta$, as well as $P+\alpha, P+\alpha+\beta, P+\alpha+2\beta$ and $P+2\alpha, P+2\alpha+\beta, P+2\alpha+2\beta$. Note that the spaces preserved by $\tilde{x}$ and $\tilde{y}$ are independent of the choice of $\tilde{x}$, $\tilde{y}$ and only depend on x and y.

LEMMA 7. *Fix a third root of unity ρ. After perhaps changing y to y^2, we have $\tilde{x}\tilde{y} = \rho\tilde{y}\tilde{x}$. The action of $\tilde{y}$ permutes the eigenspaces of x and vice versa. If V_1, V_2, V_3 are the two-dimensional spaces preserved by $\tilde{x}$ and W_1, W_2, W_3 the same for $\tilde{y}$, then points of I correspond to the nine intersections $V_i \cap W_j$. S acts transitively on I.*

PROOF. This is well known and appears in [**4**, p. 131]. We include a proof because the details of the proof will be useful. If $\tilde{x}$, $\tilde{y}$ commuted they would have the same eigenspaces and hence the same preserved two-dimensional subspaces. This proves the first statement. The second is now clear and the third statement just says, as observed above, that each of the lines preserved by x and y share one point of I. It is immediate that S acts transitively. □

Thus associated to I is a subgroup $S \subset PGL_3(\bar{F})$ where S is the image of $\tilde{S} \subset GL_3(\bar{F})$ and $\tilde{S}$ is generated by $\tilde{x}$, $\tilde{y}$ with $\tilde{x}^3 = 1$, $\tilde{y}^3 = 1$, and $\tilde{x}\tilde{y} = \rho\tilde{y}\tilde{x}$. All such $\tilde{S}$ are conjugate in $GL_3(\bar{F})$, and so all such S are conjugate in $PGL_3(\bar{F})$.

The first consequence of Lemma 7 is that S has a pairing

$$S \times S \to \langle \rho \rangle$$

defined by the map $(x, y) \to \rho^i$ where x, y lift to $\tilde{x}, \tilde{y} \in GL_3$ and $\tilde{x}\tilde{y} = \rho^i\tilde{y}\tilde{x}$. It follows that the preimage of S in $GL_3(\bar{F})$ contains the Heisenberg group of order 27. That is, if we choose $\tilde{S}$ to be generated by preimages $\tilde{x}$, $\tilde{y}$ with $1 = \tilde{x}^3 = \tilde{y}^3$, then $\tilde{S}$ is the Heisenberg group.

The identification of S with $E(I)$ via their action on I therefore amounts to an isomorphism $\psi : S \to E(I)$ where $\psi(s)$ is the equivalence class containing $(s(P), P)$. We have:

LEMMA 8. *ψ is well-defined and is a group isomorphism. Moreover, as a subset of PGL_3, S induces the trivial action on $E(I)$.*

A consequence of Lemma 8 is that $E(I)$ has a pairing inherited from S. We quote a result from [**3**, Chapter 3].

THEOREM 1. *The stabilizer of I in $\mathrm{PGL}_3(\bar{F})$ has the form*

$$S \rtimes SL_2(F_3).$$

To understand the above theorem, set $\mathrm{Aff}(I)$ to be the group of bijections of I which preserve colinearity, and we view $E(I) \subset \mathrm{Aff}(I)$ via the action of $E(I)$ on I, which preserves colinearity by Lemma 5.

LEMMA 9. *There is an exact sequence*

$$0 \to E(I) \to \mathrm{Aff}(I) \to GL_2(F_3) \to 0$$

where $\mathrm{Aff}(I) \to GL_2(F_3)$ *is induced by taking the action on* $E(I)$.

PROOF. Let $\phi : I \to I$ be a bijection that preserves the colinearity relationship. Then ϕ induces an automorphism (i.e., element of $GL_2(F_3)$) of $E(I)$. This defines the morphism above. If ϕ is the identity on $E(I)$, then $(P,Q) \sim (\phi(P), \phi(Q))$ for all P, Q so $(\phi(P), P) \sim (\phi(Q), Q)$ and there is an $\alpha \in E(I)$ containing all $(\phi(P), P)$. Thus $\phi(P) = P + \alpha$. Finally, if $A \in GL_2(F_3)$ and $P \in I$ we define $A_P : I \to I$ by letting $A_P(Q)$ be the element of I such that $(A_P(Q), P) \in A(\{(Q,P)\})$. By the lemma above A_P preserves colinearity and shows that there is a splitting $GL_2(F_3) \to \mathrm{Aff}(I)$ for every choice of P. In particular the map is onto. As we saw in Lemma 5, the action of $E(I)$ on I induces the trivial action on $E(I)$. □

Soon we will also need to consider $\mathrm{SAff}(I) \subset \mathrm{Aff}(I)$ which is the preimage of $SL_2(F_3)$.

It will be useful to notice that the action of $\mathrm{Aff}(I)$ on I induces a canonical element of $\gamma \in H^1(\mathrm{Aff}(I), E(I))$. Define $d : \mathrm{Aff}(I) \to E(I)$ by setting $d(g) = \{(g(P), P)\}$ for a choice of P. Then for $g, h \in \mathrm{Aff}(I)$, $d(gh) = \{(gh(P), P)\} = \{(gh(P), g(P))\} + \{(g(P), P)\} = g(\{(h(P), P)\} + \{(g(P), P)\} = g(d(h)) + d(g)$ and thus d is a 1 cocycle. If we define $d'(g) = \{(g(Q), Q)\}$, then $d(g) - d'(g) = g(\{(P,Q)\}) - \{(P,Q)\}$ and d and d' are cohomologous. Hence any choice of P yields the same cohomology class we call γ. If we restrict d to $E(I)$ then d is the identity, viewed as an element of $\mathrm{Hom}(E(I), E(I)) = H^1(E(I), E(I))$. Of course, if γ_S is the restriction of γ to $\mathrm{SAff}(I)$, then γ_S also restricts to the identity on $E(I)$.

PROPOSITION 1. $\gamma \in H^1(\mathrm{Aff}(I), E(I))$ *is the unique element which restricts to the identity of* $H^1(E(I), E(I)) = \mathrm{Hom}(E(I), E(I))$. *The same holds for* $\gamma_S \in H^1(\mathrm{SAff}(I), E(I))$.

PROOF. Let $G = GL_2(F_3)$ or $SL_2(F_3)$ and $H = \mathrm{Aff}(I)$ or $\mathrm{SAff}(I)$ so $H \to G$ has kernel $E(I)$. The Hochschild-Serre spectral sequence shows that there is an exact sequence $H^1(G, E(I)) \to H^1(H, E(I)) \to H^1(E(I), E(I))$, and so it suffices to show that $H^1(G, E(I)) = H^1(G, F_3 \oplus F_3) = 0$. Since $SL_2(F_3)$ has index 2 in $GL_2(F_3)$, it suffices to show $H^1(SL_2(F_3), E(I)) = 0$. As a finite group, one can calculate that $SL_2(F_3) = Q \rtimes F_3^+$ where Q is the quaternion group of order 8. Since $H^1(Q, F_3 \oplus F_3) = 0$ and $(F_3 \oplus F_3)^Q = 0$, we are done by another use of Hochschild-Serre. □

Now let $H \subset \mathrm{PGL}_3(\bar{F})$ be the stabilizer of I. We have a homomorphism $\eta : H \to \mathrm{Aff}(I)$. Furthermore, any $\phi \in \mathrm{PGL}_3(\bar{F})$ which is the identity on I must be trivial. Thus η is an injection. $S \subset H$ acts trivially on $E(I)$ and hence $\eta(S) = E(I)$. We saw above that $E(I)$ has a pairing which must be preserved by H and so $\eta(H) \subset \mathrm{SAff}(I)$. The above theorem amounts to saying that $\eta(H) = \mathrm{SAff}(I)$.

If S_t is the stabilizer of $\bar{C}$, then it must stabilize I and hence $S_t \subset H$ while $S \subset S_t$ is precisely the kernel of the action of S_t on $E(I)$.

Let $\mathcal{I}$ be the set of nine point subsets of $\mathbb{P}^2$ which are the inflections points of smooth cubic $\bar{C}$. Let $\mathcal{S}$ be the set of nine element subgroups of $PGL_3(\bar{F})$ where any $S \in \mathcal{S}$ is the image of $\tilde{S} \subset GL_3(\bar{F})$ and $\tilde{S}$ is generated by $\tilde{x}$, $\tilde{y}$ with $\tilde{x}^3 = 1$, $\tilde{y}^3 = 1$, and $\tilde{x}\tilde{y} = \rho\tilde{y}\tilde{x}$.

THEOREM 2. *The above discussion yields a one to one correspondence between* $\mathcal{I}$ *and* $\mathcal{S}$.

PROOF. We saw above that S determines I, once we observe that the description there is independent of the basis of S chosen. For sake of clarity, we review what this means. Given S, we can choose a basis x, y and preimages $\tilde{x}$ and $\tilde{y}$. Then $\tilde{x}$ has exactly three two-dimensional subspaces V_1, V_2, and V_3. Similarly, $\tilde{y}$ has W_1, W_2 and W_3. The projective points of I are the 9 lines which are the intersections $V_i \cap W_j$. Since all the elements of $\mathcal{S}$ are conjugate, the subset I must be the inflection points of some smooth cubic $\bar{C}$.

For the converse, I determines its stabilizer H, and S is the kernel of the action of H on $E(I)$. □

Note that if $\bar{G}$ is the Galois group of $\bar{F}/F$, then the above correspondence clearly commutes with the action of $\bar{G}$.

If follows from Theorem 2 that if $\phi \in PGL_3(\bar{F})$, then $\phi S \phi^{-1} = S$ if and only if $\phi(I) = I$. We have

THEOREM 3. *H is the normalizer of S in* $\mathrm{PGL}_3(\bar{F})$.

We can add to our understanding of Theorem 1 as follows. We saw above that if ϕ normalizes S then $\phi(I) = I$ and so ϕ maps to $\mathrm{SAff}(I)$. Conversely suppose that $a, b, c, d \in F_3$ are such that $ad - bc = 1$. Write $x_1 = \tilde{x}^a\tilde{y}^b$, $\tilde{y}_1 = \tilde{x}^c\tilde{y}^d$ and note that $\tilde{x}_1^3 = \tilde{y}_1^3 = 1$ and $\tilde{x}_1\tilde{y}_1 = \rho\tilde{y}_1\tilde{x}_1$. Thus setting $\tilde{x} \to \tilde{x}_1$ and $\tilde{y} \to \tilde{y}_1$ defines an automorphism of $M_3(\bar{F})$ and ϕ exists by Noether-Skolem. Thus H maps onto $\mathrm{SAff}(I)$.

For a given I, we can look at all the curves $\bar{C}$ which contain I. It is classically known that this forms a line L_I in $\mathbb{P}^9$ and all j invariants appear somewhere on a point of this line. It is instructive to outline how we know this. First of all, let $\tilde{N} \subset GL_3(\bar{F})$ be in the preimage of $H = N(S)$. Let $S = \langle x \rangle \oplus \langle y \rangle$ and let $x', y' \in \tilde{N}$ be preimages with $x'^3 = 1 = y'^3$. If $\tilde{S} \subset \tilde{N}$ is the group generated by x', y', then $\tilde{S}$ is the Heisenberg group and $\tilde{S}\bar{F}^*$ is the full preimage of S. Thus $\tilde{N}/\tilde{S}\bar{F}^* = N(S)/S = \mathrm{SL}_2(F_3)$. Let $M \subset \tilde{H}$ be the abelian group generated by x' and ρ.

Of course, $\bar{F}^3$ is a module over $\tilde{N}$. Obviously, one could describe this module but we do not need this. It suffices to note that as a $\tilde{S}$ module it is $\mathrm{Ind}_M^{\tilde{S}} L$ where M acts on $L = \bar{F}^1 v$ by setting $x'(v) = \rho v = \rho(v)$. That is, $V = \bar{F}^3$ has a basis $\{v_0, v_1, v_2\}$ where $\bar{F}v_i$ is an x' eigenspace with eigenvalue ρ^i and y permutes the v_i. Let V^* be the dual of V with dual basis v_0^*, v_1^*, v_2^*. Of course, $S^3(V^*)$ is spanned by the degree-three monomials in the v_i^*. Moreover, the inverse image of I in V are the nine lines L_{ij} where L_{ij} is the simultaneous zero of v_i^*, and $w_j^* = v_0^* + \rho^j v_1^* + \rho^{2j} v_2^*$ which is the eigenvector for y.

Of course ρ acts trivially on $S^3(V^*)$, and the $\tilde{S}$ fixed subspace is spanned by $v_0^* v_1^* v_2^*$ and $(v_0^*)^3 + (v_1^*)^3 + (v_2^*)^3$. Thus the projective space of C's fixed by $\tilde{S}$ is a projective line. This is the classical Hessian pencil and associated Hessian normal form.

Having made this computation, we are ready to prove:

THEOREM 4. *Suppose $S \subset PGL_3(\bar{F})$ is associated with $I \subset \mathbb{P}^2$. Then a cubic curve $\bar{C}$ contains I if and only if S fixes $\bar{C}$ or equivalently that $\tilde{S}$ fixes $\bar{C}$, where $\tilde{S}$ is the Heisenberg group in the preimage of S in $GL_3(\bar{F})$.*

PROOF. If $\tilde{S}$ fixes $\bar{C}$, then $\bar{C}$ is in the span of $v_0^*v_1^*v_2^*$ and $(v_0^*)^3 + (v_1^*)^3 + (v_2^*)^3$. We need to show that any such $\bar{C}$ is in the ideal (v_i^*, w_j^*) for any i and j. This implies that $I \subset \bar{C}$. We compute that $w_0^*w_1^*w_2^* = (v_0^*)^3 + (v_1^*)^3 + (v_2^*)^3 + (-3)(v_0^*v_1^*v_2^*)$ and this direction is obvious.

Conversely, suppose $I \subset \bar{C}$. We need to show that $\tilde{S}$ fixes $\bar{C}$. Note that any zero set of an f contains I if and only if $f \in \cap_{i,j}(v_i^*, w_j^*) = J$. Clearly, J and hence $J \cap S^3(V^*)$ is preserved by $\tilde{H}$. It thus suffices to show that no eigenvector for $\tilde{H}$, with nontrivial eigenvalue, is in J. We can check this one by one. $(v_0^*)^3 + \rho(v_1^*)^3 + \rho^2(v_2^*)^3$ is clearly not in (v_0^*, w_0^*) and the same holds for $(v_0^*)^3 + \rho^2(v_1^*)^3 + \rho(v_2^*)^3$. The $\tilde{H}$ span of $(v_1^*)^3v_2^*$ has basis this vector and $(v_2^*)^2v_0$, $(v_0^*)^2v_1^*$. Any eigenvector has a nonzero $(v_1^*)^3v_2^*$ term and since $(v_1^*)^3v_2^* \notin (v_0^*, w_0^*)$ it follows that this eigenvector is not in I. The same argument works for the $\tilde{S}$ span of $(v_1^*)^2v_0^*$. □

We defined L_I to the line of $\bar{C}$'s which contain I or equivalently which are fixed by S. It follows that the normalizer, $N(S)$, of S in $PGL_3(\bar{F})$ acts on $L(I)$. Since all the S's and I's are conjugate, any $PGL_3(\bar{F})$ orbit intersects L_I. If $\bar{C}, \bar{C}' \in L_I \cap O$ and $\bar{C}' = g(\bar{C})$, then both these cubics have stabilizer containing S. Thus $g \in N(S)$. The quotient of L_I by $N(S)$ is a $\mathbb{P}^1$, and the associated map $L_I \to \mathbb{P}^1$ can be viewed as taking any curve $\bar{C}$ to its j invariant. A generic element of L_I, specifically one with $j \neq 0, 1728$, has stabilizer S_t where S_t/S has order 2. Thus $L_I \to \mathbb{P}^1$ is generically a Galois cover with group $N(S)/S_t \cong SL_2(F_3)/\pm 1$ which has order 12. However, we need to be more careful. Assume first that $\bar{C}$ has $j \neq 0, 1728$. While there are 12 cubic curves $\bar{C} \supset I$, there are 24 associated maps $\iota : I \to \bar{C}$ and $b : \bar{C} \to \mathbb{P}^2(\bar{F})$, because if $s \in S_t$, we can replace ι by $s\iota$ and b by bs^{-1}, preserving $b\iota$ which is the fixed embedding of I. Of course, when $j = 0$ there are 4 curves $\bar{C}$ and when $j = 1728$ there are 6 such curves $\bar{C}$, but there are always 24 embeddings and 24 maps $\iota : I \to \bar{C}$ for some $\bar{C}$. Thus with fixed embedding $I \to \mathbb{P}^2$, choosing an embedding $\bar{C} \to \mathbb{P}^2$ is also specifying an embedding $\iota : I \to \bar{C}$. Let us make a piece of the above discussion explicit.

PROPOSITION 2. *Let $I \subset \mathbb{P}^2(\bar{F})$ be fixed. Let $\bar{C}$ be one of the smooth cubic curves containing I with embedding $\iota : \bar{C} \to \mathbb{P}^2(\bar{F})$. Suppose $\iota' : \bar{C}' \to \mathbb{P}^2(\bar{F})$ is an embedding of a smooth cubic curve isomorphic to $\bar{C}$ as curves. Then there is a unique $g \in \mathrm{PGL}_3(\bar{F})$ such that $g \circ \iota = \iota'$. If $\bar{C}' \supset I$, then $g \in H$ is the stabilizer of I. If $\bar{C}' = \bar{C}$ as subvarieties of $\mathbb{P}^2(\bar{F})$, then $g \in S_t$ is the stabilizer of $\bar{C}$, and each $g \circ \iota$ is a distinct embedding $I \to \bar{C}$. There is a one to one correspondence between such embeddings $I \to \bar{C}$ and embeddings $\bar{C} \to \mathbb{P}^2(\bar{F})$. If $g \notin S_t$, then $g(\bar{C}) \neq \bar{C}$ as subsets of $\mathbb{P}^2(\bar{F})$.*

PROOF. Most of this has been said or is trivial. If $\bar{C}' \cong \bar{C}$ is a cubic curve, then the embedding is given by $\mathcal{L}(3[0])$ and differs from $\bar{C} \subset \mathbb{P}^2(\bar{F})$ by an element of $\mathrm{PGL}_3(\bar{F})$. If $I \subset \bar{C}'$ then I must be the inflection points and $g \in H$. If $g \in S_t$ then $g \circ \iota = \iota \circ g$ by 2 and the fact that both sides agree on I. But no element of S_t acts trivially on I, so all these embeddings are distinct. □

We defined $\psi : S \to E(I)$ above. Fix an embedding $I \subset \mathbb{P}^2_{\bar{F}}$. Assume $\bar{C} \supset I$. We want to extend ψ to an identification of S and $E(I)$ with the three torsion points of $E(\bar{C}) = E$, the Jacobian of the genus one curve $\bar{C}$. Given $\iota : I \to \bar{C}$, there is an induced $E(\iota) : E(I) \to E$ given by $(P,Q) \to P - Q$. Recall that S_t was the stabilizer of $\bar{C}$. Then we remarked above that S_t/S was the automorphism group of $E(\bar{C})$ because the translation action of S on $\bar{C}$ becomes trivial on $E(\bar{C})$.

We define $\phi_C : S \to E(\bar{C})$ as $E(\iota) \circ \psi$ where $\iota : I \to \bar{C}$ is a choice of embedding. That is, $\phi_C(s) = s(\iota(P)) - \iota(P)$ for any choice of P in I.

LEMMA 10. *Assume $I \subset \bar{C} \subset \mathbb{P}^2_{\bar{F}}$ is given with ficed embeddings. Then there is an induced injective group homomorphism $\phi_C = E(\iota) \circ \psi : S \to E(I) \to E(\bar{C}) = E$ and ϕ_C preserves the pairings.*

PROOF. As a composition of homomorphisms, ϕ_C is clearly a homomorphism. That ϕ_C preserves the pairing follows from the decription of the Weil pairing from [**8**, p. 98]. ϕ_C is injective because if $\phi_C(s) = 1$ then $s(P) = P$ for one and hence all P. Since $E(I)$ acts faithfully, $s = 1$. □

Since $\psi : S \cong E(I)$ is fixed, the faithful action of $\mathrm{SL}_2(F_3)$ on $E(I)$ translates to an action on S. If $\phi, \phi' : S \to E(\bar{C})$ are two morphisms, we define $\phi \equiv \phi'$ to mean that $\phi' = \phi \circ s$ where $s \in S_t/S \subset \mathrm{SL}_2(F_3)$. Let $[\phi]$ be the equivalence class of ϕ.

LEMMA 11. *If $\bar{C} \subset \mathbb{P}^2(\bar{F})$ has some unspecified embedding $I \subset \bar{C}$, then $\bar{C}$ uniquely defines $[\phi_C]$.*

PROOF. Two choices of embeddings $\iota, \iota' : I \to \bar{C}$ differ by an element $s \in S_t$ the stabilizer of $\bar{C}$. □

Now suppose $I \subset \bar{C}$ and $I \subset \bar{C}'$ where $\bar{C} \cong \bar{C}'$, so $\bar{C}' = g(\bar{C})$ for $g \in H \subset PGL_3(\bar{F})$ the stabilizer of I and so g induces an isomorphism. There is an induced isomorphism $E(g) : E(\bar{C}) \cong E(\bar{C}')$ but as g is not unique, $E(g)$ is not unique. However, two isomorphisms $E(\bar{C}) \cong E(\bar{C}')$ can only differ by an automorphism of $E(\bar{C}) = S_t/S$.

PROPOSITION 3. *There is a one to one correspondence between equivalence classes of $\phi : S \to E(\bar{C})$ and $S \to E(\bar{C}')$. If $[\phi_C] = [\phi_{C'}]$ then $\bar{C} = \bar{C}'$ as subvarieties of $\mathbb{P}^2(\bar{F})$.*

PROOF. Since the automorphism group of $E(\bar{C})$ acts faithfully on the three torsion subgroup, and since ϕ is a homomorphism, it is clear that if $s \in S_t/S$, then $\phi \circ s = s \circ \phi$. Thus the ambiguity in the identification of $E(\bar{C})$ and $E(\bar{C}')$ leaves $[\phi]$ unaffected.

ϕ_C and $\phi_{C'}$ both preserve the pairing, as does any element of S_t/S. Thus if $\bar{C}$ and $\bar{C}'$ both contain I, ϕ_C and $\phi_{C'}$ differ by an element of $\mathrm{SL}_2(F_3)$. That is, $\phi_{C'} = s \circ \phi_C$ where $s \in \mathrm{SL}_2(F_3)$. Of course s is not uniquely determined, but it unique modulo S_t/S. On the other hand, if we choose embeddings $\iota : I \to \bar{C}$ and $\iota' : I \to \bar{C}$, then $\iota' = s' \circ \iota$ where $s' \in H$ the stabilizer of I. Thus $E(\iota') = \bar{s}' \circ E(\iota)$ where $\bar{s}' \in SL_2(F_3)$ is the image of s' as in Theorem 1. If $\phi_C = \phi_{C'}$ then $s = 1$ and s' is in S_t which is the priemage of S_t/S the automorphism group. Then $\bar{C}' = s'(\bar{C}) = \bar{C}$. □

4. General F

Now assume F is a general field of characteristic not 2 or 3. Let $\bar{F}$ be the separable closure and $\bar{G}$ the Galois group of $\bar{F}/F$. It will be useful to consider fields $F \subset K \subset \bar{F}$ where K/F is Galois with group G. If, say, C is a structure defined over F, we let C_K mean the extension of scalars. If $C \subset \mathrm{SB}(A)$ (and hence defined over F) then so is $E = E(C)$ and $I \subset \mathrm{SB}(A)$, but E maybe defined over fields where C is not. It is equally true that $I \subset \mathrm{SB}(A)$ may be defined over F and $C_K \supset I_K$ but C not defined over F. Let C above have stabilizer S_t, and distinguished $S \subset S_t$. G acts on $K \times_F \mathrm{SB}(A)$ and $K \times_F \mathrm{SB}(S^3(A)) = P^9_K$, as well as on A^*_K and A^*_K/K^*. If $K = \bar{F}$, we can identify A_K with $M_3(\bar{F})$ and A^*_K/K^* with $PGL_3(\bar{F})$ and so $M_3(\bar{F})$ and $PGL_3(\bar{F})$ have induced $\bar{G}$ actions but not the obvious ones. Rather these are the actions where $M_3(\bar{F})^{\bar{G}} = A$ and $(PGL_3(\bar{F}))^{\bar{G}} = A^*/F^*$. Similarly, $\bar{G}$ acts on $\mathbb{P}^2(\bar{F})$ with quotient $SB(A)$. If $\sigma \in \bar{G}$, $g \in PGL_3(\bar{F})$ and $P \in \mathbb{P}^2(\bar{F})$, then $\sigma(g)(\sigma(P)) = \sigma(g(P))$. Recall the line bundle $\mathcal{O}(1)_{\mathbb{P}^9}$ descended to a line bundle L on $SB(A)$ and $\bar{L}$ on $\mathbb{P}^2(\bar{F})$. Of course, the global sections of $\bar{L}$ have a $\bar{G}$ action whose invariants are the global sections of L. The curves we are studying are exactly the zeroes of global sections of L.

If $C \subset SB(A)$, then G preserves I as a set. That is, I is a zero dimensional reduced subvariety of $\mathrm{SB}(A)$. Moreover, $\bar{G}$ preserves I as a set if and only if it preserves the associated $S \subset \mathrm{PGL}_3(\bar{F})$ as a set (not elementwise). Since $\bar{G}$ preserves C, it preserves S_t, but may act nontrivially on S_t/S because $\bar{G}$ may act nontrivially on the automorphism group of $E(C)$.

We fix the embedding $I \subset \mathrm{SB}(A)$ and call it a. We also use a to denote the induced embedding $\bar{I} \subset \mathbb{P}^2(\bar{F})$. If $\sigma \in \bar{G}$, then $\sigma \circ a = a \circ \sigma$ because a is defined over F. Since the set of lines in $\mathbb{P}^2$ are preserved by G, G preserves colinearity and there is an induced map $G \to \mathrm{Aff}(I)$. If $S \subset PGL_3(\bar{F})$ is preserved by $\bar{G}$, then G also preserves the pairing $S \times S \to \langle \rho \rangle$ where $\bar{G}$ might act nontrivially on $\langle \rho \rangle$.

Recall that given $I_K \subset C_K \subset SB(K \otimes_F A)$, we have an induced $\phi_C : S \to E(C_K)$ defined as the composition $S \to E(I_K) \to E(C_K)$. We want to use ϕ_C to understand how G or $\bar{G}$ acts on the set of C_K. More precisely, we will use the Galois module structure of S and $E(C_K)$ and whether or not this is preserved by ϕ_C.

If $\bar{G}$ preserves I, then it permutes the set of $\bar{C}$'s containing I. In particular, it permutes the finite set of $\bar{C}$'s that are all isomorphic. Then $\bar{G}$ also acts on $E(I)$ and we have defined $\psi : S \to E(I)$ independent of any $\bar{C}$. That is, ψ preserves the action of $\bar{G}$. Thus ϕ_C preserves Galois actions if and only if $E(I) \to E(C_K)$ does. However, if $E(C)$ is defined over F, then G acts on $E(C)$. We prove:

PROPOSITION 4. *Assume $I \subset SB(A)$ is such that $\bar{I} \subset \mathbb{P}^2(\bar{F})$ is a set of inflection points for a smooth degree 3 curve $\bar{C} \subset \mathbb{P}^2(\bar{F})$.*

(a) $\psi \circ \sigma = \sigma \circ \psi$.
(b) *If $C \subset SB(A)$ contains I, then* $\sigma \phi_C \sigma^{-1} = \phi_C$.

PROOF. Recall that $\psi(s') = (s'(Q), Q)$ for any choice of Q. For example, we can choose $Q = \sigma(P)$. Then $\psi \circ \sigma(s) = (\sigma s \sigma^{-1}(\sigma(P), \sigma(P)) = \sigma((s(P), P)) = \sigma \circ \psi(s)$ which proves (a).

As for (b), the assumptions imply that $\iota : I \to \bar{C}$ is preserved by $\bar{G}$ and (b) follows from (a). □

In other words, if $I \subset C$ are both defined over F, which we will observe means that $\bar{G}$ preserves C, then ϕ_C preserves the Galois actions on its domain and range. It will be vital for us to prove a converse:

PROPOSITION 5. *Let $\bar{C} \supset I$ be a smooth cubic curve such that $E(\bar{C})$ is defined over F. Suppose that for all $\sigma \in \bar{G}$, $\sigma \circ \phi_C \circ \sigma^{-1} = \phi_C \circ t$ for some $t \in S_t/S$ as set maps. Then $\bar{C}$ is $\bar{G}$ invariant and defines $C \subset SB(A)$.*

PROOF. Let $\sigma \in \bar{G}$ and define $\iota' = \sigma \circ \iota \circ \sigma^{-1}$ which defines an embedding $I \to \bar{C}'$ where $\bar{C}' = \sigma(\bar{C})$ (note that σ preserves I). Since $\bar{C} \cong E(\bar{C})$ is defined over F, $\bar{C}'$ and $\bar{C}$ are isomorphic over $\bar{F}$. Let $b : \bar{C} \to \mathbb{P}^2(\bar{F})$ be the given embedding of $\bar{C}$ in $\mathbb{P}^2(\bar{F})$, so $a = b\iota$ is the given fixed embedding of $\bar{I}$ in $\mathbb{P}^2(\bar{F})$. Then a is fixed by $\bar{G}$, so $a = b\iota = \sigma \circ a \circ \sigma^{-1} = (\sigma \circ b \circ \sigma^{-1}) \circ (\sigma \circ \iota \circ \sigma^{-1})$ and $\sigma \circ b \circ \sigma^{-1}$ is an embedding for $\bar{C}'$ and defines a $\phi_{C'}$. Our assumption is that $[\phi_{C'}] = [\phi_C]$ and so $\bar{C} = \bar{C}'$. That is, $\bar{G}$ preserves $\bar{C}$ as a set. If $\bar{C}$ is the zero set of some $q \in \bar{L}$, then $\bar{G}$ preserves $\bar{F}q$ and hence, by Hilbert 90, there is a $q \in L$ whose zeroes $C \subset SB(A)$ define $\bar{C}$ over F. □

We can now state and prove our main theorem.

THEOREM 5. *Let E be an elliptic curve over F. Let A/F be a degree-three algebra. Let K/F be the Galois extension with group G obtained by adjoining all the three torsion points $E[3]$ to F. There is a $C \subset \mathrm{SB}(A)$ with $E(C) = E$ if and only if there is a subgroup $S \subset (A \otimes_F K)^*/K^*$ such that S is preserved by G, S is generated by x, y with preimage x', y' such that $x'y' = \rho y'x'$, and S is isomorphic to $E(C)[3]$ as G modules with pairing.*

PROOF. We restrict ourselves to the $PGL_3(\bar{F})$ orbit associated with E. If C is as given, then C defines $I \subset \mathrm{SB}(A)$ and so an associated $S \subset \mathrm{PGL}_3(\bar{F})$. By Lemma 4, $\phi_C : S \to E[3]$ is a G morphism. Note that, in particular, the elements of S are fixed by the Galois group of $\bar{F}/K$ and so $S \subset (A \otimes_F K)^*/K^*$.

Conversely, suppose $\phi : S \to E[3]$ is a G isomorphism preserving the pairing. Let S define I and let $\bar{C} \supset I$. We would like to prove there is a $g \in N(S)$ such that $g(\bar{C})$ comes from a curve in $SB(A)$. Let $\bar{g} = \phi_C^{-1} \circ \phi \in SL_2(F_3)$ and lift $\bar{g}$ to $g \in N(S)$. Note that $\bar{g}$ has determinant one because ϕ preserves the pairing. Now $\phi = \phi_C \circ g = \phi_{g(C)}$ preserves the $\bar{G}$ action so we are done by Proposition 5. □

Assume I is defined over F. It is clear that the elements of $\bar{G}$ act as affine transformations of I and so there is a homomorphism $\bar{G} \to \mathrm{Aff}(I)$. Let us consider the possible images of $\bar{G}$ in $\mathrm{Aff}(I)$. Set G_0 to be the image of $\bar{G}$ in $GL_2(F_3)$. If $I \subset C$ and C is defined over F, then G_0 is the Galois group of K/F where K is the field defined by adjoining all the three torsion points of $E(C)$. So if $\bar{H}$ is the kernel of $\bar{G} \to G_0$, then $\bar{H}$ is the Galois group of $\bar{F}/K$.

If the primitive third root ρ is in F, then $G_0 \subset SL_2(F_3)$. Otherwise, the image of $G_0 \to GL_3(F_3) \to F_3^*$, the second map being the determinant, is the Galois group of $F(\rho)/F$. Other than this, we cannot restrict G_0 in any way.

Of course, $\bar{H}$ maps to $S \cong E(I)$ with image we call H_0. If $\bar{J}$ is the kernel of $\bar{H} \to H_0$, then $\bar{F}^{\bar{J}}$ is $K(I)$, obtained by adjoining the inflection points themselves to K. Note that since S acts by translation, adjoining one point of I over K is equivalent to adjoining them all.

Suppose I corresponds to $S \subset A_K^*/K^*$. Let $x, y \in S$ be a basis and lift to $\tilde{x}, \tilde{y}$. Then I consists of $L_x \cap L_y$ where $L_x \subset \bar{F}^3$ is a two-dimensional space preserved by

$\tilde{x}$, and similarly for L_y. Summing over the L_y, we have that all L_x (and L_y) are defined over $K(I)$. That is, $\tilde{x}$ and $\tilde{y}$ have degree three equations which split. We will say, in this case, that x (and y) splits.

We can say this another way. Taking cubes defines an injective morphism $S/K^* \to K^*/(K^*)^3$. If $T \subset K^*/(K^*)^3$ is the image, we can form the field $K(T^{1/3})$ of all cube roots and let $T^{1/3} \subset K(T^{1/3})$ be the cube roots of all elements in T. By Kummer theory the Galois group of $K(T^{1/3})/K$ is $\mathrm{Hom}(T^{1/3}/K^*, \mu_3)$. The cube and cuberoot maps compose to form a well-defined homomorphism $S \to T^{1/3}/K^*$. Using the pairing on S and duality, we get a homomorphism $\mathrm{Hom}(T^{1/3}/K^*, \mu_3) \to S$ and the composition $\bar{H} \to \mathrm{Hom}(T^{1/3}/K^*, \mu_3) \to S$ can be easily seen to be the map above.

LEMMA 12. $K(I) = K(T^{1/3})$. *The Galois group of $K(I)/K$ has degree 1,3, or 9 and if it is not 9 then $A_K \cong M_3(K)$. If $K(I)/K$ has degree 3, then no point of I is rational over K nonetheless.*

PROOF. Over $K(T^{1/3})$ both x and y split and so I is defined over this field. Conversely, $K(T^{1/3})$ is clearly the smallest field where x splits for any $x \in S$. The Galois group of $K(I)/K$ is a subgroup of S. If $K(T^{1/3})/K$ has degree 1 or 3, then for some nontrivial $\tilde{z} = \tilde{x}^a\tilde{y}^b$ we have $\tilde{z}^3$ is a cube which forces the splitting of A_K. □

Let G_1 be the image of $\bar{G}$ in $\mathrm{Aff}(I)$, so we have an exact sequence $1 \to H_0 \to G_1 \to G_0 \to 1$ which is induced by the split exact sequence $1 \to E(I) \to \mathrm{Aff}(I) \to GL_2(F_3) \to 1$. In fact, the first sequence also splits which is almost but not quite immediate. If H_0 is trivial there is nothing to prove, and if $H_0 \cong E(I)$ the splitting is immediate. If H_0 has order three, then this extension defines an $\alpha \in H^2(G_0, H_0)$ and it also is the case that $\alpha = 0$ as follows. If nontrivial, C_3 is the 3-Sylow subgroup of G_0 and it suffices to observe that α restricts to 0 in $H^2(C_3, H_0)$. This follows because one can check that $\mathrm{Aff}(I)$ has no elements of order 9.

We further investigate the possibilities. In general we know that G_0 can be any subgroup of $GL_2(F_3)$, and if $H_0 = 1$ or $H_0 \cong E(I)$ there is nothing more to say. When H_0 has order 3 we can restrict G_0 further. Then $\tilde{S}$ is generated by $\tilde{s}$ and $\tilde{y}$ where $\tilde{y}$ is split and $\tilde{x}$ is not. Clearly A_K is split. G_0 acts on H_0 via conjugation and clearly G_0 preserves the cyclic subgroup generated by y. If $\rho \in F$, this implies that $G_0 = 1$, G_0 is a subgroup of the cyclic group C_6. If $\rho \notin F$, then G_0 is a subgroup of $G_0 = S_3 \oplus C_2$ containing the central subgroup C_2.

We want to use Theorem 5 to describe the C that appear for a $E(C)$ given by that theorem. Recall that any such C is a principal homogeneous space over $E(C)$, and all such spaces are classified by the Galois cohomology group $H^1(\bar{G}, E(\bar{C}))$. Note that this is the correct cohomology group because $E(\bar{C})$ is the group of automorphisms of $\bar{C}$ as a principal homogeneous space over $E(\bar{C})$.

More precisely, let $\bar{C}$ be such a space and let P be a point on $\bar{C}$. Define $e : \bar{G} \to E(\bar{C})$ by setting $e(\sigma) = \sigma(P) - P \in E(\bar{C})$. It is easy to check that e is a one cocycle. Note that if we pick another point Q on $\bar{C}$ and use that to define $e' : \bar{G} \to E(\bar{C})$, then $e(\sigma) - e'(\sigma) = \sigma(\alpha) - \alpha$ for $\alpha = P - Q$ the element of $E(\bar{C})[3]$. Thus $\bar{C}$ defines a class in $\gamma' \in H^1(\bar{G}, E(\bar{C}))$.

In our situation, it makes sense to define the cycle using an inflection point $P \in I \subset \bar{C}$. Furthermore, let $\phi : S \to E(\bar{C})[3]$ be the Galois preserving isomorphism. Note that ϕ is a ϕ_C from the previous section, so there is an embedding $I \subset \bar{C}$

such that $\phi(s) = s(P) - P$ for any choice of P. It is clear that any $\sigma \in \bar{G}$ defines an affine map on I, so we have a homomorphism $\Phi : \bar{G} \to \text{Aff}(I)$.

The composition $\bar{G} \to \text{Aff}(I) \to GL_2(F_3)$ defines the action of $\bar{G}$ on $E(I)$ and hence on $E(\bar{C})[3]$ and hence on S. Let $\gamma \in H^1(\text{Aff}(I), E(I))$ be the canonical class defined in Proposition 1.

THEOREM 6. *$\Phi^*(\gamma) \in H^1(\bar{G}, E(\bar{C})[3]))$ maps to $\gamma' \in H^1(\bar{G}, E(\bar{C}))$ and γ' defines $\bar{C}$ as a principal homogeneous space over $E(C)$.*

The connection between $\bar{C}$ and the algebra A is stated in [**5**] and briefly recalled next. Since $S \subset PGL_3(\bar{F})$, we have a composition $\eta : H^1(\text{Aff}(I), E(I)) \to H^1(\bar{G}, E(\bar{C})[3]) \to H^1(\bar{G}, PGL_3(\bar{F})) \to H^2(\bar{G}, \bar{F}^*)$ where, we recall, the last map is not a homomorphism because the cohomology set $H^1(\bar{G}, PGL_3(\bar{F}))$ is not a group.

PROPOSITION 6. *$\eta(\gamma) \in H^2(\bar{G}, F^*)$ defines the Brauer class of A/F. Thus if $C \subset \text{SB}(A)$, then C is defined by the image of a $\gamma' \in H^1(\bar{G}, E(\bar{C})[3])$ which maps to the Brauer class of A/F.*

PROOF. We have a given action of $\bar{G}$ on $M_3(\bar{F})$ such that the invariant ring is A and such that the quotient of the action on $\mathbb{P}^2(\bar{F})$ is $\text{SB}(A)$. If $C \subset \text{SB}(A)$ and $P \in I \subset \bar{C}$, then P defines the cocycle $e(\sigma) : \bar{G} \to E(\bar{C})[3])$. Then $e(\sigma)^{-1}\sigma$ defines a new action on $\mathbb{P}^2(\bar{F})$ which fixes P and hence has quotient variety $\mathbb{P}^2(F)$, so a new invariant ring $M_3(F)$. □

5. Examples

Now we start giving examples, including examples where an $E(C)$ does not appear. We start with the easiest corollary of Theorem 5. Let E be an elliptic curve defined of F and $K \supset F$ be the field gotten by adjoining all the three torsion points of E. Assume $K = F$ or $K = F(\rho)$ where ρ is a third root of one. The next result is proven in [**5**] and [**6**] for the $K = F$ case and in the preprint [**1**] for the $K = F(\rho)$ case, by different means.

COROLLARY 2. *If E/F is an elliptic curve where all three torsion points are defined over $F(\rho)$, and A/F is a degree-three central simple algebra, then there is an $C \subset \text{SB}(A)$ such that $E(C) = E$.*

PROOF. When $K = F$, this is immediate because A/F is cyclic. When $K = F(\rho)$, we note the following. First, if σ generates the Galois group of K/F, then E has a three torsion point P such that $\sigma(P) = P$ and a three torsion point Q such that $\sigma(Q) = -Q$. If $L \supset K$ is such that L/F is dihedral, then $L = L' \otimes_F K$ where $L' = F(a^{1/3})$ and σ fixes $a^{1/3}$. If M'/F is cyclic of degree 3 and $M = M' \otimes K$, then $M = K(b^{1/3})$ where $\sigma(b^{1/3}) \in b^{-1/3}F^*$. Again, E is the Jacobian of some $C \subset SB(A)$ by 5 and the fact that A/F is cyclic. □

Let K still be the extension field of F obtained by adjoining all the three torsion points of $E(C)$. To begin with, assume K/F is cyclic of degree three. If C exists, there is an $S \subset (A \otimes_F K)^*/K^*$ generated by images of $x', y' \in A \otimes_F K$ such that $x'y' = \rho y'x'$ and S is preserved by $G = \langle \sigma \rangle$. It follows that we can choose x', y' such that $\sigma(y') \in y'K^*$ and $\sigma(x') \in x'y'K^*$. Suppose $\sigma(y') = y'z$. Since σ has order three on $A \otimes_F K$, we have $z\sigma(z)\sigma^2(z) = 1$ and so $z = \sigma(u)/u$ for $u \in K^*$. Replacing

y' by $y'u^{-1}$, we may assume $\sigma(y') = y'$ or $y' \in A$. Now write $\sigma(x') = x'y'w$ for $w \in K^*$. Again using that σ has order three, we have

$$x' = \sigma^3(x') = \sigma^2(x'y'w) = \sigma(x'y'wy'\sigma(w)) = x'y'wy'\sigma(w)y'\sigma^2(w)$$

or $1 = y'^3 w\sigma(w)\sigma^2(w)$. That is, A has the form (a, b) where $(a, K/F)$ is trivial. Said another way, A is split by an $F(a^{1/3})$ such that $(a, K/F)$ is split.

The above condition is actually necessary and sufficient.

PROPOSITION 7. *Let A/F, E and K/F be as above. Then there is a $C \subset \mathrm{SB}(A)$ with Jacobian E if and only if A is split by $F(a^{1/3})$ where $(a, K/F)$ is trivial.*

PROOF. We saw one direction above. Suppose A is split by $F(a^{1/3})$ and $a = N_{K/F}(w)$. Then $N_{K/F}(a/w^3) = a^3/N_{K/F}(w)^3 = 1$. Thus there is a $b' \in K$ with $\sigma(b')/b' = a/w^3$. Let B'/K be the degree-three cyclic algebra generated by x', y' where $x'^3 = a$, $y'^3 = b'$, and $x'y' = \rho y'x'$. The action of σ on K extends to B' by setting $\sigma(x') = x'$ and $\sigma(y') = y'x'w^{-1}$. Then as above, $\sigma^3(y') = \sigma^2(y'x'w^{-1}) = \sigma(y'x'w^{-1}x'\sigma(w^{-1})) = y'x'w^{-1}x'\sigma(w^{-1})x'\sigma^2(w^{-1}) = y'aN_{K/F}(w^{-1}) = y'$. It follows that this extension of σ has order 3 on B' and thus the invariant algebra B is central simple of degree three over F. Since B contains x' it is split by $F(a^{1/3})$. Consider $B^\circ \otimes A$ which is also split by $F(a^{1/3})$ and hence is represented by an algebra $A' = (a, c)$ over F. Then $A \otimes_F K$ is similar to $D = B' \otimes_K (a, c) = (a, b')_K \otimes (a, c)_K \sim (a, b'c)_K$. More precisely, let $e \in F(a^{1/3}) \otimes_F F(a^{1/3})$ be the separating idempotent. Then e can be viewed as an element of D, $A_K = eDe$, and e is σ fixed. Let $x'', y'' \in (a, c)$ be the elements with $x''^3 = a$, $y''^3 = c$, and $x''y'' = \rho y''x''$. Set $\alpha = e(x' \otimes 1)e$, $\beta = e(y' \otimes y'')e$, both elements in $eDe = A_K$ (but remember e is the identity there). Clearly, $\alpha^3 = a$ because e commutes with x'. For the same reason, $\alpha\beta = \rho\beta\alpha$. Also, $\sigma(\beta) = e(\sigma(y') \otimes \sigma(y''))e = e(y'x'w^{-1} \otimes y'')e = (e(y' \otimes y'')(x'w^{-1} \otimes 1))e = e(y' \otimes y'')ee(x'w^{-1} \otimes 1)e = \beta\alpha w^{-1}$. If S is the image of α, β in $(A \otimes_F K)^*/K^*$ we are done. □

As a route to a counterexample (that is, a case where $E(C)$ does not appear), let F be a field complete with respect to a discrete valuation ring with prime π. Suppose K/F ramifies at a prime π and A/F is a division algebra unramified at π. Then any norm of K/F has the form $u^3\pi^r$ and if r is prime to 3, $F(\pi^{r/3})$ cannot be a subfield of the unramified A. Thus if K/F, and E are as in Proposition 7, E cannot appear as an $E(C)$ for $C \subset \mathrm{SB}(A)$. More generally, if F is a discrete valued field with prime π and the above hold over the completion F_π, then again E does not appear. For a specific example, according to Bruce Jordan, let E be the elliptic curve $y^2 + y = x^3 - x^2 - 10x - 20$. This is $X(11)$. If $K/\mathbb{Q}$ is obtained by adjoining the three torsion points, then $K/\mathbb{Q}$ ramifies over $\mathbb{Q}$ only at 3, and 11 and with ramification index 3 or 6. Thus there is an $F \supset \mathbb{Q}$ such that K/F is cyclic of degree three and ramifies at a prime p extending 11. Note that F must contain ρ. Furthermore, over K_p/F_p is ramified and obtained by adjoining the three torsion points of E to F_p. Now the p-adic fields have no unramified division algebras but if we adjoin an indeterminate we can create one. The degree-three algebra $A = (K_p(t)/F_p(t), t)$ over $F_p(t)$ is unramified and unsplit at p. It follows that E does not appear as $E(C)$ for a $C \subset \mathrm{SB}(A)$.

One might hope that when K/F is more general, there might be a result similar to Proposition 7, that is, a criterion with a cohomological flavor. The next result makes clear there can be an obstruction of a more arithmetic nature, involving the non-appearance of the dihedral group of order 6 as a Galois group.

COROLLARY 3. *Suppose E/F is an elliptic curve and K/F is the field obtained by adjoining all the three torsion points of E. Assume K/F is cyclic of order 2 and F contains ρ. Let D/F be a division algebra of degree three. Then there is a $C \subset \mathrm{SB}(D)$ with $E(C) = E$ if and only if there is a field L/K splitting D such that L/F is Galois with group S_3, the dihedral group of order 6.*

PROOF. Let σ generate the Galois group of K/F. Assume $S \subset (D \otimes K)^*/K^*$ exists as in Theorem 5. Then σ acts as -1 on S. If L/K is any of the cyclic field extensions coming from S, then L/F is dihedral Galois.

Conversely, suppose such an L/K exists. Then $D \otimes_F K = (a, b)_K$ where $\sigma(a) = a^{-1}z^3$ for $z \in K^*$. Since D is fixed by σ, $\sigma(b) = b^{-1}N(u)$ where $u \in L^*$ and $N : L^* \to K^*$ is the norm map. There is a surjection $K^*/(K^*)^3 \to K^*/N(L^*)$ which is a morphism of modules over the group ring $F_3[\langle\sigma\rangle]$. Let $\bar{b} \in K^*/N(L^*)$ be the image of b. Since this group ring is semisimple, there is a preimage $\tilde{b} \in K^*/(K^*)^2$ such that $\tilde{b}$ maps to $\bar{b}$ and $\sigma(\tilde{b}) = (\tilde{b})^{-1}$. That is, there is a $b' \in K^*$ such that b' maps to $\tilde{b}$ which maps to $\bar{b}$. That is, $b' = bN(u')$ for some $u' \in L^*$ and $\sigma(b') = b'^{-1}w^3$ for some $w \in K^*$. We can write $D \otimes_F K = (a, b')_K$ and so $D \otimes_F K$ contains x, y with $x^3 = a$, $y^3 = b$, $\sigma(x) = x^{-1}z$, $\sigma(y) = y^{-1}w$ and $xy = \rho yx$. If $S \subset (D \otimes_F K)^*/K^*$ is generated by the image of x and y, then S is as needed. □

The above result can be used in two ways to get interesting examples. Let K be a finite extension of some $\mathbb{Q}_p$, and E an elliptic curve over K with the following properties. First, suppose p is prime to three and K contains a primitive third root of one. In addition, assume all the three torsion points of E are rational over K. It follows that there is no L/K, an extension of fields such that the Galois group is S_3. Let K'/K be a quadratic extension of fields and E'/K the corresponding quadratic twist. Then $\mathrm{Gal}(K'/K)$ acts on the three torsion points as -1. Let D/K be either of the degree-three division algebras. $D \otimes_K K'$ cannot have the required S because of the following. Suppose S existed, and let L/K' be a cyclic extension defined by a rank one subgroup of S. Then L/K would have Galois group S_3, a contradiction. Thus E' does not appear as an $E(C)$ for $C \subset \mathrm{SB}(D)$ but E does.

The above is an example where there are two elliptic curves with the same j invariant where only one of them appears as an $E(C)$ for $C \subset \mathrm{SB}(D)$. For a different kind of example, assume F is a number field containing ρ and E/F is an elliptic curve such that the field, K, formed by adjoining all the three torsion is such that K/F is cyclic Galois of degree four with $\langle\sigma\rangle$ as Galois group. Note that this property is also true for all the quadratic twists of E over F. By density, there are infinitely many primes p of F such that $K_p = K \otimes_F F_p$ is a field where F_p is the completion. For all but finitely many of these primes p, for all finite $K' \supset F_p$, there is no Galois extension of fields L/K' with Galois group S_3. Assume $D \otimes_F F_p$ is a division algebra for one of these primes p, and K'_p is such that $F_p \subset K'_p \subset K_p$ and K_p/K'_p has Galois group $\langle\sigma^2\rangle$. If $S \subset D^*/F^*$ existed as in Theorem 5, the same would be true for $(D \otimes_F K_p)^*/K_p^*$. Since σ^2 acts on S as -1, this is a contradiction. Hence for such D and E, there are no $C \subset \mathrm{SB}(D)$ such that $E(C)$ has the same j invariant as E.

References

[1] B. Antieau and A. Auel, *Explicit descent on elliptic curves and splitting Brauer classes*, Preprint.

[2] M. Artin, *Brauer-Severi varieties*, Brauer groups in ring theory and algebraic geometry (Wilrijk, 1981), Lecture Notes in Math., vol. 917, Springer, Berlin-New York, 1982, pp. 194–210. MR657430

[3] I. V. Dolgachev, *Classical algebraic geometry: A modern view*, Cambridge University Press, Cambridge, 2012, DOI 10.1017/CBO9781139084437. MR2964027

[4] C. O'Neil, *Jacobians of genus one curves*, Math. Res. Lett. **8** (2001), no. 1-2, 125–140, DOI 10.4310/MRL.2001.v8.n2.a3. MR1825265

[5] C. O'Neil, *The period-index obstruction for elliptic curves*, J. Number Theory **95** (2002), no. 2, 329–339, DOI 10.1016/S0022-314X(01)92770-2. MR1924106

[6] C. O'Neil, *Erratum to: "The period-index obstruction for elliptic curves" [J. Number Theory* **95** *(2002), no. 2, 329–339; MR1924106]*, J. Number Theory **109** (2004), no. 2, 390, DOI 10.1016/j.jnt.2004.07.006. MR2106488

[7] D. J. Saltman, *Lectures on division algebras*, CBMS Regional Conference Series in Mathematics, vol. 94, Published by American Mathematical Society, Providence, RI; on behalf of Conference Board of the Mathematical Sciences, Washington, DC, 1999, DOI 10.1090/cbms/094. MR1692654

[8] J. H. Silverman, *The arithmetic of elliptic curves*, 2nd ed., Graduate Texts in Mathematics, vol. 106, Springer, Dordrecht, 2009, DOI 10.1007/978-0-387-09494-6. MR2514094

[9] A. Várilly-Alvarado and B. Viray, *Smooth compactifications of certain normic bundles*, Eur. J. Math. **1** (2015), no. 2, 250–259, DOI 10.1007/s40879-015-0035-7. MR3386237

Center for Communications Research, 805 Bunn Drive, Princeton, New Jersey 08540

Email address: saltman@idaccr.org

Contemporary Mathematics
Volume **800**, 2024
https://doi.org/10.1090/conm/800/16063

More on skew braces and their ideals

Agata Smoktunowicz

Abstract. This paper introduces the notion of a strongly prime ideal, and shows that the largest solvable ideal in a finite brace equals the intersection of all strongly prime ideals in this brace. This is used to generalise some well-known results from ring theory into the context of braces and pre-Lie algebras. Several open questions are also posed.

Prime ideals are an important part of noncommutative ring theory. It is known that matrix rings over prime rings are prime, and the Amitsur-Levitski theorem assures that the ring of n by n matrices over a commutative ring satisfies the standart polynomial of degree $2n$ [**4**]. Prime ideals play an important role in the theory of polynomial identities, for example, a result of Small assures that a ring satisfying a polynomial identity and the ascending chain condition on two-sided ideals satisfies the descending chain condition on prime ideals [**30**]; in particular, it holds for commutative rings [**22**]. Prime rings are used for constructing Martindale rings of quotients, and they appear in the theory of rings with generalised polynomial identities. Goldie's theorem assures that prime Noetherian rings have Artinian rings of quotients. The Faith-Utumi theorem permits even better characterisation of prime Noetherian rings, 'sandwiching' them between two matrix rings.

Subsequently, the notion of prime ideal was generalised to various other structures. For example, prime ideals in tropical mathematics have been investigated in [**21**]. In [**34,35**], Zelmanov obtained fundamental results on the structure theory of strongly prime Jordan algebras. Moreover, in [**3**] Amitsur defined the prime radical for non-associative algebras and generalised Levitsky's theorem that the prime radical equals the intersection of all prime ideals in all associative rings. In this paper, we introduce strongly prime ideals for braces and pre-Lie algebras. Our approach is similar to that of Amitsur and Levitsky, but we define N-sequences differently so as to make a better fit with braces and pre-Lie algebras. We consider strongly prime ideals instead of prime ideals for this same reason.

We consider a different model to that in [**25**], which also generalised the notion of a prime ideal from ring theory to braces. In [**25**], it was not possible to prove some structure theorems, so the notion of strongly prime braces is introduced in this paper to differentiate it from the notion of the prime brace considered there.

2020 *Mathematics Subject Classification.* Primary 16N60, 17D99, 17A65.

Key words and phrases. Braces, prime ideals, strongly prime ideals, Yang-Baxter equation.

The author is grateful for support from the EPSRC programme grant EP/R034826/1 and from the EPSRC research grant EP/V008129/1.

This allows us to show that the largest solvable ideal in a finite brace equals the interesection of all strongly prime ideals in this brace for example. Hence every finite brace without non-zero solvabe ideals can be embedded in a direct product of strongly prime braces. Recall that the notion of a solvable ideal was introduced for braces in [**6**]. An ideal I of a brace A is called solvable if $I_k = 0$ for some $k > 0$ where I_i is defined inductively by $I_1 = I$ and $I_{i+1} = I_i * I_i$. Various radicals have been introduced for braces by many authors. In [**28**], Rump introduced left and right nilpotent braces and radical chains $A^{i+1} = A * A^i$ and $A^{(i+1)} = A^{(i)} * A$ for a left brace A, where $A = A^1 = A^{(1)}$ (the original construction of Rump is for right braces, but we give the natural translation of it to left braces here).

Various other radicals in braces were subsequently introduced, in analogy with ring theory and group theory. Recall that solvable braces were introducted in [**6**], and in [**24**] the radical of a brace was introduced as the intersection of all maximal ideals in a given brace.

Various radicals have been defined for pre-Lie algebras, including the left nilpotent radical, the Koszul radical and Albert radical. Recall that an ideal I of a pre-Lie algebra A is called solvable if $I_k = 0$ for some $k > 0$, where I_i is defined inductively by $I_1 = I$ and $I_{i+1} = I_i \cdot I_i$. Another important radical in pre-Lie algebras, the Albert radical $\alpha(a)$, is defined to be the intersection of all maximal ideals M of A such that A^2 is not a subset of M. If such a maximal ideal does not exist then we let $\alpha(A) = A$ [**13**].

1. Background information

Braces were introduced by Wolfgang Rump circa 2005 to describe all involutive, non-degenerate set-theoretic solutions of the Young-Baxter equation. Subsequently, skew-braces were introduced by Guarnieri and Vendramin [**20**] to allow construction of non-involutive solutions of the Yang-Baxter equation. Braces and skew braces have also found applications in other research areas, such as Garside monoids, Hopf-Galois extensions, pre-Lie algebras, quantum integrable systems, groups with bijective 1-cocycles and regular subgroups of a holomorph, lattice theory,quasigroups, virtual knot theory; see, for example, [**5, 10, 14**–**19, 23, 29**]. Simple braces have been also investigated [**6, 12**].

It is known [**11, 28**] that if R is a Jacobson radical ring with operations $+, \cdot$, then R is a brace with the same operation $+$, and with the multiplication $a \circ b = a \cdot b + a + b$. Recall that a set B with binary operations $+$ and $*$ is a *left brace* if $(B, +)$ is an abelian group, and the following version of distributivity combined with associativity holds:

$$(a + b + a * b) * c = a * c + b * c + a * (b * c), a * (b + c) = a * b + a * c$$

for all $a, b, c \in B$. Moreover, $(B, \circ)$ is a group where we define $a \circ b = a + b + a * b$.

See [**28**] for the original definition. For a shorter equivalent definition using group theory, see [**11**]. In what follows we will use the definition in terms of operation '$\circ$' (presented in [**11**]): a set B with binary operations of addition $+$, and multiplication $\circ$ is a brace if $(B, +)$ is an abelian group, $(B, \circ)$ is a group and for every $a, b, c \in B$

$$a \circ (b + c) + a = a \circ b + a \circ c.$$

We will use the following commonly used definition of an ideal in a brace.

DEFINITION 1. [29] We say that I is an ideal in a brace $(B,+,\circ)$ if I is a subset of B and for each $a,b \in I$ and each $r \in B$ we have $a+b \in I, a-b \in I, r*a \in I, a*r \in I$.

Recall that $a \circ b = a*b+a+b$ for all $a,b \in B$.

We will use the following notation: Let I,J be subsets of a left brace A, then $I*J$ is the additive subgroup of A generated by elements $i*j$, where $i \in I$, $j \in J$.

Notice that if I,J are ideals in a brace A then $I*J$ need not to be an ideal in I; see [28] for concrete examples.

Similarly, if I,J are subsets of a pre-Lie algebra A then $I \cdot J$ is the additive subgroup of A generated by elements $i \cdot j$ where $i \in I$, $j \in J$.

DEFINITION 2. We say that I is an ideal in a pre-Lie algebra $(A, +, \cdot)$ if I is a subset of A,and for each $a,b \in I$ and each $r \in A$ we have $a+b \in I, a-b \in I, r \cdot a \in I, a \cdot r \in I$.

All braces and $\mathbb{F}$-braces considered in this paper are left braces.

1.1. Strongly prime ideals. In what follows, A is a brace with operations $+,\circ,*$, where $(A,+)$ is an abelian group, $(A,\circ)$ is a group, and, as usual, $a \circ b = a*b+a+b$. Moreover, $a*(b+c) = a*b+a*c$ for $a,b,c \in A$. We introduce a strongly prime brace.

DEFINITION 3. Let A be a brace. We say that a brace A is a strongly prime brace if every product of any number of non-zero ideals in A is non-zero. All the products use the operation $*$.

DEFINITION 4. Let A be a brace. We say that a brace A is a strongly semiprime brace if, for every non-zero ideal I in A, every product of any number of copies of I is non-zero. All the products use the operation $*$.

DEFINITION 5. Let A be a brace. We say that an ideal I in A is strongly prime if the factor brace A/I is a strongly prime brace.

DEFINITION 6. Let A be a brace. We say that an ideal I in A is strongly semiprime if the factor brace A/I is strongly semiprime.

2. N-sequences and N-ideals

Let A be a brace and $a \in A$. By $\langle a \rangle$ we will denote the smallest ideal in A containing a. Recall that for an ideal I of A, we define $I_1 = I$, and inductively $I_{n+1} = I_n * I_n$. So I is solvable if and only if $I_n = 0$ for some n.

DEFINITION 7. Let A be a brace and let $a \in A$. We define an N-sequence to be a sequence $a_1, a_2, a_3, \ldots$ of elements in A satisfying

$$a_{n+1} \in \langle a_n \rangle_{\alpha_n}$$

for some natural numbers $0 < \alpha_1 < \alpha_2 < \ldots$.

We will say that this N-sequence starts with element a_1.

Notice that

$$\langle a \rangle_1 = \langle a \rangle, \langle a \rangle_2 = \langle a \rangle * \langle a \rangle,$$
$$\langle a \rangle_3 = (\langle a \rangle * \langle a \rangle) * (\langle a \rangle * \langle a \rangle).$$

DEFINITION 8. Let A be a brace and let I be an ideal in A. We say that I is an N-ideal if every N-sequence in A starting with an element in I reaches zero.

LEMMA 9. *Let A be a brace. The sum of any number of N-ideals in A is an N-ideal in A.*

PROOF. Let I, J be two N-ideals in A. We will first show that $I + J = \{a + b : a \in I, b \in J\}$ is an N-ideal in I. Let $a_1, a_2, \dots$ be an N-sequence with $a_1 = a + b$ where $a \in I$ and $b \in J$. Consider the factor brace A/J and the N-sequence $a_1' = a_1 + J = a + J$, $a_2' = a_2 + J, \dots$ in A/J. Notice that every N-sequence in A starting with $a \in I$ will reach zero; therefore, every N-sequence in A/J will reach zero (because $(r + J) * (s + J) = r * s + J$ for all $r, s \in A$). It follows that $a_k' = 0$ for some k, hence $a_k \in J$ for some k. Observe that $a_k, a_{k+1}, a_{k+2}, \dots$ form an N-sequence in A starting with $a_1'' = a_k \in J$, and since J is an N-ideal we see that $a_n = 0$ for some n. This proves that $I + J$ is an N-ideal in A. Notice also that $I + J$ is an ideal in A by Lemma 2.7 [**25**].

Observe that this implies that the sum of a finite number of N-ideals is an N-ideal. Therefore the sum of any number of N-ideals is an N-ideal, as a_1 belongs to a sum of N-ideals if and only if it belongs to a sum of finite number of N-ideals; therefore, every N-sequence starting with a_1 reaches zero. □

2.1. Solvable ideals. Recall that the sum of a finite number of solvable ideals is a solvable ideal [**25**]. Similarly as for pre-Lie algebras [**7, 9, 13**], we can define for every finite brace the solvable radical as the largest solvable ideal, which also equals the sum of all solvable ideals in this brace. We define the prime radical of a brace A to be the largest N-ideal in A. In this section, we show that for finite braces the prime radical and the solvable radical coincide.

We recall a definition of a solvable brace from [**6**]:

DEFINITION 10. For a brace A we define $A_1 = A$ and inductively $A_{i+1} = A_i * A_i$ for $i \geq 1$. Recall that A is said to be solvable if $A_n = 0$ for some n. By induction one proves that $A_{i+1} \subseteq A_i$ for all i.

DEFINITION 11. An ideal I in a brace A is solvable, if I is a solvable skew brace.

We obtain the following result:

THEOREM 12. *Let A be a finite brace. Then the largest solvable ideal of A equals the intersection of all strongly prime ideals in A.*

PROOF. Let $a \in A$ be such that the ideal generated by a is solvable, and let P be a strongly prime ideal in A. Then $J = \langle a \rangle + P$ is an ideal in A containing P, by Lemma 2.7 [**25**]. Observe that $\{j + P : j \in J\}$ is a solvable ideal in the factor brace A/P. As the factor brace A/P is strongly prime it cannot have non-zero solvable ideals. Consequently $J \subseteq P$, so $a \in P$. It holds for every strongly prime ideal in A. Therefore, a is in the intersection of all strongly prime ideals in A.

Suppose now that a is in every strongly prime ideal in A. We will show that the ideal $\langle a \rangle$ generated by a in A is solvable ($\langle a \rangle$ is the smallest ideal in A containing a). We will first show that every N-sequence starting in a will reach zero, so $\langle a \rangle$ is an N-ideal. Suppose on the contrary that there is an N-sequence $a_1, a_2, \dots$ which has all elements non-zero and $a_1 = a$. Let I be an ideal of A which is maximal with respect to the property that I does not contain any element a_i (such an ideal exists by Zorn's lemma, and it may be the zero ideal). Observe that I is a strongly prime ideal, because if $J[1], \ldots J[n]$ are ideals containing I and larger than I, then by the maximality of I there is an element a_t from our N-sequence such that $a_t \in J[k]$ for $k = 1, 2, \ldots n$ (since if $a_i \in J[k]$ then $a_{i+1} \in J[k]$). Define $J = \bigcap_{i=1}^{n} J[i]$. Notice that $J \neq 0$ since $a_t \in J$. Therefore, for $k \geq t$

$$a_{k+1} \in \langle a_k \rangle_{\alpha_k} \subseteq J_{\alpha_k},$$

where J_{α_k} is defined as in Definition 7. So if $\alpha_k = 2$ then $J_{\alpha_k} = J_2 = J * J$, if $\alpha_k = 3$ then $J_{\alpha_k} = J_3 = (J * J) * (J * J)$, etc. Recall that α_k comes from the definition of our fixed N-sequence where we have

$$a_{k+1} \in \langle a_k \rangle_{\alpha_k}.$$

Let L be some product of ideals $J[1], \ldots, J[n]$. Then L contains some product of copies of ideal J, since $J \subseteq \bigcap_{i=1}^{n} J[i]$. Let J_m be defined as in Definition 7, so if $m = 2$ then $J_m = J_2 = J * J$, if $m = 3$ then $J_m = J_3 = (J * J) * (J * J)$, etc. It can be proved that L contains J_m for all $m > j$ for some j (with a similar proof as in Lemma 6.11 in [**25**]). Let $k > t$ be such that $\alpha_k > j$. Notice that $a_k \subseteq J$ implies

$$a_{k+1} \in J_{\alpha_k} \subseteq J_{j+1} \subseteq L.$$

Therefore all products of ideals $J_1, \ldots, J_n$ are non-zero (because all elements a_i are non-zero). Therefore I is a strongly prime ideal in A which does not contain a, a contradiction.

We will now show that $\langle a \rangle$ is a solvable ideal. Notice that $\langle a \rangle$ is a brace. It suffices to show that every finite brace B in which each N-sequence reaches zero is solvable.

We will prove that B is solvable by induction on the number of elements in B. If B has only one element then it is a zero element, as every brace contains zero, and the result holds. Suppose the result holds for all braces of cardinality smaller than i, and suppose that B has cardinality $i + 1$. We will construct an N-sequence of elements in B, let a_1 be an arbitrary non-zero element in B, and we construct our N-sequence by induction by taking as a_{i+1} an arbitrary non-zero element from $\langle a_i \rangle_i$. Because every N-sequence will reach zero, we will have $a_i \neq 0$ and $\langle a_i \rangle_i = 0$ for some i. Since $a_i \in \langle a_i \rangle$ we have that $\langle a_i \rangle$ is a non-zero solvable ideal in A. Notice that $\langle a_i \rangle$ is an ideal in brace B. Since every N-sequence in brace B reached zero, then every N-sequence in the factor brace $B/\langle a_i \rangle$ reaches zero. By the inductive assumption $B/\langle a_i \rangle$ is a solvable brace. Notice that $\langle a_i \rangle$ and $B/\langle a_i \rangle$ are solvable braces, so by Lemma 6.4 from [**25**] B is a solvable brace. □

Notice that the proof of the above results uses similar methods as in the proof of Levitsky's theorem, namely that the prime radical in an associative ring equals the intersection of all prime ideals.

LEMMA 13. *A finite (non-trivial) brace is strongly semiprime if and only if it has no non zero solvable ideals.*

PROOF. It suffices to show that the solvable radical of A is not zero if and only if A is not strongly semiprime. If A is not strongly semiprime, then there is some non-zero ideal I such that some product L of some number of copies of I is zero. It can be proved by induction (on the number of copies of I) that any product of some copies on I contains I_n, for some n. Therefore, I is a solvable ideal (since $I_n \subseteq L = 0$ for some n).

Assume now that the solvable radical of A is not zero, and let I be a non-zero solvable ideal in A, then $I_n = 0$ for some n. As the product I_n of some copies of I is zero then A is not strongly semiprime. □

Let $A_1, A_2, \ldots, A_n$ be braces, then the direct product $A_1 \oplus A_2 \oplus \cdots \oplus A_n$ is a brace consisting of elements $(a_1, \ldots, a_n)$ with $a_i \in A_i$ with the addition

$$(a_1, \ldots, a_n) + (a_1', \ldots, a_n') = (a_1 + a_1', \ldots, a_n + a_n')$$

and multiplication

$$(a_1, \ldots, a_n) * (a_1', \ldots, a_n') = (a_1 * a_1', \ldots, a_n * a_n').$$

We obtain the following classification of strongly semiprime braces. We use similar methods as in related results in noncommutative ring theory.

THEOREM 14. *Every finite strongly semiprime brace A embeds as a brace in the direct product of some strongly prime braces $A_1, \ldots, A_n$, for some n. Moreover, A contains the direct product of some braces $L_1, \ldots, L_n$, where each L_i is a non-zero ideal in the brace A_i.*

PROOF. Because A is finite, we can find the minimal possible number n such that there exist strongly prime ideals $P_1, \ldots, P_n$ in A such that $\bigcap_{i=1}^n P_n = 0$. A embeds in a direct product of strongly prime braces $A/P_1, A/P_2, \ldots, A/P_n$ under the mapping $f(a) = (a + P_1, \ldots, a + P_n)$ (because the kernel of the map f is zero, since $\bigcap_{i=1}^n P_n = 0$).

By the minimality of n, for every i there is an $a_i \in A$ such that $a_i \in P_j$ for $j \neq i$ and $a \notin P_i$. Let L_i' be the smallest ideal of A containing a_i, then $L_i' \subseteq P_j$ for $j \neq i$. Notice that $L_i = \{l + P_i : l \in L_i'\}$ is an ideal in A/P_i. It follows that the direct product of ideals L_i is contained in the image of A under the map $f(a) = (a + P_1, \ldots, a + P_n)$, as required. □

REMARK 15. The above results can be generalised to the case of pre-Lie algebras by substituting the words pre-Lie algebra for the word brace in all occurrences of the word brace.

3. The Baer radical

In [**25**], the Baer radical of a brace was defined. It was shown that the Baer radical equals the intersection of all prime ideals in A. Recall that an ideal I in a brace A is prime if for any two ideals $I \subseteq J, I \subseteq K$ in A, if $J \neq I$, and $I \neq K$ then $J * K$ is not a subset of I. In other words, the factor brace A/I is a prime brace, i.e., a brace in which the product of any two non-zero ideals is non-zero. To construct the Baer radical, the following definition of n-sequence was defined, see [**25**]. Notice that this is a different definition of N-sequence than in this paper but both generalise the ring theoretical notion of n-sequence.

Let A be a brace, the sequence of elements $a_1, a_2, \ldots$ from A is an n-sequence in A if for every i, $a_{i+1} \in \langle a_i \rangle * \langle a_i \rangle$.

We finish with the following open questions:

QUESTION 1. Is every finite strongly prime brace a simple brace?

It is known that there are finite braces which are prime and not simple [**6**]. Recall that a brace A is simple if it has no proper non-zero ideals. There are also finite braces which are prime and not strongly prime [**27**].

QUESTION 2. Let A be a finite brace in which right multiplication operators R_x are nilpotent. Can A be a prime brace, strongly prime brace or simple brace? Recall that $R_x(y) = y * x$ for $x, y \in A$.

QUESTION 3. Let A be a finite pre-Lie algebra over a field of characteristic $p > 0$, in which right multiplication operators R_x are nilpotent. Can A be a prime pre-Lie algebra, strongly prime pre-Lie algebra or simple pre-Lie algebra? Recall that $R_x(y) = y \cdot x$ for $x, y \in A$.

There exist pre-Lie algebras over a field of characteristic zero which are finitely dimensional, right nil and simple [**7, 9**].

References

[1] S. A. Amitsur, *A general theory of radicals. I. Radicals in complete lattices*, Amer. J. Math. **74** (1952), 774–786, DOI 10.2307/2372225. MR50563

[2] S. A. Amitsur, *A general theory of radicals. II. Radicals in rings and bicategories*, Amer. J. Math. **76** (1954), 100–125, DOI 10.2307/2372403. MR59256

[3] S. A. Amitsur, *A general theory of radicals. III. Applications*, Amer. J. Math. **76** (1954), 126–136, DOI 10.2307/2372404. MR59257

[4] A. S. Amitsur and J. Levitzki, *Minimal identities for algebras*, Proc. Amer. Math. Soc. **1** (1950), 449–463, DOI 10.2307/2032312. MR36751

[5] D. Bachiller, *Counterexample to a conjecture about braces*, J. Algebra **453** (2016), 160–176, DOI 10.1016/j.jalgebra.2016.01.011. MR3465351

[6] D. Bachiller, F. Cedó, E. Jespers, and J. Okniński, *Asymmetric product of left braces and simplicity; new solutions of the Yang-Baxter equation*, Commun. Contemp. Math. **21** (2019), no. 8, 1850042, 30 pp., DOI 10.1142/S0219199718500426. MR4020748

[7] C. Bai, *An introduction to pre-Lie algebras*, Algebra and applications 1—non-associative algebras and categories, ISTE, London, 2020, pp. 245–273. MR4484613

[8] T. Brzeziński, *Trusses: between braces and rings*, Trans. Amer. Math. Soc. **372** (2019), no. 6, 4149–4176, DOI 10.1090/tran/7705. MR4009388

[9] D. Burde, *Left-symmetric algebras, or pre-Lie algebras in geometry and physics*, Cent. Eur. J. Math. **4** (2006), no. 3, 323–357, DOI 10.2478/s11533-006-0014-9. MR2233854

[10] F. Catino, I. Colazzo, and P. Stefanelli, *Regular subgroups of the affine group and asymmetric product of radical braces*, J. Algebra **455** (2016), 164–182, DOI 10.1016/j.jalgebra.2016.01.038. MR3478858

[11] F. Cedó, E. Jespers, and J. Okniński, *Braces and the Yang-Baxter equation*, Comm. Math. Phys. **327** (2014), no. 1, 101–116, DOI 10.1007/s00220-014-1935-y. MR3177933

[12] F. Cedó, E. Jespers, and J. Okniński, *Every finite abelian group is a subgroup of the additive group of a finite simple left brace*, J. Pure Appl. Algebra **225** (2021), no. 1, Paper No. 106476, 10 pp., DOI 10.1016/j.jpaa.2020.106476. MR4122077

[13] K. S. Chang, H. Kim, and H. C. Myung, *On radicals of a left-symmetric algebra*, Comm. Algebra **27** (1999), no. 7, 3161–3175, DOI 10.1080/00927879908826619. MR1695315

[14] F. Chouraqui, *Garside groups and Yang-Baxter equation*, Comm. Algebra **38** (2010), no. 12, 4441–4460, DOI 10.1080/00927870903386502. MR2764830

[15] I. Colazzo, E. Jespers, and Ł. Kubat, *Set-theoretic solutions of the pentagon equation*, Comm. Math. Phys. **380** (2020), no. 2, 1003–1024, DOI 10.1007/s00220-020-03862-6. MR4170296

[16] C. Dietzel, *Braces of order p^2q*, J. Algebra Appl. **20** (2021), no. 8, Paper No. 2150140, 24 pp., DOI 10.1142/S0219498821501401. MR4297324

[17] A. Doikou, *Set-theoretic Yang-Baxter equation, braces and Drinfeld twists*, J. Phys. A **54** (2021), no. 41, Paper No. 415201, 21, DOI 10.1088/1751-8121/ac219e. MR4318629

[18] A. Doikou and A. Smoktunowicz, *Set-theoretic Yang-Baxter & reflection equations and quantum group symmetries*, Lett. Math. Phys. **111** (2021), no. 4, Paper No. 105, 40 pp., DOI 10.1007/s11005-021-01437-7. MR4294149

[19] T. Gateva-Ivanova, *Set-theoretic solutions of the Yang-Baxter equation, braces and symmetric groups*, Adv. Math. **338** (2018), 649–701, DOI 10.1016/j.aim.2018.09.005. MR3861714

[20] L. Guarnieri and L. Vendramin, *Skew braces and the Yang-Baxter equation*, Math. Comp. **86** (2017), no. 307, 2519–2534, DOI 10.1090/mcom/3161. MR3647970

[21] Z. Izhakian and L. Rowen, *Ideals of polynomial semirings in tropical mathematics*, J. Algebra Appl. **12** (2013), no. 2, 1250143, 35, DOI 10.1142/S0219498812501435. MR3005595

[22] A. V. Jategaonkar, *A counter-example in ring theory and homological algebra*, J. Algebra **12** (1969), 418–440, DOI 10.1016/0021-8693(69)90040-4. MR240131

[23] P. Jedlička, A. Pilitowska, and A. Zamojska-Dzienio, *The retraction relation for biracks*, J. Pure Appl. Algebra **223** (2019), no. 8, 3594–3610, DOI 10.1016/j.jpaa.2018.11.020. MR3926229

[24] E. Jespers, Ł. Kubat, A. Van Antwerpen, and L. Vendramin, *Radical and weight of skew braces and their applications to structure groups of solutions of the Yang-Baxter equation*, Adv. Math. **385** (2021), Paper No. 107767, 20 pp., DOI 10.1016/j.aim.2021.107767. MR4256133

[25] A. Konovalov, A. Smoktunowicz, and L. Vendramin, *On skew braces and their ideals*, Exp. Math. **30** (2021), no. 1, 95–104, DOI 10.1080/10586458.2018.1492476. MR4223285

[26] A. Konovalov, A. Smoktunowicz, and L. Vendramin, *Erratum to the paper "On skew braces and their ideals"*, Exp. Math. **31** (2022), no. 1, 346, DOI 10.1080/10586458.2021.1980466. MR4399130

[27] D. Puljić, A. Smoktunowicz, and K. Nejabati Zenouz, *Some braces of cardinality p^4 and related Hopf-Galois extensions*, New York J. Math. **28** (2022), 494–522. MR4395578

[28] W. Rump, *Braces, radical rings, and the quantum Yang-Baxter equation*, J. Algebra **307** (2007), no. 1, 153–170, DOI 10.1016/j.jalgebra.2006.03.040. MR2278047

[29] W. Rump, *The brace of a classical group*, Note Mat. **34** (2014), no. 1, 115–144. MR3291816

[30] L. W. Small, *Prime ideals in Noetherian* PI*-rings*, Bull. Amer. Math. Soc. **79** (1973), 421–422, DOI 10.1090/S0002-9904-1973-13196-9. MR313307

[31] A. Smoktunowicz, *On Engel groups, nilpotent groups, rings, braces and the Yang-Baxter equation*, Trans. Amer. Math. Soc. **370** (2018), no. 9, 6535–6564, DOI 10.1090/tran/7179. MR3814340

[32] A. Smoktunowicz, *Algebraic approach to Rump's results on relations between braces and pre-Lie algebras*, J. Algebra Appl. **21** (2022), no. 3, Paper No. 2250054, 13 pp., DOI 10.1142/S0219498822500542. MR4391819

[33] A. Smoktunowicz and L. Vendramin, *On skew braces (with an appendix by N. Byott and L. Vendramin)*, J. Comb. Algebra **2** (2018), no. 1, 47–86, DOI 10.4171/JCA/2-1-3. MR3763907

[34] E. I. Zel′manov, *Goldie theorems for Jordan algebras* (Russian), Sibirsk. Mat. Zh. **28** (1987), no. 6, 44–52, 217. MR928338

[35] E. I. Zel′manov, *Goldie theorems for Jordan algebras. II* (Russian), Sibirsk. Mat. Zh. **29** (1988), no. 4, 68–74, 223, DOI 10.1007/BF00969864; English transl., Siberian Math. J. **29** (1988), no. 4, 567–573 (1989). MR969104

School of Mathematics, University of Edinburgh, Peter Guthrie Tait Road, Edinburgh, EH9 3FD, Scotland

Email address: `A.Smoktunowicz@ed.ac.uk`

Selected Published Titles in This Subseries

Israel Mathematical Conference Proceedings (IMCP) is a publication, part of the Contemporary Mathematics Series, devoted to the proceedings of conferences, symposia and seminars. Collections of papers focusing on a certain subject will also be published. Prospective volumes may be submitted to any member of the editorial board. Each volume has an editor (or editors) responsible for its preparation. In order to ensure inexpensive and timely distribution, authors are requested to submit to the Editor of the volume an electronic TEX file of their manuscript in AMS-LATEX, using the Contemporary Mathematics style file which can be downloaded at `http://www.ams.org/tex/author-info.html`. For further information, contact the Managing Editor, IMCP, Department of Mathematics, Bar-Ilan University, Ramat-Gan 52900, Israel; e-mail: `vishne@macs.biu.ac.il`.

800 **Avinoam Mann, Louis H. Rowen, David J. Saltman, Aner Shalev, Lance W. Small, and Uzi Vishne, Editors,** Amitsur Centennial Symposium, 2024

726 **Eugene Plotkin, Editor,** Groups, Algebras and Identities, 2019

699 **Mark L. Agranovsky, Matania Ben-Artzi, Catherine Bénéteau, Lavi Karp, Dmitry Khavinson, Simeon Reich, David Shoikhet, Gilbert Weinstein, and Lawrence Zalcman, Editors,** Complex Analysis and Dynamical Systems VII, 2017

667 **Mark L. Agranovsky, Matania Ben-Artzi, Greg Galloway, Lavi Karp, Dmitry Khavinson, Simeon Reich, Gilbert Weinstein, and Lawrence Zalcman, Editors,** Complex Analysis and Dynamical Systems VI: Part 2: Complex Analysis, Quasiconformal Mappings, Complex Dynamics, 2016

653 **Mark L. Agranovsky, Matania Ben-Artzi, Greg Galloway, Lavi Karp, Dmitry Khavinson, Simeon Reich, Gilbert Weinstein, and Lawrence Zalcman, Editors,** Complex Analysis and Dynamical Systems VI: Part 1: PDE, Differential Geometry, Radon Transform, 2015

636 **Simeon Reich and Alexander J. Zaslavski, Editors,** Infinite Products of Operators and Their Applications, 2015

619 **Gershon Wolansky and Alexander J. Zaslavski, Editors,** Variational and Optimal Control Problems on Unbounded Domains, 2014

591 **Mark L. Agranovsky, Matania Ben-Artzi, Greg Galloway, Lavi Karp, Vladimir Maz'ya, Simeon Reich, David Shoikhet, Gilbert Weinstein, and Lawrence Zalcman, Editors,** Complex Analysis and Dynamical Systems V, 2013

568 **Simeon Reich and Alexander J. Zaslavski, Editors,** Optimization Theory and Related Topics, 2012

554 **Mark Agranovsky, Matania Ben-Artzi, Greg Galloway, Lavi Karp, Simeon Reich, David Shoikhet, Gilbert Weinstein, and Lawrence Zalcman, Editors,** Complex Analysis and Dynamical Systems IV: Part 2. General Relativity, Geometry, and PDE, 2011

553 **Mark Agranovsky, Matania Ben-Artzi, Greg Galloway, Lavi Karp, Simeon Reich, David Shoikhet, Gilbert Weinstein, and Lawrence Zalcman, Editors,** Complex Analysis and Dynamical Systems IV: Part 1. Function Theory and Optimization, 2011

514 **Arie Leizarowitz, Boris S. Mordukhovich, Itai Shafrir, and Alexander J. Zaslavski, Editors,** Nonlinear Analysis and Optimization II, 2010

513 **Arie Leizarowitz, Boris S. Mordukhovich, Itai Shafrir, and Alexander J. Zaslavski, Editors,** Nonlinear Analysis and Optimization I, 2010

489 **David Ginzburg, Erez Lapid, and David Soudry, Editors,** Automorphic Forms and L-functions II, 2009

488 **David Ginzburg, Erez Lapid, and David Soudry, Editors,** Automorphic Forms and L-functions I, 2009

For a complete list of titles in this series, visit the
AMS Bookstore at **www.ams.org/bookstore/conmseries/**.